Soil Abiotic and Biotic Interactions and Impact on the Ecosystem and Human Welfare

Preface

Physical, chemical, and biological processes are not independent but interactive processes in soil environments. Soil can be defined as complex interactive biogeochemical reactors, reservoirs of organisms (mainly microorganisms), and a major compartment of terrestrial ecosystems under the influence of anthropogenic activities. To improve our scientific knowledge of soil resources and its application to remediation and long-term management, it is of major importance and interest to study soil organization and function not only through the subdisciplines of Soil Sciences, but also through interactive and integrative approaches. The study of the interactions between soil constituents and soil organisms has to be considered at different scales, namely from the molecular level to field / landscape systems and is, indeed, essential to stimulating further research to uncover the dynamics and mechanisms of reactions and processes in soil and related environments. Therefore, Commission 2.5 has officially been created in the scientific structure of the International Union of Soil Sciences (IUSS) to deal with physical / chemical / biological interfacial interactions, i.e., abiotic and biotic interfacial interactions in the soil and related environments. These fundamental interactive processes have enormous impacts on ecosystem productivity, services, integrity, and human welfare.

The 21 chapters in this book were largely selected from papers presented at Symposium 47 "Soil Mineral-Organic Component-Microorganism Interactions and the Impact on the Ecosystem and Human Welfare" of IUSS Working Group MO and Symposium 06 "Frontiers of Soil Chemistry and Biochemistry of the Soil Rhizosphere" of IUSS Commission II, the 17th World Congress of Soil Science, Bangkok, Thailand, August 14–21, 2002. Part I overviews the subject and addresses abiotic and biotic interactions and the impact on restoration of terrestrial ecosystems and human welfare. Parts II, III, and IV deal with the roles of abiotic and biotic interactions in the transformations of (a) natural organics and xenobiotics, (b) nitrogen, phosphorous, sulfur, and boron, and (c) metals and metalloids respectively. Part V addresses the issue of rhizosphere processes, the bottleneck for sustaining biological productivity and protecting the human food chain. It is hoped that this book will stimulate research and education in this extremely important and exciting area of science for years to come.

All the chapters in this book have been critically reviewed by external referees and editors. We are grateful to the authors for their contributions and to the external referees for their invaluable inputs that helped maintain the quality of this volume. We also thank Ms. Joy Drohan and Dr. A.T. Rambo for English editing some chapters. Our sincere appreciation is extended to the University of Saskatchewan for awarding a Publication Fund grant. We also express our gratitude to the Univesity of Naples, Italy for providing a grant to Professor Antonio Violante to enable his visit to the University of Saskatchewan to facilitate completion of this book project.

This book is an essential reference for chemists and biologists studying environmental systems, as well as for earth, soil, and environmental scientists. It will serve as a useful reference for professors, students, consultants, and others in Environmental Science, Soil Sciences, Ecology, and Ecotoxicology.

P.M. Huang
A. Violante
J.-M Bollag
P. Vityakon

Contents

PART III. NITROGEN, PHOSPHORUS, SULFUR AND BORON

Contributors

Adriano, D.C.
University of Georgia
Savannah River Ecology Lab.
Drawer E, Aiken, SC 29802,
USA

Agnelli, A.
Istit. Studio degli Ecosistemi
Sezione di Firenze, CNR
Piazzale delle Cacine 28
50144 Firenze, Italya

Arora, Sanjay
Division of Soil Science and
Agricultural Chemistry
Sher-e-Kashmir University of
Agriculture
Main Campus-Chatha
Jammu (J & K) 180 009
India

Berthelin, J.
Centre National de la Recherche
Scientifique
Centre de Pedologie Biologique
17, rue Notre Dame des Pauvers
B.P. 5–54501 Vandoeuvree-Les-Nancy
Cedex, France

Bolan, N.S.
Soil and Earth Sciences
Massey University
New Zealand

Bollag, J.-M.
Lab. Soil Biochemistry
Pennsylvania State University
University Park, PA 16802,
USA

Borggaard, O.K.
Chemistry Dept.
Royal Veterinary and Agricultural
University
40 Thorvaldensvej
DK-1871 Frederiksberg C, Denmark

Chen, W.
College Resources and Environment
Huazhong Agricultural University
Wuhan 430070 P.R. China

Cuter, J.N.
Canadian Light Source Inc.
University of Saskatchewan
Saskatoon SK S7N 0X4, Canada

D'Acqui, L.P.
Istit per lo Studio degli Ecosistemi
Sezione di Firenze, CNR
Via Madonna del Piano Ed. C/D, 50019,
Sesto Fiorentino, Italia

Dassonville, F.
Inst. National Recherche Agronomique
Unité Climat, Sol et Environnement
Domaine Saint-Paul, Site Agroparc
84914 Avignon Cedex 9, France

Del Gaudio, S.
Dipart, Scienze del Suolo
della Pianta e dell'Ambiente
Università di Napoli Federico II
80055 Portici (Napoli), Italya

Demanet, R.
Dept. Agriculture Production
Universidad de La Frontera
P.O. Box 54-D Temuco, Chile

Doerr, S.H.
Dept. Geography
University of Wales Swansea
Singleton Park, Swansea, SA2 8PP
UK

Douglas, P.
Dept. Chemistry
University of Wales Swansea
Singleton Park, Swansea, SA2 8PP
UK

Ferreira, A.J.D.
Centro das Zonas Costeiras e do Mar
Depart. Ambiente e Ordenamento
Universidade de Aveiro
P-3810-193 Aveiro, Portugal

Gimsing, A.L.
Chemistry Dept.
Royal Veterinary and Agricultural University
40 Thorvaldsensvej
DK-1871 Frederiksberg C, Denmark

Guggenberger, G.
Institute of Soil Science and Plant Nutrition
Martin Luther University of Halle-Wittenberg 06108 Halle (Saale)
Germany

Guo, X.
College Resources and Environment
Huazhong Agricultural University
Wuhan 430070
P.R. China

Haider, K.
Kastanienallee 4
D-82041 Deisenhofen
Germany

Haskins, C.
Dept. Chemistry
University of Wales Swansea
Singleton Park, Swansea, SA2 8PP
UK

He, J.Z.
College Resources and Environment
Huazhong Agricultural University
Wuhan 430070
P.R. China

Hu, H.
College Resources and Environment
Huazhong Agricultural University
Wuhan 430070
P.R. China

Huang, P.M.
Dept. Soil Science
University of Saskatchewan
51 Campus Drive
Saskatoon SK S7N 5A8,
Canada

Huang, Q.
College Resources and Environment
Huazhong Agricultural University
Wuhan 430070
P.R. China

Johnsey, L.
Dept. Chemistry
University of Wales Swansea
Singleton Park, Swansea, SA2 8PP
UK

Jokic, A.
Dept. Soil Science
University of Saskatchewan
51 Campus Drive
Saskatoon, SK S7N 5A8
Canada

Kiryushin, V.I.
Timiryazev Agricultural Academy
Russian Academy Sciences
ul. Timiryazeva 49
Moscow 127550
Russia

Li, X.Y.
College Resources and Environment
Huazhong Agricultural University
Wuhan 430070 P.R. China

Ling, W.T.
Key Lab. Subtropical Soil Resources and Environment
Ministry of Agriculture
Huazhong Agricultural University
Whuan 430070
P.R. China

Liu, C.
Kuo Testing Labs Inc.
Agricultural, Environmental, Industrial
337 South 1st Avenue
Othello, WA 99344, USA

Liu, F.
College Resources and Environment
Huazhong Agricultural University
Wuhan 430070
P.R. China

Llewellyn, C.T.
Dept. Chemistry
University of Wales Swansea
Singleton Park, Swansea, SA2 8PP
UK

Mainwaring, K.A.
Dept. Chemistry
University of Wales Swansea
Singleton Park, Swansea, SA2 8PP
UK

Malawska, M.
Inst. Botany
Dept. Plant Systematics and Geogaphy
Warsaw University
00-478 Warszawa
Al, Ujazdowskie 4, Poland

Mora, M.L.
Chemistry Dept.
Universidad de La Frontera
P.O. Box 54-D
Temuco, Chile

Morley, C.P.
Dept. Chemistry
University of Wales Swansea
Singleton Park, Swansea, SA2 8PP
UK

Onyatta, J.O.
National Council Science Technology
P.O. Box 30623
Code 00200, City Square
Nairobi, Kenya

Pigna, M.
Dipart. Scienze del Suolo
della Pianta e dell'Ambiente
Università di Napoli Federico II
80055 Portici (Napoli), Italya

Rao, M.A.
Dipart. Scienze del Suolo,
della Pianta e dell'Ambiente
Università di Napoli Federico II
80055 Portici (Napoli)
Italya

Rasmussen, L.H.
Chemistry Dept.
Royal Veterinary and Agricultural
University
40 Thorvaldsensvej
DK-1871 Frederiksberg C
Denmark

Renault, P.
Inst. National Recherche Agronomique
Unité Climat, Sol et Environnement
Domaine Saint-Paul, Site Agroparc
84914 Avignon Cedex 9, France

Ricciardella, M.
Dipart. Scienze del Suolo
della Pianta e dell'Ambiente
Università di Napoli Federico II
80055 Portici (Napoli), Italya

Ritsema, C.J.
Alterra, Land Use and Soil Processes
Team
P.O. Box 47
6700 AA Wageningen,
The Netherlands

Schnitzer, M.
Eastern Cereal and Oilseed Research
Center
Central Experimental Farm
Agriculture and Agri-Food Canada
Ottawa, ON, Canada

Schulten, H.-R.
Inst. Soil Science
University of Rostock
Justus-von-Liebitz-Weg 6
D-18051 Rostock, Germany

Shene, C.
Dept. Chemical Engineering
Universidad de La Frontera
P.O. Box 54-D
Temuco, Chile

Santi, C.A.
Istit. Studio degli Ecosistemi
Sezione di Firenze, CNR
Piazzale delle Cascine 28
50144 Firenze Italya

Sparvoli, E.
Istit. Studio degli Ecosistemi
Sezione di Firenze, CNR
Piazzale delle Cascine 28
50144 Firenze
Italya

Stagnitti, F.
Deakin University
School Ecology and Environment
PO Box 423
Warrnambool, Victoria, Australia 3280

Szilas, C.
Chemistry Dept.
Royal Veterinary and Agricultural University
40 Thorvaldsensvej
DK-1871 Frederiksberg C, Denmark

Tani, M.
Dept. Agro-Environmental Science
Obihiro University of Agriculture and Veterinary Medicine
Obihiro, Japan

Violante, A.
Dipart. Scienze del Suolo
della Pianta e dell'Ambiente
Università di Napoli Federico II
80055 Portici (Napoli), Italya

Wang, H.
Inst. Soil and Water Resources and Environmental Science
Zhejiang University
Hangzhou 310029
P.R. China

Wilkomirski, B.
Inst. Botany
Dept. Plant Systematics and Geography
Warsaw University
00-478 Warszawa
Al, Ujazdowskie 4, Poland

Xie, Z.
Inst. Soil and Water Resources and Environmental Science
Zhejiang University
Hangzhou 310029
P.R. China

Xu, J.
Inst. Soil and Water Resources and Environmental Science
Zhejiang University
Hangzhou 310029
P.R. China

Ziogas, A.K.
Dept. Civil Engineering
Demokritus University of Thrace
67100 Xanthi, Greece

External Referees

Adriano, Domy C.
University of Georgia
Savannah River Ecology Lab
Drawer, E, Aiken, SC 29802, USA

Amarger, Noelle
Laboratoire de Microbiologie des Sols
Inst. National Recherche Agronomique
21034 Dijon Cedex, France

Buurman, Peter
Dept. Environmental Sciences
University of Wageningen
P.O. Box 6700, Wageningen
The Netherlands

Dec, Jerzy
Lab. Soil Biochemistry
129 Land and Water Research
Pennsylvania State University
University Park, PA 16802, USA

Doner, H.E.
Division of Ecosystem Sciences
Dept. Environmental Science,
Policy, and Management
College of Natural Resources
University of California
Berkeley, CA 94720-3110 USA

Gianfreda, L.
Dipart. Scienze del Suolo
della Pianta e dell'Ambiente
Università di Napoli Federico II
80055 Portici (Napoli) Italya

Goh, K.M.
Soil, Plant and Ecological Science
Division
Lincoln University
Canterbury, New Zealand

Goldberg, S.
George E. Brown, Jr.
Salinity Laboratory
Riverside CA 92507 USA

Govere, Ephraim M.
Center for Bioremediation and
Detoxification
129 Land and Water Research
Pennsylvania State University
University Park, PA 16802 USA

Haider Konard
Kastanienallee 4
D-82041 Deisenhofen
Germany

Hue, N.V.
Dept. Tropical Plant and Soil Science
University of Hawaii
1910 East-West Road
Honolulu, Hawaii 96822 USA

Krishnamurti, G.S.R.
313-855 West 16th Street
North Vancouver, BC V7P 1R2 Canada

Miano, Teodoro
Dipart. Biologia e Chimica
Agro-forestale ed Ambientale
Università degli Studi
Via Amendola 165/a
Bari, Italy

Page, A.L.
Dept. Environmental Sciences
University of California
Riverside, CA 92521-0424
USA

Peak, D.
Dept. Soil Science
University of Saskatchewan
51 Campus Drive
Saskatoon SK S7N 5A8 Canada

Piccolo, Alessandro
Dipart. Scienze del Suolo
della Pianta e dell'Ambiente
Università di Napoli Federico II
80055 Portici (Napoli) Italya

Ruaysoongnern, Sawaeng
Dept. Land Resources & Environment
Khon Kaen University
Khon Kaen, Thailand

Ruggiero, Pacifico
Dipart. Biologia e Chimica
Agro-forestale ed Ambientale
Università degli Studi
Via Amendola 165/a
Bari, Italy

Saha, U.K.
Dept. Soil Science
University of Saskatchewan
51 Campus Drive
Saskatoon, SK S7N 5A8 Canada

Singh, Swaranjit
Inst. Microbial Technology
Sector 39-A
Chandigarh 160 036 India

Strynar, Mark J.
US Environmental Protection Agency
ORD/NERL
Research Triangle Park, NC 27709 USA

Tani, M.
Dept. Agro-Environmental Science
Obihiro University of Agriculture
and Veterinary Medicine
Obihiro, Japan

Torrent, José
Depart. Ciencias y
Recursos Agricolas y Forestales
Unsiversidad de Còrdoba
Apdo 3048
14080 Córdoba, Spain

Violante, Antonio
Dipart. Scienze del Suolo
della Pianta e dell'Ambiente
Università di Napoli Federico II
80055 Portici (Napoli) Italya

Part I
Overview

1

Soil Minerals and Organic Components: Impact on Biological Processes, Human Welfare, and Nutrition

K. Haider* *and* **G. Guggenberger**

Abstract

Soil structure is an important soil property and mediates many biological and physical processes. The mineralogical characteristics of a soil can influence soil stability and the relationship between organic matter contents and soil fertility. Many of the soil C simulation models incorporate soil texture and especially clay contents as important factors that control soil C, N, and P stabilization and mineralization.

Separation of soil into primary organomineral complexes shows that soil organic matter is associated with mineral particles that differ in size, structure, and function. A trend towards lower C/N, C/P_{org}, and C/S ratios in the finer size particles has been observed.

The spatial arrangement of solid particles in the soil ecosystem results in a complex and discontinuous pattern of various sized and shaped of pore spaces that are filled either with water or with air. They form the habitat of micro-organisms in microenvironments characterized by a variety of physical and chemical conditions and involve simultaneous strongly interrelated biological, chemical and physical processes.

Studies have demonstrated that the spatial complexity and heterogeneity of microenvironments within soil aggregates form sites of accumulation of organic residues through physical protection mechanisms by spatial isolation of substrates from decomposer cells and enzymes resulting from their interaction with soil inorganic materials and from their association with biologically resistant humic substances. Microporous properties of soil materials are also considered

Corresponding author: Dr. K. Haider, Kastanienallee 4, D-82041 Deisenhozen, Germany.
Email: konrad.haider@arcomail.de

important for the physical sequestration of organic or inorganic contaminants and influence the risk assessment of chemicals in the environment.

Bacteria and fungi synthesize extracellular polymers, such as polysaccharide or proteins. These polymers adhere microorganisms to organic and inorganic particles, thereby preventing their leaching from soil and the spread of contagious organisms into the environment. Water quality is controlled by the ability of soils to retain organic matter and to prevent it from entering the aqueous phase.

Roots growing through soil are exposed to infections by pathogens, especially of fungal origin. The specialized vascular wilt fungi such as *Fusarium* sp. and *Verticillium* sp. attack juvenile roots. Once inside the roots, they cause great damage to the plants. A high clay content of soil can be suppressive to these diseases, but our understanding of this suppression is not yet complete.

1 INTRODUCTION

Soils contain organomineral complexes of varying size, structure, and organization. The role of mineral colloids in the stabilization of soil organic matter (SOM) and its associated nutrients is of particular importance. Generally, more than 99% C, 95% N and S, and between 20 and 90% P in soil are found in SOM. Any biochemically controlled process of organic carbon transformation is tightly associated with mobilization of N, P, and S.

Depending on climate and management conditions, the C/N ratios vary between 10 and 15, C/P ratios between 50 and 100, and C/S ratios between 60 and 200 (Roberts et al., 1989). Important biotic and abiotic factors controlling the cycling of organic C, N, P, and S are the temperature and water regimes because they control the amount of carbon fixed by photosynthesis and the rate of decomposition and mineralization of organic compounds. Among the numerous other factors influencing the ability of the decomposer community to utilize organic compounds, the textural composition of soil and the soil structure are probably the second most important.

The role of the mineral phase in soil on the interaction with primary particles (i.e., clay minerals and Fe- and Al-oxyhydroxides) and on the interactions of organic materials and microorganisms in the stabilization of soil aggregates has been reviewed by van Veen and Kuikman (1990) Ladd et al. (1996).

The spatial arrangement of the solid particles in the soil environment forms a complex and discontinuous pattern of habitats characterized by a variety of physical and chemical conditions involving simultaneous biological, chemical, and physical processes that are strongly interrelated. Physical protection by physical separation of substrates from the decomposer community and their enzymes results partly from interaction with soil mineral matter and partly from association with resistant humic substances.

The microporous properties of organic and inorganic soil particles are also important for the physical sequestration of contaminants or microorganisms (i.e., formation of bound residues). Sorption capacity is markedly related to the contents of humic materials and clay in the soil. Humic materials influence risk assessment in the environment and preclude leaching processes into ground or surface waters. Water quality in catchment areas is controlled by the ability of soils to retain organic matter, contaminants, and microorganisms, thereby arresting their entry into the aqueous phase.

Roots growing through soil are exposed to infection by pathogens, especially by fungi. Vascular wilt fungi, once inside the roots, can cause great damage to plants. Certain mont-morillonitic soils are suppressive to these diseases, but our understanding of this suppression is still rather poor (Höper et al., 1995).

2 IMPACTS OF SOIL MINERALS ON CYCLING OF CARBON AND NUTRIENTS

Models based on correlations between soil C and clay content in large soil C databases indicate that organic C increases with precipitation and clay content and decreases with temperature (Burke et al., 1989). When other factors such as vegetation type, mean annual temperature, and drainage class are relatively constant, clay content, namely that of 2:1 type minerals and allophane, correlates positively with organic carbon content (Fig. 1.1).

Kaiser et al. (2002) explained the preferential association of SOM with clay-size particles with increasing soil depth by the fact that a major pathway for entry SOM into subsoils is via dissolved organic matter (DOM). Reactive

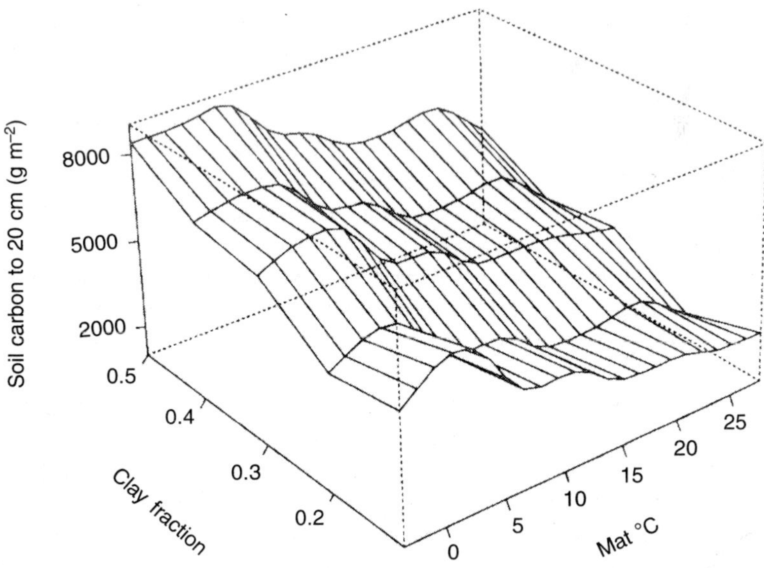

Fig. 1.1: Model prediction of soil carbon contents as function of clay content and mean annual temperature (MAT) (according to Burke et al., 1989).

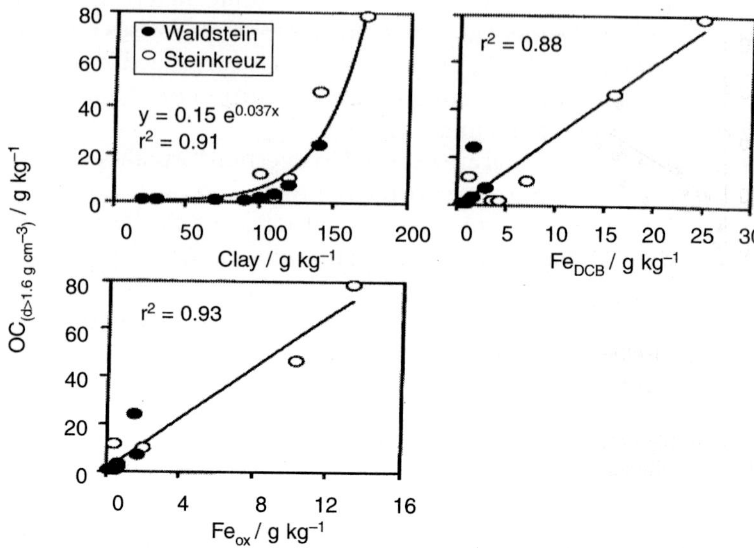

Fig. 1.2: Relationship between organic carbon (OC) in the density fraction d >1.6 g cm^{-3} and the soil content of clay and extractable iron (according to Kaiser et al., 2002).

minerals of the clay fraction, such as aluminum and iron hydrous oxides, are the most effective sorbents for DOM (Kaiser and Zech, 2000). The proportion of mineral-bound SOM correlated well but nonlinearly with the clay content and linearly with the content of dithionite-citrate-bicarbonate-extractable iron and oxalate-extractable iron (Fe_{DCB} and Fe_{Ox}; Fig. 1.2).

Hassink (1997) reported a significant correlation between clay plus silt content and organic C and N contents (Fig. 1.3). The protective effects of clay on soil C, N, P, and S have also been included to describe the turnover of soil organic C and associated nutrients (e.g., van Veen et al., 1985; Parton et al., 1987; Percival et al., 2000).

Introduction of this clay factor in models is based on the fact that clay soils have a greater capacity to preserve or protect microbial biomass and provide an environment for closer interaction between microorganisms and the products of decay. This results in a larger proportion of C and N within the microbial biomass of soils and promotes a higher efficiency of use of metabolic products by the soil biota.

To compare concentrations of SOM in the different size separates independent of the different soil carbon levels, Christensen (1992) defined the enrichment factor, *E*. It relates the content of an organic matter component in a particular size fraction to the content of that component in the whole soil:

$$E_{C,N,P,S} = [(\text{mg C, N, P, S g}^{-1}\text{ separate})/(\text{mg C, N, P, S g}^{-1}\text{ whole soil})].$$

Generally, the highest enrichment of C was found in fine silt and in clay-size separates, with E_C ranging from 1.3 to 3.9 for silt and greater than 10 for

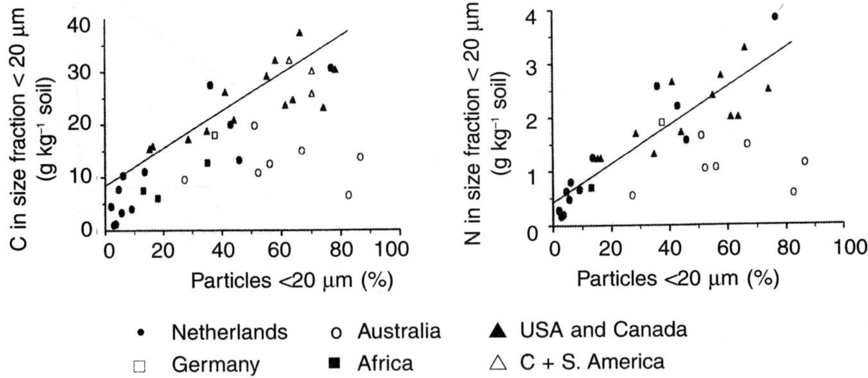

Fig. 1.3: Relationship between organic C and N in the particle-size fraction < 20 μm (clay + silt in g kg^{-1} soil) in uncultivated and grassland soils of temperate and tropical regions (from Hassink, 1997).

clay (Christensen, 1992). Similar trends of enrichment factors for organic C (E_C), N (E_N), P_{org} (E_P), and S (E_S) among the size fractions suggest that particle-size fractionation yields characteristic pools not only for C accumulation, but also for the accumulation of N, P, and S (Fig. 1.4).

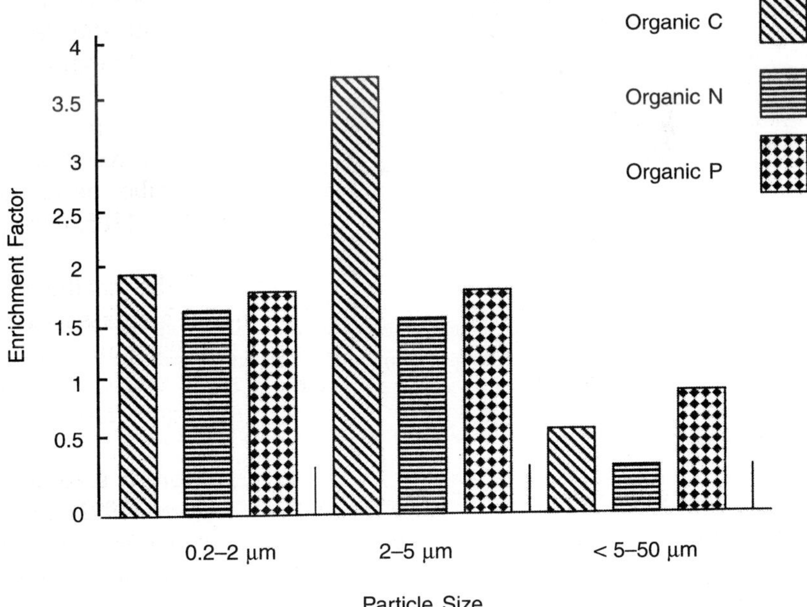

Fig. 1.4: Enrichment factor of organic C, N, and P in particle-size separates of topsoil (extracted from Table 6 of Guggenberger and Haider, 2002).

3 IMPACT OF SOIL MINERALS ON SOIL STRUCTURE

The mineralogical characteristics and management of a soil influences aggregate stability and its relationship with the organic matter content. Six et al. (2000) investigated aggregates from four different long-term agricultural field experiments with native vegetation (NV), no tillage (NT), and conventional tillage (CT) treatments. The authors showed a general decrease in mean weighted diameter of aggregates with increase in cultivation intensity (Fig. 1.5). Soils 1 to 3 are dominated by 2:1 clays and their SOM is the critical binding agent. Contrarily, in soil 4, characterized by a mixed (2:1 and 1:1) clay mineralogy, aggregate stability results from the electrostatic interactions between kaolinite, oxides, and vermiculite and is not as dependent on soil organic matter content as in soils 1 to 3 (Six et al., 2000).

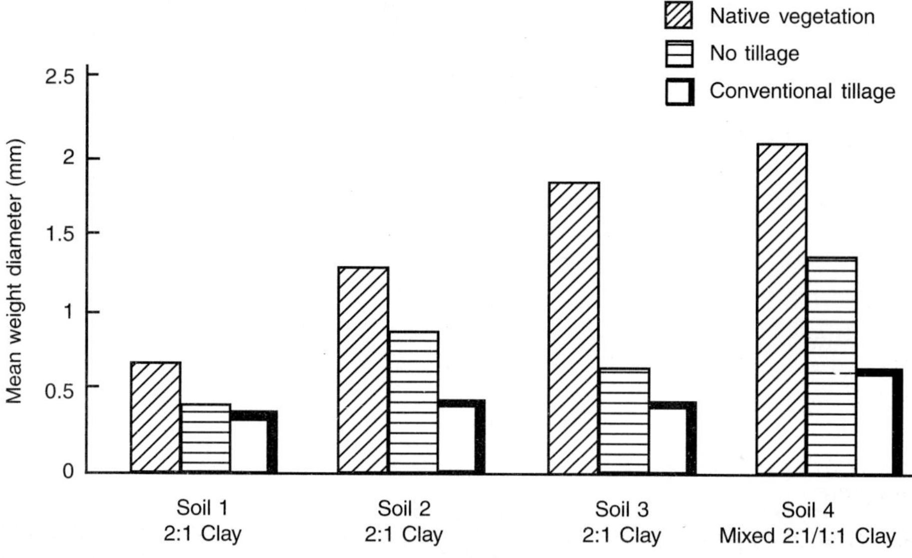

Fig. 1.5: Mean weighted diameters of soil aggregates of soils with different clay characteristics from fields under different types of long-term management (extracted from Six et al., 2000).

4 C/N, C/P_{ORG} AND C/S RATIOS IN SOM ASSOCIATED WITH SOIL PARTICLES

A trend toward lower C/N ratios in finer size particles has often been reported (Christensen, 1992; Anderson et al., 1981); C/N ratios in the clay-size separates are mostly less than 10. This trend also holds true for the C/P_{org} ratios but less so for the C/S ratios.

C/N, C/P_{org} and C/S ratios found for sand-size separates are at the same range of values reported for plant litter. Those of the clay fractions contrarily are within or close to the range of ratios found in microbial biomass.

Microorganisms have C/N ratios of 7 to 12, C/P_{org} ratios of 14 to 36, and C/S ratios of 57 to 85 (Hedges and Oades, 1997; Sumann et al., 1998).

The weakest relationship with soil texture was found for the C/S ratios (Fig. 1.6).

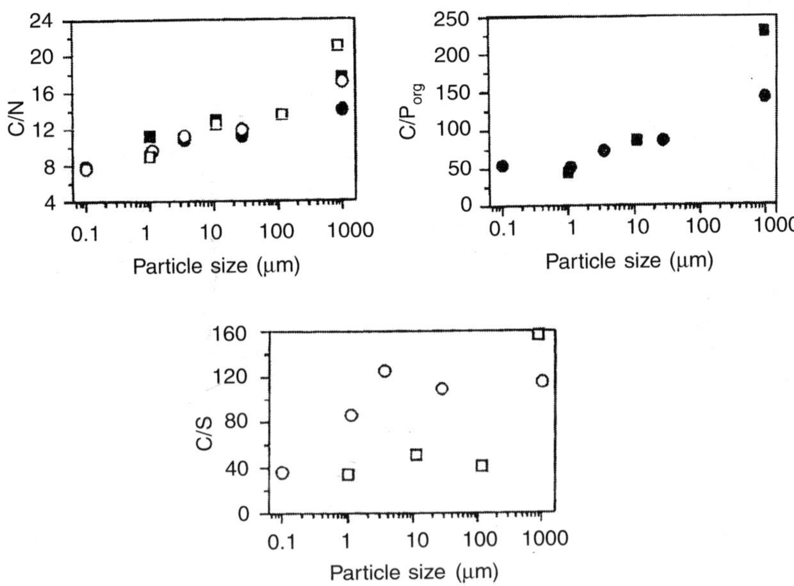

Fig. 1.6: Lower C/N-, C/OP- and C/S ratios at lower soil-particle sizes (Christensen 1992; Anderson et al., 1981).

5 COMPOSITION AND ORIGIN OF SOM ASSOCIATED WITH SOIL PARTICLES

Golchin et al. (1994) observed mostly large, undecomposed root and plant fragments in the 500 to 2,000 μm fractions and more decomposed materials in the 10 to 100 μm fractions. According to Amelung et al. (1998), organic matter in the coarse sand fraction comprises mainly slightly decomposed plant residues, whereas SOM associated with fine sand consists of fairly altered organic debris and fine root particles. Microbial cells in primary particle-size fractions cannot be detected by microscopy, probably due to their destruction by the high-energy sonication used for complete soil dispersion (Stemmer et al., 1998). CP/MAS ^{13}C NMR spectroscopy (for details of this technique, see Skjemstad et al., 1996) provides important insight into the chemical structure of SOM associated with different primary particle fractions (Fig. 1.7).

In general, O-alkyl resonances derived from carbohydrates or from ether bonds in degraded lignin predominate across the entire textural range but are minimized in the silt fraction (Baldock et al., 1992; Guggenberger et al., 1995). The strongest O-alkyl signals are always observed in the sand fraction, and

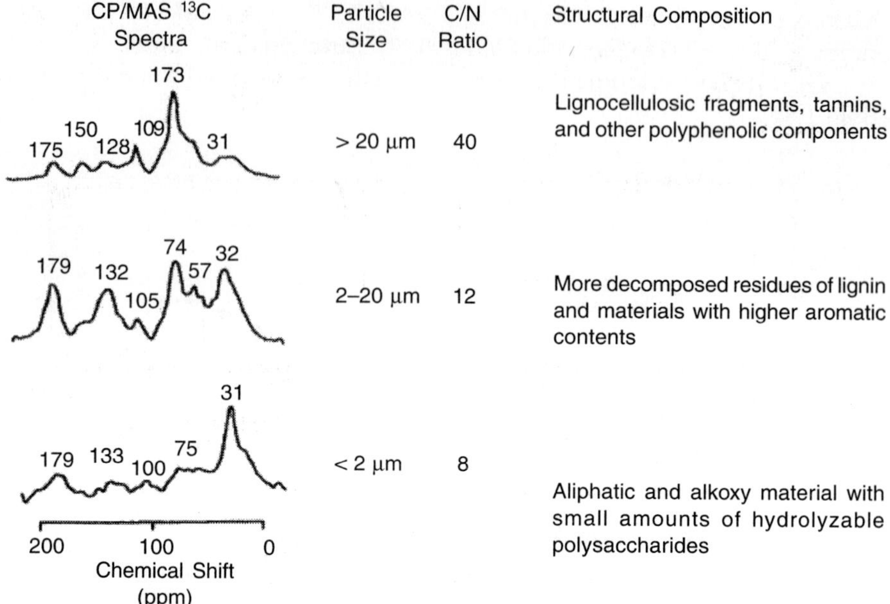

Fig. 1.7: Structural composition of the organic matter associated with decreasing particle-size fractions (Baldock et al., 1992).

spectra of the organic matter associated with sand strongly resemble those of fresh plant litter. Aromatic C, and in particular the C-substituted aromatic C, maximizes in silts but is lowest in the clay fraction(s). Alkyl C steadily increases with decreasing particle size and represents in the clay fraction the major or second major C species. Oades et al. (1987) identified long-chain material as the dominating form of alkyl C and considered a contribution of natural waxes.

Solid-state ^{15}N NMR spectroscopy of nitrogen-containing compounds of the fine particle size fractions (<20 μm) of two podzols showed that 60 to 90% of the nitrogen in this fraction is assigned to amides and a smaller percentage to amino groups (Knicker et al., 2000). This confirms earlier studies demonstrating that peptide-like material plays an important role in formation of the refractory nitrogen in SOM (Knicker and Lüdemann, 1995; Knicker et al., 1997). Moreover, substantial proportions of peptide materials were also observed in hydrolysis residues from clay and fine silt, and it can be hypothesized that pedogenic oxides of Al and Fe contribute to the stabilization of peptides in nonhydrolyzable forms (Leinweber and Schulten, 2000). The previous hypothesis of major contributions from N-containing heterocyclic aromatic compounds to the stabilized N fractions seems to be of minor importance (Knicker et al., 2000).

6 STABILIZATION OF SOM AND XENOBIOTICS IN MICROPOROUS STRUCTURES OR ON HYDROPHOBIC SURFACES

The spatial arrangement of solid particles in the soil ecosystem results in a complex and discontinuous pattern in pore spaces of various size and shape.

Pore neck sizes determine the accessibility by microorganisms and soil enzymes. Microbial access to substrates located in small pores or sorbed on solid surfaces may be limited and the substrates thereby protected from rapid decomposition (de Jonge et al., 2000). Bacteria enclosed in smaller pores may also be protected for a long time against grazing by soil animals.

Within pores of less than 8 μm (mesopores) dissolved organic matter is even inaccessible to exoenzymes (Arai and Norde, 1990; Fig. 1.8). Carbohydrates and nitrogenous substances can interact with soil colloids by sorption, polymerization, or entrapment in voids of inorganic or organic soil constituents (Kaiser and Guggenberger, 2003).

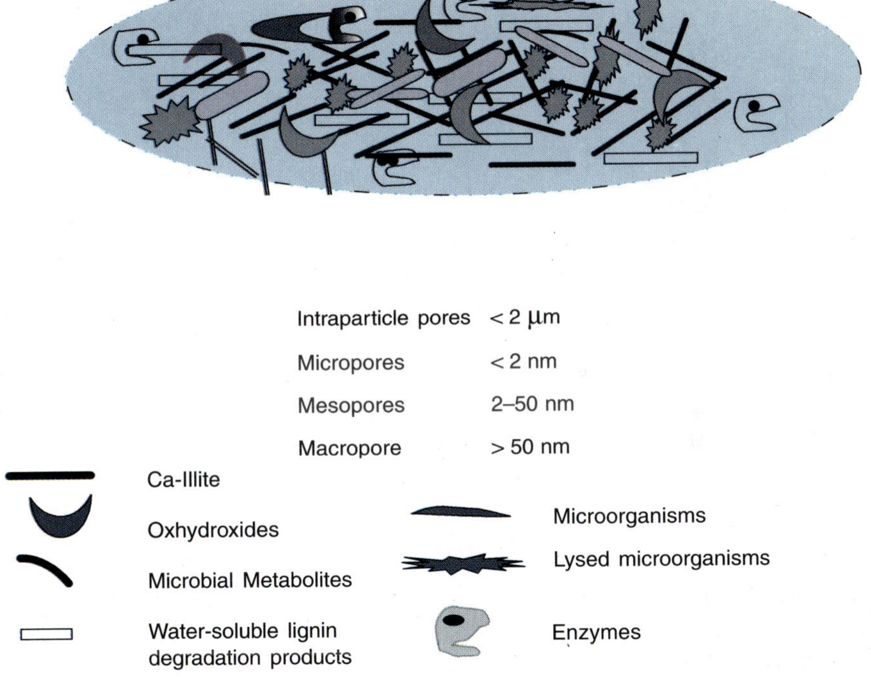

Fig. 1.8: Model for the stabilization of dissolved organic matter in soil pore system (Kaiser and Guggenberger, 2003).

Mineralization of low concentrations (10 ppm or below) of xenobiotics, phenols, or phenolic acids, amino acids, or carbohydrates can be strongly retarded because of sorptive interaction with pores or surfaces may be strongly retarded (Martin and Haider, 1979; Swinnen and van Veen, 1992). Xenobiotics such as pesticides or polyaromatic hydrocarbons can be stabilized in the microporous structure of minerals or soil organic matter. This sorption of chemicals in soil can be rapid or may require weeks, months, or even years to reach equilibrium (Alexander, 1995; Hatzinger and Alexander, 1995; Dec and Bollag, 1997). Figure 1.9 indicates a declining bioavailability of phenanthrene and similar polyaromatic hydrocarbons with increasing in time of contact in a sterile, organic-rich soil.

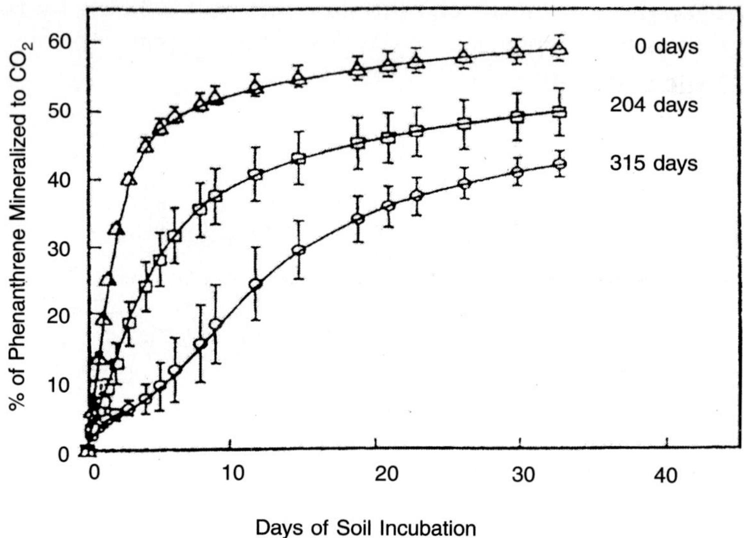

Fig. 1.9: Biodegradation of phenanthrene (10 ppm in soil) after progressive aging for 0, 204, and 315 days (Hatzinger and Alexander, 1995).

7 ORGANOMINERAL INTERACTIONS AND WATER QUALITY

The minimal leaching of bacteria from soils and the difficulty in separating microorganisms from soils is due to the attachment (adhesion) of microorganisms to solid particles or entrapment in soil aggregates or narrow pores. Review articles about the interactions of microorganisms with soil particles are numerous (e.g., Stotzky, 1986; Robert and Chenu, 1992; Chenu and Stotzky, 2002). Emphasis, however, should be given to the fact that adsorption and adhesion inhibit the movement of microorganisms, and mainly that of pathogenic organisms in soil, precluding their entry into surface water or ground water.

Water quality in catchment areas is controlled by the ability of soils to retain organic matter and prevent its entering the aqueous phase. The adsorption capacity of the soil is related to the clay content of the A horizon or the surface area of the soil. Figure 1.10 shows that the DOC of water in different catchments decreases from about 20 mg L^{-1} in catchments with sandy A horizons to about 5 mg L^{-1} in those with loamy A horizons (Nelson et al., 1990, 1993). The water quality with respect to organic matter is related to the adsorptive capacity of the clay.

The various management regimes in catchments are of minor influence but sorption of organic materials is also influenced by the salt content and sodium adsorption ratios of the water (Oades, 1995).

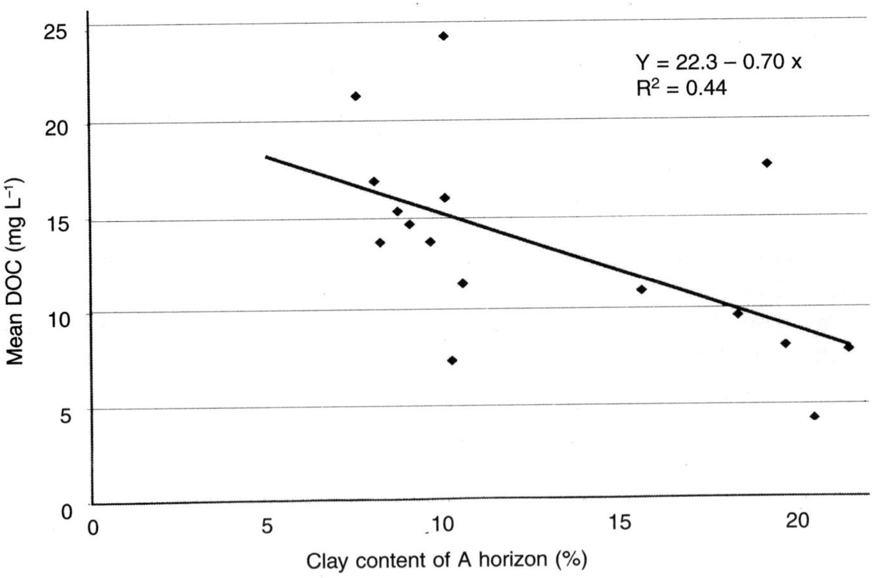

Fig. 1.10: Influence of clay content in the A horizon on adsorption of dissolved organic carbon (DOC) (Nelson et al., 1990).

8 INVOLVEMENT OF CLAY MINERALS IN ROOT-DISEASE SUPPRESSION

Roots growing through soil are exposed to infection by pathogens, especially by fungi. The specialized vascular wilt fungi such as *Fusarium* sp. and *Verticillium* sp. attack juvenile roots. Once inside the roots, they cause great damage. Various studies suggest that abiotic and biotic soil properties influence plant diseases and hence the expressions "suppressive" and "conducive" soils were coined (Alabouvette, 1986).

The composition and activity of an antagonistic soil microflora are not independent of abiotic soil properties (Höper et al., 1995; Dominguez et al., 2001). Soils that showed low incidence of fusarium wilt disease even after a long production time contained montmorillonitic clays, whereas soils with severe disease symptoms did not (Stotzky and Martin, 1963). The influence of clay type has been experimentally demonstrated by finding that the addition of montmorillonite or of kaolinite together with illite and montmorillonite to a wilt-conducive sandy soil (Amir and Alabouvette, 1993), with a simultaneous increase in soil pH to 7.0 had the strongest effect (Höper et al., 1994). Studies by Dominguez et al. (2001), however, showed that a high clay content of soils did not always appear to be related as such to the development of disease resistance: other factors such as extractable Fe and pH also seem to play an important role for the development of a suppressive or conducive soil.

9 CONCLUSIONS

1. The mineralogical characteristics of a soil influence soil organic matter content, its fertility, and its structure.
2. The spatial arrangement of soil particles results in a complex and discontinuous pattern of pore spaces of various size and shape, which are important for water and air supply.
3. The microporous structure forms habitats for microorganisms, protects humus from degradation, and favors its sequestration. It also diminishes the risks of chemicals causing environmental damage.
4. Clay contents and sequestration within small pores protect ground and surface waters from contamination through pathogens, xenobiotics, and dissolved humic compounds.
5. Certain clay minerals in soil can suppress specific fungal root diseases.

References

Alabouvette C. 1986. Fusarium-wilt suppressive soils from the Chateaurenard region: a review of a 10-year study. *Agronomie* 6: 273–284.

Alexander M. 1995. How toxic are toxic chemicals in soil? *Environ. Sci. Tech.* 29: 2713–2717.

Amelung W., Zech W., Zhang X., Follett R.F., Tiessen H., Knox E., and Flach K.-W. 1998. Carbon, nitrogen, and sulfur pools in particle-size fractions as influenced by climate. *Soil Sci. Soc. Amer. J.* 62: 172–181.

Amir H. and Alabouvette C. 1993. Involvement of soil abiotic factors in the mechanism of soil suppressiveness to fusarium wilts. *Soil Biol. Biochem.* 25: 157–164.

Anderson D.W., Saggar S., Bettany J.R., and Stewart J.W.B. 1981. Particle size fractions and their use in studies of soil organic matter: I. The nature and distribution of forms of carbon, nitrogen and sulfur. *Soil Sci. Soc. Amer. J.* 45: 767–772.

Arai T. and Norde W. 1990. The behavior of some model proteins at solid-liquid interfaces, I. Adsorption from single protein solutions. *Colloids and Surfaces* A 51: 1–15.

Baldock J.A., Oades J.M., Waters A.G., Peng X., Vassallo, A.M., and Wilson, M.A. 1992. Aspects of the chemical structure of soil organic materials as revealed by solid-state ^{13}C NMR spectroscopy. *Biogeochem.* 16: 1–42.

Burke I.C., Yonker C.M., Parton W.J., Cole C.V., Flach K., and Schimel D.S. 1989. Texture, climatic and cultivation effects on soil organic matter content in U.S. grassland soils. *Soil Sci. Soc. Amer. J.* 53: 800–805.

Chenu C. and Stotzky G. 2002. Bio-physico-chemical interactions between microorganisms and soil particles. In: *Interactions between Soil Particles and Microorganisms and Their Impact on the Terrestrial Environment.* P.M. Huang, J.-M. Bollag, and N. Senesi (eds.). IUPAC Series on Analytical and Physical Chemistry of Environmental Systems. Wiley-Interscience, New York, NY, pp. 3–40.

Christensen B.T. 1992. Physical fractionation of soil and organic matter in primary particle size and density separates. *Adv. Soil Sci.* 20: 1–90.

de Jonge H., de Jonge L.W., and Mittelmeijer-Hazelegger M.C. 2000. The microporous structure of organic and mineral soil materials. *Soil Sci.* 165: 99–108.

Dec J. and Bollag J.-M. 1997. Determination of covalent interactions and noncovalent binding between xenobiotic chemicals and soil. *Soil Sci.* 162: 858–874.

Dominguez J., Negrin M.A., and Rodriguez C.A. 2001. Aggregate water-stability, particle-size and soil solution properties in conducive and suppressive soils to Fusarium wilt of banana from Canary Islands (Spain). *Soil Biol. Biochem.* 33: 449–455.

Filip Z., Haider K., and Martin J.P. 1972. Influence of clay minerals on the formation of humic substances. *Soil Biol. Biochem.* 4: 148–154.

Golchin A., Oades J.M., Skjemstad J.O., and Clarke P. 1994. Soil structure and carbon cycling. *Austr. J. Soil Res.* 32: 1043–1064.

Gregorich E.G., Kachanoski R.G., and Voroney R.P. 1988. Ultrasonic dispersion of aggregates: Distribution of organic matter in size fractions. *Can. J. Soil Sci.* 68: 395–403.

Guggenberger G. and Haider K. 2002. Effects of mineral colloids on biogeochemical cycling of C, N, P, and S in soil. In: *Interactions between Soil Particles and Microorganisms and Their Impact on the Terrestrial Environment.* P.M. Huang, J.-M. Bollag, and N. Senesi (eds.). IUPAC Series on Analytical and Physical Chemistry of Environmental Systems. Wiley-Interscience, New York, NY, pp. 267–322.

Guggenberger G., Zech W., Haumaier L., and Christensen B.T. 1995. Land-use effects on the composition of organic matter in particle-size separates of soil: II. CPMAS and solution ^{13}C NMR analysis. *Eur. J. Soil Sci.* 46: 147–158.

Hassink J. 1997. The capacity of soils to preserve organic C and N by their association with clay and silt particles. *Plant Soil* 191: 77–87.

Hatzinger P.B. and Alexander M. 1995. Effect of aging of chemicals in soil on their biodegradability and extractability. *Environ. Sci. Tech.* 29: 537–545.

Hedges J.I. and Oades J.M. 1997. Comparative organic geochemistries of soils and marine sediments. *Org. Geo.* 27: 319–361.

Höper H., Steinberg C., and Alabouvette C. 1995. Involvement of clay type and pH in the mechanisms of soil suppressiveness to Fusarium wilt of flax. *Soil Biol. Biochem.* 27: 955–967.

Jenkinson D.S. 1988. Soil organic matter and its dynamics. In: *Russell's Soil Conditions and Plant Growth.* 11th edition. A. Wild (ed.). Longman, New York, NY, pp. 564–607.

Jin Y., Pratt E., and Yates M.V. 2000. Effect of mineral colloids on virus transport through saturated sand columns. *J. Environ. Qual.* 29: 532–539.

Kaiser K. and Guggenberger G. 2000. The role of DOM sorption to mineral surfaces in the preservation of organic matter in soils. *Org. Geochem.* 31: 711–725.

Kaiser K. and Guggenberger G. 2003. Mineral surfaces and soil organic matter. *Eur. J. Soil Sci.* 54: 219–236.

Kaiser, K. and Zech, W. 2000. Dissolved organic matter sorption by mineral constitutents of subsoil clay fractions. *J. Plant Nutr. Soil Sci.* 163: 531–535.

Kaiser K., Eusterhues K., Rumpel C., Guggenberger G., and Kögel-Knabner I. 2002. Stabilization of organic matter by soil minerals—investigations of density and particle-size fractions from two acid forest soils. *J. Plant Nutr. Soil Sci.* 165: 451–459.

Knicker H. and Lüdemann H.-D. 1995. N-15 and C-13 CPMAS and solution NMR studies of N-15 enriched plant material during 600 days of microbial degradation. *Org. Geochem.* 23: 329–341.

Knicker H., Lüdemann H.-D., and Haider K. 1997. Incorporation studies of NH_4^+ during incubation of organic residues by ^{15}N-CPMAS-NMR-specroscopy. *Eur. J. Soil Sci.* 48: 431–441.

Knicker H., Schmidt M.W.I., and Kögel-Knabner I. 2000. Nature of organic nitrogen in fine particle size separates of sandy soil of highly industrialized areas revealed by NMR spectroscopy. *Soil Biol. Biochem.* 32: 241–252.

Ladd J.N., Foster R.C., Nannipieri P., and Oades J.M. 1996. Soil structure and biological activity. In: *Soil Biochemistry.* G. Stotzky and J.-M. Bollag (eds.). Marcel Dekker, New York, NY, vol. 9, pp. 23–78.

Leinweber P. and Schulten H.-R. 2000. Nonhydrolyzable forms of soil organic nitrogen: Extractability and composition. *J. Plant Nutr. Soil Sci.* 163: 433–439.

Martin J.P. and Haider K. 1979. Effect of concentration on decomposition of some ^{14}C-labelled phenolic compounds, benzoic acid, glucose, cellulose, wheat straw, and Chlorella protein in soil. *Soil Sci. Soc. Amer. J.* 43: 917–920.

Nelson P.L., Baldock J.A., and Oades J.M. 1993. Concentration and composition of dissolved organic carbon in streams in relation to catchment soil properties. *Biogeochem.* 19: 27–35.

Nelson P.N., Cotsaris E., Oades J.M., and Buursill D.B. 1990. The influence of soil clay content on dissolved organic matter in stream waters. *Austr. J. Marine Freshw. Res.* 41: 761–768.

Oades J.M. 1995. Recent advances in organomineral interactions: Implications for carbon cycling and soil structure. In: *Environmental Impact of Soil Component Interactions.* P.M. Huang et al. (eds.). CRC Press, Boca Raton, FL, pp. 119–134.

Oades J.M., Vassallo A.M., Waters A.G., and Wilson M.A. 1987. Characterization of organic matter in particle size and density fractions from a red-brown earth by solid-state ^{13}C N.M.R. *Austr. J. Soil Res.* 25: 71–82.

Parton W.J., Schimel D.S., Cole C.V., and Ojima D.S. 1987. Analysis of factors controlling soil organic matter levels in Great Plains grasslands. *Soil Sci. Soc. Amer. J.* 51: 1173–1179.

Percival H.J., Parfitt R.L., and Scott N.A. 2000. Factors controlling soil carbon levels in New Zealand grasslands: is clay content important? *Soil Sci. Soc. Amer. J.* 64: 1623–1630.

Robert M. and Chenu C. 1992. Interactions between soil minerals and microorganisms. In: *Soil Biochemistry.* G. Stotzky and J.-M. Bollag (eds.). Marcel Dekker, New York, NY, vol. 7, pp. 307–404.

Roberts T.L., Bettany J.R., and Stewart J.W.B. 1989. A hierarchical approach to the study of organic C, N, P and S in western Canadian soils. *Can. J. Soil Sci.* 69: 739–749.

Six J., Elliot E.T., and Paustian K. 2000. Soil structure and soil organic matter: II. A normalized stability index and the effect of mineralogy. *Soil Sci. Soc. Amer. J.* 64: 1042–1049.

Skjemstad J.O., Clarke P., Taylor J.A., Oades J.M., and McClure S.G. 1996. The chemistry and nature of protected carbon in soil. *Austr. J. Soil Res.* 34: 251–171.

Stemmer M., Gerzabek M.H., and Kandeler E. 1998. Organic matter and enzyme activity in particle-size fractions of soils obtained after low-energy sonication. *Soil Biol. Biochem.* 30: 9–17.

Stotzky, G. 1985. Mechanisms of addesion & Clays, with reference to soil systems. In: *Bacterial Adhesion.* D.C. Savage and M. Fletcher (eds.). Plenum Press, New York. p. 195.

Stotzky G. 1986. Influence of soil mineral colloids on metabolic processes, growth, adhesion, and ecology of microbes and viruses. In: *Interactions of Soil Minerals with Natural Organics and Microbes.* P.M. Huang and M. Schnitzer (eds.). SSSA Spec. Publ. 17: 305–428.

Stotzky G. and Martin T. 1963. Soil mineralogy in relation to the spread of Fusarium wilt of banana in Central America. *Plant Soil* 18: 317–337.

Sumann M., Amelung W., Haumaier L., and Zech W. 1998. Climatic effects on soil organic phosphorus in the North American Great Plains identified by phosphorus-31 nuclear magnetic resonance. *Soil Sci. Soc. Amer. J.* 62: 1580–1586.

Swinnen J. and van Veen J. 1992. Unterscheidung von Wurzel- und mikrobieller Atmung im Boden durch Exudatmarkierung. *Berichte über Landwirtschaft, Humushaushalt* 206: 114–116.

Tiessen H. and Stewart J.W.B. 1983. Particle-size fractions and their use in studies of soil organic matter: II. Cultivation effects on organic matter composition in size fractions. *Soil Sci. Soc. Amer. J.* 47: 509–514.

van Veen J.A. and Kuikman P.J. 1990. Soil structural aspects of decomposition of organic matter by micro-organisms. *Biogeochem.* 11: 213–234.

van Veen J.A., Ladd J.N., and Amato M. 1985. Turnover of carbon and nitrogen through the microbial biomass in a sandy loam and a clay soil incubated with [^{14}C(U)]glucose and [^{15}N]$(NH_4)_2SO_4$ under different moisture regimes. *Soil Biol. Biochem.* 17: 257–274.

2

Modeling Soil Microbial and Geochemical Processes under Anaerobic Conditions

F. Dassonville* *and* **P. Renault**

Abstract

Anaerobiosis in soil can have undesirable environmental effects on the soil, the deep vadose zone, the aquifer, and the atmosphere. These effects include mobilization of metals and emission of greenhouse gases resulting from microbial processes in close interaction with abiotic geochemical transformations. The main anaerobic microbial catabolism pathways include respiration, fermentation, acetogenesis, and methanogenesis. On the one hand the soil's geochemical properties affect microbial activities through substrate availability, thermodynamic regulations, and inhibitions. On the other hand microbial activities, by modulating adsorption/desorption and dissolution/precipitation reactions, affect the pH of the soil solution, its reduction/oxidation level, and the complexation of metals with volatile fatty acids and solid particles. However, some of the solid phases (e.g., calcite, metal oxides, and oxyhydroxides) can partially buffer changes in the solution. Abiotic transformations can also partly regenerate the terminal electron acceptors used by microorganisms. Few models combine microbial and geochemical transformations. Of those that do, most were initially intended to describe anaerobic biodigesters. For this environment they legitimately limit the geochemical transformations they model to acid/base reactions and ignore reactions between solids and the solution. A new modeling approach that explicitly describes anaerobic microbiology and soil geochemistry has recently been proposed. It was tested with batch experiments and proved to reflect the experimental data.

Corresponding author: Fabrice Dassonville, CIRAD Amis Programme Agronomie Equipe Most TA 40/01, Avenue Agropolis 34398 Montpellier Cedex 05 France, E-mail: Fabrice.dassonville@cirad.Fr

A number of environmental problems could benefit from improved biogeochemical models, but such models must explicitly distinguish between microbial and abiotic transformations with their different kinetic behaviors and causal factors.

1 OVERVIEW

Soils are usually well aerated, in which case anaerobic microbial activities are restricted to small anoxic sites for short periods (Parry et al., 2000; Parry et al., 1999; Hojberg et al., 1994; Zausig et al., 1993; Tiedje et al., 1984). Situation in which anaerobiosis predominates include (i) flooded soils (rice paddies, wetlands, and peatlands) (Chidthaisong and Conrad, 2000; Liesack et al., 2000; Chidthaisong et al., 1999; Krylova et al., 1997), and (ii) soils that have received heavy applications of organic amendment, under which conditions transient moisture excesses occur (De Cockborne et al., 1999; Glissman and Conrad, 1999).

Anaerobiosis can have an impact on soil functions as well as on the deep vadose zone, underlying aquifer, and atmosphere. It affects root growth by modifying the bioavailability of nutrients, e.g., P (Bertrand, 1998) and O_2 for roots without aerenchyma, i.e., those that cannot transport gases from shoot to root (Drew, 1997; Visser et al., 1997). It also affects all aerobic microbial activities. Anaerobic microbial activities can lead to an accumulation of toxic transient products (e.g., lactate, propionate, and butyrate) (De Cockborne et al., 1999; Aguilar et al., 1995; Fuzukaki et al., 1990a,b; Van Den Heuvel et al., 1988), can mobilize metals (e.g. Al or Mn), and can generate simple complexing organic molecules (De Cockborne et al., 1999; Förstner, 1987). Consequently, anaerobic microbial activities play a part in transporting Mn or Al into groundwater and/ or mobilizing them within the aquifer by supplying this environment with volatile fatty acids from the soil (Förstner, 1987). Other anaerobic microbial activities contribute to N_2O and CH_4 emissions into the atmosphere (Chidthaisong et al., 1998; Inubushi et al., 1997; Peters and Conrad, 1996; Prieme, 1994; Bouwman, 1990).

On the positive side, anaerobiosis can be used to eliminate excess NO_3^- in surface water (Curmi et al., 1997) a condition necessary for the emergence of young rice plants (Forde, 2000). It also stimulates the microbial dissimilatory reduction of FeIII involved in the degradation of some organic contaminants such as nonchlorinated organics and a wide variety of monoaromatic compounds; these are converted into harmless CO_2 and organic byproducts that can be easily metabolized further (Lovley and Coates, 2000; Lovley, 1995). Anaerobiosis can enhance the microbial dissimilatory reduction of toxic metals (e.g., Pb(II), Se(VI), Cr(VI), Hg(II)) and radionuclides (e.g., U(VI), Tc(VII)), which can then be further converted into insoluble forms (Pb(O), Se(O), Cr(III), U(IV), Tc(IV), etc.) or volatile forms (Hg(o)) for easy removal from contaminated environments (Lovley and Coates, 2000).

Anaerobic microbial activities are closely linked to geochemical reactions in the solution and at solid-solution interfaces (Reddy et al., 1998; Stumm and

Morgan, 1996; Zehnder and Stumm, 1988). On the one hand, they depend on the geochemical characteristics of the solution and on transient changes in these characteristics, which affect (i) substrate availability for microorganisms in the solution and at solid/solution interfaces (Brown et al., 1999; Bousserrhine et al., 1998; Stumm and Morgan, 1996; Roden and Zachara, 1996; Lefebvre-Drouet and Rousseau, 1995), and (ii) the feasibility of microbial reactions, through inhibition by particular chemical species in solution (NO_2^-, volatile fatty acids, etc.) (Klüber and Conrad, 1998; Aguilar et al., 1995; Fukuzaki et al., 1990a,b; Van Den Heuvel et al., 1988) and thermodynamic regulation (H_2) (Schink, 1997; Stams, 1994). On the other hand, microbial activities affect the geochemical characteristics of the solution, including (i) the pH, via net microbial production of H^+ (e.g. during the reduction of NO_2^- to N_2O), CO_2, NH_4^+, HS^-, organic acids, etc. (Stumm and Morgan, 1996), (ii) soil reduction/oxidation level, by influencing redox couples including NO_3^-/NO_2^-, Fe^{3+}/Fe^{2+}, H^+/H_2, SO_4^{2-}/HS^-, CO_2/CH_4 (Peters and Conrad, 1996; Glinski et al., 1995), and (iii) the amount of simple complexing fatty acids, including acetate, propionate, and butyrate (Glissman and Conrad, 1999; Chidthaisong et al., 1998; Krylova et al., 1997; Chin and Conrad, 1995; Küsel and Drake, 1995). Microbial activities also affect solid phases, for example: (i) adsorption/desorption reactions via microbiological byproducts (H^+ and other cations) that interact with solid phases according to their solid characteristics (Stumm and Morgan, 1996; Robert and Chenu, 1992), (ii) dissolution of minerals (metal oxides and oxyhydroxides, calcite, etc.) that further modify soil solution composition (Stumm and Morgan, 1996; Lefebvre-Drouet and Rousseau, 1995), and (iii) precipitation of other minerals such as green rust (Refait et al., 1999; Trolard and Bourrié, 1999; Abdemoula et al., 1998; Génin et al., 1998; Trolard et al., 1997; Schwertmann and Fechter, 1994). Microbial effects may be partly neutralized by geochemical reactions at the solid-solution interface, including dissolution of calcite (Stumm and Morgan, 1996) and/or metal oxides (goethite, etc.) and precipitation of other minerals (siderite, green rust) (Refait et al., 1999; Abdemoula et al., 1998). Abiotic geochemical transformations may also partly regenerate some microbial substrates (e.g., chemical oxidation of HS^- into S^0, with concomitant reduction of Fe^{3+} to Fe^{2+} (Murase and Kimura, 1997) or indirectly enhance other transformations. For example, precipitation of Fe^{2+} from FeII minerals (siderite, polysulfur species) or FeII–III minerals (green rust) may thermodynamically facilitate reduction of Fe^{3+} and hence the dissolution of FeIII minerals (goethite, hematite, ferrihydrite) (Brown et al., 1999).

The main anaerobic microbial activities in soil have been extensively studied with regard to (i) the reduction of various terminal electron acceptors during respiratory processes, including N, Mn, Fe, and S oxides (Lovley, 1995; Widdel, 1988), (ii) fermentation and acetogenic reactions (Chin and Conrad, 1995; Stams, 1994), and (iii) methanogenesis (Oremland, 1988). So far, few studies have focused on lipid and protein degradation (Jones, 1999; Finnerty, 1989). Description of the diversity and structure of microbial consortia has improved with recent use of molecular biology techniques (Head et al., 1998). Geochemical reactions have generally been well characterized. However, adequate description is still lacking for (i) the solid and solute speciations of some compounds, including Fe (Brown et al., 1999), (ii) kinetics between solid particles and solution (Stumm and Morgan, 1996), and (iii) the properties of organic ligands, which vary with the soil.

A number of modeling studies have been carried out on anaerobic environments, including the deep vadose zone and the aquifer (Hunter et al., 1998; Salvage and Yeh, 1998), anaerobic waste biodigesters (Vavilin et al., 1998; Jeyseelan, 1997; Vavilin et al., 1997a; Vavilin et al., 1996a,b,c; Shin and Song, 1995; Vavilin et al., 1995a,b,c; Vavilin et al., 1994a,b; Angelidacki et al., 1993; Vasiliev et al., 1993), and the rumen (Susmel et al., 1999; Dijkstra et al., 1998). They have led to some detailed models (see also Tables 2.1 and 2.2 below). Unfortunately, such models cannot be applied directly to soil environments because (i) soil microbial communities are varied and therefore the relative contributions of the metabolic pathways to oxidation of organic matter are varied (Liesack et al., 2000), (ii) models established for rumen and biodigesters ignore chemical and biological interactions between solid particles and the solution (see Table 2.1), and (iii) models established for the aquifer mostly account for geochemical processes and do not accurately describe microbial activities (Hunter et al., 1998).

A few models explicitly combine microbial processes and abiotic geochemical transformations within the soil; of these, most limit their consideration of abiotic transformations to describing acid/base reactions (Vavilin et al., 2000; Lokshina and Vavilin, 1999; Vavilin et al., 1997b). A new model accurately describes anaerobic microbial activities (glucose metabolism) and soil geochemical transformations (acid/base reactions, complexation reactions, reduction/oxidation reactions, dissolution/precipitation, and adsorption/desorption) and has proven capable of reflecting experimental data (Dassonville et al., 2004).

2 ANAEROBIC MICROBIAL ACTIVITIES AND ORGANIC MATTER DECOMPOSITION

2.1 Main Pathways Involved in Anaerobic Decomposition of Organic Compounds

Both aerobic and anaerobic types of microbial catabolism first of all involve hydrolysis of organic polymers (lignin, polysaccharides, lipids, proteins, and other complex compounds), followed by oxidation of the monomers so produced (Zehnder and Stumm, 1988). The products of the initial hydrolysis are identical, but aerobic and anaerobic conditions involve different microbial communities and different mechanisms (Brune et al., 2000; Zehnder and Svensson, 1986), giving different reaction rates (Kristensen et al., 1995). For example, Leschine (1995) showed that cellulose hydrolysis results from fungal activity under aerobic conditions and from bacterial activity under anaerobic conditions. The products of the initial hydrolysis are (i) mainly glucose, some other pentoses and hexoses

Table 2.1: Main microbial anaerobic processes and geochemical transformations accounted for in models dealing with soil, anaerobic digester reactors, and deep vadose zone

			Microbial activities										Geochemical reactions				
Application	References	Year	Hydrolysis	N-oxide respiration	Fermentation	Mn-respiration	S-respiration	Fermentation	Acetogenesis	Methanogenesis	Pop.dynamics	Biogeochemical interactions	Acid/base	Reduction/ oxidation	Complexation	Adsorption/ desorption	Dissolution precipitation
Soil	Dassonville et al.,	2004	•	•	•		•	•	•		•	•	•	•	•	•	•
	Segers and Kengen,	1998		*1	*1	*1	*1			•	•						
	Di Chio et al.,	1999		*2	*2	*2	*2	*2	*2	*2	•						
	Di Chio et al.,	2000		*2	*2	*2	*2	*2	*2	*2	•						
	Grant,	1998						•	•	•	•						
	Blagodatsky et al.,	1998		*2	*2	*2	*2	*2	*2	*2	•						
	Blagodatsky and Richter,	1998		*2	*2	*2	*2	*2	*2	*2	•						
	Cao and Dent,	1995								•							
	Vavilin et al.,	2000	•				•	•	•	•	•	•	•				
	Lokshina andVavilin,	1999	•				•	•	•	•	•	•	•				
	Vavilin et al.,	1997a	•				•	•	•	•	•	•	•				
	Vavilin et al.,	1997b	•				•	•	•	•	•	•	•				
	Grant et al.,	1993a	•	*2	*2	*2	*2	*2	*2	*2	•						
	Grant et al.,	1993b	•	*2	*2	*2	*2	*2	*2	*2	•						

Contd.,

(Table 2.1 Cont'd)

Anaerobic reactor digester																	
	Vavilin et al.,	1998	•				•	•	•	•	•	•					
	Vavilin et al.,	1997a	•				•	•	•	•	•	•					
	Jeyseelan,	1997						•	•	•	•						
	Vavilin et al.,	1996a	•				•	•	•	•	•	•	•				
	Vavilin et al.,	1996b	•				•	•	•	•	•	•	•				
	Vavilin et al.,	1996c	•				•	•	•	•	•	•	•				
	Vavilin et al.,	1995a	•				•	•	•	•	•	•	•				
	Vavilin et al.,	1995b	•				•	•	•	•	•	•	•				
	Vavilin et al.,	1995c	•				•	•	•	•	•	•	•				
	Shin and Song,	1995	•					•		•							
	Vavilin et al.,	1994a	•				•	•	•	•	•	•	•				
	Vavilin et al.,	1994b	•				•	•	•	•	•	•	•				
	Vasiliev et al.,	1993	•				•	•	•	•	•	•	•				
	Angelidacki et al.,	1993	•					•	•	•	•	•	•				
Deep vadose zone	Hunter and Van Capellen,	1998	•	•	•	•	•			•		•	•	•		•	•
	Salvage and Yeh	1998		*3	*3	*3	*3	*3	*3	•	•	•	•	•	•	•	•

*1 The authors consider an unspecified alternative electron acceptor.

*2 The authors describe anaerobic microbial activities via N and C turnover in soil. No anaerobic microbial activity is specified.

*3 The authors assume that their model allows for an unlimited number of microbial species, electron acceptors, and microbial reactions as long as a sufficient number of reactions are specified to solve the number of unknowns.

Table 2.2: Main chemical compounds accounted for in models dealing with soil, anaerobic digester reactors, and deep vadose zone

			Solutes														
		Year	Organic S.						-Mineral S.			-Gases			Solids		
Application	References		Formula	4cetate	Propionate	Butyrate	Lactate	Alcohols	Other Organic S.	N-compounds	S-compounds	CO_2	H_2	H_2S	CH_4	Calcite	Metal-Oxide
Soil	Dassonville et al.,	2004	•	•		•		•		•	•	•	•	•	•	•	•
	Segers and Kengen,	1998		•								•			•		
	Di Chio et al.,	1999							•	•		•					
	Di Chio et al.,	2000							•	•		•					
	Grant,	1998		•					•	•		•			•		
	Blagodatsky et al.,	1998							•	•		•					
	Blagodatsky and Richter,	1998							•	•		•					
	Cao and Dent,	1995							•			•			•		
	Vavilin et al.,	2000		•	•	•			•	•	•	•	•	•	•		
	Lokshina et Vavilin,	1999		•	•	•			•	•	•	•	•	•	•		
	Vavilin et al.,	1997a		•	•	•			•	•	•	•	•	•	•		
	Vavilin et al.,	1997b		•	•	•			•	•	•	•	•	•	•		
	Grant et al.,	1993a							•	•		•					
	Grant et al.,	1993b							•	•		•					

(Contd.,)

(Table 2.2 cont'd)

			C1	C2	C3	C4	C5	C6	C7	C8	C9	C10	C11	C12	C13	C14	C15
Anaerobic digester reactor	Vavilin et al.,	1998		•	•	•			•	•	•	•	•	•	•		
	Vavilin et al.,	1997a		•	•	•			•	•	•	•	•	•	•		
	Jeyseelan,	1997		•	•	•			•			•			•		
	Vavilin et al.,	1996a		•	•	•			•	•	•	•	•	•	•		
	Vavilin et al.,	1996b		•	•				•	•	•	•	•	•	•		
	Vavilin et al.,	1996c		•	•				•	•	•	•	•	•	•		
	Vavilin et al.,	1995a		•	•				•	•	•	•	•	•	•		
	Vavilin et al.,	1995b		•	•				•	•	•	•	•	•	•		
	Vavilin et al.,	1995c		•	•				•	•	•	•	•	•	•		
	Shin and Song,	1995							•						•		
	Vavilin et al.,	1994a		•	•				•	•	•	•	•	•	•		
	Vavilin et al.,	1994b		•	•				•	•	•	•	•	•	•		
	Vasiliev et al.,	1993		•	•				•	•	•	•	•	•	•		
	Angelidacki et al.,	1993		•	•	•			•	•		•	•		•		
Deep vadose zone	Hunter et al.,	1998							•	•	•	•	•	•	•	•	•
	Salvage and Yeh	1998	*1	*1	*1	*1	*1	*1	*1	*1	*1	*1	*1	*1	*1	*1	*1

*1 The authors assume that their model allows for an unlimited number of microbial species, electron acceptors, and microbial reactions as long as a sufficient number of reactions are specified to solve for the number of unknowns.

from polysaccharides (Colberg, 1988), (ii) free organic acids and other simple compounds such as glycerol and galactose from lipids (McInerney, 1988), and (iii) amino acids with various functional groups (including –COOH, –OH, $-NH_2$, –SH, and aromatic cycles) from proteins (McInerney, 1988).

The resultant monomers are then oxidized with the transfer of electrons and H^+ to cofactors (NAD^+, FAD^+, etc.), which are regenerated by reduction/oxidation reactions, including anaerobic respiration and fermentation. Transient compounds and final products depend on the microorganisms involved and the environmental conditions (Zehnder and Svensson, 1986). Organic monomers are oxidized through various pathways (Fig. 2.1):

—catabolism, producing CO_2 (as does aerobic catabolism) when there are terminal electron acceptors such as N-oxides (NO_3^-, NO_2^-, N_2O) (Moodie and Ingledew, 1990; Stouthamer, 1988) to totally oxidize various monomers. Fe- and Mn-oxides can also act significantly as terminal electron acceptors in microbial respiration (Nealson and Little, 1997; Lovley, 1995; Nealson and Saffarani, 1994; Nealson and Myers, 1992). Toxic metals (Cr(VI)), metalloids (As(V), Se(VI)),

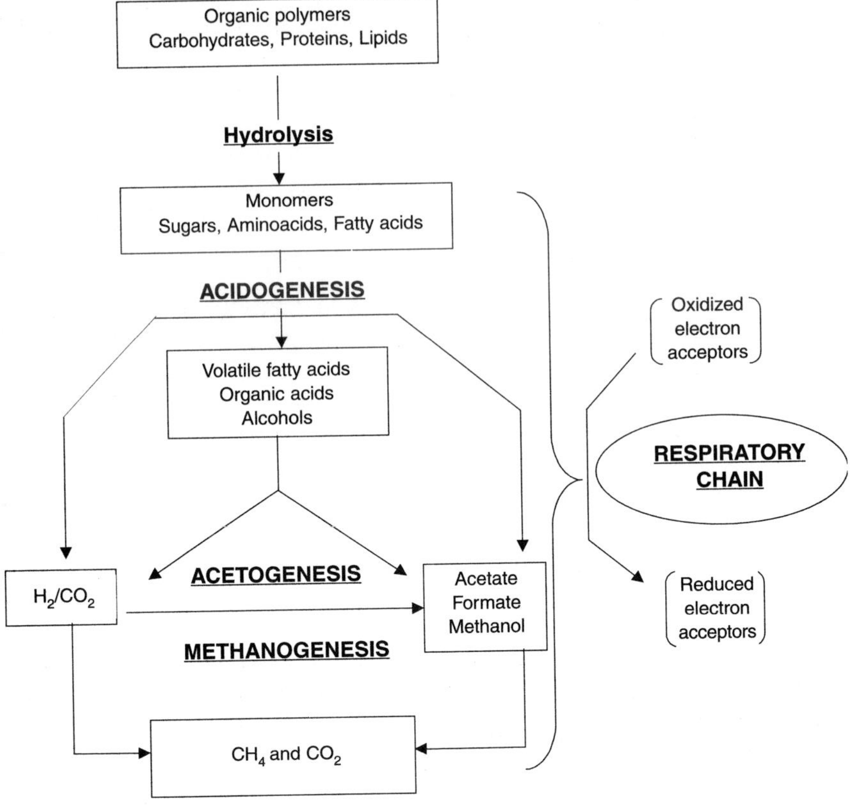

Fig. 2.1: Main pathways involved in the anaerobic degradation of organic matter.

and radionuclides (U(VI)) can also be used as electron acceptors in contaminated environments (Lovley and Coates, 2000);

—fermentation pathways, including acidogenesis (Zehnder and Stumm, 1988), in which various carbohydrates and amino acids are metabolized to alcohols and organic acids. Carbon dioxide and H_2 can be produced simultaneously from the lysis of formate (Pelmont, 1993). During fermentation, microorganisms use some of the substrates or transient organic compounds as electron acceptors;

—acetogenesis (Schink, 1997; Stams, 1994; Dolfing, 1988), using fermentation products such as volatile fatty acids, alcohols, amino acids and, in some case, aromatic compounds as substrates for acetate and H_2 production;

—reduction of oxidized S compounds to oxidize fermentation products (acetate, propionate, butyrate, etc.) with SO_4^{2-} and SO_3^{2-}, $S_2O_3^{2-}$, or S as the final electron acceptors (Hedderich et al., 1999; Wind and Conrad, 1995; Moodie and Ingledew, 1990; Legall and Fauque, 1988). This respiration produces either acetate or H_2S and CO_2, depending on the functional microbial populations concerned (Widdel, 1988);

—methanogenesis (Schink, 1997; Oremland, 1988), producing CH_4 either from substrates such as acetate (acetoclastic pathway) or from CO_2 and H_2 (hydrogenoclastic pathway). During carbohydrate degradation, the acetoclastic pathway generally accounts for 2/3 of CH_4 production and the hydrogenoclastic pathway for 1/3 (Wogel, 1988). The remaining few percent are produced from other compounds, e.g. methanol and formate (Oremland, 1988).

2.2. Acetate and H_2 as Key Intermediates in Organic Monomer Degradation

Anaerobic microbial metabolism involves key intermediates such as acetate and H_2, which can play a major role in the final steps of anaerobic metabolism. As demonstrated for forest and litter soil (Küsel and Drake, 1999; Krzyszowska et al., 1996; Wagner et al., 1996; Küsel and Drake, 1995), rice paddies (Glissman and Conrad, 1999; Chidthaisong et al., 1998), and other flooded soils (Tsusuki and Ponnamperuma, 1987), acetate is generally the main volatile fatty acid produced in anaerobic metabolism. It may be the final product of anaerobic metabolism when (i) H_2 accumulates due to environmental conditions that reduce acetate consumption (Chin and Conrad, 1995; Stams, 1994; Conrad and Weeter, 1990) and/or (ii) there is no inorganic electron acceptor (NO_3^-, for example) to allow acetate to be completely oxidized. Acetate may be further consumed under conditions that enhance H_2 consumption, i.e., conditions promoting sulfate-reduction (Schink, 1997; Widdel, 1988), homoacetogenesis (Chin and Conrad, 1995; Diekert and Wohlfarth, 1994; Conrad et al., 1989), and methanogenesis (Conrad, 1999; Chin and Conrad, 1995; Conrad et al., 1989; Oremland, 1988). Many authors mention the major role of H_2 in the reduction of oxidized metals (Lovley, 1995), the final steps of anaerobic metabolism (Schink, 1997; Stams, 1994; Dolfing, 1988; Wolin, 1988), and the degradation of many compounds such as volatile fatty acids (Schink, 1997; Stams, 1994) (Fig. 2.2) and amino acids (Nanninga and Gottschal, 1985). The presence of species that consume H_2 modifies H_2 metabolism (Wolin, 1988; Zehnder and Stumm, 1988).

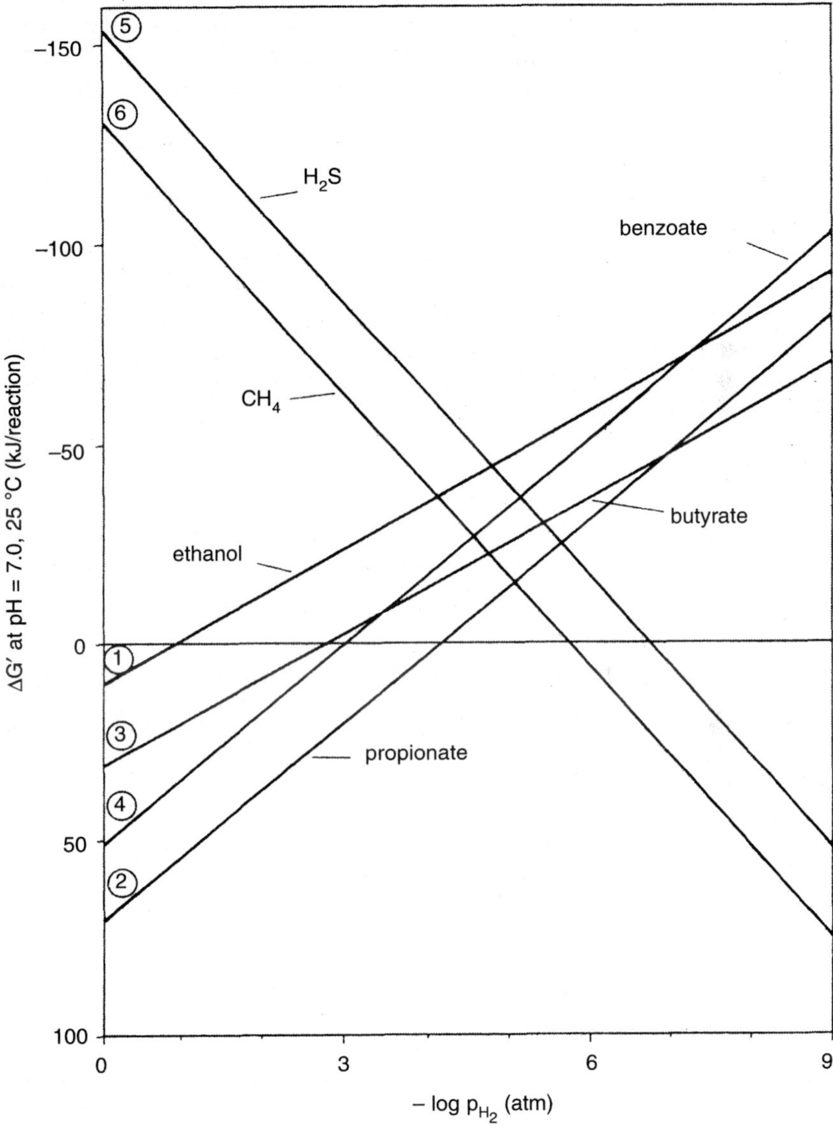

Fig. 2.2: Effect of H_2 partial pressure on free energy (ΔG′) for acetogenetic reactions involving ethanol, propionate, butyrate, and benzoate, for sulfate reduction using H_2 substrate, and for hydrogenoclastic methanogenesis. Calculations were made assuming 1 mM of either ethanol, acetate, propionate, butyrate, or benzoate, 100 mM of bicarbonate, equimolar concentrations of SO_4^{2-} and HS^-, and 0.5 atm partial pressure of methane (Dolfing, 1988).

2.3 Functional Anaerobic Microbial Communities

In any given environment, a number of bacterial species predominate and drive specific catabolic pathways (Schimel and Gulledge, 1998; Schimel, 1995). One can define functional microbial communities according to their anaerobic catabolic pathways. However, many microorganisms can be simultaneously or successively involved in several microbial pathways, depending on the environmental conditions. Several Fe-, Mn-, and sulfate-reducers (Widdel, 1988) also possess NO_3^- reductase in their respiratory chain. In the case of fermentation pathways, some microorganisms may favor those that provide the most energy (Pelmont, 1993). Several respiratory bacteria develop fermentation behavior when no inorganic electron acceptor other than CO_2 is available (Hedderich, 1999; Widdel, 1988). In methanogenic communities, it is unusual for a number of catabolic pathways to be involved simultaneously or successively, but some communities can drive both hydrogenoclastic and acetoclastic pathways (Oremland, 1988). It is therefore important when modeling complex soil biogeochemical behavior to understand the rule that determines reaction priorities and their respective contributions to energy requirements.

Microbial dynamics is specific to each functional microbial community and contribute to that community's capacity to adapt to a particular environment. It may be linked to substrate consumption and the level of ATP production, which differs between catabolic pathways and between respiratory chains (Moodie and Ingledew, 1990). The overall microbial dynamics can be characterized through biomass quantification (Witt et al., 2000) or microorganism enumeration (Küsel et al., 1999). The problem of extracting and cultivating microorganisms can be bypassed using epifluorescent microscopy (acridine orange counts, DAPI) (Kepner and Pratt, 1994). However, the overall microbial dynamics reflects only the balance between death and growth in a functional microbial community (viz. communities that can survive in unfavorable conditions by suspending their vital functions). The dynamics of specific microbial populations or groups can often be monitored nowadays with molecular biological techniques, including fluorescent in-situ hybridization (FISH) (Weber et al., 2001) (Fig. 2.3) or dot blot hybridization using specific probes targeting the 16S rRNA gene (Weber et al., 2001; Schimel, 1995). Several studies have used these techniques to determine the composition of bacterial groups (Weber et al., 2001, Teske et al., 1996); some of these studies have combined molecular measurements with measurements of particular processes and environmental parameters (Ramsing et al., 1996). Other methods providing a rapid overview of a complex microbial community and assessing community fluctuations have been developed, such as denaturing or temperature gradient gel electrophoresis (D/ TGGE) (Weber et al., 2001) terminal restriction fragment length polymorphism (Scheid and Stubner, 2001), and single-strand conformation polymorphism (Delbès et al., 2001; Zumstein et al., 2000).

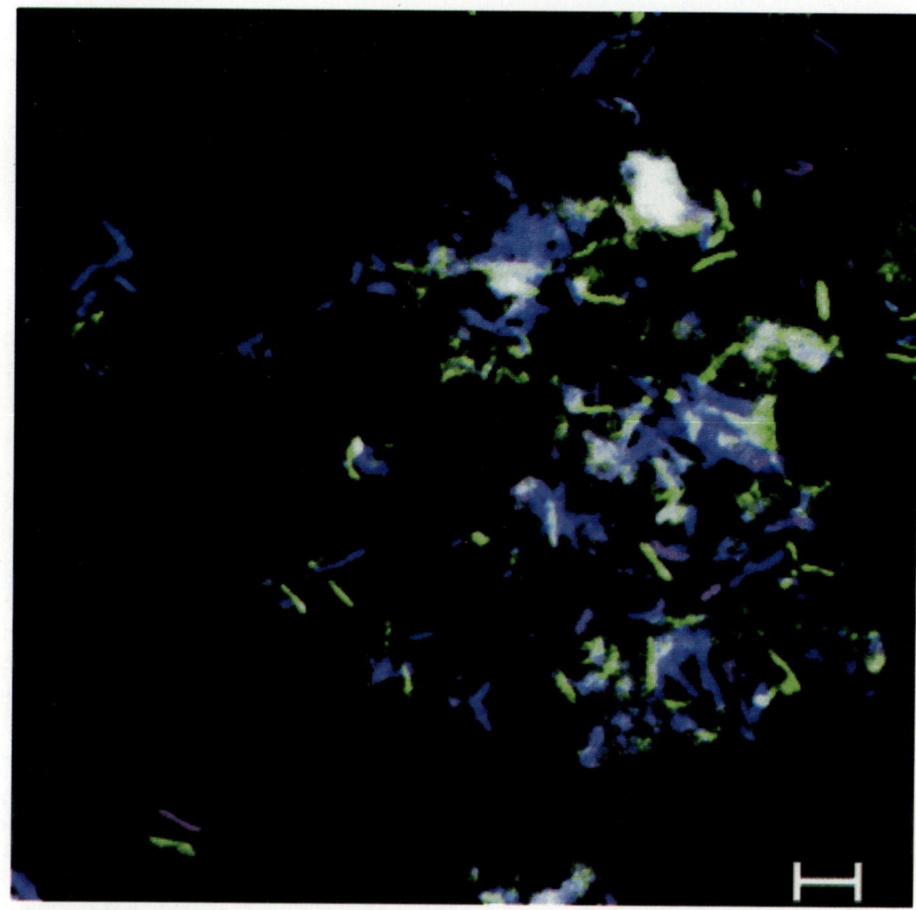

Fig. 2.3: FISH of fixed cells detached mechanically from straw incubated for 8 days in soil slurries. Three 16S rDNA probes were used simultaneously. CY5-labeled bacterial-domain probe Eub338 (blue) was used together with CY3-labeled probe Clost XIVa (red), specific for parts of clostridial cluster XIVa, and FITC-labeled probe Clost XIVa-c (green), designed from 16S rDNA sequence data. The overlay of probes Eub338 and Clost XIVa resulted in pink cells; the overlay of Eub338 and Clost XIVa-c resulted in green-to-turquoise cells. The number of cells hybridized simultaneously with the group-specific probes Clost XIVa and Clost XIVa-c (resulting in white cells) was less then 1% of all Eub338-detected cells. Scale bars, 5 μm (Weber et al., 2001).

3 INTERACTIONS BETWEEN MICROBIAL ACTIVITIES AND GEOCHEMICAL TRANSFORMATIONS

Anaerobic microbial activities interact closely with abiotic geochemical transformations in the solution and at solid-solution interfaces (Reddy et al., 1998; Stumm and Morgan, 1996; Glinski et al., 1995). Soil geochemical characteristics affect the nature and intensity of microbial processes (Reddy et al., 1998), while anaerobic microbial processes modify geochemical properties (Reddy et al., 1998; Glinski et al., 1995). However, reactions at solid/solution interfaces can partly buffer changes to the soil solution (Stumm and Morgan, 1996).

3.1 Effects of Geochemical Characteristics on Anaerobic Microbial Activities

First and foremost, soil geochemistry affects substrate availability for microorganisms in the solution and at solid/solution interfaces. The rate of anaerobic catabolism is often limited by the rate of initial hydrolysis of organic polymers (Colberg, 1988; Jash and Gosh, 1996; Leschine, 1995). This hydrolysis is affected by substrate structure (or substrate accessibility), including molecular ordering (Leschine, 1995), particle geometry (Vavilin et al., 1996a), and surface availability for microorganism fixation (Vavilin et al., 1996a). For example, cellulose has a crystalline structure (Leschine, 1995) and its hydrolysis rate depends on its degree of crystallinity (generally ranging from 60 to 90%). Similar accessibility problems occur during the degradation of other polymers such as lignin (Colberg, 1988). Polymer hydrolysis is also affected by the release of extracellular enzymes such as amylase (Pelmont, 1993), cellobiose (Leschine, 1995), lipase (McInerney, 1988), and protease (McInerney, 1988). Extracellular enzyme activity is greatly affected by solid/liquid interfaces: adsorption of extracellular enzyme on minerals leads to conformational changes that reduce enzyme catalytic activity (Quiquampoix, 1987a, 1987b).

Structural characteristics such as specific surface area (Brown et al., 1999; Roden and Zachara, 1996), contact with microorganisms (Brown et al., 1999), and crystalline-chemistry (Ghiorse, 1988; Legall and Fauque, 1988), including the presence of substituents in the structure (Bousserrhine et al., 1998), are also of great importance to the microbial reduction of metal oxides and oxyhydroxides. In rice paddies where ferrihydrite, lepidocrocite, goethite, and hematite Fe-oxides are abundant, studies have shown that anaerobic microorganisms preferentially reduce ferrihydrite and lepidocrocite, which are metastable and have low crystallinity (Karim, 1984; Bacha and Hossner, 1977). Contact between microorganisms and metal oxides and oxyhydroxides is not essential, since (i) humus can act as an electron shuttle, for example, between FeIII-reducing organisms and FeIII oxides (Lovley, 1995), and (ii) microorganisms can release chelating systems such as siderophores (Lovley, 1995).

The availability of substrates in solution depends on the pH of the solution, which affects (i) the permeability of the microbial membrane (Pelmont, 1993), and (ii) the speciation of chemical species that can act both as substrates and as inhibitors for microbial activities (e.g., volatile fatty acids are inhibitors in an undissociated form (Vavilin and Lokshina, 1996b)) in solid phases, solution, and the solid/solution interface. Several chemical compounds in soil solution influence the intensity and/or orientation of microbial pathways. For example, NO_3^- amendment prevents fermentation processes (Jugsujinda et al., 1995).

Other soil solution chemical compounds can inhibit microbial activities. We can distinguish between (i) general types of inhibition that affect all microbial activities, for example, undissociated volatile fatty acids (propionate, lactate, and butyrate) that can have a bacteriostatic effect (Aguilar et al., 1995, Fuzukaki et al., 1990a,b; Van Den Heuvel et al., 1988) and NO_2^- (Klüber and Conrad, 1998), and (ii) specific types of inhibition of defined microbial activities such as the inhibitory effect of NO_3^- on N_2O reduction during denitrification (Klüber and Conrad, 1998). There are also specific thermodynamic types of inhibition that influence acetogenesis, sulfate reduction, and methanogenesis and are exergonic or endergonic, depending on the partial pressure of H_2 in the soil (Schink, 1997; Stams, 1994) (Fig. 2.2).

3.2 Effects of Anaerobic Microbial Activities on Geochemical Characteristics of a Soil

Microbial activities directly affect the geochemical characteristics of the solution. Anaerobic microbial activities first affect the pH and the reduction/oxidation level of the solution (Chidthaisong and Conrad, 2000; Reddy et al., 1998; Stumm and Morgan, 1996; Ponnamperuma, 1972). The initial decrease in the reduction/oxidation level results from the sequential and/or partly simultaneous reduction of terminal electron acceptors. Their order of occupation can be affected by the pH (Stumm and Morgan, 1996). According to thermodynamic theory, the order in which electron acceptor reduction occurs, from the most oxidized state to the most reduced state, is as follows: NO_3^-, Mn^{4+}, Fe^{3+}, SO_4^{2-}, CO_2 (Chidthaisong and Conrad, 2000; Peters and Conrad, 1996) and finally organic compounds that act as electron acceptors during respiratory processes (e.g. fumarate) (Fig. 2.4). Nevertheless, (i) a number of respiratory processes can take place simultaneously because soil microorganisms include numerous populations, each of which can use only one or a few terminal electron acceptors successively or simultaneously, (ii) sequential reduction may partly result from the production of transient compounds (including volatile fatty acids and H_2) that are necessary for sulfate reduction and methanogenesis (Stams, 1994), and (iii) actual reduction processes can take place only if microorganisms with adequate physiological potentials are actually present in sufficient numbers (Liesack et al., 2000; Peters and Conrad, 1996). As an example, reduction of NO_3^-, Mn_4^+, Fe^{3+}, and SO_4^{2-} can take place simultaneously (Peters and Conrad, 1996). Because abiotic reductions/oxidations are generally slow transformations and because microbial reductions/oxidations are kinetic transformations that do not ensure thermodynamic equilibrium between redox couples, we prefer to use the imprecise expression "reduction/oxidation level" rather then "redox potential."

Soil solution pH variations depend on initial conditions. Microbial activities modulate the pH via the net microbial production of H^+ (e.g. during reduction of NO_2^- to N_2O), CO_2, NH_4^+, HS^-, organic acids, etc. (Stumm and Morgan, 1996). In acidic soils, the pH increment results from dissolution of metal oxides and oxyhydroxides (Ponnamperuma, 1972). In basic soils (e.g. calcareous soils),

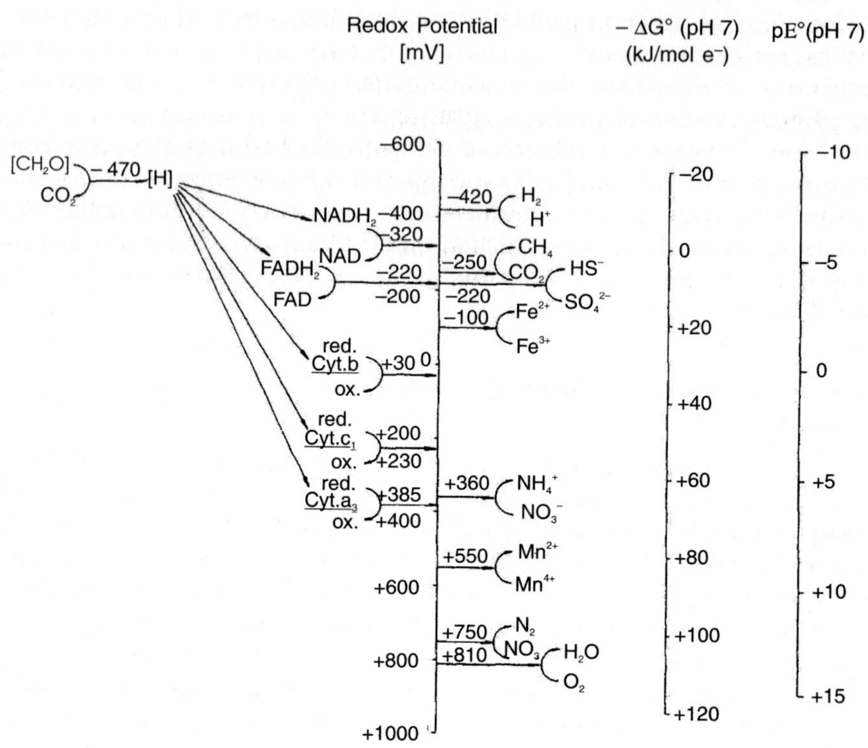

Fig. 2.4: Sequential reduction diagram for the biological redox sequence with organic carbon (CH_2O) acting as the electron donor. Calculated for standard conditions at pH 7 (CH_2O represents one-sixth of glucose) (Zehnder and Stumm, 1988).

the initial pH decrease results from an increase in CO_2 concentration and the transient production of volatile fatty acids (Ponnamperuma, 1972). This pH decrease may be minimized by the direct consumption of H^+ during the reduction of NO_2^- to N_2O (Moodie and Ingledew, 1990). The subsequent consumption of volatile fatty acids (Schink, 1997; Stams, 1994), dissolution of metal oxides and oxyhydroxides (Brown et al., 1999), and partial consumption of CO_2 during homoacetogenesis (Diekert and Wohlfarth, 1994) and methanogenesis (Oremland, 1988) tend to lead to a late increment in soil pH. Another important consequence of anaerobic microbial activities in solution is the production of volatile fatty acids that act as ligands for metal complexation (De Cockborne et al., 1999; Stumm and Morgan, 1996; Förstner, 1987). It has already been shown that these ligands can greatly increase the mobilization of metals under anaerobic conditions (Cambier and Charlatchka, 1999). Microbial activities can also generate inorganic anions such as S_2^-, CO_3^{2-}, OH^-, etc., which also act as ligands, with the nature of the ligand depending on redox conditions:

for example, S_2^- can be considered a major inorganic ligand under anoxic conditions (Stumm and Morgan, 1996).

Changes in the geochemical characteristics of the solution induced by microbial activities lead to other changes at the solid/solution interface level (adsorption/desorption reactions and precipitation/dissolution reactions). Microbial anaerobic activities can alter solid phases such as clays and their specific cation exchange capacity, which can be modified as a result of microbial byproducts (such as H^+ and other cations) interacting with the specific clay surface (specific exchange sites, specific surface area, specific electrical charge), leading to a change in the distribution of species between solution and solid phases (Stumm and Morgan, 1996; Robert and Chenu, 1992). For example, (i) H^+ released by microbial activities competes with metals to form surface complexes with ligands (Robert and Chenu, 1992) and can result in the release of Al in solution (Guibaud and Ayele, 2000) and (ii) microorganisms can reduce the FeIII bound in the clay structure, thereby decreasing clay swelling, causing the phyllosilicate structure to collapse, decreasing surface area, and increasing the surface charge density as a function of cation exchange capacity (Kostka et al., 1999). Microbial anaerobic activities can also alter mineral phases by changing the pH and so lead to the dissolution of minerals such as calcite, metal oxides, and oxyhydroxides (Stumm and Morgan, 1996; Legall and Fauque, 1988). Changes in soil geochemical characteristics can also cause precipitation and the formation of new minerals, these processes being affected by (i) partial pressure of CO_2, which acts on the pH of the solution and plays a direct part in the precipitation of carbonate minerals, and (ii) the reduction/oxidation level of the soil. We can distinguish between transformation into (i) sulfurs and polysulfur species, (ii) carbonates such as siderite ($FeCO_3$) or hydroxides ($Fe(OH)_2$), and (iii) minerals that incorporate FeII and FeIII, such as green rust (Refait et al., 1999; Abdemoula et al., 1998; Génin et al., 1998; Trolard et al., 1997).

3.3 Buffering Effects of Solid Phases

Soil can buffer the variations caused by anaerobic microbial processes. In calcareous soils, an increase in the partial pressure of CO_2 induces calcite dissolution and so minimizes the dissociation of H_2CO_3 into H^+ and HCO_3^- (Stumm and Morgan, 1996):

$$CaCO_3 + CO_2 + H_2O \rightleftarrows Ca^{2+} + 2HCO_3^- \qquad \text{... (1)}$$

Soil acidification induces metal oxide and oxyhydroxide dissolutions that can buffer a pH decrease. For example, when goethite dissolves, hydroxide ions are released:

$$FeOOH + H_2O \rightleftarrows Fe^{3+} + 3OH^- \qquad \text{... (2)}$$

Dissolution/precipitation reactions increase abiotic reduction/oxidation reactions in the soil solution and so improve the regeneration of terminal electron acceptors for microbial respiration. Consequently, HS^- and S^0 may be partially reoxidized into S^0 and SO_4^{2-} via abiotic and biotic reduction/oxidation reactions respectively, with concomitant reduction of Fe^{3+} to Fe^{2+} (Murase and Kimura, 1997) (Fig. 2.5). One practical consequence of this process is that when dissolution of FeIII oxides and oxyhydroxides occurs simultaneously with precipitation of FeII- or FeII–III-oxides and oxyhydroxides, SO_4^{2-} regeneration can be much greater than in situations wherein no FeII oxides precipitate (Fig. 2.6). There are various FeII oxides, including polysulfur species (FeS), siderite ($FeCO_3$), and other oxides such as green rusts, for which the oxidation level is between II and III. The recent identification of green rust oxides (Abdemoula et al., 1998) may lead to changes in our understanding of SO_4^{2-} regeneration because green rust is less soluble than amakinite ($Fe(OH)_2$).

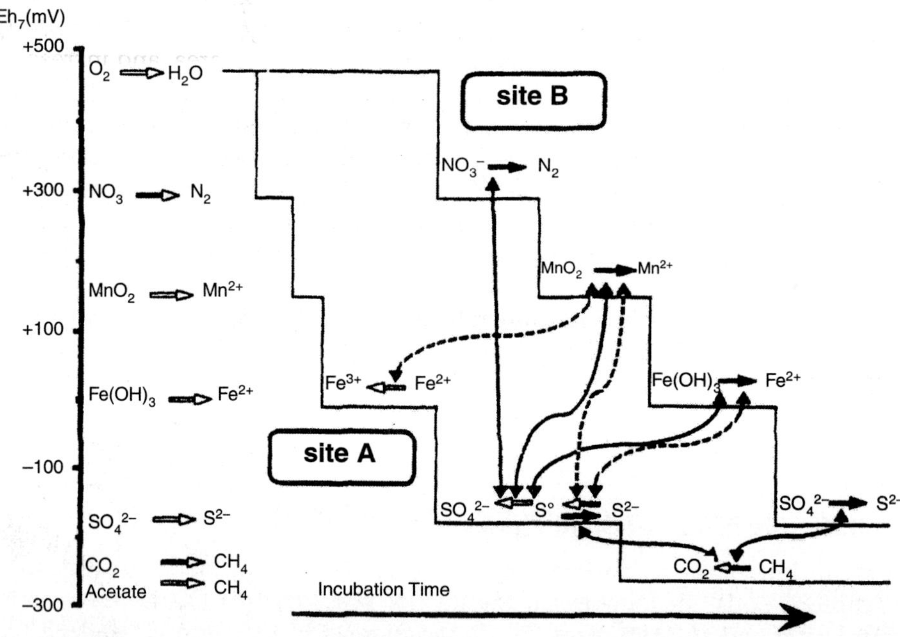

Fig. 2.5: A dynamic reduction model in submerged paddy soils. Two sites with different reduction states are considered; a site in which the reduction process advances further (site A) and a site in which the process is slow (site B) (Murase and Kimura, 1997).

4 MODELS COMBINING ANAEROBIC MICROBIAL PROCESSES AND GEOCHEMICAL TRANSFORMATIONS

4.1 Distinction between Microbial and Abiotic Transformations

Geochemical models that have attempted to describe geochemical reactions in solutions and at solution-solid interfaces include WHAM (Tipping, 1994),

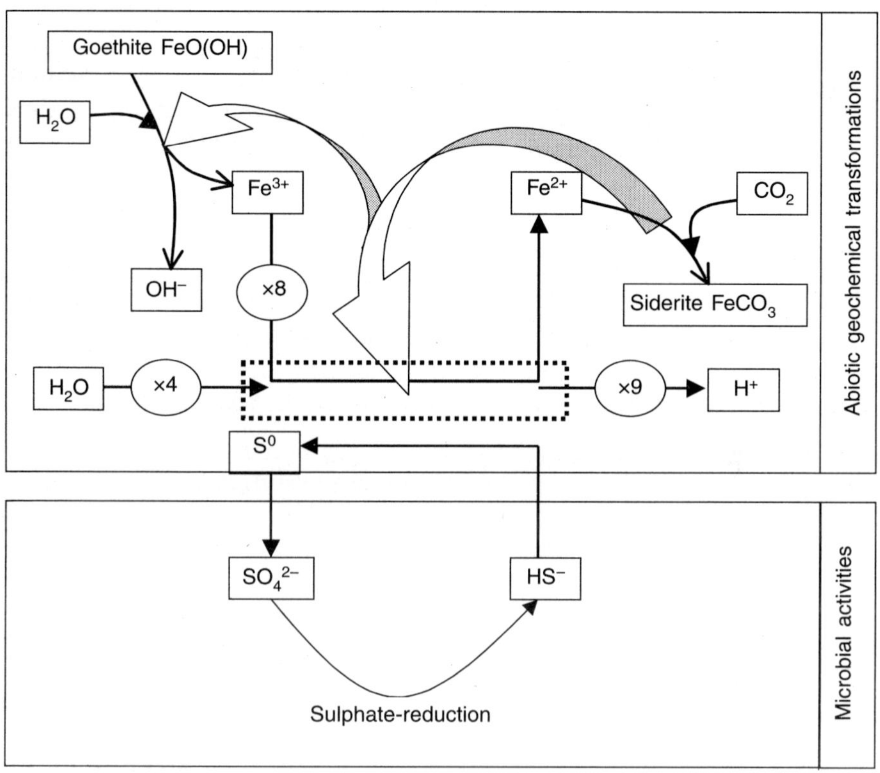

Fig. 2.6: Abiotic regeneration of SO_4^{2-} with concomitant reduction of Fe^{3+} into Fe^{2+} under anoxic conditions; Amplification of regeneration with precipitation of siderite and dissolution of goethite.

EQUIL(T) (Fritz, 1981), MIN3P (Mayer, 1999), PHREEQC (Parkhurst and Appelo, 1995), and AQUA (Vallès and Bourgeat, 1988). These models generally take no account of the microbial or abiotic origins of the transformations; they assume that transformation rates are high enough to ensure thermodynamic equilibrium (in thermodynamic models) or to minimize Gibbs free energy in the reactions involved (in kinetic models). Some of these models do roughly account for microbial effects. For example, (i) a CO_2 partial pressure that reflects microbial activity was included in the work of Marlet (1996) in AQUA, and (ii) PHREEQC has been applied to simple systems involving the biological degradation of nitrolotriacetate (NTA) in water.

Although the rates of several transformations, including acid/base and complexation reactions, are high enough to use thermodynamic approximations and ignore the microbial or abiotic origin of such transformations, in many cases it is necessary to distinguish explicitly between microbial and abiotic transformations for the following reasons:

— microbial reduction/oxidation transformations do not ensure thermodynamic equilibrium between redox couples (Pankow, 1991; Stumm and Morgan, 1996), and abiotic reduction/oxidation transformation rates are generally too low to minimize the deviations between the thermodynamic equilibria (Pankow, 1991);

— microbial activities explain most of the changes in organic materials that are involved in organomineral interactions (including metal complexation) and also provide the energy supply for heterotrophic microorganisms (Zehnder and Stumm, 1988);

— the presence/absence and the importance of some functional microbial communities may explain a number of behavioral differences between soils (Liesack et al., 2000; Schimel and Gulledge, 1998; Schimel, 1995).

More generally, redox chemistry in the soil solution involves kinetic reactions as well as adsorption/desorption and precipitation/dissolution reactions, whereas acid/base and complexation reactions may, as a first approximation, be regarded as thermodynamic (Stumm and Morgan, 1996). Although geochemical transformations of all five types may be either microbially or abiotically mediated, some reactions are preferentially microbial (e.g. oxidation of S^0 into SO_4^{2-} with concomitant reduction of Fe^{3+} to Fe^{2+}; Moodie and Ingledew, 1990), whereas others are preferentially abiotic (e.g. oxidation of S^{2-} into S^0 with concomitant reduction of Fe^{3+} to Fe^{2+}). The respective contributions of abiotic and microbial processes to net production of compounds may therefore be of the same order of magnitude: e.g. NO_3^- reduction to NH_4^+ through (i) dissimilatory NO_3^- reduction (Hansen et al., 1996) and (ii) concomitant abiotic oxidation of sulfate green rust can occur at similar rates.

4.2 Modeling Abiotic Geochemical Transformations and Anaerobic Microbial Activities

As summarized in Table 2.1, numerous models of anaerobic microbial activities that sometimes interact with abiotic geochemical transformations have been developed for anaerobic environments, including the deep vadose zone and aquifer (Hunter et al., 1998, Salvage and Yeh, 1998) anaerobic waste biodigesters (Vavilin et al., 1998; Jeyseelan, 1997; Vavilin et al., 1997a; Vavilin et al., 1996a,b,c; Shin and Song, 1995; Vavilin et al., 1995a,b,c; Vavilin et al., 1994a,b; Angelidacki et al., 1993; Vasiliev et al., 1993), and soils (Dassonville et al., 2004; Di Chio et al., 2000; Vavilin et al., 2000; Di Chio et al., 1999; Lokshina and Vavilin, 1999; Blagodatsky et al., 1998; Blagodatsky and Richter, 1998; Grant, 1998; Segers and Kengen, 1998; Vavilin et al., 1997a,b; Cao and Dent, 1995; Grant et al., 1993a,b). Only models proposed during the last decade are discussed here. Models dealing with rumen and sediments are not described because they are the least developed.

4.2.1 Models established for biodigesters

Most of the models dealing with anaerobic microbial activities have been established for waste biodigesters (Table 2.1, 2.2). The most advanced of these is the METHANE model. First proposed in 1993 (Vasiliev et al., 1993), this model has since been applied to other environments, including soils (Vavilin et al., 2000; Lokshina and Vavilin, 1999) (Fig. 2.7). With respect to biodigestion, this model has been applied in studies on (i) industrial and municipal anaerobic degradation processes used in the food industry and wastewater degradation (Vavilin et al., 1994a), (ii) municipal sewage sludge degradation (Vavilin et al., 1997a), (iii) cattle manure degradation (Vavilin et al., 1998), and (iv) anaerobic degradation of complex organic matter (Vavilin et al., 1994a). The METHANE model has been widely applied in analyses of the phases of anaerobic digestion (Vavilin et al., 1996c), kinetics of volatile fatty acid degradation (Vavilin and Lokshina, 1996b), kinetics of particulate organic matter hydrolysis (Vavilin et al., 1996a), inhibition of anaerobic metabolism (including biogas toxicity) (Vavilin et al., 1995c), inhibition caused by NH_4^+ and H_2S (Vavilin et al., 1995c), and the

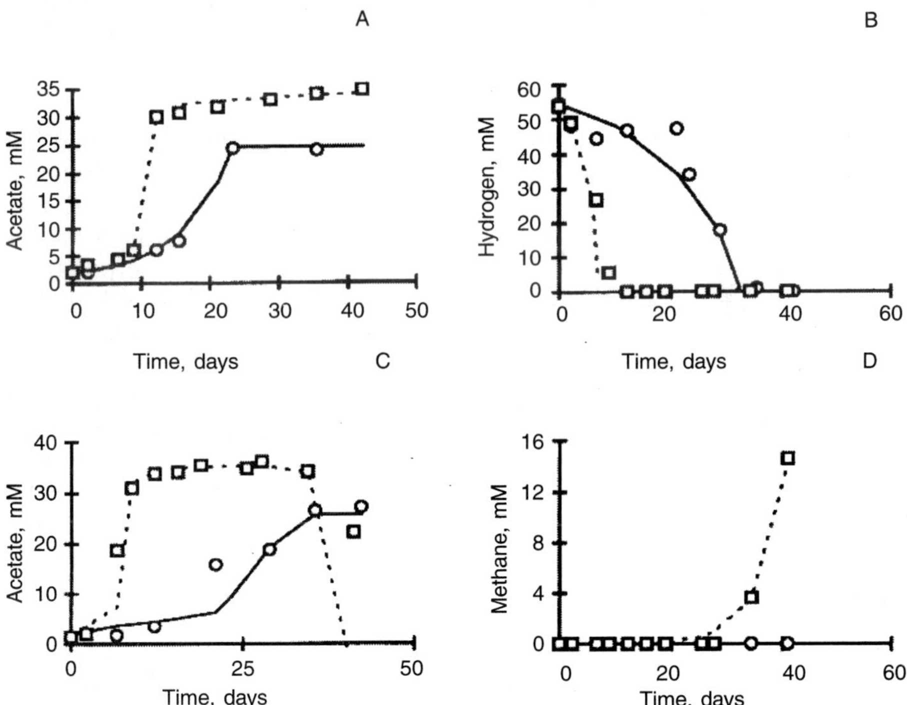

Fig. 2.7: METHANE model predictions (lines) and experimental data (points) of acetate and methane formation from endogenous matter of Geisberger forest soil samples (Germany) (A) and from samples of the same supplemented H_2/CO_2 (B, C, D). Dotted lines and squares depict the processes at 30°C. Solid lines and circles depict the processes at 15°C (Lokshina and Vavilin, 1999). Experimental data are taken from Küsel and Drake (1995).

effect of SO_4^{2-} on methanogenesis (Fomichev and Vavilin, 1997; Vavilin et al., 1994b). In our opinion, this model offers a good description of anaerobic metabolism for anaerobic biodigesters. However, it is of limited use for studies of anaerobic soil metabolism because (i) it does not account for anaerobic respiration in the presence of N-, Fe-, and Mn-oxides, (ii) ignores biological and physicochemical interactions between solid particles and the solution, (iii) does not combine microbiological and geochemical soil processes, except for acid/base reactions, and (iv) the predominant microbial species in biodigesters differ from those in soils, so the relative contributions of the various metabolic pathways are not the same.

Other anaerobic waste biodigester models are less advanced than the METHANE model for describing anaerobic metabolism, which they simplify. For example, Shin and Song (1995) consider anaerobic metabolism to be a series of acidification and methanation reactions occurring under optimum environmental conditions and with a constant concentration of active organisms. Furthermore, none of these models combine microbiological and geochemical processes, except for acid/base reactions in the work by Angelidacki et al. (1993), which includes a detailed pH description to accurately simulate NH_4^+ inhibition (Table 2.1, 2.2). Such models are therefore applicable only to studies dealing with anaerobic biodigesters.

4.2.2 Models established for the deep vadose zone and the aquifer

Models established for the deep vadose zone and the aquifer predict chemical changes within contaminated aquifers by combining degradation reactions, geochemistry, and chemical transport (Hunter et al., 1998; Salvage and Yeh, 1998). These models describe geochemical processes comprehensively (acid/base, reduction/oxidation, complexation, dissolution/precipitation, adsorption/desorption) but do not emphasize anaerobic microbial activities like the models for anaerobic biodigesters do. Although they describe most of the anaerobic respiration steps (N-, Fe-, Mn-oxides, CO_2), they give only an overall description of the fermentative and acetogenic processes that are important in soil. Nor do they specify the functional microbial communities involved in microbiological activities. Apart from these models, the BIOKEMOD model (Salvage and Yeh, 1998) seems to be the most developed. This is a general model for simulating geochemical and microbiological reactions in batch aqueous solutions and may be coupled with hydrologic transport codes to simulate chemically and biologically reactive transport. Unfortunately, it has been applied only to simple systems involving biological and chemical reactions, such as (i) substrate utilization and microbial growth (consumption of $N\text{-}NO_2^-$ by *Nitrobacter winogradski*) and (ii) mixed microbiological and chemical kinetics (biodegradation of cobal-nitrolotriacetate by *Chelatobacter heintzii*). To our knowledge, this model is too general since it gives only an overall governing equation describing biogeochemical reaction systems and does not define specific microbiological processes (fermentations, etc.) or geochemical processes and their associated parameters. Furthermore, it concerns only aqueous solutions and might not be easy to transpose to a complex system such as soil.

4.2.3 Models established for soil

A few models describe anaerobic metabolism in soils (Table 2.1, 2.2). Some models are used to describe CH_4 emission in soils but describe only the final steps of anaerobic metabolism (Grant, 1998; Segers and Kengen, 1998; Vavilin et al., 1997b, Cao and Dent, 1995). Others describe anaerobic metabolism via C and N turnover in soils, but overlook most of the anaerobic metabolic steps (Di Chio et al., 2000; Di Chio et al., 1999; Blagodatsky et al., 1998; Blagodatsky and Richter, 1998; Grant et al., 1993a,b). They are mainly used to study disturbances due to changes in the soil C/N ratio (Di Chio et al., 1999; Di Chio et al., 2000) and to describe microbial growth associated with C and N turnover in soil (Blagodatsky et al., 1998; Blagodatsky and Richter, 1998).

Until recently, only the METHANE model accounted for geochemical transformations (acid/base reactions only). Now, however, a new model has been proposed that integrates anaerobic microbial activities with soil geochemical transformations (Dassonville et al., 2004). This model describes the dynamics of six functional microbial communities, their decomposition after death, and the catabolism of carbohydrates through denitrification, dissimilatory NH_4^+ production, Fe(III) reduction, fermentation, acetogenesis, and SO_4^{2-} reduction. It has been combined with a model that thermodynamically describes acid/base, reduction/oxidation, and complexation reactions in solution, and kinetic precipitation and dissolution. The model was tested with an anaerobically incubated soil amended with (i) glucose or (ii) glucose and NO_3^- and reflected the experimental microbial data (Fig. 2.8) and geochemical characteristics including pH and reduction/oxidation potential (Fig. 2.9). However, some hypotheses, microbiological processes (e.g. dynamics of microbial functional communities), and geochemical transformations (e.g. actual speciation of solid FeII and FeIII oxyhydroxides) remained unproven. Furthermore, this model was tested in extreme environmental conditions (high initial glucose and NO_3^- levels in batch experiments), which do not prevail in most anaerobic soils. It therefore needs to be tested under environmental conditions commonly found in anaerobic soils to assess its potential for adaptation to in-situ conditions.

4.3 Model Parameterization

In simulating the biogeochemical behavior of soils under anaerobic conditions, numerous biogeochemical transformations need to be accounted for simultaneously; hence estimates of numerous parameters must be obtained. Little information is available on the parameterization of most of the models presented in Tables 2.1 and 2.2. We can distinguish between approaches in which parameter values have been taken from the literature (Di Chio et al., 1999; Di Chio et al., 2000; Grant, 1998; Segers and Kengen, 1998; Jeyseelan, 1997; Shin and Song, 1995; Grant et al., 1993a) and when data were lacking, fitted to obtain

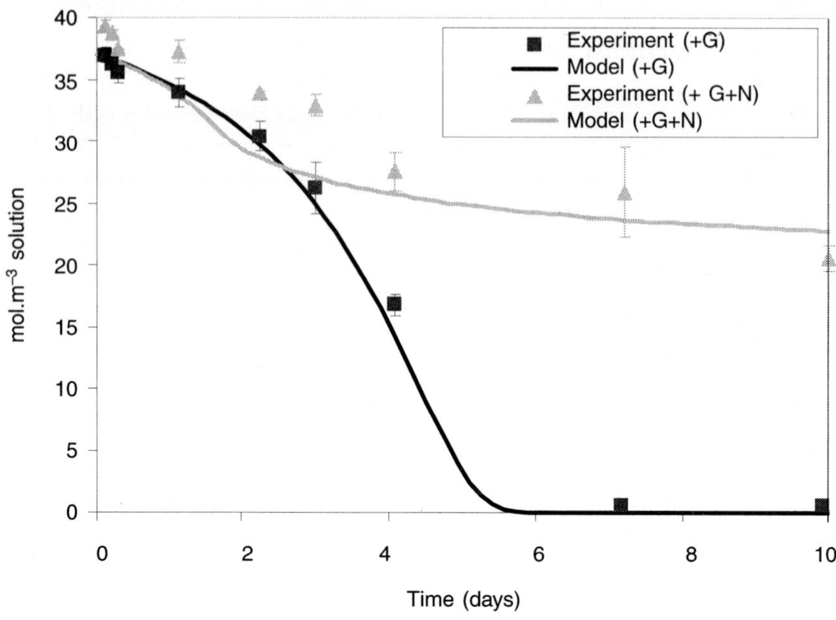

Fig. 2.8: Experimental and simulated variations in glucose concentration over time for an anaerobic incubated soil amended with glucose (+G), and glucose + NO_3^- (+G+N) (Dassonville et al., 2004).

a reasonable range for the parameter values (Vavilin et al., 1997a; Vavilin et al., 1996a,b,c; Cao and Dent, 1995; Vavilin et al., 1995a), from those approaches in which parameter values were estimated to fit simulations (Vavilin et al., 2000; Lokshina et Vavilin, 1999; Blagodatsky et al., 1998; Blagodatsky and Richter, 1998; Hunter et al., 1998; Salvage and Yeh, 1998; Vavilin et al., 1998; Fomichev and Vavilin, 1997; Vavilin et al., 1997a; Vavilin et al., 1996a,b,c; Cao and Dent, 1995; Vavilin et al., 1995a; Vavilin et al., 1994a,b; Angelidacki et al., 1993). Unfortunately, for most of these models, neither the procedures used nor the problems resulting from these procedures (e.g. correlations between parameter estimates, etc.) have been reported. The report on the model proposed by Dassonville et al. (2004) gave an accurate description of the parametrization procedure. Some parameter values were taken from experimental data, other values were estimated by a procedure to fit simulations and, when data were lacking, fitted to obtain a reasonable range for parameter values. The potential effects of some fitted parameters are also discussed. This made it much easier to obtain estimates of numerous parameters required for modeling soil biogeochemical processes. However, it is still difficult to distinguish between

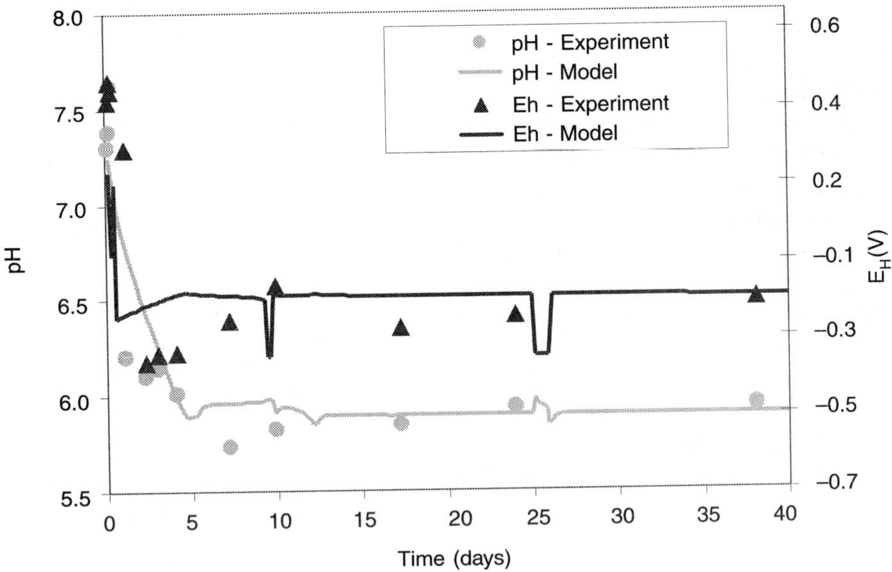

Fig. 2.9: Experimental and simulated variations in E_H and pH over time for an anaerobic incubated soil amended with glucose (+G), and glucose + NO_3^- (+G+N). Overestimation of FeII in solution for two periods (days 9–10, days 25–27) led to a decrease in E_H (Dassonville et al., 2004).

major and lesser processes and to assess the degree of accuracy required for parameter estimates.

5 CONCLUSIONS

Anoxic conditions greatly affect the geochemical characteristics of soil as well as the properties of the deep vadose zone, aquifer, and atmosphere. Undesirable effects include metal mobilization (e.g. Mn and Al), greenhouse gas emissions (e.g. CH_4 and N_2O), and solid phase alterations (clays, minerals, etc.); positive effects include the reduction of NO_3^- pollution and the immobilization of some radionuclides (e.g. U and Tc). When soil conditions shift from oxic to anoxic, various anaerobic microbial activities start up simultaneously and/or successively, consuming the limited numbers of various terminal electron acceptors through N-, Fe-, Mn., S- and CO_2-respiration or fermentative and acidogenic reactions. Accumulation of products (including CO_2 and volatile fatty acids) produces further modifications in the geochemical characteristics of the soil. Microbial processes interact closely with abiotic geochemical transformations. On the one hand, soil geochemical characteristics affect microbial processes through substrate availability, thermodynamic regulation, and inhibition. On the other hand, microbial activities affect the pH of the soil solution, its reduction/oxidation level, complexation of metals with volatile fatty acids, and the alteration of solid particles (clays, minerals, etc.). Abiotic transformations also partially buffer the geochemical changes caused by microbial activities and partly regenerate the terminal electron acceptors used by microorganisms.

As a rule, microbial and abiotic reduction/oxidation reactions do not ensure thermodynamic equilibrium: the reaction rates are not high enough to maintain Gibbs free energy in reactions at around 0 and ensure thermodynamic equilibria between redox couples. Some abiotic reduction/oxidation reactions can be ignored (e.g. oxidation of S^0 into SO_4^{2-} with concomitant reduction of Fe^{3+} to Fe^{2+}), whereas the rates of others may be of the same order of magnitude as microbial transformations (e.g. reduction of NO_3^- to NH_4^+ with concomitant oxidation of sulfate green rust). By contrast, several instances of biological reduction/oxidation can be ignored (e.g. oxidation of S^{2-} into S^0 with concomitant reduction of Fe^{3+} to Fe^{2+}). Factors affecting microbial reactions differ from those that affect abiotic ones. Microbial contributions to reduction/ oxidation reactions largely depend on the presence or absence of microbial populations able to produce these reactions. But this sequential use of electron acceptors also depends on the availability of electron donors, including volatile fatty acids and H_2, to reduce S oxides or CO_2. Consideration of other types of reactions also requires distinguishing between microbial and abiotic reactions: for example, although abiotic acid/base reactions may be regarded as instantaneous reactions, the pH of soil solutions largely depends on the partial pressure of CO_2, which results mainly from microbial lysis of formate under fermentative conditions and therefore depends on factors affecting cell membrane permeability.

Given this kind of consideration, models combining microbial and abiotic geochemical transformations should generally distinguish between biotic and abiotic processes and explicitly account for interactions between these two types of reaction. Few models to date have combined microbial and geochemical transformations. Of those that have, most were initially proposed for anaerobic biodigesters; for this environment they legitimately limit the geochemical transformations they model to acid/base reactions and ignore reactions between solids and the solution. Aquifer models describe geochemical processes that may differ from those prevailing in soils and do not emphasis on microbial activities. Most of the models established for soil still lack satisfactory descriptions of key microbiological and geochemical processes. With regard to other media, a feature of soil reactors is the presence of reactive solid phases (especially calcite and metal oxyhydroxides) that help to minimise changes (pH and reduction/oxidation level), favor microbial populations that can utilize these solid phases (e.g. Fe^{3+} reducers), and limit the bioavailability of several substrates. Soil reactors also have their particular functional microbial biodiversity, partly resulting from the existence of these solid phases and partly from soil history, including impact of any previous anoxic events. A new model was recently proposed in which biotic and microbial transformations interact closely. This model was tested with batch experiments and was able to reflect the experimental data. Despite these first satisfactory results, some processes and hypotheses remain unproven. Furthermore, the model was checked in extreme environmental conditions, which do not prevail in most anaerobic soils. Nonetheless, it seems that this model could be usefully applied to anaerobic conditions in which microbial and geochemical processes closely interact.

Given the great complexity of soil biogeochemical processes, future models will necessarily be simplified representations of the real situation. These simplified models will have to describe (i) anaerobic metabolism steps associated with key intermediates such as H_2 and acetate (production and consumption, importance of the relative consumption pathways), (ii) the functioning and dynamics of a number of functional bacterial communities that simultaneously or successively drive specific catabolic pathways (respiration, fermentation, acetogenesis and methanogenesis), and (iii) numerous interactions between anaerobic microbial activities and geochemical transformations in the solution and at solid-solution interfaces, as well as the buffering capacity of soil geochemistry.

Acknowledgments

This research was conducted at INRA, unité "Climat, Sol et Environnement," Avignon (France). We are grateful to Harriet Coleman, for reviewing the English version of the report.

References

Abdemoula M., Trolard F., Bourrié G., and Génin J.M.R. 1998. Evidence for the Fe(II)–Fe(III) Green Rust "Fougerite" mineral occurrence in a hydrodynamic soil and its transformation with depth. *Hyperfine Interactions* 112: 235–238.

Aguilar A., Casas C., and Lema J.M. 1995. Degradation of volatile fatty acids by differently enriched methanogenic cultures: Kinetics and inhibition. *Water Res.* 29: 505–509.

Angelidacki I., Ellegard L., and Arhing B.K. 1993. A mathematical model for dynamic simulation of anaerobic digestion of complex substrates: Focusing on ammonia inhibition. *Biotech. Bioeng.* 42: 159–166.

Bacha R.E. and Hossner L.R. 1977. Characteristics of coating formed on rice roots as affected by iron and manganese additions. *Soil Sci. Soc. Amer. J.* 41: 931–935.

Bertrand I. 1998. Importance de la libération de protons sur la mobilisation du phosphore minéral et du fer par les racines. Etude des minéraux modèles calcite et goethite. Thesis, University Aix-Marseille III, Marseille, France.

Blagodatsky S.A. and Richter J. 1998. Microbial growth in soil and nitrogen turnover: A theorical model considering the activity state of microorganisms. *Soil Biol. Biochem.* 30: 1743–1755.

Blagodatsky S.A., Yevdokimov I.V., Larionova A.A., and Richter J. 1998. Microbial growth in soil and nitrogen turnover: Model calibration with laboratory data. *Soil Biol. Biochem.* 30: 1757–1764.

Bousserrhine N., Gasser U., Jeanroy E., and Berthelin J. 1998. Effects of aluminium substitution on free-reducing bacteria activity and dissolution of goethites. *C. R. Acad. Sci. Série II. Science Terre Planètes* 326 : 617–624.

Bouwman A.F. 1990. *Soils and the Greenhouse Effects*. John Wiley & Sons, New York, USA.

Brown G.E., Henrich V.E., Clark D.L., Eggleston C. et al. 1999. Metal oxides and their interactions with aqueous solutions and microbial organisms. *Chem. Rev.* 99: 77–174.

Brune A., Frenzel P., and Cypionka H. 2000. Life at the oxic-anoxic interface: Microbial activities and adaptations. *FEMS Microb. Ecol.* 24: 691–710.

Cambier P. and Charlatchka R. 1999. Influence of reducing conditions on the mobility of divalent trace metals in soils. In: *Fate and Transport of Heavy Metals in the Vadose Zone*. H.M. Selim and I.K. Iskandar (eds.). Lewis Publ., Boca Raton, FL, USA, pp. 702–765.

Cao M. and Dent J.B. 1995. Modeling methane emissions from rice paddies. *Global Biogeochemical Cycles* 9: 183–195.

Chidthaisong A. and Conrad R. 2000. Turnover of glucose and acetate coupled to reduction of nitrate, ferric iron and sulfate to methanogenesis in anoxic rice field soil. *FEMS Microb. Ecol.* 31: 73–86.

Chidthaisong A. and Conrad R. 2000. Pattern of non-methanogenic and methanogenic degradation of cellulose in anoxic rice field soil. *FEMS Microb. Ecol.* 31: 87–94.

Chidthaisong A., Obata H., and Watanabe I. 1998. Methane formation and substrate utilisation in anaerobic rice soils as affected by fertilisation. *Soil Biol. Biochem.* 31: 135–143.

Chidthaisong A., Rosenstock B., and Conrad R. 1999. Measurement of monosaccharides and conversion of glucose to acetate in anoxic rice field soil. *Appl. Environ. Microb.* 65: 2350–2355.

Chin K.J. and Conrad R. 1995. Intermediary metabolism in methanogenic paddy soil and the influence of temperature. *FEMS Microb. Ecol.* 18: 85–102.

Colberg P.J. 1988. Anaerobic degradation of cellulose, lignin, oligolignols, and monoaromatic lignin derivatives. In: *Biology of Anaerobic Microorganisms*. A.J.B. Zehnder (ed.). John Wiley & Sons, New York, NY, pp. 333–373.

Conrad R. 1999. Contribution of hydrogen to methane production and control of hydrogen concentrations in methanogenic soils and sediments. *FEMS Microb. Ecol.* 28: 193–202.

Conrad R. and Weeter B. 1990. Influence of temperature on energetics of hydrogen metabolism in homoacetogenic, methanogenic, and other anaerobic bacteria. *Arch. Microb.* 155: 94–98.

Conrad R., Bak F., Seitz H.J., Thebrath B., Mayer H.P., and Schültz H. 1989. Hydrogen turnover by psychotrophic homoacetogenic and mesophilic methanogenic bacteria in anoxic paddy soil and lake sediment. *FEMS Microb. Ecol.* 62: 285–294.

Curmi P., Bidois J., Bourrie G., Cheverry C. et al. 1997. Rôle du sol sur la circulation et la qualité des eaux au sein de paysages présentant un domaine hydromorphe. Incidence sur la teneur en nitrates des eaux superficielles d'un bassin versant armoricain. *Etude gestion sols* 4: 95–114.

Dassonville F., Renault P., and Vallès V. 2004. A model describing the interactions between anaerobic microbiology and geochemistry in a soil amended with glucose and nitrate. *Eur. J. Soil Sci.* 55(1): 29–45.

De Cockborne A.M., Vallès V., Bruckler L., Sévenier, G. et al. 1999. Environmental consequences of apple waste deposition. *J. Environ. Qual.* 28: 1031–1037.

Delbès C., Leclerc M. Zumstein E., Godon J.J., and Moletta R. 2001. A molecular method to study population and activity dynamics in anaerobic digestors. *Water Sci. Tech.* 43: 51–57.

Di Chio D., Potenz D., and Righetti E. 1999. Degradation of organic substances in the soil: Proposal for a mathematical model. *Bioresources Tech.* 67: 267–278.

Di Chio D., Potenz D., and Righetti E. 2000. Test of a mathematical model for degradation kinetics of components of cultivated soil. Its use to determine optimal nutrient dosage required to maintain stable soil conditions. *Bioresources Tech.* 74: 249–255.

Diekert G. and Wohlfarth G. 1994. Metabolism of homoacetogens. *Antony Van Leeuwenhoek* 66: 209–221.

Dijkstra J., France J., and Tamminga S. 1998. Quantification of the recycling of microbial nitrogen in the rumen using a mechanistic model of rumen fermentation processes. *J. Agric. Sci.* 130: 81–94.

Dolfing J. 1988. Acetogenesis. In: *Biology of Anaerobic Microorganisms*. A.J.B. Zehnder (ed.). John Wiley & Sons, New York, NY, pp. 333–373.

Drew M.C. 1997. Oxygen deficiency and root metabolism: Injury and acclimation under hypoxia and anoxia. *Ann. Rev. Plant Physiol. Plant Molec. Biol.* 48: 223–250.

Finnerty W.F. 1989. Microbial lipid metabolism. In: *Microbial Lipids.* C. Ratledge and S.G. Wilkinson (eds.). Acad. Press, London, UK, vol. II, pp. 525–66.

Fomichev A.O. and Vavilin V.A. 1997. The reduced model of self-oscillating dynamics in an anaerobic system with sulfate-reduction, *Ecol. Mod.* 95: 133–144.

Forde B.G. 2000. Nitrate transporters in plants: structure, function and regulation. *Biochim. Biophys. Acta* 1465: 219–235.

Förstner U. 1987. *Metals Speciation, Separation, and Recovery,* Lewis Publ., Chelsea, UK.

Fritz B. 1981. Etude thermodynamique et modélisation des réactions hydrothermales et diagéniques. *Sci. Géol. Mém.* 65: 197 pp.

Fukuzaki S., Nishio N., and Nagai S. 1990a. Kinetics of the methanogenic fermentation of acetate. *Appl. Environ. Microb.* 56: 3158–3163.

Fukuzaki S., Nishio N., Shobayashi M., and Nagai S. 1990b. Inhibition of the fermentation of propionate to methane by hydrogen, acetate, and propionate. *Appl. Environ. Micro.* 56 719–723.

Génin J.-M., Bourrié G., Trolard F., Abdemoula M. et al. 1998. Thermodynamic equilibria in aqueous suspensions of synthetic and natural Fe(II)–Fe(III) green rusts: Occurrences of the mineral in hydromorphic soils. *Environ. Sci. Tech.* 32: 1058–1068.

Ghiorse W.C. 1988. Microbial reduction of manganese and iron. In: *Biology of Anaerobic Microorganisms.* A.J.B. Zehnder (ed.). John Wiley & Sons, New York, NY, pp. 305–333.

Glinski J., Stahr K., Stepniewska Z., and Brzezinska M. 1995. Changes of redox and pH conditions in a flooded soil amended with glucose and manganese oxide or iron oxide under laboratory conditions. *Zeitschrift Pflanzen. Boden.* 159: 297–304.

Glissman K. and Conrad R. 1999. Fermentation pattern of methanogenic degradation of rice straw in anoxic paddy soil. *FEMS Microb. Ecol.* 31: 117–126.

Grant R.F. 1998. Simulation of methanogenesis in the mathematical model "ecosys". *Soil Biol. Biochem.* 30: 883–896.

Grant R.F., Juma N.G., and McGill W.B. 1993a. Simulation of carbon and nitrogen transformations in soil: Mineralization. *Soil Biol. Biochem.* 25: 1317–1329.

Grant R.F., Juma N.G., and McGill W.B. 1993b. Simulation of carbon and nitrogen transformations in soil: Microbial biomass and metabolic products. *Soil Biol. Biochem.* 25: 1331–1338.

Guibaud G. and Ayele J. 2000. pH and ionic strength effect on release of aluminium by limousin acidic brown earth soils—impact on natural water pollution. *Environ. Tech.* 21: 257–269.

Hansen H.C., Koch C.B., Nancke-Krogh H., and Sorensen J. 1996. Abiotic nitrate reduction to ammonium: Key role of green rust. *Environm. Sci. Tech.* 30: 2053–2056.

Head I.M., Saunders J.R., and Pickup R.W. 1998. Microbial evolution, diversity, and ecology: A decade of ribosomal rRNA analysis of uncultivated microorganisms. *Microbial. Ecol.* 35:1–21.

Hedderich R., Klimmek O., Kröger A., Dirmeier R. et al. 1999. Anaerobic respiration with elemental sulfur and with disulfides. *FEMS Microb. Rev.* 22: 353–381.

Hojberg O., Revsbech N.P., and Tiedje J.M. 1994. Denitrification in soil aggregates analyzed with microsensors for nitrous oxide and oxygen. *Soil Sci. Soc. Amer. J.* 58: 1691–1698.

Hunter K.S., Wang Y., and Van Cappellen P. 1998. Kinetic modeling of microbially-driven redox chemistry of subsurface environments: coupling transport, microbial metabolism and geochemistry. *J. Hydrology* 209: 53–80.

Inubushi K., Hori K., Matsumoto S., and Wada H. 1997. Anaerobic decomposition of organic carbon in paddy soil in relation to methane emission to the atmosphere. *Water Sci. Tech.* 36: 523–530.

Jash T. and Ghosh D.N. 1996. Studies of the solubilization kinetics of solid organic residues during anaerobic biomethanation. *Energy* 21: 725–730.

Jeyaseelan S.A. 1997. A simple mathematical model for anaerobic digestion process. *Water Sci. Tech.* 35: 185–191.

Jones D.L. 1999. Amino acid biodegradation and its potential effects on organic nitrogen capture by plants. *Soil Biology Biochemistry* 31: 613–622.

Jugsujinda A., Delaune R.D., and Lindau C.W. 1995. Influence of nitrate on methane production and oxydation in a flooded soil. *Comm. Soil Sci. Plant Analysis* 26: 2449–2459.

Karim Z. 1984. Formation of aluminium-substituted goethite in seasonally waterlogged rice soils. *Soil Sci. Soc. Amer. J.* 48: 410–413.

Kepner R.L. and Pratt J.R. 1994. Use of fluorochromes for direct enumeration of total bacteria in environmental samples: Past and present. *Microb. Rev.* 58: 603–615.

Klüber H.D. and Conrad R. 1998. Effects of nitrate, nitrite, NO and N_2O on methanogenesis and other redox processes in anoxic rice field soil. *FEMS Microb. Ecol.* 25: 301–318.

Kostka J.E., Wu J., Nealson K.H., and Stucki J.W. 1999. The impact of structural Fe(III) reduction by bacteria on the surface chemistry of smectite clay minerals. *Geochem. Cosmochim. Acta* 63: 3705–3713.

Kristensen E., Ahmed S.I., and Devol A.H. 1995. Aerobic and anaerobic decomposition of organic matter in marine sediments: Which is fastest ? *Limnol. Oceanog.* 40 (1995): 1430–1437.

Krylova N.I., Janssen P.H., and Conrad R. 1997. Turnover of propionate in methanogenic paddy soil. *FEMS Microb. Ecol.* 23: 107–117.

Krzyszowska A.J., Blaylock M.J., Vance G.F., and David M.B. 1996. Ion-chromatographic analysis of low molecular weight organic acids in spodosol forest floor solutions. *Soil Sci. Soc. Amer. J.* 60: 1565–1571.

Küsel K. and Drake H.L. 1995. Effects of environmental parameters on the formation and turnover of acetate by forest soil. *Appl. Environ. Microb.* 61: 3667–3675.

Küsel K. and Drake H.L. 1999. Microbial turnover of low molecular weight organic acids during leaf litter decomposition. *Soil Biol. Biochem.* 31: 107–118.

Küsel K., Wagner C., and Drake H.L. 1999. Enumeration and metabolic product profiles of the anaerobic microflora in the mineral soil and litter of a beech forest. *FEMS Microb. Ecol.* 29: 91–103.

Lefebvre-Drouet E. and Rousseau M.F. 1995. Dissolution de différents oxyhydroxydes de fer par voie chimique et par voie biologique: Importance des bactéries réductrices. *Soil Biol. Biochem.* 27: 1041–1050.

Legall J. and Fauque G. 1988. Dissimilatory reduction of sulfur compounds. In: *Biology of Anaerobic Microorganisms*. A.J.B. Zehnder (ed.). John Wiley & Sons, New York, NY, pp. 587–639.

Leschine S.B. 1995. Cellulose degradation in anaerobic environments. *Ann. Rev. Microb.* 49: 399–426.

Liesack W., Schnell S., and Revsbech N.P. 2000. Microbiology of flooded rice paddies. *FEMS Microb. Ecol.* 24: 625–645.

Lokshina L.Ya. and Vavilin V.A. 1999. Kinetic analysis of low temperature methanogenesis. *Ecol. Mod.* 285–303.

Lovley D.R. and Coates J.D. 2000. Novel forms of anaerobic respiration of environmental relevance. *Curr. Opin. Microb.* 3: 252–256.

Lovley D.R. 1995. Microbial reduction of iron, manganese, and other metals. *Adv. Agron.* 54: 54–231.

Marlet S. 1996. Alcalinisation des sols dans la vallée du fleuve Niger. Modélisation des processus physico-chimiques et évolution des sols sous irrigation, Thesis, University of Montpellier, Montpellier, France.

Mayer K.U. 1999. A numerical model for multicomponent reactive transport in variability saturated porous media. Thesis, Dept. Earth Sciences, University of Waterloo, Waterloo, Ontario, Canada.

McInerney M.J. 1988. Anaerobic hydrolysis and fermentation of fats and proteins, In: *Biology of Anaerobic Microorganisms*. A.J.B. Zehnder (ed.). John Wiley & Sons, NY, pp. 373–417.

Moodie A.D. and Ingledew J. 1990. Microbial anaerobic respiration. *Adv. Microbial. Physiol.* 31: 225–269.

Murase J. and Kimura M. 1997. Anaerobic reoxydation of Mn^{2+}, Fe^{2+}, S^0 and S^{2-} in submerged paddy soils. *Biol. Fert. Soil* 25: 302–306.

Nanninga H.J. and Gottschal J.C. 1985. Amino acid fermentation and hydrogen transfer in mixed cultures. *FEMS Microb. Ecol.* 31: 261–269.

Nealson K.H. and Myers C.R. 1992. Microbial reduction of manganese and iron: new approaches to carbon cycling. *Appl. Environ. Microb.* 58: 439–443.

Nealson K.H. and Saffarani D. 1994. Iron and manganese in anaerobic respiration: Environmental significance, physiology, and regulation. *Ann. Rev. Microb.* 48: 311–343.

Nealson K.H. and Little B. 1997. Breathing manganese and iron: Solid-state respiration. *Adv. Appl. Microb.* 45: 213–239.

Oremland R.S. 1988. Biogeochemistry of methanogenic bacteria. In: *Biology of Anaerobic Microorganisms.* A.J.B. Zehnder (ed.). John Wiley & Sons, New York, NY, pp. 641–707.

Pankow J.F. 1991. Thermodynamic principles. In: *Aquatic Chemistry Concepts.* Lewis Publishers, Boca Raton, FL, USA, pp. 17–51.

Parkhurst D.L. and Appelo C.A.J. 1995. User's guide to PHREEQC—A computer program for speciation, reaction-path, advective-transport, and inverse geochemical calculations. US Geological survey Water-Resources Investigations. Report 99–4259, 143 pp.

Parry S., Renault P., Chenu C., and Lensi R. 1999. Denitrification in pasture and cropped soil clods as affected by pore space structure. *Soil Biol. Biochem.* 31: 493–501.

Parry S., Renault P., Chadoeuf J., Chenu C., and Lensi. R. 2000. Particulate organic matter as a source of denitrification variability in soil clods. *Eur. J. Soil Sci.* 51: 271–281.

Pelmont J. 1993. *Bactéries et environnement: adaptations physiologiques.* Presses Universitaires de Grenoble, Grenoble, France.

Peters V. and Conrad R. 1996. Sequential reduction processes and initiation of CH_4 production upon flooding of oxic upland soils. *Soil Biol. Biochem.* 28: 371–382.

Ponnamperuma F.N. 1972. The chemistry of submerged soils. *Adv. Agron.* 24: 29–88.

Prieme A. 1994. Production and emission of methane in a brackish and a freshwater wetland. *Soil Biol. Biochem.* 26: 7–18.

Quiquampoix H. 1987a. A stepwise approach to the understanding of extracellular enzyme activity in soil, I. Effect of electrostatic interactions on the conformation of a β-D-glucosidase adsorbed on different mineral surfaces. *Biochimie* 69: 753–763.

Quiquampoix H. 1987b. A stepwise approach to the understanding of extracellular enzyme activity in soil, II. Competitive effects on the adsorption of β-D-glucosidase in mixed mineral or organo-mineral systems. *Biochimie* 69: 765–771.

Ramsing N.B., Fossing H., Ferdelman T.G., Andersen F., and Thampdrup B. 1996. The distribution of bacterial populations in a stratified Fjord (Mariager Fjord, Denmark) quantified by in situ hybridization and related to chemical gradients in the water column. *Appl. Environ. Microb.* 62: 1391–1404.

Reddy K.R., D'Angelo E.M., and Harris W.G. 1998. Biogeochemistry of wetlands. In: *Handbook of Soil Science.* M.E. Summer (ed.). CRC Press, Boca Raton, FL, USA, pp. 89–119.

Refait P., Bon C., Simon L., Bourrié G. et al. 1999. Chemical composition and gibbs standard free energy of formation of Fe(II)–Fe(III) hydroxysulfate green rust and Fe(II) hydroxide. *Clay Minerals* 34: 499–510.

Robert M. and Chenu C. 1992. Interactions between soil minerals and microorganisms. In: *Soil Biochemistry.* G. Stotzky and J-M. Bollag (eds.). Marcel Dekker, New York, NY, USA, vol. I, pp. 307–403.

Roden E.E. and Zachara J.M. 1996. Microbial reduction of crystalline iron(III) oxides: Influence of oxide surface area and potential for cell growth. *Environ. Sci. Tech.* 30: 1618–1628.

Salvage K.M. and Yeh G.T. 1998. Development and application of a numerical model of kinetic and equilibrium microbiological and geochemical reactions. *J. Hydrology* 209: 27–52.

Scheid D. and Stubner S. 2001. Structure and diversity of Gram-negative sulfate-reducing bacteria on rice roots. *FEMS Microb. Ecol.* 36: 175–183.

Schimel J. 1995. Ecosystem consequences of microbial diversity and community structure. *Ecol. Studies* 113: 239–254.

Schimel J. and Gulledge J. 1998. Microbial community structure and global trace gases. *Global Change Biol.* 4: 745–758.

Schink B. 1997. Energetics of syntrophic cooperation in methanogenic degradation. *Microb. Molec. Biol. Rev.* 61: 262–280.

Schwertmann U. and Fechter H. 1994. The formation of green rust and its transformation to lepidocrocite. *Clay Minerals* 29: 87–92.

Segers R. and Kengen S.W.M. 1998. Methane production as a function of anaerobic carbon mineralization: A process model. *Soil Biol. Biochem.* 30: 1107–1117.

Shin H.S. and Song Y-C. 1995. A model for evaluation of anaerobic degradation characteristics of organic waste : Focusing on kinetics, rate-limiting step. *Environ. Sci. Tech.* 16: 775–784.

Stams J.M. 1994. Metabolic interactions between anaerobic bacteria in methanogenic environments. *Antonie Van Leeuwenhoek* 66: 271–294.

Stouthamer A.H. 1988. Dissimilatory reduction of oxidized nitrogen compounds. In: *Biology of Anaerobic Microorganisms*. A.J.B. Zehnder (ed.). John Wiley & Sons, New York, NY, pp. 245–303.

Stumm W. and Morgan J.J. 1996. Aquatic chemistry. In: *Chemical Equilibria and Rates in Natural Waters*. J.L. Shnoor, A.J.B. Zehnder (eds.). John Wiley & Sons, New York, NY, USA.

Susmel P., Sphangero M., and Stefanon B. 1999. Interpretation of rumen biodegradability of concentrate feeds with a Gompertz model. *Ann. Feed Sci. Tech.* 79: 223–237.

Teske A., Wawer C., Muyzer G., and Ramsing N.B. 1996. Distribution of sulfate-reducing bacteria in a stratified Fjord (Mariager Flord, Denmark) as evaluated by most-probable-number counts and denaturing gradient gel electrophoresis of PCR-amplified ribosomal DNA fragments. *Appl. Environ. Microb.* 62: 1405–1415.

Tiedje J.M., Sexstone A.J., Parkin T.B., and Revsbech N.P. 1984. Anaerobic processes in soil. *Plant Soil* 76: 197–212.

Tipping E. 1994. WHAM—A chemical equilibrium model and computer code for waters, sediments, and soils incorporating a discrete site / electrostatic model of ion-binding by humic substances. *Computers & Geosci.* 20: 973–1020.

Trolard F. and Bourié G. 1999. Influence of green rusts on oxido-reduction sequences in soils. *C. R. Acad. Sci. Série 2*, 329: 801–806.

Trolard F., Génin J.M.R., Abdemoula M., Bourrié G., Humbert B., and Herbillon A. 1997. Identification of a green rust mineral in a reductomorphic soil by Mössbauer and Raman spectroscopies. *Geochim. Cosmochima Acta* 61: 1107–1111.

Tsusuki K. and Ponnamperuma F.N. 1987. Behavior of anaerobic decomposition products in submerged soils. Effects of organic material amendment, soil properties, and temperature. *Soil Sci. Plant Nutr.* 33: 13–33.

Vallès V. and Bourgeat F. 1988. Geochemical determination of gypsum requirement of cultivated sodic soil. I. Chemical interactions. *Arid Soil Res. Rehab.* 165–177.

Van Den Heuvel J.C., Beeftink H.H., and Verschuren P.G. 1988. Inhibition of the acidogenic dissimilation of glucose in anerobic continuous cultures by free butyric acid. *Appl. Environ. Microb.* 56: 719–723.

Vasiliev V.B., Vavilin V.A., Rytov S.V., and Ponomarev A.V. 1993. Simulation model of anaerobic digestion of organic matter by a microorganism consortium: Basic equations. *Water Ressources Res.* 20: 633–643.

Vavilin V.A., Vasiliev V. B., Ponomarev A.V., and Rytov S.V. 1994a. Simulation model 'methane' as a tool for effective biogas production during anaerobic conversion of complex organic matter. *Bioresources Tech.* 48: 1–8.

Vavilin V.A., Vasiliev V. B., Ponomarev A.V., and Rytov S.V. 1994b. Self-oscillating coexistence of methanogens and sulfate-reducers under hydrogen sulfide inhibition and the pH-regulating effect. *Bioresources Tech.* 49: 105–119.

Vavilin V.A., Rytov S.V., and Lokshina L.Ya. 1995a. Modelling hydrogen partial pressure change as a result of competition between the butyric and propionic groups of acidogenic bacteria. *Bioresources Tech.* 54: 171–177.

Vavilin V.A., Vasiliev V.B., and Rytov S.V. 1995b. Modelling of gas pressure effects on anaerobic digestion. *Bioresources Tech.* 52: 25–32.

Vavilin V.A., Vasiliev V.B., Rytov S.V., and Ponomarev A.V. 1995c. Modeling ammonia and hydrogen sulfide inhibition in anaerobic digestion. *Water Resources* 29: 827–835.

Vavilin V.A., Rytov S.V., and Lokshina L.Ya. 1996a. A description of hydrolysis kinetics in anaerobic degradation of particulate organic matter. *Bioresource Tech.* 56: 229–237.

Vavilin V.A. and Lokshina L.Ya. 1996b. Modeling of volatile fatty acids degradation kinetics and evaluation of microorganism activity. *Bioresource Tech.* 57: 69–80.

Vavilin V.A., Vasiliev V.B., and Rytov S.V. 1996c. Simulation of constituent processes of anaerobic degradation of organic matter by the methane model. *Antonie Van Leeuwenhoek* 69: 15–23.

Vavilin V.A., Rytov S.V., and Lokshina L.Ya. 1997a. A balance between hydrolysis and methanogenesis during the anaerobic digestion of organic matter. *Microbiology* 66: 846–851.

Vavilin V.A., Lokshina L.Ya., Rytov S.V., Kotsyurbenko O.R., Nozhevnikova A.N., and Parshina S.N. 1997b. Modelling methanogenesis during anaerobic conversion of complex organic matter at low temperatures. *Water Resources Tech.* 36: 531–538.

Vavilin V.A., Lokshina L.Ya., Rytov S.V., Kotsyurbenko O.R., and Nozhevnikova A.N. 1998. Modelling low-temperature methane production from cattle manure by an acclimated microbial community, *Bioresource Tech.* 63: 159–171.

Vavilin V.A, Lokshina L.Ya., Rytov S.V., Kotsyurbenko O.R., and Nozhevnikova A.N. 2000. Description of two-step kinetics in methane formation during psychrophilic H_2/CO_2 and mesophilic glucose conversions. *Bioresource Tech.* 71: 195–209.

Visser E.J.W., Nabben R.H.M., and Blom C.W.P.M. 1997. Elongation by primary lateral roots and adventitious roots during conditions of hypoxia and high ethylene concentrations. *Plant Cell Environ.* 20: 647–653.

Wagner C., Grießhammer A., and Drake H.L. 1996. Acetogenic capacities and the anaerobic turnover of carbon in a Kansas prairie soil. *Appl. Environ. Microb.* 62: 494–500.

Weber S., Stubner S., and Conrad R. 2001. Bacterial populations colonizing and degrading rice straw in anoxic paddy soil. *Appl. Environ. Microb.* 67: 1318–1327.

Widdel F. 1988. Microbiology and ecology of sulfate and sulfur-reducing bacteria, In: *Biology of Anaerobic Microorganisms*. A.J.B. Zehnder (ed.). John Wiley & Sons, New York, NY, pp. 469–587.

Wind T. and Conrad R. 1995. Sulfur compounds, potential turnover of sulfate and thiosulfate, and numbers of sulfate-reducing bacteria in planted and unplanted paddy soil. *FEMS Microb. Ecol.* 18: 257–266.

Witt C., Gaunt J.L., Galicia C.C., Ottow J.C.G., and Neue H-U. 2000. A rapid chloroform-fumigation extraction method for measuring soil microbial biomass carbon and nitrogen in flooded rice soils. *Biol. Fert. Soils* 30: 510–519.

Wogel F. 1988. Biochemistry of methane production In: *Biology of Anaerobic Microorganisms*. A.J.B. Zehnder (ed.). John Wiley & Sons, New York, NY, pp. 469–587.

Wolin M.J. 1988. Hydrogen transfer in microbial communities, In: *Microbial Interactions and Communities*. A.T. Bull and J.H. Slater (eds.). Acad. Press, London, UK, vol. I, pp. 323–356.

Zausig J., Stepniewski W., and Horn R. 1993. Oxygen concentration and redox potential gradients in saturated model soil aggregates. *Soil Sci. Soc. Amer. J.* 57: 908–916.

Zehnder A.J.B. and Svensson B.H. 1986. Life without oxygen: What can and what cannot? *Experientia* 42: 1197–1205.

Zehnder A.J.B. and Stumm W. 1988. Geochemistry and biogeochemistry of anaerobic habitats. In: *Biology of Anaerobic Microorganisms*. A.J.B. Zehnder (ed.). John Wiley & Sons, New York, NY, pp. 1–39.

Zumstein E., Moletta R., and Godon J.J. 2000. Examination of two years community dynamics in an anaerobic bioreactor using fluorescence polymerase chain reaction (PCR) single-strand conformation polymorphism analysis. *Environ. Microb.* 2: 69–78.

3

Impact of Soil Mineral-Organic Component-Microorganism Interactions on Restoration of Terrestrial Ecosystems

J.-M. Bollag*, J. Berthelin, D.C. Adriano, *and* **P.M. Huang**

Abstract

Soil is a dynamic system in which continuous interaction takes place between soil minerals, organic matter, and organisms. Each of these three major soil components influences the physicochemical and biological properties of terrestrial systems. Interactions between the mineral, organic, and biological factors have enormous impact on terrestrial processes critical to environmental quality and ecosystem health; they control the cycling and bioavailability of nutrients, metals, and xenobiotic substances in the environment through physical, chemical, biochemical and biological processes.

Human activities and industrialization development generate byproducts and waste that must be disposed of in a way that shall not affect the environment. Even when such byproducts are used on agricultural lands as a resource, e.g. organic waste that may contribute to maintain or increase the organic matter and nutrient content in the soil, there are growing concerns about the fate of undesirable constituents they may contain. A wide variety of naturally occurring toxic and recalcitrant organic compounds exist on earth; however, nature has evolved ways of mineralizing many of them without human intervention. Microorganisms are ultimately responsible for mineralizing most organic matter to carbon dioxide, water, and inorganic components. However, soil mineral colloids also play a significant role in abiotic transformation of organic compounds through catalysis. Furthermore, soil mineral colloids substantially influence microbial activity and enzymatic reactivity and the subsequent transformation of organic components.

**Corresponding author:* Dr. J.-M. Bollag, Soil Biochemistry, Pennsylvania State University, University Park, PA 16802, USA. E-mail: jmbollag@psu.edu

Over a very long period of time, natural degradation might remove many of the organic contaminants, but could possibly result in an accumulation of heavy metals. Affordable novel technologies are needed to enhance natural remediation processes (natural attenuation) and reduce health risks by restoring natural balances. Therefore, ecosystem restoration can best be achieved through development of innovative management strategies involving biotic and abiotic interactive processes.

1 INTRODUCTION

Considerable work has been done to explain the mechanisms of microbial degradation of xenobiotics and other contaminants (Alexander, 1999). Extensive studies have also been carried out to determine the effects of minerals and soil organic matter on the fate of xenobiotics (Huang and Bollag, 1998; Berthelin et al., 1999; Haider 1999; Dec et al., 2002) and metals (McBride, 1994; MacLaughlin et al., 1998; Huang and Germida, 2002; Sparks, 2003). Relatively little has been done, however, to integrate the information obtained to evaluate biotic and abiotic interactions in relation to environmental pollution.

Soils are an integral compartment of the environment. Cycling of materials, including nutrients essential to humans, animals, and plants, and toxic pollutants in the pedosphere can influence thus significantly the dynamics and fate of these materials in terrestrial and aquatic environments. The impact of the interactions among minerals, organic components and microorganisms in the soil on material cycling in the pedosphere and on environmental sustainability warrants increased attention.

The microbial and soil component interactions in the pedosphere affect the metabolic transformation of natural and xenobiotic organic compounds and the fate of metals and other inorganic components. These interactions substantially influence changes in the form and composition of materials located in this zone and their subsequent translocation and cycling in the pedosphere.

2 COMPLEXITY OF SOIL AND TRANSFORMATION OF XENOBIOTICS AND METALS

Soil is undoubtedly the most complex of all microbial habitats. Largely because of this complexity, there is insufficient information on how and where most microbial activity occurs in situ and which microorganisms are the most important participants. There is considerable evidence that certain particulate soil components, especially some types of clay minerals, significantly affect microbial life (Stotzky, 1986; Huang, 1990; Theng and Orchard, 1995). Clay minerals appear to exert their primary influence by modifying the physicochemical characteristics of microbial habitats; this either enhances or attenuates the growth and metabolism of individual microbial populations, which, in turn, influence the growth and activity of other populations. In contrast to these indirect effects of clay minerals, relatively little is known about the mechanisms of direct surface interactions (e.g. adhesion) between clays and microorganisms.

The characteristics of soils allow for a unique environment that encourages biotic and abiotic interactions. These interactions influence the transformation, dynamics, and toxicity of organic compounds and metals.

2.1 Xenobiotics

The transformation of xenobiotics in a multicomponent system, such as soil, is a result of the combined activity of microorganisms, extracellular enzymes, mineral colloids and humic materials (Huang and Bollag, 1998). Each of these important components of the soil system not only participates in the transformation of xenobiotics, but also modifies the activity of the other components.

Humic substances associated with mineral colloids add an additional aspect to the process of xenobiotic biodegradation in soil environments. Humic substances are strong adsorbents themselves, and the adsorption of microorganisms, enzymes and xenobiotics on the organic fraction may be difficult to distinguish from that occurring on the mineral surfaces. Therefore, it is difficult to determine the effect of each of these fractions on biodegradation. Humic acids are believed to catalyze certain transformation reactions, but their major impact is their ability to adsorb (microorganisms, enzymes and chemicals) or to be adsorbed (on soil minerals). Minerals exhibit mixed functions as well. They transform naturally occurring and xenobiotic substrates abiotically; at the same time, they act as sorbents, thus altering the impact of microorganisms, enzymes, and chemicals. Adsorption and other binding interactions that occur on both mineral and humic surfaces are believed to reduce the bioavailability of xenobiotics.

2.2 Metals

The transformation of metals is governed by abiotic and biotic processes in soil and related environments (Huang, 2000; Huang and Germida, 2002; Sparks, 2003). Abiotic processes include solution complexation, adsorption-desorption, precipitation-dissolution, redox reactions, and catalysis.

A series of complexation reactions in the soil solution affect metal transformation. Metals vary often act as the central components that attract neutral or negatively charged ligands. If the interaction between the cation and the ligand is strong enough for the ligand to displace the inner hydration sphere, the resultant complexation is termed the formation of an inner-sphere complex. If the interaction between the cation and the ligand is not strong enough to displace the inner hydration sphere, the association is termed the formation of an outer-sphere complex or an ion pair. Complexation reactions of metals with ligands in the soil solution are significant in determining the chemical behavior and toxicity of metals (Hayes and Traina, 1998). In view of the occurrence of organic and inorganic ligands in soil environments (Szmigielska et al., 1996; Cieslinski et al., 1998; MacLaughlin et al., 1998; Marschner, 1998) and the stability constants of the complexes of metals with these ligands (NIST, 1997) a large fraction of soluble metal ions in the soil solution may be actually in complex forms.

Adsorption-desorption reactions significantly influence dynamics and equilibria of metal ions in soil environments. (McBride, 2000). Soils are remarkable for their ability to remove metal ions from the soil solution by adsorption reactions. Metal oxides and edges of phyllosilicates, organic matter, and organomineral complexes are surface reactive sites of soils for chemisorption of metals. Few studies have investigated the adsorption-desorption reactions in the rhizosphere. Excretion products of roots include a variety of low-molecular-mass organic acids (LMMOAs), many of which are capable of forming complexes with metal ions (Robert and Berthelin, 1986; Huang and Germida, 2002). Krishnamurti et al. (1997) reported that the increase in Cd release in the presence of LMMOAs can be explained by the surface complexation of the particulate-bound Cd in soil with LMMOAs, which is reflected in increased release of Cd from the soils with increase in stability constant of Cd-LMMOA complexes (Fig. 3.1).

Chemical precipitation may influence the concentration of metals in soil solutions. For many of the more abundant elements such as Al, Fe and Mn, precipitation of mineral forms is common and may control their solubility. For

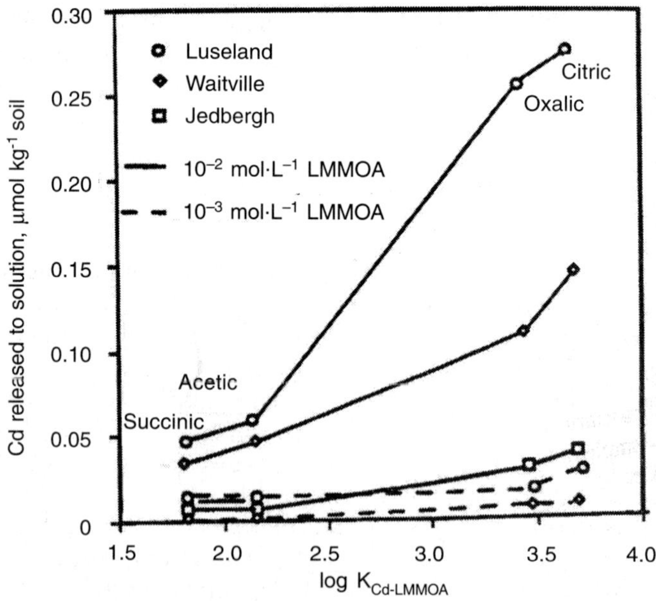

Fig. 3.1: Relationship between Cd released from soils by selected low molar mass organic acids (LMMOAs) (acetic, citric, oxalic, and succinic acids) during the reaction period of 0.25 h and the logarithm of the stability constant (log K) of Cd-LMMOA complexes (Krishnamurti et al., 1997).

most of the trace metals, direct precipitation from solution through homogeneous nucleation appears to be less likely than adsorption-desorption by virtue of the low concentrations of these metals in the soil solutions in well aerated dry-land soils. However, in soil environments, heterogeneous nucleation is more likely than homogeneous nucleation due to the presence of mineral, organic, and microbial surfaces which can catalyze the nucleation set of precipitation (McBride, 2000; Huang and Germida, 2002). The energy barrier to nucleation is reduced or removed by these surfaces. When soils become heavily polluted, metal solubility may reach a level to cause precipitation. In addition, precipitation may occur in the immediate vicinity of the phosphate fertilizer zone where the concentration of heavy metals present as impurities may be sufficiently high. In reduced environments where the sulfide concentration is sufficiently high, precipitation of trace elements as sulfides may have a significant role of metal transformation (Robert and Berthelin, 1986).

Redox reactions are important in controlling the chemical speciation and toxicity of a number of contaminant metals, notably As, Se, Cr, Pu, Co, Pb, Ni, and Cu (McLaughlin et al., 1998; Sparks, 2003). Redox reactions are also important in controlling the transformation and reactivity of Mn and Fe oxides in soils, which have enormous capacities to adsorb metal pollutants and are the major sinks of these pollutants. Furthermore, reduction of sulfate to sulfide in aerobic environments may also affect metal speciation and solubility. Masschelyn and Patrick (1994) summarized the critical redox potentials for transformation of some contaminants (Fig. 3.2). There has been little study of how soil redox potential changes in the rhizosphere could affect contaminant chemistry (MacLaughlin et al., 1998).

Abiotic catalytic reactions in soils and associated environments are more common than previously thought. Mn oxides and Fe-bearing minerals have the ability of catalyze the transformation of metals, metalloids, and other inorganic (Huang, 2000). Manganese oxides are effective catalysts in promoting many reactions such as the transformation of Cr (III) to Cr (VI) (Bartlett and

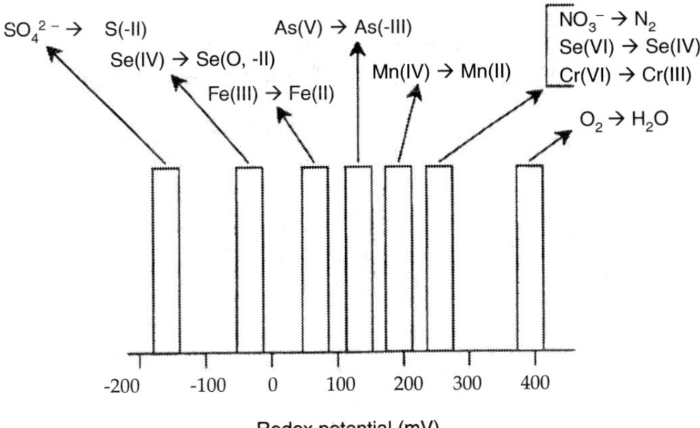

Fig. 3.2: Critical redox potentials for some contaminant ions (Masschelyn and Patrick, 1994).

James, 1979; Amacher and Baker, 1982), As (III) to As (V) (Oscarson et al., 1981, 1983), Fe (II) to Fe (III) (Krishnamurti and Huang, 1987), Pu (III) to Pu (IV) (Cleveland 1970; Amacher and Baker, 1982), the autooxidation of Mn (II) (Ross and Bartlett, 1981) and the oxidation of nitrite to nitrate (Bartlett, 1981). As an example, the conversion of more toxic As (III) to much less toxic As (V) by catalysis of Mn (IV) oxide is shown in Table 3.1. Heterogeneous catalytic reactions involving electron transfer between transition metals and Fe-bearing minerals have also been demonstrated (Wehrli and Stumm, 1989; Ilton and Veblen, 1994; Peterson et al., 1996; White and Peterson, 1996. Abiotic catalysis thus, merits close attention in understanding the transformation of metals, metalloids, and other inorganics and the impact on system restoration.

Table 3.1: Oxidation of As (III) and sorption of As by Mn (IV) oxide (Oscarson et al., 1981)

As(III) or As (V added)	As (III)	As (V)	Mn	Final pH
μ mL^{-1}		——μ mL^{-1} in solution——		
100 As (III)	ND*	83.5 ± 1.4^{+}	0.41 ± 0.12	7.1
300 As (III)	63.2 ± 7.0	186 ± 5	8.08 ± 0.36	7.1
500 As (III)	213 ± 4	205 ± 1	6.06 ± 0.60	7.3
1,000 As (III)	665 ± 5	216 ± 4	4.16 ± 0.86	7.5
300 As (V)	ND	298 ± 1	0.06 ± 0.02	7.5

*ND = not detectable

$^{+}$Mean + SD; n = 3

Microbial activity also has a very important role in influencing dynamics of metals in the terrestrial ecosystem. Microbes can dissolve minerals by direct or indirect action under aerobic and anaerobic conditions (Kurek, 2002). When oxidized metal compounds such as Fe (III), Mn (IV) or As (V) act as terminal election acceptors, anaerobic respiration becomes an example of direct dissolving action under anaerobic conditions. Oxidation of ferrous iron or sulfur entities of metal sulfides to obtain energy is an example of direct dissolving action under aerobic conditions. Indirect dissolution of minerals can be the result of microbial activity connected with the productions of organic and inorganic acids, and oxidizing agents which can influence soil conditions including changes in pH and Eh. Metals can also be mobilized from minerals by complexation with biomolecules in microbial metabolites. Volatilization of metals and metalloids or biomethylation of metals and metalloids from the soil into the atmosphere can be a mechanism of detoxification of toxic elements such as Hg, As, and Se for microbes.

Microbial interactions with metals can lead to the formation of fine-grained minerals and thus have an impact on bioremediation strategies (McLean et al., 2002). The mechanism by which bacteria initiate the formation of minerals in bulk solution vary widely between species. There may be a combination of biochemical and surface mediated reaction. Microbial Mn (II) oxidation is a major process that can produce Mn oxide coatings on soil particles 10^5 times faster then abiotic oxidation (Tebo et al., 1997). Bacteria such as *Leptothrix* sp. precipitate Mn oxide on their outermost structure called a sheath (Fig. 3.3). Manganese oxides are highly reactive mineral phases and help control the mobility of metals in soils and related environments through adsorption on their surfaces. Biogenic Mn oxides formed by *L. discophora SS-1* have significantly larger surface area and higher Pb adsorption capacity than abiotically precipitated Mn oxides (Nelson et al., 1999).

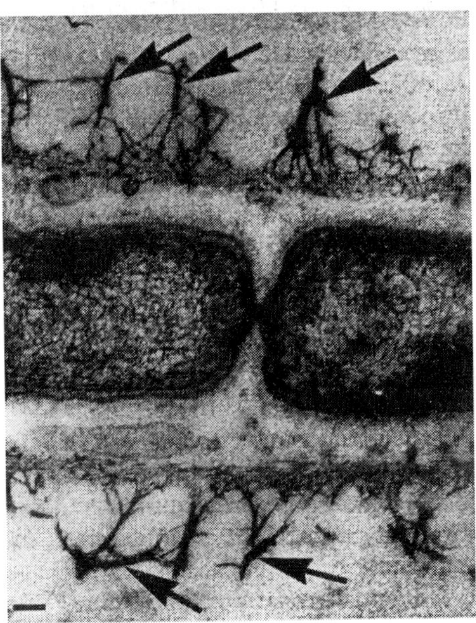

Fig. 3.3: Thin section of *Leptothrix* sp. which is precipitating manganese oxide on its outermost structure called a sheath. Arrows point to the manganese mineral phase identified by EDS. Scale bar = 150 nm (McLean et al., 2002).

3 PROBLEM OF SOIL POLLUTION

The soil has become increasingly subjected to various chemical stresses, not only because of our need for more food and fiber, but also because of ever-increasing industrialization. Various anthropogenic substances, either organic or inorganic in nature, upon entering the soil may not only adversely affect its productivity potential, but may also compromise the quality of the food chain and groundwater. This situation may require risk assessment and evaluation of remedial techniques in order to restore the quality of the soil so that safe food products and clean groundwater and air may be obtained once again.

The human activities related to industrialization generate large amounts of byproducts and waste that must be disposed of in an environmentally benign way. Sometimes byproducts of industrial processes are used on agricultural lands as a resource, e.g. organic waste that should maintain or increase the organic matter and nutrient content of a soil, but there are growing concerns about the fate of undesirable constituents they may contain.

A wide variety of naturally occurring toxic and recalcitrant organic compounds exist on earth. In addition, various man-made materials have been dumped on land adjacent to industrial plants in landfills and on unregulated dumping grounds. As a result, the soils at many of these sites contain a complex mixture of contaminants, such as petroleum products, organic solvents, metals, acids, bases, brines, and radionuclides. Over a very long period of time, natural degradation activities may eventually destroy most of these organic contaminants. However, affordable technologies are needed to speed up the natural remediation processes. Furthermore, natural degradation activities would not solve the problems of metal contaminants. Therefore, risk management through remediation is essential to reducing health risks and restoring natural balances. The treatments currently used to remove or destroy contaminants include physical, chemical and biological technologies.

4 REMEDIATION OF CONTAMINANTS

The cleanup of soils polluted by hazardous man-made materials has become a matter of urgent public concern. Traditional methods of waste handling, such as landfilling or incineration, often exchange one problem for another. For instance, landfilling merely confines the pollution while doing little to remove it; certain kinds of garbage are known to remain intact for decades in a landfill. Incineration removes wastes, but disposal of ash or residues still remains and new concerns about air pollution are created. As a result, society has turned to technology to devise better methods of disposal; one such method is bioremediation.

Microorganisms are ultimately responsible for degrading most organic matter to carbon dioxide, minerals, and water. Bioremediation utilizes the natural potential of microorganisms to cause transformation, mineralization or complexation by directing those capabilities toward environmental pollutants.

The enhancement of microbial degradation as a means of bringing about the *in situ* clean up of contaminated soils has spurred much research (Adriano et al., 1999). The most common methods to stimulate degradation rates include supplying inorganic nutrients and oxygen, but the addition of degradative microbial inocula or enzymes as well as the use of plants (phytoremediation) should also be considered (Bollag et al., 1994). The successful implementation of soil remediation requires interdisciplinary cooperation among soil biology, soil chemistry, and engineering experts. Basic research is needed to better understand the biological, chemical, and physical factors affecting transformation pathways and reaction rates.

4.1 Natural Remediation Processes

Nature has evolved ways of mineralizing many contaminants or xenobiotics without human intervention. Soil organic matter is recycled by a diverse array of soil organisms including bacteria, fungi, actinomycetes, protozoa, earthworms, and insects. Microorganisms are usually responsible for mineralizing most organic matter to carbon dioxide, water, and inorganic components. However, soil mineral colloids also play a significant role in abiotic transformation of organic compounds through catalysis (Huang, 2000). In addition, soil mineral colloids influence microbial growth, metabolism, survival and enzymatic reactions (Stotzky, 1986; Theng and Orchard, 1995).

Over a very long period of time, natural degradation (also known as natural attenuation) might remove many of the organic contaminants in soil, but may result in an accumulation of heavy metals. Affordable novel technologies are needed to enhance natural remediation processes (natural attenuation) and restore healthy environmental conditions.

Natural remediation usually influences the extent of bioavailability (or mobility) of contaminants in the soil complex (Adriano et al., 2003). The different natural remediation-bioavailability processes influencing contaminant dynamics are depicted in Figure 3.4. In the initial phase (A) the dynamics of contaminants between the aqueous phase and solid phase are regulated by various biogeochemical processes, the extent of which determines metal partitioning. Some of these processes include such basic ones like desorption/adsorption, precipitation/dissolution, complexation, redox reactions, etc. In essence, the nature of these processes determines the kinetics of natural remediation that can be parameterized by bioavailability. The role of the rhizosphere in

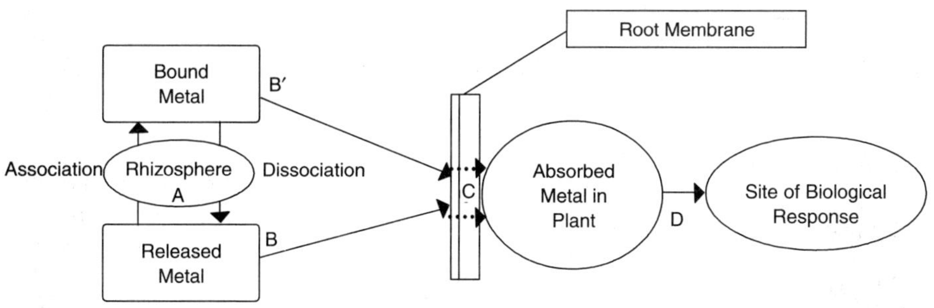

Bioavailability processes (A, B, B′ and C)

Metal interactions between phases (A)	Transport of metals to plant (B, B′)	Passage across root membrane (C)	Circulation within plant, accumulation in target tissue and toxic effects (D)

Fig. 3.4: Bioavailability processes in soils. Note the delineation of the whole scheme into "Natural remediation" (left) and "Bioavailability" (right). Modified from the National Research Council (2003).

contaminant transformation and partitioning can be very important and is discussed below. The next phase (B, B′) involves the transport of contaminants to organisms. The contaminant can be transported in soluble, colloidal, and/or particulate form. The particulate form can play a significant role in humans and animals. For example, ingestion of soil particles by children is an important exposure pathway for lead. By the same token, ingestion of soil particles by grazing livestock is an important exposure pathway for animals feeding on contaminated pastures. From the aqueous phase, metals can be mobilized either in soluble or colloidal form, the latter being viewed as facilitated transport. Usually inorganic and OM-based colloids are highly reactive thus they characteristically contain much higher metal concentrations compared with those in the solution. The next phase (C) involves passing through a biological membrane, which in many instances can serve as a biofilter for contaminants. In plants, this is represented by the root membrane; in humans, it is represented by the GI (gastrointestinal) tract. Then the last phase (D) involves circulation and assimilation in the metabolic machinery of the organisms, culminating in some form of biological response. This usually refers to the end point of interest.

The potential role of higher plants in accelerating natural remediation in contaminated soils has recently focused on the rhizosphere, which is characterized by distinct physical, chemical, and biological conditions (Figs. 3.4 and 3.5). These are created by the plant roots and their microbial associations. These rhizosphere-related biogeochemical processes are unstable because they are considerably influenced by edaphic and climatic conditions. The edaphic influence is in turn modified by physical, mineralogical, chemical, and biological features of the soils. Due to the limited areal extension of the rhizosphere, special tools and techniques are required to study its characteristics and processes (e.g. Wenzel et al., 2001).

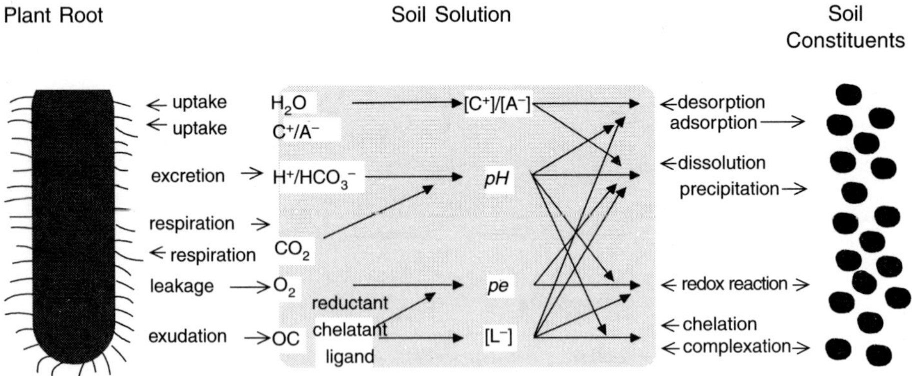

Fig. 3.5: Schematic rhizosphere, showing the various exudates and how they can influence abiotic factors and mechanisms in the soil-solution interface. Legends: OC = organic carbon; C^+ = cation; A- = anion; L^- = ligand; re = redox potential. (modified from Hinsinger, 2001).

The rhizosphere is spatially defined as the few millimeters of soil surrounding the plant roots and influenced by their activity as well as the microbial assemblages associated with the roots. The rhizosphere can also be functionally defined as a highly dynamic, solar / plant-driven micro-environment characterized by feedback loops of interactions between root processes, soil characteristics, and the dynamics of the associated microbial population (Wenzel et al., 2001). Soil characteristics such as pH and redox potential, nutrient status, presence of contaminants, and physical properties all influence root growth and the dynamics of the microbial population. On the other hand, rhizosphere soil characteristics are modified by plants by means of root exudation, uptake of nutrients and contaminants, and root growth. The energy that drives these changes is provided by sunlight through photosynthesis. A key component of this system is represented by root exudates. It has been calculated that between 10 and 40% of the total net C assimilated by crops is released in the form of soluble root exudates and insoluble materials such as cell walls and mucilage. Plants release root exudates that stimulate microbial activity and biochemical transformations and / or enhancement of mineralization of metals in the rhizosphere (Anderson et al., 1993).

The chemical and biological reactions occurring in the rhizosphere play an important role in the natural remediation and subsequent bioavailability of contaminants to plants. Plant roots may change the physical, chemical, and biological conditions of the soil in the rhizosphere which, compared to bulk soil, is enriched with organic substances of plant and microbial origin, including organic acids (OAs), sugars, amino acids, lipids, coumarins, flavonoids, proteins, enzymes, aliphatics, aromatics, carbohydrates, etc. (Pomilio et al., 2000). Among them the OAs are the most abundant and chemically very reactive with soil constituents. The commonly found OAs in the rhizosphere are acetic, butyric, citric, fumaric, lactic, malic, malonic, oxalic, propionic, tartaric, and succinic acids (Szmigielska et al., 1996; Koo, 2001). In soils, OAs are involved in biogeochemical processes that release the not-so-readily available toxic metals such as cadmium (Krishnamurti et al., 1997; Onyata and Huang, 2003) and nutrients such as phosphorus, iron, and other micronutrients (Awad and Römheld, 2000). Organic acids in the rhizosphere affect the dynamics of metals in soils via their effect on acidification, metal chelation and complexation, precipitation, redox reactions, microbial activity, rhizosphere physical properties, and root morphology (Huang and Germida, 2002).

Several environmental factors can influence the rate of biodegradation: pH, temperature, soil moisture, oxygen, nutrient contents, and types of bacteria present or introduced into the soil (Alexander, 1999). Bioremediation can involve biostimulation by changing one or more of these environmental conditions or by adding a limiting nutrient.

Soils contaminated with recalcitrant xenobiotics or environments hostile to indigenous microorganisms may benefit from microbial inoculation. Using classic enrichment techniques or gene transfer technology, large populations of bacteria may be developed, then inoculated into the contaminated soil (Bollag et al., 1994).

4.2 Biotic Processes

Microorganisms are the primary factors that influence the fate of organic xenobiotics in soil. Environmentally, the most desired result of microbial activity is the complete degradation (mineralization) of toxic chemicals of both natural and synthetic origin. Some anthropogenic compounds may be used by microorganisms as a source of energy and nutrients for growth. Another important process of microbial transformation, however, is cometabolism, in which microorganisms transform the xenobiotic molecules, but are unable to use these molecules as a source of carbon or energy.

The cause of microbial transformation is the activity of intracellular and extracellular enzymes produced by microorganisms. Xenobiotics having chemical structures similar to organic compounds that occur in nature are usually more susceptible to biodegradation than those whose structures bear little resemblance to natural products. This is because the microbial enzymes that specifically degrade toxic chemicals of natural origin may also be able to degrade structurally analogous xenobiotics. On the other hand, when microorganisms are exposed to structurally different xenobiotics, they usually need a prolonged acclimation period during which they may undergo genetic mutations, resulting in the production of novel degrading enzymes (van der Meer et al., 1992; Singh et al., 1999).

Microbial enzymes are involved in complex relationships with other components of the soil system; these relationships may considerably modify their ability to mediate the transformation of anthropogenic compounds (Burns, 1986; Naidja et al., 2000). The efficiency of intracellular enzymes depends on the conditions that microorganisms face. As a result, the activity of cellular enzymes is frequently identified as the activity of the microorganisms which produce them. Consequently, microorganisms tend to be viewed as the subject matter of investigatory insight, whereas the role of the enzymes tends to be underrated. In the case of extracellular enzymes, however, the perspective is different. Once released from their producers, extracellular enzymes can begin to function as a sole cause of xenobiotic biodegradation.

Our present knowledge of the microbial degradation of xenobiotics has been accumulated through numerous in vitro experiments in which xenobiotic compounds were exposed to specific microorganisms isolated from soil or other natural environments. Monitoring the transforming or degradative activities of microorganisms under field conditions may generate ambiguous results, because it is difficult to distinguish between physico-chemical and microbial transformation reactions. The ability of microorganisms to transform xenobiotics relies largely on four major processes: oxidation, reduction, hydrolysis, and synthetic reactions, which are all catalyzed by microbial enzymes.

It is of great importance to also consider the fact that the microbial activity in soil is strongly influenced by mineral colloids and humates that bind organic chemicals, inorganic ions, and water films to their surfaces (Huang, 1990). In addition, extracellular enzymes are rapidly sorbed at clay and organic surfaces. Immobilized through these events, enzymes usually acquire greater stability but may modify their activity. Some immobilized enzymes in soils are associated with the humic fraction by the formation of enzyme-phenolic copolymers during the generation of humic substances. Enzymes are vital biotic catalysts in the transformation of organic pollutants (Bollag, 1992). Compared with abiotic catalysts, they differ in capacity and kinetics in mediating oxidative coupling reactions (Pal et al., 1994).

Much research has been conducted on the use of bioremediation as a tool to cleanup many environmental contaminants such as polycyclic aromatic hydrocarbons, explosives, pesticides, polychlorinated biphenyls, and other organic toxicants. However, use of bioremediation technology for metal and other inorganic remediation is just becoming recognized as an efficient and effective strategy. Cadmium, lead, zinc, nickel, cobalt, and other base metals generally exist as cations in aqueous phase, and adsorption to the electronegative sites of the surface layers of bacteria can remove these toxic metals very effectively (Mullen et al., 1989; Volesky, 1990). Highly toxic and mobile oxyanions of uranium, chromium, arsenic, and selenium are of particular concern in many soils and groundwaters throughout this world. Their speciation in the environment and hence their reactivity, bioavailability, toxicity, and mobility are influenced significantly by microbe-mediated reactions. Metal oxyanions such as chromates, arsenates and selenates do not bind to any significant extent to the electropositive sites of the bacterial surface or through metal ligand bridging. Bioremediation of these metaloxyanions is predominantly based on microbially catalyzed redox conversions to insoluble forms (McLean et al., 2000).

Application of bioremediation technology to decontaminate polluted sites is still a developing science. The mechanisms driving microbial activity and degradation pathways of specific pollutants need to be further elucidated before successful and better controlled site-specific treatments can be applied. Recent advances in biotechnology have enabled modification of organisms at the molecular level for improved degradative performance. This approach has already contributed new tools for the analysis and monitoring of complex environmental processes.

4.3 Abiotic Processes

Abiotic factors such as the catalytic activity of mineral and organic colloids, may be a substantial source for the degradation of organic compounds in soil. To date, however, reports of abiotic transformation have been limited to just a relatively narrow group of xenobiotics.

The significance of soil mineral-catalyzed abiotic transformation of naturally occurring and xenobiotic compounds in the environment has become widely recognized only in recent years (Huang and Bollag, 1998). Clay minerals, with their high concentration in soils, their large surface area, and relatively high charge density, contribute to the overall xenobiotic transformation at least as much as does the organic matter. Even in soil environments where biological activity is intense, abiotic transformation is of importance.

In addition to clay minerals with well-ordered crystalline structures, soils contain mineral colloids, such as metal oxides, aluminosilicates and carbonates, that exist in poorly ordered forms. The contribution of these short-range ordered colloids to the transformation of organic chemicals on mineral surfaces warrants in-depth research (Huang, 2000). The poorly ordered colloids appear to play a dual role in humification and binding interactions between xenobiotics and soil. First, they may be directly involved in the oxidation of organic chemicals of natural and anthropogenic origin with the formation of desorption-resistant organic residues. Second, they may influence the efficiency of enzymes that contribute to the same reactions by immobilizing and thus stabilizing them (Naidja et al., 2000). Therefore, abiotic catalytic reactions resulting in the transformation of toxic organics and their impact on ecosystem remediation, deserve increased attention.

Besides addressing the presence of toxic organics, soil scientists, especially soil chemists, are addressing the issue of remediation of soils which have been contaminated with metals, metalloids, and other inorganics. The role of soil chemists in soil remediation can be demonstrated by citing a few of the simpler approaches using soil ameliorants. Many ameliorants have been used to remediate polluted soils (Table 3.2). These soil ameliorants are rather inexpensive and are readily available on a global scale. Application of limestone to elevate soil pH to about 6.5 is recommended when agricultural soils are amended with sewage sludge high in metals, especially cadmium (Page et al., 1983; Adriano et al., 1986; Pierzynski et al., 1994). The addition of certain clay minerals, such

Table 3.2: Selected ameliorants adapted to metal-contaminated soils (Adriano et al., 1998)

Technique	Target contaminants	Soil processes involved	Constraints
Limestone	Metals, radionuclides	Precipitation, sorption	Ineffective for oxyanions; certain crops (lettuce, spinach, tuber, and others); short term
Zeolite	Metals, radionuclides	Ion exchange, sorption, fixation	Insufficient data; short term
Apatite	Metals	Sorption, precipitation, complexation	Selective; insufficient data
Clay minerals	Metals, radionuclides	Ion exchange, sorption, fixation	Type of clay; short term

as vermiculite, to contaminated soils may be effective in immobilizing metals and radionuclides such as ^{137}Cs and ^{90}Sr (Adriano et al., 1997; Mench et al., 1997). Zeolites, which serve as molecular sieves, have been used as soil ameliorants. Addition of zeolites to soils contaminated with metals in Poland significantly reduced the concentrations of Cd and Pb in the roots and shoots of a number of pot-grown crops indicating the potential applicability of this ameliorant in metal-contaminated soil cleanup (Gworek, 1992a,b). Use of P and K fertilizers may have some beneficial effects on certain contaminants. In practice, when mixed with lead hydroxyapatite, $Ca_{10}(PO_4)_6(OH)_2$, will dissolve and precipitate as hydroxypyromorphite $Pb_{10}(PO_4)_6(OH)_2$ (Ma et al., 1993). Use of K fertilizer is based on its similar geochemical and physiological behavior towards Cs. As such they compete for the same absorption sites on plant roots, causing antagonistic effects. The ameliorants in Table 3.2 have been applied in agricultural soils but may provide only interim solutions in stabilizing contaminants. Their efficacy over the long term still needs to be investigated (Chlopecka and Adriano, 1996; Adriano et al., 1998). Furthermore, interactions of abiotic and biotic processes, especially in the rhizosphere, in influencing the stabilization of contaminants and the efficacy of ameliorants remain to be investigated as discussed below.

4.4 Interactions of Abiotic and Biotic Processes

Abiotic and biotic reactions are not independent but rather interactive processes in soil environments. Literature data indicate that the transformation of organic pollutants in soils is a result of the combined action of microorganisms, extracellular enzymes, mineral colloids, and humic substances (Dec et al., 2002). Each of these four major components of the soil system not only participates in the transformation of organic pollutants, but also modifies the participation of the other three. Mixed biotic/abiotic mechanisms by which organic pollutants are degraded in soils and related environments will be an important area of future investigation.

Abiotic and biotic catalysts coexist in soils. Abiotic catalysts can influence microbial formation of enzymes and enzymatic activity (Claus and Filip, 1990; Gianfreda and Bollag, 1994). Furthermore, many soil abiotic catalysts also influence the activity of desorbed enzymes. The influence of interactions of abiotic and biotic catalysts on the transformation, mobility, bioavailability, and toxicity of organic pollutants and their impact on the development of remediation strategies for terrestrial ecosystem, is thus an issue to be intensely studied for years to come (Huang, 2000).

Interactions of abiotic and biotic processes are not only important in the transformation of organic pollutants but also in the dynamics and fate of metals, metalloids, and other inorganics in soils, especially at the soil-root interface (MacLaughlin et al., 1998). In the rhizosphere, the kinds and concentration of substrates differ from those in bulk soil due to root exudation (Lynch, 1990). This leads to colonization by various populations of bacteria, fungi, protozoa, and nematodes. Plant-microbe interactions result in durable biological processes in the rhizosphere. These interactions, in turn, affect physicochemical reactions in the rhizosphere. The rhizosphere environment is governed by an interacting soil-plant-microbe trinity. Therefore, the reactions and processes of metals and related inorganics can only be investigated satisfactorily with interdisciplinary approaches.

The rhizosphere in soil environments is the bottleneck for contamination of the terrestrial food chain. The impact of physiochemical and biological processes in the rhizosphere on speciation, mobility, bioavailability, and toxicity of environmental pollutants and restoration of ecosystem health merit increased attention (Huang and Germida, 2002).

5 SUMMARY AND FUTURE PROSPECTS

Minerals, organic matter, and microorganisms are intimately associated in soils and sediments and closely interact in environmental processes. These interactions are especially important in the soil rhizosphere and the sediment-water interface, where low-molecular weight biochemicals are abundant and microbial activity intense.

Current research is driven by the widespread contamination of soils with organic pollutants, metals, metalloids and other inorganics and the desire for practical applications in the field of the knowledge acquired during earlier basic research. These explorations not only increase our understanding of the interactions between microorganisms and other soil components, but also assist us in establishing and applying methods for waste remediation and the cleanup of contaminated soils.

As the importance of bioavailability has been recognized and is being considered in regulatory decision-making, a greater understanding of the abiotic and biotic interactive processes and factors affecting the bioavailability of contaminants to plants and organisms is necessary. Increased knowledge will provide more realistic information about how to take bioavailability and toxicity into consideration for risk assessment and site remediation.

Research into the interactions between microorganisms, humic materials, and minerals in the soil continues to impact the fields of environmental remediation and regulation through the development of new techniques and an expanding base of knowledge about the relationships between microorganisms and soil particles. Mixed biotic/abiotic mechanisms by which xenobiotic chemicals, organics, metals, metalloids and other inorganics are transformed and degraded in soil environments will be an important area of future investigations.

Acknowledgements

Fundings from the US Environmental Protection Agency (EPA; Grant No. R-823847) and a Discovery Grant (Huang 2383) from the Natural Sciences and

Engineering Research Council of Canada, and the publication fund of the University of Saskatchewan are very much appreciated.

References

Adriano D.C. 1986. *Trace Elements in the Terrestrial Environment*. Springer-Verlag, New York, NY.

Adriano D.C., Chlopecka A., and Kaplan D.I. 1998. Role of soil chemistry in soil remediation and ecosystem conservation. In: *Soil Chemistry and Ecosystem Health*. P.M. Huang, D.C. Adriano, T.J. Logan, and R.T. Checkai (eds). SSSA Special Publ. no. 52. Soil Sci. Soc. Amer., Madison, WI, USA, pp. 361–386.

Adriano D.C., Albright J., Whicker F.W., and Iskandar I.K. 1997. Remediation of metal- and radionuclide-contaminated soil. In: *Remediation of Metal-Contaminated Soils*. I.K. Iskandar and D.C. Adriano (eds). Science Review Northwood, England, pp. 27–45.

Adriano D.C., Bollag J.-M., Frankenberger W.T., and Sims R.C. (eds.). 1999. Bioremediation of Contaminated Soils. Agron. Monograph. Amer. Soc. Agron., Madison, WI, USA. 772 pp.

Adriano D.C., Vangronsveld J., Bolan N.S., and Wenzel W.W. 2005. Accelerated natural remediation—a powerful tool in environmental cleanup. *Geoderma* (in press).

Alexander M. 1999. *Biodegradation and Bioremediation*, Acad. Press, Inc., San Diego, CA, USA (2nd ed.).

Amacher M.L. and Baker D.E. 1982. Redox reactions involving chromium, plutonium, and manganese in soils. DOE/DP/04515-1. Pennsylvania State University, University Park, PA

Anderson T.A., Guthrie E.A., and Walton B.T. 1993. Bioremediation in the rhizosphere. *Environ. Sci. Tech.*, 27: 2630–2636.

Awad F. and Römheld V. 2000. Mobilization of heavy metals from a contaminated calcareous soil by plant borne and synthetic chelators and their uptake by wheat plants. *J. Plant. Nutr.* 23: 1847–1855.

Bartlett R.J. 1981. Non-microbial nitrite-to-nitrate transformation in soils. *Soil Sci. Soc. Am. J.* 45: 1054–1058.

Bartlett R.J. and James B. 1979. Behavior of chromium in soils. III. Oxidation. *J. Environ. Qual.* 8: 31–35.

Berthelin J., Huang P.M., Bollag J.-M., and Andreux F. (eds.). 1999. *Effect of Mineral-Organic-Microorganism Interactions on Soil and Freshwater Environments*. Kluwer Acad./Plenum Publ., New York, NY, 378 pp.

Bollag J.-M. 1992. Decontaminating soil with enzymes: an in-situ method using phenolic and anilinic compounds. *Environ. Sci. Tech.* 25: 1876–1881.

Bollag J.-M., Mertz T., and Otjen L. 1994. Role of microorganisms in soil bioremediation. In: *Bioremediation through Rhizosphere Technology*. ACS Symp. Series 563. T.A. Anderson and J.R. Coats (eds.). Amer. Chem. Soc., Washington, DC, pp. 2–10.

Burns R.G. 1986. Interactions of enzymes with soil mineral and organic colloids. In: *Interaction of Soil Minerals with Natural Organics and Microbes*. P.M. Huang and M. Schnitzer (eds). SSSA Spec. Publ. No. 17. Soil Sci. Soc. Amer. Madison. WI, USA, pp. 429–451.

Chlopecka A. and Adriano D.C. 1996. Mimicked *in situ* stabilization of metals in a cropped soil. *Environ Sci. Tech.* 30: 3294–3303.

Cieslinski G., Van Rees K.C.J., Szmigielska A.M., Krishnamurti G.S.R., and Huang P.M. 1998. Low-molecular-weight organic acids in rhizosphere soils of durum wheat and their effects on cadmium bioaccumulation. *Plant Soil* 203: 109–117.

Claus H. and Filip Z.K. 1990. Effects of clays and other solids on the activity of phenoloxidases produced by some fungi and actinomycetes. *Soil Biol. Biochem.* 22: 483–488.

Cleveland J.M. 1970. *The Chemistry of Plutonium*. Gordon and Breach, New York, NY, USA.

Dec J. and Bollag J.-M. 2001. Use of enzymes in bioremediation. In: *Pesticide Biotransformation in Plants and Microorganisms: Similarities and Divergences*. ACS Symp. Series 777. J.C. Hall, R.E. Hoagland, and R.M. Zablotowicz (eds.). Amer. Chem. Soc., Washington DC, pp. 182–193.

Dec J., Huang P.M., Senesi N., and Bollag J.-M. 2002. Impact of interactions of microorganisms and soil colloids on the transformation of organic pollutants. In: *Interactions between Soil Particles and Microorganisms and the Impact on the Terrestrial Ecosystem*. P.M. Huang, J.-M. Bollag, and N. Senesi (eds). John Wiley & Sons, Chichester, England, pp. 323–378.

Gianfreda L. and Bollag J.-M. 1994. Effect of soils on the behavior of immobilized enzymes. *Soil Sci. Soc. Amer. J.* 58: 1672–1681.

Gworek B. 1992a. Land inactivation in soils by zeolites. *Plant Soil* 143: 71–74.

Gworek B. 1992b. Inactivation of cadmium in soil by synthetic zeolites. *Environ. Pollut.* 75: 269–271.

Haider K. 1999. Microbe-soil-organic contaminant interactions. In: *Bioremediation of Contaminated Soils.* D.C. Adriano, J.-M. Bollag, W.T. Frankenberger and R.C. Sims (eds.). Agron. Monograph Ser. 37. Madison, WI, USA, pp. 33–51.

Hayes K.F. and Traina S.J. 1998. Metal ion speciation and its significance in ecosystem health. In: *Soil Chemistry and Ecosystem Health*. In: P.M. Huang, D.C. Adriano, T.J. Logan, and R. T. Checkai (eds). SSSA Spec. Publ. no. 52. Soil Sci. Soc. Amer., Madison, WI, USA, pp. 45–84.

Hinsinger P. 2001. Bioavailability of metals as related to root-induced chemical changes in the rhizosphere. In: *Trace Elements in the Rhizosphere.* G.R. Gobran, W.W. Wenzel, and E. Lombi (eds.). CRC Press, Boca Raton, FL, USA, pp. 25–41.

Huang P.M. 1990. Role of soil minerals in transformations of natural organics and xenobiotics in the environment. In: *Soil Biochemistry*. J.-M. Bollag, and G. Stotzky (eds.). Marcel Dekker, New York, NY, vol. 6, pp. 29–115.

Huang P.M. 2000. Abiotic catalysis. In: *Handbook of Soil Science*. M.E. Sumner (editor-in-chief). CRC Press, Boca Raton, FL, USA, pp. B303–332.

Huang P.M. and Germida J.J. 2002. Chemical and biological processes in the rhizosphere: Metal pollutants. In: *Interactions between Soil Particles and Microorganisms.* P.M. Huang, J.-M. Bollag, and N. Senesi (eds). John Wiley & Sons, Chichester, UK, pp. 381–438.

Huang P.M. and Bollag J.-M. 1998. Minerals-organics-microorganisms interactions in the soil environment. In: *Structure and Surface Reactions of Soil Particles*. P.M. Huang, N. Senesi and J. Buffle (eds.), John Wiley & Sons, Chichester, England, pp. 3–39.

Ilton E.S. and Veblen D.R. 1994. Chromium sorption by phlogopite and biotite in acidic solutions at 25°C: Insights from X-ray photoelectron spectroscopy and electron microscopy. *Geochim. Cosmochim. Acta* 58: 2777–2788.

Koo B.J. 2001. Assessing bioavailability of metals in biosolid-treated soils: root exudates and their effects on solubility of metals. PhD diss., Univ. of California, Riverside, CA.

Krishnamurti G.S.R. and Huang P.M. 1987. The catalytic role of birnessite in the transformation of iron. *Can. J. Soil Sci.* 67: 533–543.

Krishnamurti G.S.R., Cieslinski G., Huang P.M., and Van Rees K.C.J. 1997. Kinetics of cadmium release from soils as influenced by organic acids: Implications in cadmium availability. *J. Environ. Qual.* 26: 271–277.

Kurek E. 2002. Microbial mobilization of metals from soil minerals under aerobic conditions. In: *Interactions between Soil Particles and Microorganisms*. P.M. Huang, J.-M. Bollag and N. Senesi (eds). John Wiley & Sons, Chichester, UK, pp. 189–225.

Lynch J.M. 1990. *The Rhizosphere.* John Wiley & Sons, Chichester, UK, 458 pp.

Ma Q.Y., Logan T.J., Ryan J.A., and Traina S.J. 1993. *In situ* lead immobilization by apatite. *Environ. Sci. Tech.* 27: 1803–1810.

MacLaughlin M.J., Smolden E., and Merck R. 1998. Soil-root interface: Physicochemical processes: In: *Soil Chemistry and Ecosystem Health*. P.M. Huang, D.C. Adriano, T.J. Logan, and R.T. Checkai (eds.). SSSA Special Publ. no. 52. Soil Sci. Soc. Amer., Madison, WI, USA, pp. 233–277.

Marschner M. 1998. Soil-root interface: Biological and biochemical processes. In: *Soil Chemistry and Ecosystem Health*. P.M. Huang, D.C. Adriano, T.J. Logan, and R.T. Checkai (eds). SSSA Spec. Publ. no. 52. Soil Sci. Soc. Amer., Madison, WI, USA, pp. 191–231.

Masschelyn P.H. and Patrick W.H. 1994. Selenium, arsenic and chromium redox chemistry in wetland soils and sediments. In: *Biogeochemistry of Trace Elements*. D.C. Adriano et al. (eds). Sci. Tech. Lett., Northwood, pp. 615–625.

McBride M.B. 1994. *Environmental Chemistry of Soils*. Oxford Univ. Press, Oxford, UK.

McBride M.B. 2000. Chemisorptions and precipitation reactions. In: *Handbook of Soil Science*. M.E. Sumner (editor-in-chief). CRC Press, Boca Raton, FL, USA, pp. B265–B302.

McLean J.S., Lee J.U., and Beveridge T.J. 2002. Interactions of bacteria and environmental metals, fine-grained mineral development and bioremediation strategies. In: *Interactions between Soil Particles and Microorganisms*. P.M. Huang, J.-M. Bollag, and N. Senesi (eds). John Wiley & Sons, Chichester, UK, pp. 227–261.

Mench M., Amans V., Sappin-Didier V., Forgues S. et al. 1997. A study of additives to reduce availability of Pb in soil to plants. In: *Remediation of Metal-contaminated Soils. Iskandar I.K. and Adriano D.C. (eds).* Science Rev., Northwood, England, pp 185–202.

Mullen M.D., Wolf D.C., Ferris F.G., Beveridge T.J., Flemming C.A., and Bailey G.W. 1989. Bacterial sorption of heavy metals. *Appl. Environ. Microbiol*. 55: 3143–3149.

Naidja A., Huang P.M., and Bollag J.-M. 2000. Enzyme-clay interactions and their impact on transformations on natural and anthropogenic organic compounds in soil. *J. Environ. Qual.* 29: 677–691.

National Research Council (NRC). 2003. Bioavailability of contaminants in soils and sediments: Processes, tools and applications. In: *Report of Committee on Bioavailability of Contaminants in Soils and Sediments*. NRC. Water Sci. Tech. Board (WSTB), Washington DC, USA, 432 pp.

Nelson Y.M., Lion L.W., Ghiorse W.C., and Shuler M.L. 1999. Production of biogenic Mn oxides by *Leptothrix discophora* SS-1 in a chemically defined growth medium and evaluation of their Pb adsorption characteristics. *Appl. Environ. Microb.* 65: 175–180.

NIST. 1997. Standard Reference Database 46. *Critically Selected Stability Constants of Metal Complexes*, Version 5.0 NIST, US Dept. Commerce, Gaithersburg, MD, USA.

Onyatta J.O. and Huang P.M. 2003. Kinetics of cadmium release from selected tropical soils from Kenya by low-molecular-weight organic acids. *Soil Sci.* 168: 234–252.

Oscarson D.W., Huang P.M., Defosse C., and Herbillon A. 1981. Oxidative power of Mn (IV) and Fe (III) oxides with respect to As (III) in terrestrial and aquatic environments. *Nature* 291: 50–51.

Oscarson D.W., Huang P.M., Liaw W.K., and Hammer U.T. 1983. Kinetics of oxidation of arsenite by various manganese dioxides. *Soil Sci. Soc. Amer. J.* 47: 644–648.

Page A.L., Gleason T.L., Smith J.E., Iskandar I.K., and Sommers L.E. 1983. *Utilization of Municipal Wastewater and Sludges on Land*. Univ. California, Riverside, CA, USA.

Pal S., Bollag J.-M., and Huang P.M. 1994. Role of abiotic and biotic catalysts in the transformation of phenolic compounds through oxidative coupling reactions. *Soil Biol. Biochem*. 26: 813–820.

Peterson M.L., Brown G.E., and Parks G.A. 1996. Direct XAFS evidence for heterogenous redox reaction at the aqueous chromium/magnetite interface. *Colloid Surf. A*. 107: 77–88.

Pierzynski G.M., Sims J.T., and Vance G.F. 1994. *Soil and Environmental Quality*. Lewis Publ., Boca Raton, FL, USA.

Pomilio A.B., Leicach S.R., Grass M.Y., Ghersa C.M, Santoro M., and Vitale A.A. 2000. Constituents of the root exudates of *Avena fauta* grown under far-infrared-enriched light. *Phytochem. Anal.* 11: 304–308.

Robert M. and Berthelin J. 1986. Role of biological and biochemical factors in soil mineral weathering. In: *Interactions of Soil Minerals with Natural Organics and Microbes*. P.M. Huang and M. Schnitzer (eds). SSSA Spec. Publ. No. 17. Soil Sci. Soc. Amer., Madison, WI, USA, pp. 453–495.

Ross D.S. and Bartlett R.J., 1981. Evidence for non-microbial oxidation of manganese in soil. *Soil Sci.* 132: 153–160.

Singh B.K., Kuhad R.C., Singh A., Lal R., and Tripathi K.K. 1999. Biochemical and molecular basis of pesticide degradation by microorganisms. *Crit. Rev. Biotech*. 19: 197–226.

Sparks D.L. 2003. *Environmental Soil Chemistry*. Acad. Press, San Diego, CA, USA.

Stotzky G. 1986. Influence of soil mineral colloids on metabolic processes, growth, adhesion, and ecology of microbes and viruses. In: *Interactions of Soil Minerals with Natural Organics and*

Microbes. P.M. Huang and M. Schnitzer, (eds.). SSSA Spec. Publ. No. 17. Soil Sci. Soc. Amer., Madison, WI, pp. 305–428.

Szmiegielska A.M., Van Rees K.C.J., Cieslinski G., and Huang P.M. 1996. Low molecular weight dicarboxylic acids in rhizosphere soils of durum wheat. *J. Agric. Food Chem.* 44: 1036–1040.

Tebo B.M., Ghiorse W.C., van Waasbergen L.G., Siering P.L., and Caspi R. 1997. Bacterially mediated mineral formation: Insights into manganese (II) oxidation from molecular genetic and biochemical studies. In: *Geomicrobiology: Interactions between Microbes and Minerals*. J.F. Banfield and K.H. Nealson (eds.). Mineral. Soc. Amer., Washington DC, USA, vol. 35, 225 pp.

Theng B.K.G. and Orchard V.A. 1995. Interactions of clays with microorganisms and bacterial survival in soil: A physicochemical perspective. In: *Environmental Impact of Soil Component Interactions.* Vol. II: *Metals, other Inorganics, and Microbial Activities.* CRC/Lewis Publ., Boca Raton, FL, USA, pp. 123–143.

van der Meer, J.R. de Vos, W.W., Harayama S., and Zehnder A.J.B. 1992. Molecular mechanisms of genetic adaptation to xenobiotic compounds. *Microb. Rev.* 56: 677–694.

Volesky B. 1990. Removal and recovery of heavy metals by biosorption. In: *Biosorption of Heavy Metals*. B. Volesky (ed). CRC Press, Boca Raton, FL, USA, pp. 7–43.

Wehrli B. and Stumm W. 1989. Vanadyl in natural waters: Adsorption and hydrolysis promote oxygenation. *Geochim. Cosmochim. Acta* 53: 69–77.

Wenzel W.W., Wieshammer G., Fitz W.J., and Puschenreiter M. 2001. Novel rhizobox design to assess rhizosphere characteristics at high spatial resolution. *Plant Soil* 237: 37–45.

White A.F. and Peterson M.L. 1996. Reduction of aqueous transition metal species on the surfaces of Fe (II) containing oxides. *Geochim. Cosmochim. Acta* 60: 3799–3814.

Part II
Natural Organics and Xenobiotics

4

Soil and Organic Wastes: Transformation of Olive Oil Mill Waste Water and Clay Particle Aggregation

L.P. D'Acqui*, E. Sparvoli, A. Agnelli, *and* **C.A. Santi**

Abstract

In the Mediterranean area olive mill wastes (OW) reach an amount to 30×10^6 t y^{-1} and constitute a serious organic pollution risk for surface water. Olive mill waste water (OW) has been applied to cultivated soils to reduce their environmental impact and to exploit their amendment properties. Positive effects on fertility have generally been reported but less attention paid to the effect of OW on soil physical properties. This study investigated organic transformations, aggregate formation, and their stability in water in two different clays incubated with fresh OW. For this purpose, OW was added to a Zettlitz kaolinite and an Arizona montmorillonite. Clays and OW were incubated for 90 days and subjected to wetting and drying cycles. The organic matter transformation was characterized by solid state CP/MAS ^{13}C NMR spectroscopy. Aggregation was assessed on selected sizes of clay aggregates obtained after incubation by pore size distribution (Hg intrusion + N_2 adsorption techniques) and by aggregate stability in water measurements. Organic matter from incubated OW clay samples of both clay types has shown enrichment of alkyl groups and decrease of O-alkyl. Incubation of OW with kaolinite promoted a substantial increase of aggregation and aggregate stability in water. The mineralogy of clays can substantially influence both the transformation of organic matter and the aggregation processes that, in turn, could affect the amendment properties of OW applied to soils.

**Corresponding author:* Dr. LPD'Acqui. Istituto per lo studio degli Ecosistemi, Sezione di Furenze, CNR, via Madonna del Piano Ed. C/D, 50019, Sesto Fiorentino, Italia. E-mail: dacqui@ise.cnr.it

1 INTRODUCTION

Reduction of soil organic matter (SOM) through cultivation decreases the chemical and physical fertility of soil. Concomitantly a surplus of organic wastes is produced by the agroindustry. These opposite trends should be coupled in a unique sustainable natural cycle.

Olive oil is a typical and valuable agroindustrial product in all Mediterranean countries. Olive mill wastes in the Mediterranean area reach 30×10^6 t y^{-1}. In particular, olive mill waste water (OW) can be a serious pollution risk for surface water because of the high (150–200 g L^{-1}) chemical oxygen demand (COD) and the presence of phytotoxic and antibacterial polyphenols (1–10 kg t^{-1}) (Levi-Minzi et al., 1992). However, because of its valuable content of nutrient elements and organic matter (1 t contains 150 kg SOM, 1 kg N, 5 kg K_2O, 0.5 kg P_2O_5), OW can be a useful natural resource as soil amendment (Lopez et al., 1992). Experiments carried out directly applying fresh OW on soil have given generally positive results (Paris, 1998). A two-year experiment carried out with lysimeters and in the field, showed that a clay soil rich in calcium carbonate evenly treated with 10,000 t ha^{-1} y^{-1} of OW, can adsorb all organic and inorganic constituents present in the amendment (Cabrera et al., 1996). At lower rates, close to that allowed by the Italian environmental laws (80 t ha^{-1} on bare soil, 40 t ha^{-1} on standing crops), positive effects have been observed in crop yields and residual soil fertility (Bonari et al., 1993).

In spite of much interest paid to the interactions of OW and soil productivity, the effects of OW on soil structure, particle aggregation, and aggregate stability in water, that in turn can affect overall fertility, have not received sufficient attention (Cox et al., 1997).

The study investigated the organic transformations and the dynamics of clay particle aggregation occurring when a kaolinite and a montmorillonite are incubated with OW. Interactions between clay particles (the more active soil mineral fraction) and these organics provided a simple model for assessing the physical fertility improvement when OW is used as a soil amendment.

2 MATERIALS AND METHODS

2.1 Description of Materials

Fresh OW was collected in January 2000 from an olive mill near Pisa (Italy); the material was maintained at 4°C for 10 days and freeze dried. Two standard clays were used for the experiment: a montmorillonite from Arizona, USA (Van Olphen and Fripiat, 1979), and a kaolinite from Zettlitz, Czech Republic (Grim, 1968).

2.2 Experimental Setup and Design

A lab trial was conducted with two treatments, i.e. two clay types (kaolinite and montmorillonite) and two levels of OW amendment (0 g kg^{-1} and 80 g kg^{-1}

of clay). They were arranged in randomized complete block design with three replicates.

As for the OW treatment, 2.4 g of freeze dried OW (524.1 mg g^{-1} of total organic C and 15.0 mg g^{-1} of N) were added to 30 g of air-dried clays screened through a 0.053 mm sieve. They were thoroughly mixed and successively wetted with deionized water to reach 50% of the water-holding capacity. An amount of 2.4 g of OW was used to obtain a content of about 4% C in the organomineral mixtures. This C concentration is necessary for acquiring a relatively high solid state CP-MAS ^{13}C NMR signal. This amount of OW is comparable with the values reported in the literature. The C and N content measured in the organomineral mixtures obtained were 38.8 mg g^{-1} and 1.05 mg g^{-1} respectively. The mixtures were treated with 1 ml of an inoculum of soil microbial biomass obtained by a water extract from a natural clay soil (soil/water ratio 1:2.5) and incubated in open Petri dishes for 90 days at room temperature (24°–26°C). Each sample was rewetted every time the moisture level fell below 20%. Eleven wetting/drying cycles were imposed during the incubation. As controls, clays were incubated without adding OW. After 90 days the sample from each Petri dish was gently dry-sieved through 10, 8, 6, 4, 3, 2, 1, 0.5, 0.2, 0.1, 0.053 mm sieves, the aggregates separated, ranked in dimensional classes, and weighed. The weight of aggregates of each dimensional class was expressed as the mean of three replicates and standard error was calculated. The error propagation technique (Skoog et al., 1996) was used to calculate the standard errors of the values obtained by calculation.

2.3 Physical Analyses of Incubated Samples

Clay microstructure

Microstructure studies, over the range of 1 to 20 nm equivalent cylindrical diameter, were carried out on dried (18 h at 70°C under vacuum) aggregates of about 10 mm in diameter. Specific surface area and pore size distribution (PSD) were determined using N_2 adsorption at 77° K with a Carlo Erba Sorptomatic 1900 apparatus. Pore size distribution over the range of 20 to 2×10^5 nm diameter, was evaluated by Hg porosimetry with a Carlo Erba Porosimeter 2000. The PSD curves obtained by N_2 adsorption analysis and Hg intrusion technique were combined (Sparvoli et al., 1989) to get a continuous PSD curve over the range of 2 to 2×10^5 nm diameter, according to Gregg and Sing (1982). The ordinate (mm^3 g^{-1}) of each point of the PSD curve represents, according to Fiès and Bruand (1990), the slope of the cumulative pore volume curve between two successive values of diameter, calculated with the expression

$$\Delta V\, [\Delta\, (\log d)]{-1}$$

where V is the pore volume and d the pore diameter.

The microstructure was also observed using a Philips 505 scanning electron microscope after coating the samples with gold.

Aggregate stability

Aggregate stability in water tests were undertaken on the aggregates with a diameter between 4 and 8 mm, representing the more frequent class obtained from the clays treated and untreated with OW. The aggregates were placed in a plastic container with water (liquid/solid ratio 15:1) and shaken for 1 h with a Techne Shaking Bath SB-16. The shaking speed was 5.5, equivalent to 72 cycles min^{-1}. After treatment, the samples were wet-sieved through 2, 1, 0.5, 0.2, 0.1, and 0.053 mm sieves. The resultant aggregate fractions were dried at 60°C and weighed.

2.4 Organic Transformation

Total organic C and N contents were measured by a flash combustion technique (1800°C under O_2 flow), using a Carlo Erba NA 1500 CNS Elemental Analyzer. Each sample was run in triplicate, the values expressed as the mean of three replicates, and the standard error calculated.

To observe the organic transformation of the OW during incubation, representative aliquots of incubated OW, mont. + OW and kaol. + OW samples were analyzed by solid state CP-MAS ^{13}C NMR spectroscopy.

3 RESULTS AND DISCUSSION

3.1 Clay Microstructure

The combined (N_2+Hg) PSD of kaolinite and montmorillonite incubated samples, treated and untreated with OW, are shown in Figures 4.1 and 4.2. The PSD of treated and untreated clay samples exhibited a strong peak at 200 nm diameter for kaolinite (Fig. 4.1) and 4 nm for montmorillonite (Fig. 4.2). Moreover, montmorillonite samples also showed a smaller second peak around 8.5×10^3 nm diameter.

The two main peaks are characteristic of the two clay minerals and are related to the arrangement of the different types of morphological units (crystallites for kaolinite and quasicrystal for montmorillonite) at their primary organizational level (Tessier, 1984; Oades, 1986).

In the case of kaolinite, as a consequence of OW addition, the characteristic peak shifted slightly toward larger pores with an increase in corresponding volume. In the case of montmorillonite, the sample treated with OW showed a marked reduction of the typical pore peak with a decrease in related volume. Modifications of pore peak characteristics of the two clays are attributed to the interaction of organic molecules with clay morphological units. In kaolinite the shifting of the peak toward larger pores, observed in the PSD curve may have been by a tangential orientation of kaolinite crystallites around the organic material, such as organic residues and hyphae, that induced a fairly open structure (Chenu, 1989; Dorioz et al., 1993). Similar behavior has been reported

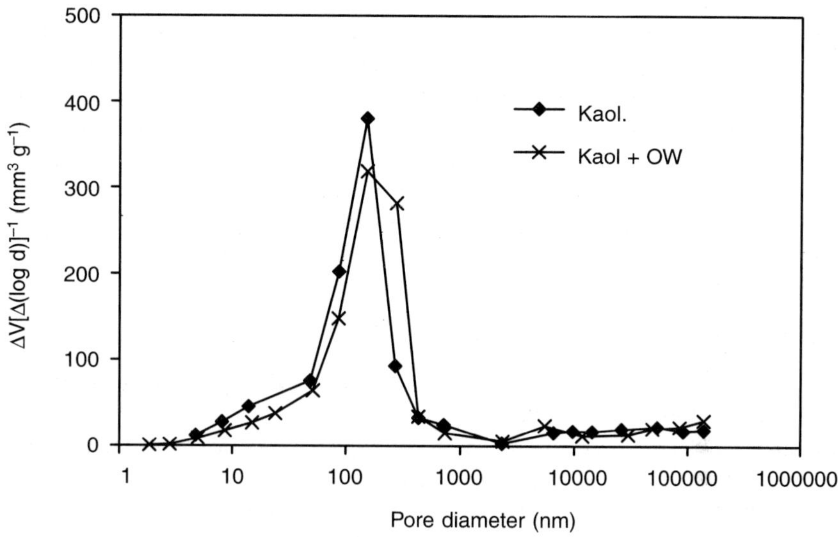

Fig. 4.1: Pore size distribution (PSD) of kaolinite samples treated with OW and the control. The samples underwent wet/dry cycles for 90 days. The values of the ordinate were calculated according to the expression: $\Delta V\ [\Delta(\log d)]^{-1}$, where V= pore volume ($mm^3\ g^{-1}$); d= pore diameter (nm).

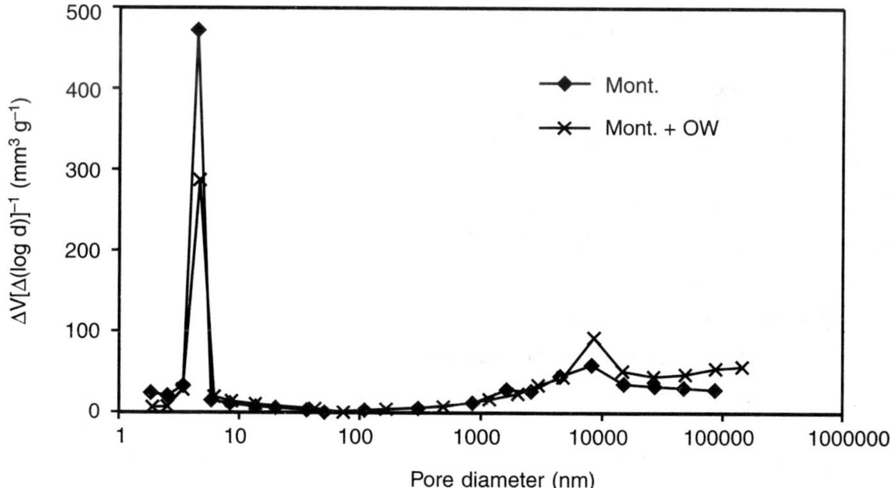

Fig. 4.2: Pore size distribution (PSD) of montmorillonite samples treated with OW and the control. The samples underwent wet/dry cycles for 90 days. The values of the ordinate were calculated according to the expression: $\Delta V\ [\Delta(\log d)]^{-1}$; where V= pore volume ($mm^3 g^{-1}$); d= pore diameter (nm).

by D'Acqui et al. (1999). In montmorillonite the strong reduction of pore volume is associated with a marked decrease (about 62%) in the specific surface areas (SSA) (Table 4.1). This reduction could be explained by the presence of organic molecules of a particular size, probably produced by the decomposition of the organic material, that clog the clay micropores and prevent the penetration of N_2 (D'Acqui et al., 1999). In this specific case, the phenomenon could be magnified by the presence of aliphatic substances, the more representative organic compounds of the OW treated samples as shown by NMR analysis (see below). Adsorption of aliphatic moieties on the primary particles of the montmorillonite could not only could prevent N_2 adsorption entry, but even order the turbostratic arrangement of clay primary particles whereby successive layers exhibit random displacements and rotations relative to each other (Farmer, 1978) with a further reduction in SSA, due to the hydrophobic extended interaction (Calamai et al., 2000).

Structural reorganization of the two treated clays does not affect the pore volume detected by N_2 + Hg measurements (Table 4.1).

In treated kaolinite, an increased volume of pores in the range of 20 nm to 2×10^5 nm diameter (measured by the Hg intrusion technique), due to the tangential orientation of crystallites, was compensated by a reduction in micropores (measured by the N_2 adsorption technique) (Table 4.1). In treated montmorillonite the reduction of pore volume measured by N_2 adsorption technique was compensated by an increase in larger pores. The porosity in this range, attributable to the wetting/drying-induced self-mulching phenomenon, is enhanced by the interaction with the organics. This confirms the role of organic matter in aggregate formation (Stevenson, 1982).

In treated montmorillonite, the presence of aliphatic compounds and other organic substances could affect the N_2 measurements by reducing the N_2 sites of adsorption and, as a consequence, reduction in the SSA and micropore volume (Table 4.1). However, the effect of OW on the SSA was smaller in kaolinite than in montmorillonite, since kaolinite morphological units are larger and well stacked with low interaction surfaces (Oades, 1986).

3.2 Aggregate Stability Tests

The dimensional distribution and the mean weight diameter (MWD) of aggregates, separated by a gentle dry sieving after incubation, are presented in Table 4.2. Both kaolinite and montmorillonite, OW-treated and untreated, showed the formation of aggregates larger than the starting dimension (Ø < 0.053 mm). The eleven wetting-drying cycles applied during incubation and the consequent shrink and swell processes could induce this phenomenon. Formation of larger aggregates was particularly evident in montmorillonite because of the strong self-mulching activity characteristic of this mineral. Moreover, treatment of both clays with OW produced an appreciable increase of larger aggregates, as evidenced by the higher values of the MWD of the OW-treated samples.

Table 4.1: Distribution of pore volume and total C and N of OW-treated kaolinite and montmorillonite and the control. The samples underwent wet/dry cycles for 90 days.

Samples	Specific surface area	Pore volume		Total pore volume	Total C	Total N
		by n_2(a) (A)	by Hg(b) (B)	(A) + B)		
	$m^2 g^{-1}$	$mm^3 g^{-1}$			$g kg^{-1}$	
OW					38.8 (± 0.3)	1.5 (± 0.12)
Kaolinite	17.6 (± 0.3)	43.1 (± 0.8)	132.1 (± 4.6)	175.2 (± 4.7)	—	—
Kaol. + OW	13.4 (± 0.4)	24.6 (± 1.1)	164.0 (± 8.2)	188.6 (± 8.3)	28.6 (± 0.8)	0.88 (± 0.20)
Montmorillonite	102.6 (± 2.0)	168.3 (± 5.1)	70.0 (± 1.2)	238.8 (± 5.2)	—	—
Mont. + OW	38.3 (± 0.9)	104.6 (± 4.3)	105.9 (± 3.1)	210.5 (± 5.3)	25.9 (± 0.3)	0.75 (± 0.10)

Values are means of three replicates (± standard error).
[a]Specific volume measured by nitrogen adsorption in the pore range of 1–20 nm diameter.
[b]Specific volume measured by mercury intrusion in the pore range of 20 nm–2 × 10^5 nm diameter.

Table 4.2: Distribution of dry-sieved aggregate-size fractions and mean weight diameter of OW-treated kaolinite and montmorillonite samples and controls. The samples underwent wet/dry cycles for 90 days

Sample	Aggregate size-fractions (mm)*												Mean weight diameter
	> 10	10-8	8-6	6-4	4-3	3-2	2-1	1-0.5	0.5-0.2	0.2-0.1	0.1-0.053	< 0.053	
						(g kg^{-1})							
Kaolinite	0.0	0.0	81.6 ± 5.6	127.1 ± 11.2	254.1 ± 8.4	242.1 ± 8.9	152.4 ± 3.7	60.8 ± 1.7	17.6 ± 0.7	9.4 ± 0.2	4.8 ± 0.2	1.9 ± 0.3	**3.00** ± 0.16
Kao. + OW	29.0 ± 8.5	140.4 ± 13.7	314.2 ± 23.1	203.0 ± 20.5	90.8 ± 3.9	84.9 ± 3.1	68.9 ± 1.7	34.3 ± 0.5	18.1 ± 0.4	9.7 ± 0.3	4.7 ± 0.3	1.9 ± 0.3	**5.49** ± 0.02
Montmorillonite	0.0	155.3 ± 9.8	200.5 ± 20.0	218.5 ± 12.3	185.0 ± 7.1	105.4 ± 9.3	67.0 ± 5.2	29.6 ± 6.4	26.2 ± 3.7	9.5 ± 1.0	3.0 ± 0.2	0.0	**4.94** ± 0.24
Mont. + OW	340.1 ± 74.1	203.1 ± 16.6	208.7 ± 28.1	96.2 ± 10.6	46.2 ± 12.0	44.5 ± 5.8	34.6 ± 4.9	13.3 ± 2.1	7.0 ± 1.3	3.1 ± 0.4	1.8 ± 0.1	1.4 ± 0.2	**8.19** ± 0.56

*Values are means of three replicates (± standard error).

Results of the aggregate water stability test for the 4 to 8 mm aggregates showed that the untreated kaolinite samples are almost completely dispersed (< 0.053 mm in size), while untreated montmorillonite samples maintained 80% of aggregates larger than 0.053 mm in size (Figs. 4.3 and 4.4). Incubation of kaolinite with the organic material led to formation of stable aggregates (about 60% by weight) (Fig. 4.3). In the case of montmorillonite, the self-mulching phenomenon and the addition and transformation of OW stabilized about 70% of the sample in aggregates larger than 0.053 mm (Fig. 4.4). Survival of aggregates

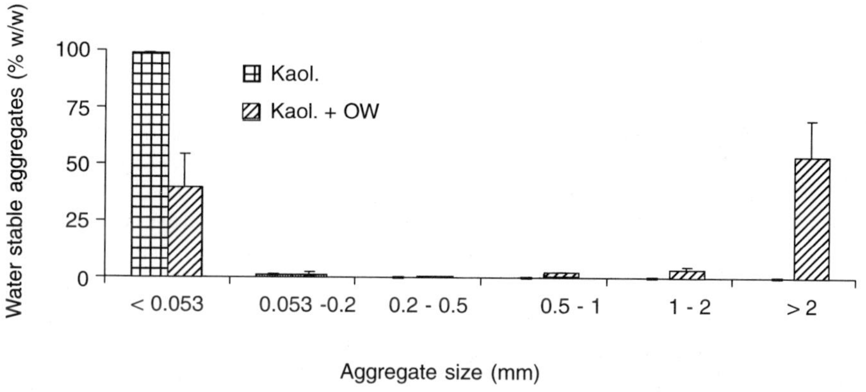

Fig. 4.3: Distribution of aggregate size-fractions of kaolinite after a test of aggregate stability in water on 4-8 mm aggregates. Vertical bars represent standard error of the mean (n = 3).

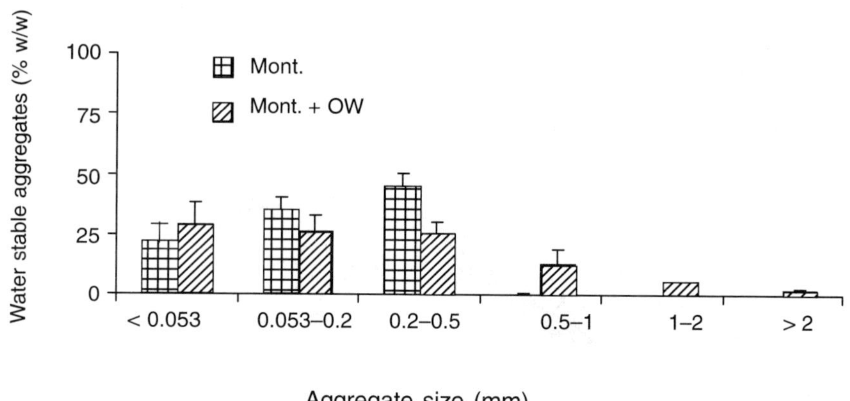

Fig. 4.4: Distribution of aggregate-size fractions of montmorillonite after a test of aggregate stability in water on 4-8 mm aggregates. Vertical bars represent standard error of the mean (n = 3).

larger than 0.5 mm (about 20% by weight) (Fig. 4.4), practically absent in the untreated samples, revealed the effects of OW and its transformation products on the stabilisation of montmorillonite aggregates. The presence of this type of aggregate and their corresponding interspaces, according to the model of hierarchic aggregates organization (Oades, 1984), can explain the increased volume for pores larger than 5×10^3 nm for the treated montmorillonite samples (Fig. 4.2). In the case of kaolinite, about 60% of the stable aggregates are larger than 2 mm in size; therefore the interaggregates pores are beyond the range measured by the Hg intrusion technique.

SEM images from the largest aggregates, separated by dry sieving, confirm these data (Photos 1 and 2). Aggregates of the OW-treated clays are surrounded

a) **Photo 1** b)

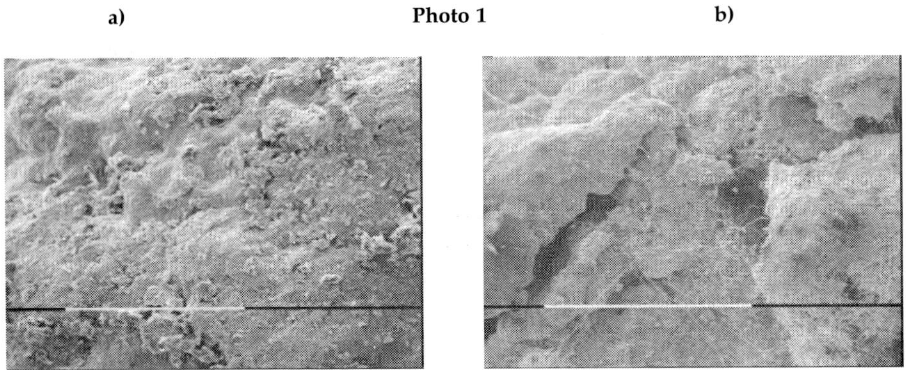

Photo 1: (**a**) Kaolinite samples incubated for 90 days without the addition of OW, no pores were observed; (**b**) Kaolinite samples treated with OW and incubated for 90 days, clay appears organized into discrete aggregates. Note the hyphae network. Scale: each section of the bar =1000 μm.

a) **Photo 2** b)

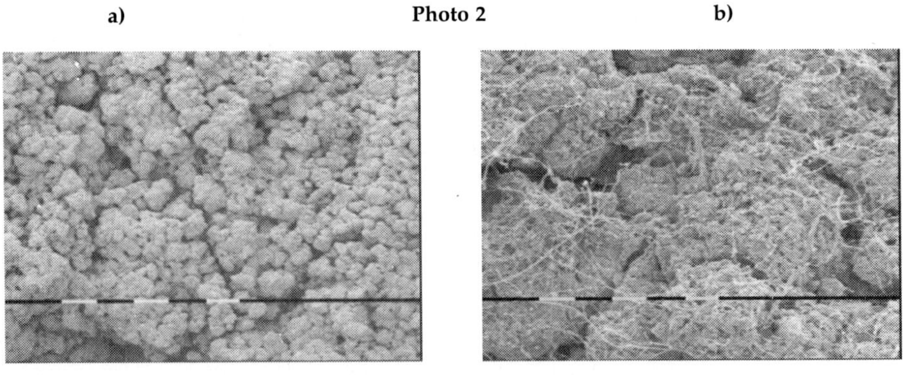

Photo 2. (**a**) Montmorillonite samples incubated for 90 days without the addition of OW; (**b**) Montmorillonite samples treated with OW and incubated for 90 days. Clay appears organized into discrete aggregates. Note the hyphae network. Scale: each section of the bar =100 μm.

by an extended network of fungal hyphae. It appears, however, that kaolinite aggregates have larger subunits than those of montmorillonite.

The hyphae network could furthermore be responsible, together with the presence of hydrorepellent long aliphatic chains, for the stability of clay macroaggregates in water.

3.3 Organic Transformations

After 90 days of incubation the reduction of C in kaol. + OW was about 26% and for mont. + OW about 33%. (Table 4.1). Similar results were observed in a previous study on the interaction between clays and organic amendments, such as plant leaves (D'Acqui et al., 1999). Other workers have found that the composition of SOM associated with the clay fractions depends on the type of clay minerals present (Wattel-Koekkoek et al., 2001).

In the case of N, no significant differences between the two clays were observed after the incubation period due to variability of the N data (Table 4.1).

Solid state ^{13}C NMR spectroscopy showed that during the incubation period the organic matter derived from OW was subjected to chemical transformations. In both kaolinite and montmorillonite there was a dramatic reduction of O-alkyl C signals (40–130 ppm region), representing polysaccharide structures (Fig. 4.5). The weak and broad signal in this region suggests the presence of denaturalized cellulose, rather than the crystalline form, and altered hemicelluloses. Reduction of the O-alkyl peak was consistent with other decomposition studies (e.g. Skjemstad et al., 1997). Polysaccharides, which decompose rapidly, are the first materials to be transformed, while the alkyl, aromatic, and carbonyl materials are slower to decompose. The net effect is a relative increase in these latter compounds as the decomposition proceeds. The incubated samples showed a strong signal, in the 0–40 ppm region, which was more pronounced in kaolinite than in montmorillonite. This signal corresponds to $(CH_2)_n$ groups (Lindeman and Adams, 1971); it can be attributed to an increase in cross-linking of long-chained alkyl-C moieties occurring during SOM transformation, and also to a selective preservation of this material (Theng et al., 1992; Kögel-Knabner et al., 1992). Moreover, kaol. + OW samples showed a well-defined peak at 130 ppm, in the region corresponding to the signals of the aromatic spectral range (115–165 ppm). Taking into account the peak shape and the presence of unsaturated fatty acids in OW, we attribute this signal mostly to olefinic carbons in acyl chains (Almendros et al., 2001). In mont. + OW samples, more heterogeneous signals occurred in the aromatic regions (110–165 ppm range) and a small band, attributable to aldehydes, ketones or quinones, can be observed in the carbonyl region (165–200 ppm range), which were not present in the original OW. The small band would indicate that a more extended oxidative process occurred in montmorillonite than in kaolinite samples. These data would agree with the lower C content measured in mont. + OW after incubation with respect to the kaol. + OW samples.

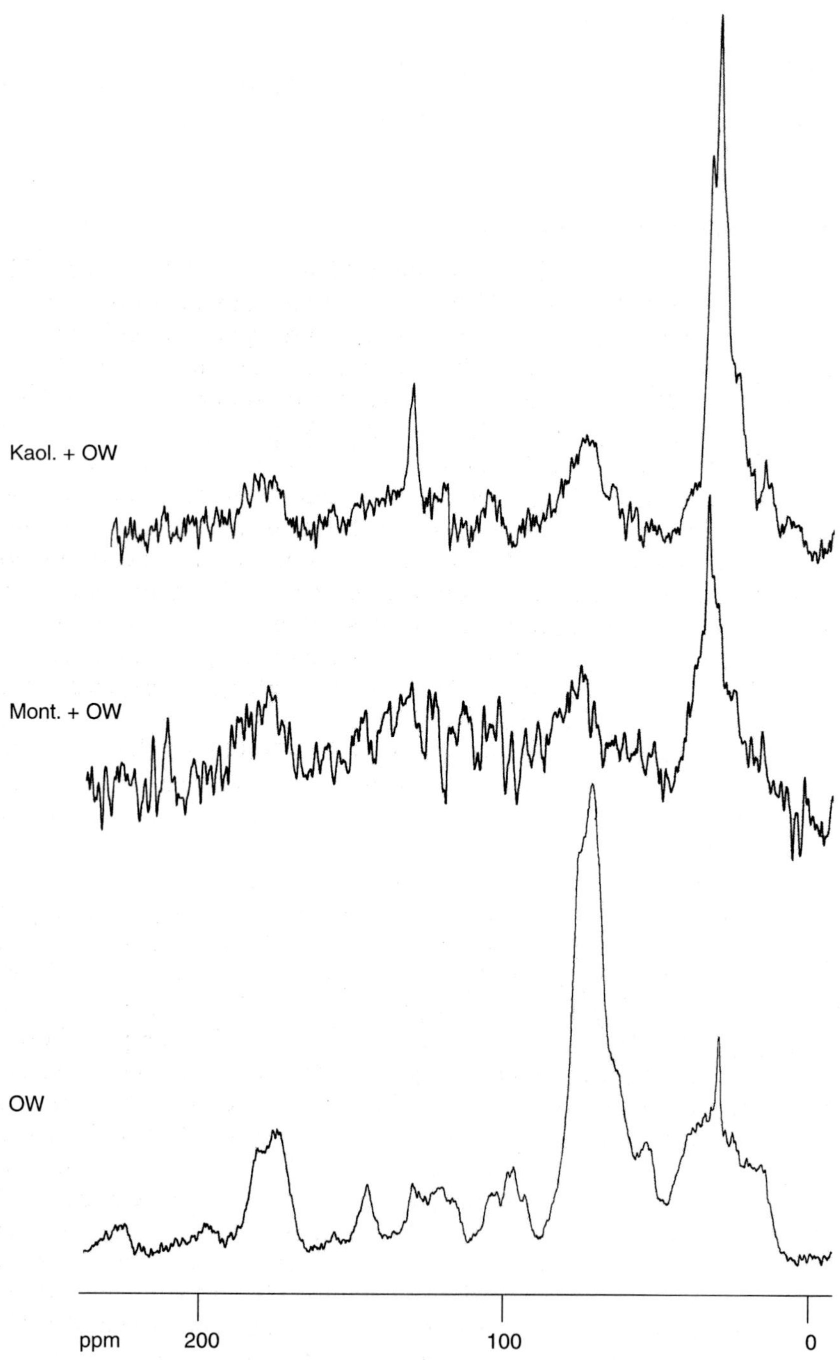

Fig. 4.5: ^{13}C CP-MAS NMR spectra of OW and OW-treated kaolinite and montmorillonite samples.

4 CONCLUSIONS

In the absence of OW, wetting and drying cycles induced better aggregation in montmorillonite than in kaolinite. Samples treated with OW led to the formation for both clays of water stable macroaggregates. The presence of OW promoted formation of larger and more stable aggregates. In fact, transformation of organic material and its interaction with clays increased aggregate water stability, fostering microbial activity at the aggregate surface and then, developing a thick mat of hyphae that bind the aggregates. Moreover, the presence of hydrophobic substances enhanced protection of formed aggregates against water dispersing action.

These findings point to the positive effect of OW used as a soil amendment. However, different clay minerals markedly affected both the organic material transformations and aggregation processes. Hence, the amendment properties of OW applied to the soil depend on the nature of the clay fraction.

Acknowledgments

We thank Prof. F. Ugolini for assisting with preparation of the manuscript, Prof. A. Degl'Innocenti and Dr. G. Almendros for their advice and helpful suggestions with regards to NMR spectra, and Mr. A. Dodero for the C and N analyses.

This work was funded by the Italian Ministry of University and Research: "Progetto di Ricerca-Reflui del Sistema Agricolo-Industriale-Settore Ambiente-Legge 95/'95".

References

Almendros G., Tinoco P., González-Vila F. J., Lüdemann H.D., Sanz J., and Velasco F. 2001. ^{13}C-NMR of forest soil lipids. *Soil Sci.* 166 (3): 186–196.

Bonari E., Macchia M., Angelini L.G., and Ceccarini L. 1993. The waste waters from olive oil extraction: their influence on the germinative characteristics of some cultivated and weed species. *Agric. Mediter.* 123 (4): 273–280.

Cabrera F., Lopez R., Martinez-Bordiu A., Lome E.D., and de Murillo J.M. 1996. Land treatment of olive mill wastewater. *Int. Biodeter. Biodegrad.* 38 (3–4): 215–225.

Calamai L., Lozzi I., Stotzky G., Fusi P., and Ristori G.G. 2000. Interaction of catalase with montmorillonite homoionic to cations with different hydrophobicity: effect on enzymatic activity and microbial utilization *Soil Biol. Biochem.* 32: 815–823.

Chenu C. 1989. Influence of a fungal polysaccharide, scleroglucan, on clay microstructures. *Soil Biol. Biochem.* 21: 299–305.

Cox L., Celis R., Hermosin M.C., Becker A., and Cornejo J. 1997. Porosity and herbicide leaching in soils amended with olive mill waste water. *Agric., Ecosyst. Environ.*65 (2): 151–161.

D'Acqui L.P., Daniele E., Fornasier F., Radaelli L. and Ristori G.G. 1999. Interaction between clay microstructure, decomposition of plant residues and humification. *Eur. J. Soil Sci.* 49: 579–587.

Dorioz J.M., Robert M. and Chenu C. 1993. The role of roots, fungi and bacteria on clay particle organization. An experimental approach. *Geoderma* 56: 179–194.

Farmer V.C. 1978. Water on particle surface. In: *The Chemistry of Soil Constituents.* D.J. Greenland and M.H.B. Hayes (eds.). John Wiley & Sons, Chichester, UK, pp. 405–448.

Fiès J.C. and Bruand A. 1990. Textural porosity analysis of a silty clay soil using pore volume balance estimation, mercury porosimetry and quantified backscattered electron scanning image (BESI). *Geoderma* 47: 209–219.

Gregg S.J., and Sing K.S.W. 1982. *Adsorption, Surface Area and Porosity*. Acad. Press Inc. London, UK, 303 pp.

Grim, R.E. 1968. *Clay Mineralogy*. McGraw-Hill Book Co., New York, NY, 596 p.

Kögel-Knabner I., de Leeuw J.W., and Hatcher P. G. 1992. Nature and distribution of alkyl carbon in forest soil profiles: implication for the origin and humification of aliphatic macromolecules. *Sci. Total Environ.* 117/118: 175–185.

Levi-Minzi R., Saviozzi A., Riffaldi R., and Falzo L. 1992. Land application of vegetable water. Effects on soil properties. *Olivae* 40: 20–25.

Lindeman L.P., and Adams J.L. 1971. Carbon-13 nuclear magnetic resonance spectrometry: chemical shifts for the paraffins through C_9. *Anal. Chem.* 43: 1245–1251.

Lopez R., Martinez-Bordiu A., Dupuy-de Lome E., Cabrera F., and Murillo J.M. 1992. Land treatment of liquid wastes from olive oil industry. *Fresenius Environ. Bull.* 1 (2): 129–134.

Oades, J.M. 1984. Soil organic matter and structural stability: mechanisms and implications for management. *Plant Soil* 76: 319–367.

Oades J.M. 1986. Association of colloidal materials in soils. *Trans. 13th Int. Cong. Soil Sci.* Hamburg, vol. 6, pp. 660–674.

Paris P. 1998. Agronomic perspectives for land disposal of waste from food processing industry. *Riv. di Agron.* 32 (4): 196–220.

Robert M., and Chenu C. 1992. Interactions between soil minerals and microorganisms. *In: Soil Biochemistry*. G. Stotzky and J.-M. Bollag (eds.), Marcel Dekker, New York, NY, vol. 7, pp. 307–404.

Skjemstad J.O., Clarke P., Golchin A., and Oades J.M. 1997. Characterization of soil organic matter by Solid-state ^{13}C NMR Spectroscopy. In: *Driven by Nature: Plant Litter Quality and Decomposition*. G. Cadisch and K. E. Giller (eds.). CAB Int., UK. pp. 253–271.

Skoog D. A., West D. M., and Holler J. J. 1996. *Fundamentals of Analytical Chemistry*, Saunders College Publ., Philadelphia, PA USA, 870 pp. (ed.)

Sparvoli E., Ristori G.G., and D'Acqui L. 1989. Porosité des sols argileux en fonction de l'hydratation. Influence du type d'argile et des ciments sur leur stabilité structurale. *C.R. Acad. Sci. Paris*, Série II 308: 327–333.

Stevenson F.J. 1982. *Humus Chemistry*. John Wiley & Sons, New York, NY, 443 pp.

Tessier D. 1984 Etudes expérimentales de l'organization des matériaux argileux. Hydratation, gonflement et structuration au cours de la desication et de la réhumectation. Thèse d'Etat, Université Paris VII, 361 pp.

Theng B.K.G., Tate K.R., and Becker-Heidmann P. 1992. Towards establishing the age, location and identity of the inert soil organic matter of a Spodosol. *Zeitschrift Pflanz.Boden.* 155: 181–184.

Van Olphen H. and Fripiat J.J. 1979. *Data Handbook for Clay Materials and Other Non-metallic Minerals*. Pergamon Press, Toronto, Ont., CAN, 346 pp.

Wattel-Koekkoek E.J.W., van Genuchten P.P.L., Buurman P., and van Lagen B. 2001. Amount and composition of clay-associated soil organic matter in a range of kaolinitic and smectictic soils. *Geoderma* 99: 27–49.

5

Dynamics of Extractable and Bound Residues of ^{14}C-Metsulfuron-Methyl in Soil

J. XU*, H. Wang, *and* **Z. Xie**

Abstract

A laboratory study was conducted to determine the dynamics of extractable and bound residues of ^{14}C-metsulfuron-methyl in soil, and the distribution of ^{14}C-bound residues in soil humic substances during a 224-day incubation period. The extractable residues of ^{14}C-metsulfuron-methyl decreased continuously with time, while the bound residues increased rapidly during the initial 28 days, and then did not change significantly over time. Degradation of extractable residues could be described using a first-order equation, with a half-degradation time ($T_{1/2}$) ranging from 75.1 to 90.1 days. The Main portion of ^{14}C-bound residues was found in the fulvic acid (FA) fraction followed by the humin fraction, and the humic acid (HA) fraction. The amount of FA-bound residues increased during the initial 28 days, then decreased with time. The HA-bound and humin-bound residues did not change significantly with time from 28 to 224 days after application. The fact that most of the ^{14}C-bound residue existed in the FA fraction provides evidence that the bound residues of metsulfuron-methyl in soil could still cause high phytotoxicity to sensitive rotational crops.

1 INTRODUCTION

Herbicides are now applied extensively in crop production. Although they are designed to be biologically active, herbicides are commonly observed in plants, soils, surface and groundwater. Increasing use of herbicides has created serious concern about the adverse effects of herbicide residues on the environment, agricultural production, and human health.

**Corresponding author:* Dr. Jianming Xu, Inst. Soil and Water Resources and Environment Science, Zhejiang University, Hangzhou 310029, P.R. China. E-mail: jmxu@zju.edu.cu

In soil environments, herbicide can be divided into extractable and bound (nonextractable). Soil extractable residues are those extractable by organic solvents, having higher bioavailability, and susceptible to degradation. Soil bound residues, operationally referred to as compounds including the parent substance and its metabolites not extractable from soil after repeated extraction over organic solvents, are not detected in normal residue analysis. However, increasing evidence indicates that bound residues of herbicides could be released over time from soil and absorbed by plants and earthworms (Gevao et al., 2000; Hague et al., 1982). Metsulfuron-methyl (methyl 2-[[(4-methoxy-6-methyl-1, 3,5- triazin-2- yl)-aminocarbonyl]-aminosulphonyl] benzoate) is a sulfonylurea herbicide widely used for pre- and post-emergence control of broad-leaf weeds in cereal crops. Even extremely low residues of metsulfuron-methyl in soil would cause high phytotoxicity to sugar beet, maize, cotton, sunflower, soybean, rape and rice seedlings (Brown, 1990; Kotoula-syka et al., 1993; Junnila et al., 1994; Yao et al., 1997). Pot experiments indicated that the bound residues of metsulfuron-methyl in soil still developed toxicity in rape seedling (Ye et al., 2002).

This study was undertaken to evaluate the dynamics of extractable and bound residues of ^{14}C-metsulfuron-methyl in soil and to examine the distribution of bound residues in soil humic fractions.

2 MATERIALS AND METHODS

2.1 Soils

The five paddy soil samples (0~20 cm) used in this study were collected from Zhejiang Province in southeastern China. Some selected properties of these soils are listed in Table 5.1.

Table 5.1: Selected properties of paddy soils used in this study

Paddy soil	pH	CEC cmol (+) kg^{-1}	OC g kg^{-1}	Clay %	Silt %	Sand %
1	6.20	25.1	23.6	35.3	60.6	4.1
2	6.22	28.5	18.3	40.0	57.0	3.0
3	6.00	9.0	13.5	29.0	32.3	38.7
4	5.36	13.7	9.1	39.0	41.1	19.9
5	6.32	30.6	22.9	45.6	43.9	10.5

2.2 Chemicals

[triazine-4-^{14}C]-labeled metsulfuron-methyl was obtained from the Chinese Institute of Atomic Energy, Beijing. Its specific activity was 4.55×10^4 Bq mg^{-1}, and radiochemical purity is more than 97.3%. Nonlabeled metsulfuron-methyl was purchased from Chem Service (West Chester, PA) and its chemical purity was >99.0%.

2.3 Incubation Experiments

A spiking solution was prepared by dissolving ^{14}C-labeled and nonlabeled metsulfuron-methyl in a 1:1 methanol:water mixture to give a concentration of 1,000 μg ml^{-1}. Fresh soil samples (60 g oven-dried weight equivalent) were weighed into 250-mL Erlenmeyer flasks and soil moisture adjusted to 50% of the soil maximum water-holding capacity by addition of distilled water. The soil sample in each flask was added with 0.6 ml of the spiking solution, giving an initial soil concentration of 10 μg g^{-1} soil. It is possible that such high concentrations of herbicides do occur in some spots in real fields due to nonuniform distribution, especially for metsulfuron-methyl which is used at very low rates. Our preliminary experiment indicated the residual methanol from the spiking solution did not influence the microbial biomass size. After complete mixture of soil and metsulfuron-methyl, a vial containing 10 ml of 0.5 mol L^{-1} NaOH was put in each flask to trap the $^{14}CO_2$ evolved during the incubation period. The flasks were then sealed, incubated at 25 ± 1°C in darkness and aerated weekly, with water loss supplemented by distilled water. Soil samples were sampled in aliquots of 6.0 g (oven-dried weight equivalent) removed from the sample flasks at 0, 7, 14, 28, 56, 112 and 224 days after application (DAA) of the herbicide and subjected to extraction and analysis as described below. Three replications were done for each soil sample.

2.4 Extraction of Radioactivity

Soil samples (6.0 g, oven-dried weight equivalent) were shaken for 2 h with 40 ml methanol and then centrifuged at 845 g for 15 min. The extraction procedure was repeated until ^{14}C radioactivity was no longer detected in the supernatant. The supernatants were then mixed together for analysis of total extractable residues of metsulfuron-methyl in each soil.

Now the soil residues were extracted with a mixture of 60 ml 0.1 mol L^{-1} NaOH and 0.1 mol L^{-1} $Na_4P_2O_7$ (pH 13) to separate the soluble organic fractions, including humic acid (HA) and fulvic acid (FA), from the insoluble humin fraction. To the brief, the soil residues were shaken with the extractant mixture overnight on a platform shaker at room temperature, and centrifuged at 845 g for 15 min to collect the supernatant. This step was repeated three times and all the supernatants were combined to form the soluble fraction. The soluble fraction was then acidified to pH 1.5 to 2 with 2 mol L^{-1} HCl to precipitate the HA fraction. The separation of precipitated HA and soluble FA was conducted by centrifugation at 845 g for 15 min. The HA precipitate was redissolved in 0.5 mol L^{-1} NaOH solution for radioactivity determination. After extraction, the residual soil (humin fraction) was left in a fume hood to be air dried.

2.5 Radioactivity Determination

The radioactivity in $^{14}CO_2$-trapped alkali solution and in HA and FA fractions was measured by a Wallack 1414 liquid scintillation counter (LSC) (Wallack; Turku, Finland) using 0.5 ml aliquots of sample solution mixed with 10 ml of scintillation cocktail I [2'5-Diphenyl Oxazole (PPO): 1'4-bis-(5-phenyl-oxazoyl-2)-benzene (POPOP): toluene: 2-ethoxyethanol = 5g: 0.4g: 600 ml: 400 ml]. Aliquots of 1 g of residual soil (humin fraction) were combusted for 5 min in an OX-600 biological sample oxidizer (R.J. Harvey Instrument Corp., Hillsdale, NJ). The $^{14}CO_2$ evolved from combustion was trapped in 15 ml scintillation cocktail II (PPO : POPOP : ethanolamine : 2-ethoxyethanol : toluene = 5g : 0.4 g : 175 ml : 350 ml : 475 ml), and radioactivity determined by LSC. The efficiency of $^{14}CO_2$ recovery, as determined by combusting ^{14}C standards, was > 92%. The total recovery of radioactivity in $^{14}CO_2$ traps and in extractable and nonextractable residues ranged from 90.1% to 100.8%.

3 RESULTS AND DISCUSSION

3.1 Extractable Residues in Soil

The extractable ^{14}C radioactivity of metsulfuron-methyl in soils significantly decreased with increase in incubation time (Table 5.2). Immediately after

Table 5.2: Extractable and bound residues of ^{14}C-metsulfuron-methyl (%) in soil

		Days after application						
Soil	Residues	0	7	14	28	56	112	224
1	Extractable	92.2 ± 0.6	81.5 ± 0.7	62.7 ± 0.5	53.5 ± 1.2	35.8 ± 0.4	21.5 ± 0.5	11.3 ± 0.2
	Bound	4.1 ± 0.1	13.5 ± 0.1	30.7 ± 0.4	37.6 ± 1.0	35.3 ± 0.4	35.2 ± 0.7	26.9 ± 0.5
2	Extractable	92. ± 0.8	78.6 ± 1.3	65.4 ± 0.4	52.2 ± 0.6	43.1 ± 1.3	20.4 ± 0.4	12.0 ± 0.2
	Bound	3.3 ± 0.5	12.7 ± 0.1	23.9 ± 0.7	36.3 ± 0.8	35.6 ± 0.6	32.4 ± 0.8	26.2 ± 0.1
3	Extractable	93.5 ± 0.8	68.2 ± 0.5	57.7 ± 0.7	35.6 ± 1.1	32.5 ± 0.3	16.5 ± 0.5	10.1 ± 0.2
	Bound	2.7 ± 0.1	23.5 ± 0.2	29.9 ± 0.2	49.9 ± 0.5	40.8 ± 0.6	35.1 ± 0.5	28.7 ± 0.1
4	Extractable	88.7 ± 0.9	71.5 ± 0.7	47.6 ± 0.8	28.4 ± 0.3	28.2 ± 0.7	19.9 ± 0.8	12.3 ± 0.3
	Bound	5.1 ± 0.4	20.1 ± 0.3	41.3 ± 0.3	55.7 ± 1.5	39.7 ± 0.9	29.2 ± 0.2	24.8 ±0.2
5	Extractable	90.7 ± 1.1	71.8 ± 0.6	61.1 ± 1.1	49.7 ± 1.4	35.1 ± 1.1	16.1 ± 0.2	14.1 ± 0.3
	Bound	2.9 ± 0.1	18.8 ± 0.2	28.6 ± 0.1	38.8 ± 0.2	36.8 ± 0.1	27.7 ± 0.6	20.9 ±0.3

application of ^{14}C-metsulfuron-methyl into soil, 88.7%~93.5% of applied ^{14}C activity was found in the methanol extractable fraction. At the end of the incubation experiment (224 DAA), the extractable residues ranged from 10.1% to 14.1% of the amount applied. Degradation of extractable ^{14}C activity over time resulted from the combination of microbial and chemical processes in soil (Beyer et al., 1988; Hemmamda et al., 1994; Ismail and Lee, 1995; Sarmah et al., 1998). The decrease pattern of extractable ^{14}C activity over time in soil could well be described using a first-order equation with a significant correlation coefficient ranging from 0.88 to 0.97 (Table 5.3). The half-degradation time ($T_{1/2}$) of extractable ^{14}C residues ranged from 75.1 to 90.1 days, which is much longer than values previously reported (James et al., 1995; Pons and Barriuso, 1998; Ye et al., 2002). The reason is that the extractable ^{14}C residues comprise not only ^{14}C metsulfuron-methyl but also its ^{14}C metabolites. Bossi et al. (1999) identified two major metabolites, 2-amino-4-methoxy-6-methyl-1,3,5-triazine and 2-(aminosulfonyl)-methyl ester benzoic acid, in extractable residues of metsulfuron-methyl from soil. Although soil properties would apparently influence the degradation of metsulfuron-methyl (Walker et al., 1989; Abdullah et al., 2001). We found no significant relation between $T_{1/2}$ and any soil property considered in this study.

Table 5.3: First-order parameters for the degradation of extractable residues of ^{14}C-metsulfuron-methyl in soils

Soil	Exponential equation	r	$T_{1/2}$ (d)
1	LnC=4.307–0.00915t	–0.970**	75.7
2	LnC =4.321–0.00898t	–0.971**	77.2
3	LnC =4.151–0.00923t	–0.939**	75.1
4	LnC =4.028–0.00769t	–0.882**	90.1
5	LnC =4.206–0.00829t	–0.920**	83.6

** Significant at the 0.01 probability levels.

3.2 Bound Residues in Soil

The bound residues increased rapidly during the initial 28 days, then gradually decreased (Table 5.2). At 28 DAA, the bound residues peaked, varying from 36.3.4% to 55.7% of applied radioactivity. The steep increase in bound residues coincided with the rapid decrease of extractable residues during the initial 28 days, which indicated a faster incorporation of extractable residues into bound residues. From 28 to 224 DAA, the amount of bound residues gradually decreased with time, which indicated that bound residues of ^{14}C-metsulfuron-methyl could be released or mineralized. The activity of microorganisms is thought to be the primary factor responsible for the release of bound pesticide residues (Gevao et al., 2000). Bound residues at 224 DAA ranged from 20.9% to 28.7% of applied radioactivity. Formation of bound residues was interpreted as resulting from the interaction between ^{14}C-metsulfuron-methyl and its metabolites with soil organic components. The possible binding mechanisms included ionic bonding, charge transfer complexes, hydrogen bonding, and hydrophobic sorption (Pons and Barriuso, 1998; Walker and Jurado, 1998; Gevao et al., 2000; Wang et al., 2002).

3.3 Bound Residues in Soil Humic Fractions

Of the bound residues of ^{14}C metsulfuron-methyl in soil, the main portion was found in the FA fraction followed by the humin fraction, then the HA fraction (Table 5.4), i.e. 60.8–78.3%, 18.1–35.7%, and 3.5–4.8% of total bound residues, respectively. The distribution of ^{14}C activity of bound residues in the various humic substances might be related to the amount of their active functional groups, which was usually higher in FA than in HA and humin.

Table 5.4: Bound residues of ^{14}C-metsulfuron-methyl (%) in soil humic fractions

Soil	Bound residues	Days after application						
		0	7	14	28	56	112	224
1	FA	3.9 ± 0.1	10.8 ± 0.1	24.4 ± 0.1	27.3 ± 0.2	24.2 ± 0.1	23.2 ± 0.1	16.4 ± 0.2
	HA	0.1 ± 0.01	0.4 ± 0.1	0.7 ± 0.02	1.2 ± 0.03	1.8 ± 0.02	2.7 ± 0.01	0.9 ± 0.01
	Humin	0.1 ± 0.01	2.4 ± 0.03	5.5 ± 0.1	9.1 ± 0.2	9.4 ± 0.2	9.4 ± 0.5	9.6 ± 0.3
2	FA	3.2 ± 0.1	11.1 ± 0.1	20.4 ± 0.2	30.6 ± 0.3	28.0 ± 0.2	23.7 ± 0.2	18.1 ± 0.2
	HA	0.1 ± 0.01	0.3 ± 0.01	0.9 ± 0.02	1.4 ± 0.02	1.8 ± 0.02	2.4 ± 0.02	1.3 ± 0.01
	Humin	0.0	1.4 ± 0.01	2.6 ± 0.1	4.3 ± 0.1	5.9 ± 0.2	6.3 ± 0.06	6.8 ± 0.2
3	FA	2.6 ± 0.1	21.4 ± 0.2	26.6 ± 0.5	44.2 ± 0.4	34.5 ± 0.2	28.1 ± 0.5	22.2 ± 0.2
	HA	0.1 ± 0.01	0.4 ± 0.01	0.9 ± 0.1	1.6 ± 0.02	1.7 ± 0.01	2.0 ± 0.02	1.2 ± 0.01
	Humin	0.1 ± 0.01	1.7 ± 0.2	2.3 ± 0.1	4.1 ± 0.3	4.7 ± 0.2	4.9 ± 0.2	5.4 ± 0.2
4	FA	5.0 ± 0.2	17.8 ± 1.0	38.3 ± 0.2	51.1 ± 0.4	34.1 ± 0.3	23.2 ± 0.3	19.4 ± 0.5
	HA	0.1 ± 0.01	0.6 ± 0.01	0.7 ± 0.01	1.4 ± 0.05	1.4 ± 0.01	1.5 ± 0.01	0.9 ± 0.01
	Humin	0.1 ± 0.01	1.7 ± 0.2	2.4 ± 0.1	3.2 ± 0.04	4.3 ± 0.01	4.5 ± 0.03	4.5 ± 0.1
5	FA	2.8 ± 0.6	17.0 ± 0.2	26.3 ± 0.2	34.8 ± 0.5	31.6 ± 0.3	22.1 ± 0.5	15.7 ± 0.2
	HA	0.0	0.5 ± 0.02	0.5 ± 0.02	1.2 ± 0.01	1.4 ± 0.01	1.4 ± 0.02	0.9 ± 0.01
	Humin	0.1 ± 0.1	1.4 ± 0.1	1.7 ± 0.3	2.8 ± 0.1	3.9 ± 0.03	4.2 ± 0.1	4.3 ± 0.3

The change of ^{14}C activity in the FA fraction over time followed a pattern similar to that of total bound residues. The highest amount of bound residues in the FA fraction was also observed at 28 DAA, after which ^{14}C activity decreased. The amount of bound residues in HA and humin fractions did not vary significantly with time from 28 to 224 DAA.

The distribution of bound residues in the various humic fractions would play a significant role in the movement and bioreactivity of herbicides in soil. It is well known that FA is a naturally occurring water-soluble low-molecular-weight humic fraction, regarded as the dominant soluble organic fraction in soil solution under field conditions (Khan, 1982). FA can function and transport as a carrier bound residues in soil, water and the crop system. Compared with HA and humin, FA-bound residues are expected to be more bioavailable to plants and soil fauna, and more liable to leaching. Most likely the bound residues of metsulfuron-methyl in the FA fraction cause high phytotoxicity to rotation crops. Further studies on the phytotoxicity and mobility of residual metsulfuron-methyl in the FA fraction should be conducted under field conditions.

4 CONCLUSIONS

Soil bound residues of ^{14}C-metsulfuron-methyl rapidly increased as the extractable residues decreased during the initial 28 days after application, and thereafter tended to decrease until the end of the experiment (at 224 DAA). This indicated that bound residues of ^{14}C-metsulfuron-methyl in soil could be gradually released or mineralized by soil biological processes or chemical and physical processes. The major portion of bound residues of ^{14}C activity were associated with the FA fraction followed by humin and HA fractions. The phytotoxicity of metsulfuron-methyl residues to sensitive rotational crops probably results from the binding of most bound residues of ^{14}C-metsulfuron-methyl into mobile soil FA fraction.

Acknowledgments

This work was supported by the Teaching and Research Award Program for Outstanding Young Teachers in Higher Education Institutions of China, the National Natural Science Foundation of China (Grant No. 40171051), and Zhejiang Provincial Natural Science Foundation of China (Grant No. RC99032).

References

Abdullah A.R., Sinnakkannu S., and Tahir N.M. 2001. Adsorption, desorption, and mobility of metsulfuron-methyl in Malaysian agricultural soils. *Bull. Environ. Contam Toxicol.* 66: 762–769.

Beyer Jr. E.M., Duffy M.J., Hay J.V., and Schlueter D.D. 1988. Sulfonylureas. In: *Herbicides: Chemistry, Degradation and Mode of Action*. P.C. Kearny and D.D. Kaufman (eds.). Marcel Dekker, New York, NY, pp. 117–189.

Bossi R., Seiden P., Andersen S.M., Jacobsen C.S., and Streibig J.C. 1999. Analysis of metsulfuron-methyl in soil by liquid chromatography/tandem mass spectrometry. Application to a field dissipation study. *J. Agric. Food Chem.* 47: 4462–4468.

Brown H.M. 1990. Mode of action, crop selectivity and soil relations of the sulfonylurea herbicides. *Pest. Sci.* 29: 263–281.

Gevao B., Semple K.T., and Jones K.C. 2000. Bound pesticide residues in soils: a review. *Environ. Pollut.* 108: 3–14.

Hague A., Schuphan I. and Ebing W. 1982. Bioavailability of conjugated and soil-bound [^{14}C] hydroxymonolinuron-β-D-glucoside residues to earthworms and ryegrass. *Pest. Sci.* 13: 219–228.

Hemmamda S., Calmon M., and Calmon J.P. 1994. Kinetics and hydrolysis mechanism of chlorsulfuron and metsulfuron-methyl. *Pest. Sci.* 40: 71–76.

Ismail B.S. and Lee H.J. 1995. Persistence of metsulfuron-methyl in two soils. *J. Environ. Sci. Health B* 30: 485–497.

James T.K., Klaffenbach P., Holland P.T., and Rahman A. 1995. Degradation of primisulfuron-methyl and metsulfuron-methyl in soil. *Weed Res.* 35: 113~120.

Junnila S., Heinonoren-tanski H., Revio L.R., and Laitinen P. 1994. Phytotoxic persistence and microbiological effects of chlorsulfuron and metsulfuron in Finnish soils. *Weed Res.* 34: 413–423.

Khan S.U. 1982. Distribution and characteristics of bound residues of prometryn in an organic soil. *J. Agric. Food Chem.* 30: 175–179.

Kotoula-syka E., Eleftherohorinos I.G., and Gagianas A.A. 1993. Phytotoxicity and persistence of chlorsulfuron, metsulfuron-methyl in three soils. *Weed Res.* 33: 355–367.

Pons N. and Barriuso E. 1998. Fate of metsulfuron-methyl in soils in relation to pedo-climatic conditions. *Pest. Sci.* 53: 311–323.

Sarmah A.K., Kookana R.S., and Alston A.M. 1998. Fate and behaviour of triasulfuron, metsulfuron-methyl, and chlorsulfuron in the Australian soil environment: a review. *Austr. J. Agric. Res.* 49: 75–790.

Walker A. and Jurado M. 1998. Adsorption of isoproturon, diuron and metsulfuron-methyl in two soils at high soil: solution ratios. *Weed Res.* 38: 229–238.

Walker A., Cotterill E.G., Welch S.J. 1989. Adsorption and degradation of chlorsulfuron and metsulfuron-methyl in soils from different depths. *Weed Res.* 29: 281–287.

Wang H.Z., Xu J.M., Xie Z.M., and Ye Q.F. 2002. Dynamics of bound residues of metsulfuron-methyl in soil humus *Acta Sci. Circum*stant. 22: 256–260. (in Chinese).

Yao D.R., Song X.L., and Chen J. 1997. Studies on sensitivity of aftercrops to metsulfuron residue in wheat fields *Jiangsu J. Agric. Sci.* 13: 171–175. (in Chinese).

Ye Q.F., Wu J.M., and Sun J.H. 2002. ^{14}C-extractable residue, ^{14}C-bound residue and mineralization of ^{14}C-labeled metsulfuron-methyl in soils. *Environ. Sci.* 23: 62–68. (in Chinese).

6

A Study of Geochemical Parameters and Polycyclic Aromatic Hydrocarbons (PAHs) Concentration in Peat of Selected Peatlands in Northeastern Poland

M. Malawska* *and* **B. Wilkomirski**

Abstract

The concentration of polycyclic aromatic hydrocarbons (PAHs) was analyzed in samples of peat collected from four peatlands in the northeast of Poland. Peat samples were collected from different depths according to the stratigraphic profile of the peat bogs. The total concentration of the 16 anthropogenic PAHs was between 70 and 439 ng g^{-1}. Most of the PAHs were 3-ring compounds, 4-ring ones slightly fewer and the least common 5-ring compounds. In some peat samples, the concentration of perylene, which comes from a natural source, largely exceeds the concentration of all the other PAHs investigated. The concentration of anthropogenic PAHs and perylene is connected not only with pollution but also with the hydrological and geological conditions during peatland formation. The results obtained complete our knowledge about the sources of PAHs in biogenic sediments.

1 INTRODUCTION

Most wetlands in Poland are located in three provinces: the southern Baltic seashore province of high-moor peatlands and the northeastern Poland and Polish northern German provinces of low-moor and high-moor mires. These provinces include young glacial areas shaped by Baltic glaciation. The diversity of the region's geographic setting, hydrographic conditions, and landscape significantly influences wetland distribution and specificity.

**Corresponding author:* Dr. M. Malawska, Dept. of Plant Systematics and Geography, Warsaw University, 00-478, Warszawa, Al, Ujazdowskie 4, Poland. E-mail: bowi@biol.uw.edu.pl.

Assessment of historical trends in atmospheric deposition of organic contaminants by analyzing peat samples has been done many times (Aamot et al., 1987; Bracewell et al., 1993; Himberg and Pakarinen, 1994; Rapaport and Eisenreich, 1986; Sanders et al., 1995). Depositional media such as sediment cores, polar ice, and peat profiles have all been used in historical pollutant monitoring studies, as the depositional processes leading to the accumulation of organic compounds (Howsam and Jones, 1998). Peat cores represent an almost ideal medium for recording temporal changes in organic contaminant deposition rates, because peat from bogs is typically 97–98% organic matter and the sorption of hydrophobic organic contaminants to the matrix is strongly enhanced, minimizing the potential for postdepositional mobility or diagenetic alteration following deposition (Sanders et al., 1995). Bogs are archives of the changing fluxes of organic compounds from natural as well as anthropogenic sources.

PAHs are widespread environmental contaminants, suspected to be carcinogenic (Harvey, 1985). They have been detected in various concentrations not only in polluted industrial and urbanized environments, but also in soil, water, sediments, and derivative rocks in areas of various anthropogenic pressures (Bradley et al., 1994; Ollivon et al., 1995; Coleman et al., 1997). The vast majority of polycyclic aromatic hydrocarbons get to the environment from anthropogenic sources, such as the combustion and processing of fossil fuels, the coke industry, oil refineries, household and communal waste, and car and airplane engines (Harvey, 1998; Howsam and Jones, 1998; Lichtfouse et al., 1997). Changes in their qualitative distribution suggest that the sources of PAHs have shifted from biomass burning to fossil fuels combustion in the last 200 years.

Some PAHs occur in nature in coal, peat, and crude petroleum and can also be formed during processing of these materials in industrial plants. It was postulated that they should be attributed to early stages of transformation of nonaromatic constituents of higher plants such as diterpenoids, triterpenoids, steroids, and hopanoids. These compounds are called aromatic biomarkers. Their presence in geological records can be evidence of natural processes of PAH formation (Neilson, 1998).

The hydrocarbon created most frequently as a result of natural biogeochemical processes in sediments is perylene (Silliman et al., 1998; Wilcke et al., 1999). Perylene has been widely found found in both marine and freshwater sediments.

The presence of organic matter and anoxic conditions seems to be essential for the formation of perylene. However, the type of organic matter that is present may not be as crucial for perylene formation as is the process involved in forming it. Lack of notable correlation between delivery of algal and land-plant organic matter and perylene abundance in sediment implies that this PAH does not have a specific precursor and that it may originate from either aquatic or land-derived organic matter. Such a situation would imply that perylene is an indicator of diagenetic conditions and its genesis may be process-governed and independent of the type of organic matter within the sediment (Silliman et al., 1998; Pereira et al., 1999; Thiele and Brummer, 2002).

The literature contains very few papers concerning the presence of PAHs in peat. In Poland, there are about 51,000 peat bogs covering a total surface area of slightly more than 13,000 km^2 in regions with very little anthropogenic pollution. The aim of this study was to investigate the concentrations of 16 main PAHs from the U.S. Environmental Protection Agency's (U.S. EPA) list (with the exception of naphthalene), plus perylene, in different types of peat deposits formed in various geological and hydrological conditions. Moreover, we sought to test the hypothesis that the composition and quantity of PAHs depend not only on the genus (botanical composition) of peat, but also on its geochemical properties. One can suppose that in some types of peat the processes of natural formation of PAHs should be more intensive than in others.

2 MATERIALS AND METHODS

2.1. Study Sites

Field investigations were carried out in 2000 and 2001. The study was conducted in natural or nearly natural wetland communities after previous identification of a peat deposit. Regional and altitudinal variations of the wetlands in the northeastern region of Poland were taken into account in the complete identification of the stratigraphic profile of the peatlands.

Peat samples were collected in four peat bog complexes: Jarka (A), Grądy Węgorzewskie (B), Żegary (C), and Bagna Skieblewskie (D). These are situated in areas created by the Baltic glaciation. Compared to the rest of Poland, there is a considerable variety of site conditions and a high degree of largely undisturbed ecosystems. Table 6.1 lists the sampling sites and briefly describes the study areas mapped in Figure 6.1.

Table 6.1: Sampling sites and description of the area

Site no.	Sampling area	Plant communities	Total area [km^2]
A	Jarka (raised bog)	*A. Ledo-Pinetum Kobendza*	1.64
B	Grądy Węgorzewskie		
	B1. raised bog	B1. *Vaccinio uluginosi-Pinetum*	3.13
	B2. Fens	B2. *Dryopteridi-Betuletum pubescentis*	1.12
	B3. Fens	B3. *Betulo-Salicetum repentis*	0.7
	Bagna Skieblewskie		
C	C1. raised bog	C1.*Vaccinio uliginosi-Pinetum*	9.41
	C2. transitional bog	C2. *Sphagnetum magellanici boreale*	4.70
D	Żegary		
	D1 raised bog	*Ledo-Pinetum* Kobendza	4.70
	D2 transitional bog	*Alneto Betuletum*	7.0

Fig. 6.1: Main wetlands in Poland and localization in the study
A—Jarka raised bog, B—Grądy Węgorzewskie wetland, C—Żegary wetland, D—Bagna Skieblewskie wetland

The Jarka raised bog (Mechacz Wielki Reserve) and Grądy Węgorzewskie bogs are located in the Masurian Lake District in the terminal moraine landscape of main Baltic glaciation corresponding to the Pomerania moraines. The accumulation of sediments took place after repeated oscillations of the glacier.

The Żegary and Bagna Skieblewskie marshes differ from each other in terms of plant communities, as well as type of peat deposit (lowland and transitional bog respectively). Żegary marsh lies among moraine hills built of rock formations. It fills a depression surrounded by moraines and ose. The bog occupies the watershed between streams flowing to Marycha River and to nearby lakes. Żegary marsh is situated in the eastern part of the Prussian-Mazowian basin. Its main morphological elements are elevated areas (remains of the terminal moraine) and lowland areas covered with moor. The bog is located on the boundary between the basins of the Czarna Hańcza and Biebrza rivers.

2.2 Peat Sampling

Peat samples were collected in June and September 2000. Subsequent description of peat deposit character and distribution in the peatlands was based on specialist peat documentation (geological characterization and valuation of wetlands and grasslands in Poland). Geological documentation is in the computer database of Institute the for Land Reclamation and Grassland Farming (Warsaw).

Points for collection of biogenic sediment cores were selected on the basis of representative stratigraphic profiles of each peat bog deposit. Peat samples were collected from various depths of the stratigraphic profile of the investigated bogs. The peat samples were stored at –20°C until analysis was begun.

2.3 Sample Preparation and Soil Analyses

The genus of the peat was classified according to the Lang method (Lang, 1994), which takes into account its botanical origin, dominant species, and decomposition rate.

The decomposition rate (DR) of peat was determined by the von Post method (Grosse-Bauckmann, 1990).

The peat samples were air dried and the fraction >2 mm removed by dry sieving. The samples were then characterized by conventional standard procedures:

1. percentage ash content (AC) (samples were burnt in an oven at 500°C for 4 h);
2. pH determined in water and 1 M KCl at a peat: solution ratio of 1:2.5 with a potentiometric glass electrode;
3. content of extractable metal cations (Na^+, K^+, Mg^{+2}, Ca^{+2})—using 1 M ammonium acetate (NH_4OAc) reagent and measured by atomic absorption spectrophotometry (Sumner and Miller, 1996);
4. percentage of soil organic carbon (C_{org}) and humic acid organic carbon (C_{hum}) by Tiurin's dichromate method (Tiurin and Tiurina, 1940). Humic substances were extracted according to the method described by Kononowa (1968); and
5. percentage of total nitrogen (N) measured by Kjeldahl automatic analyzer.

2.4 Identification of PAHs in Peat

Samples were dried at room temperature for 2 days, then at 50°C for 2–3 h (in an oven). They were sieved through a 2-mm mesh to remove large particles and stored at 5°C prior to analysis.

We quantified 17 unsubstituted PAHs: acenaphthene (ACE), acenaphtylene (ACY), fluorene (FLU), fluoranthene (FLA), phenanthrene (PHE), anthracene (ANT), pyrene (PYR), benzo[a]anthracene (BAA), chrysene (CHRY), benzo[b]fluoranthene (BBFLA), benzo[k]fluoranthene (BKFLA); benzo[e]pyrene (BEP), benzo[a]pyrene (BAP), indeno[1,2,3-cd]pyrene (IND), dibenzo [ah]anthracene (DBAH); benzo[ghi]perylene (BGHI), and perylene (PER).

Liquid-solid extraction was performed using a water-acetone-hexane solution in the volume proportion 1:1:1 in a Soxhlet apparatus. All samples were purified used an SPE (solid phase extraction) column filled with Florisil for the cleanup of Soxhlet extracts. Details of the purification procedure are given by Maliszewska-Kordybach and Oleszek (1994). Quantitative analysis was performed using the external standard method in which the certified PM-612 standard (ULTRA Scientific Ltd.) was applied.

PAH content analysis in peat material was performed using a Hewlett-Packard gas chromatograph (5890 Series II) equipped with a Hewlett-Packard 5-MS mass selective detector and a nonpolar capillary column HP-5 (length 24 m, diameter 0.2 mm, 0.33 μm diphenyl-95% dimethylpolysiloxane film). The following temperature programming was applied: 70°C (10°C/min^{-1}) to 200°C (2.5°C min^{-1}) to 300°C (7 min). The detector temperature was 280°C. The detection limit was ca. 0.1 ng g^{-1} dry weight, assuming 10 g of sample.

2.5 Quality Assurance and Quality Control (QA/QC)

The procedures described above have been checked for recoveries and reproducibility according to the standards of PAH determination (Neilson, 1998). Prior to extraction, the extraction was investigated by spiking peat samples with four increasing amounts of standards. Recovery results were in the range of 81–96% for all the compounds analyzed. Reproducibility was calculated on replicate analyses, giving an error between 3.1 and 8.4%. Ten percent samples were extracted and analyzed in duplicate. After analysis of every 10 samples, standard samples with a known PAH content were analyzed. A solvent blank was analyzed after every 20 samples as a check on the response of gas chromatography.

3 RESULTS AND DISCUSSION

In the four sites selected a total number of 37 peat samples were collected. Botanical classification of peat genus revealed different generic types of various botanical origin and decomposition rates, namely:

— 10 samples of sedge-moss peat (*Carici-Bryaleti*), belonging to Hypnum moss peat (Bryalo-Parvocaricioni), moderately decomposed form (DR = 15 – 45%).

— 3 samples of reed –sedge peat *(Carici-Phragmiteti*) and 6 samples of sedge-peat *(Cariceti*) belonging to Sedgeous Peat (*Magnocaricioni*), both classified as reed-sedge peat. Reed-sedge peat (*Carici-Phragmiteti*) and sedge peat *(Cariceti*) were moderately decomposed (DR = 30 – 45%) and (DR = 30 – 50%) respectively. Various classifications of peat have been developed. In this study the classification based on phytosociological features was used (Lang, 1994).

— 2 samples of wood peat (*Alnioni*) originating mainly from alder (*Alneti*), osier (*Saliceti*), and mixed alder-birch (*Alno-Betuleti*) species (alder peat, alder brushwood peat, forest wood peat). The alder peats were strongly decomposed (DR = 45–50%).

— 2 samples of *Sphagno-Cariceti* peat belonging to *Minero-Sphagnioni* were decomposed (DR = 25–30%).

— 9 samples of high-moor peat belonging to poorly decomposed *Cuspidato-Sphagneti* (1 sample), *Eriophoro-Sphagneti* (7 samples), and *Pino-Sphagneti*

(1 sample). All these peat were classified as *Sphagnum* moss peat (*Ombro-Sphagnioni*).

— 5 samples of totally decomposed Gyttja (detrital, calcareous, algal, and clayey), typical of raised bogs.

Figures 6.2 to 6.7 present some chemical characteristics of peat deposits in the investigated wetlands. The decomposition rate of the peat seems to depend both on the peat species and depth of the peat deposit (Fig. 6.2). Low moors (B2, B3) are characterized by strongly decomposed sedgeous peat (*Bryalo-Parvocaricioni* and *Magnocaricioni*). Upper layers of raised bogs are built of poorly decomposed *Sphagnum* peat (*Ombro-Sphagnioni*), while the lower layers are formed by the sedgeous peat with higher decomposition rates.

In all the wetlands investigated the ash content of the peat increased with depth. This can be explained by diagenesis of wetlands and the input of water (rain water or groundwater retention). The raised bogs were formed on the low moor, which originated by swamping the areas adjacent to the river valley and by groundwater retention.

The acidity of peat observed was typical for peat species (Fig. 6.3). The *Sphagnum* peats occurring in the top layers of raised bogs were highly acidic (pH < 3), while the sedgeous peats have a pH from 3 to 5. The acidity of peat decreases with depth (Fig. 6.3).

The percentage concentration of total nitrogen, organic carbon, and humic acid varied in different species of peat and did not depend on depth (Figs. 6.4, and 6.5). The peat deposits of raised bogs (C1, C2) had relatively high concentrations of C and N but the highest concentrations of these elements were found in peats in the low moor and transitional bogs.

Figures 6.6 and 6.7 illustrate depth profiles of the distribution of exchangeable cations Na^+, K^+, Mg^{2+}, and Ca^{2+}. The highest concentrations of Na^+ and K^+ cations are observed in peat bogs A, B1, and C2, and the lowest in the low moor B3.

The lack of correspondence between the Na^+ and K^+ concentrations in the various genera of peat suggests that the cation concentrations in it are not controlled by a simple chemical equilibrium with the peat. Damman (1978) suggested that the K^+ concentration in peat correlates with the ash content at the same depths and reflects strong biological uptake by the plants. The concentration of Mg^{2+} and Ca^{2+} in the peat profiles exhibits similar trends but the concentration per se is much higher. These results generally agree with the observation that the exchangeable cation pool in peats from blanket bogs is dominated by Mg^{2+} and Ca^{2+} (Gore and Allen, 1956; Boatman, 1957). Shotyk (1997) investigated the content of trace elements in ombrotrophic peat bogs in Scotland and the Shetland Islands. The cation concentrations in peat profiles showed very little change with depth, were consistent with an ombrotrophic interpretation on the cores, and provided no evidence of any chemical influence by the rocks surrounding the bogs. Shotyk suggested that the cation

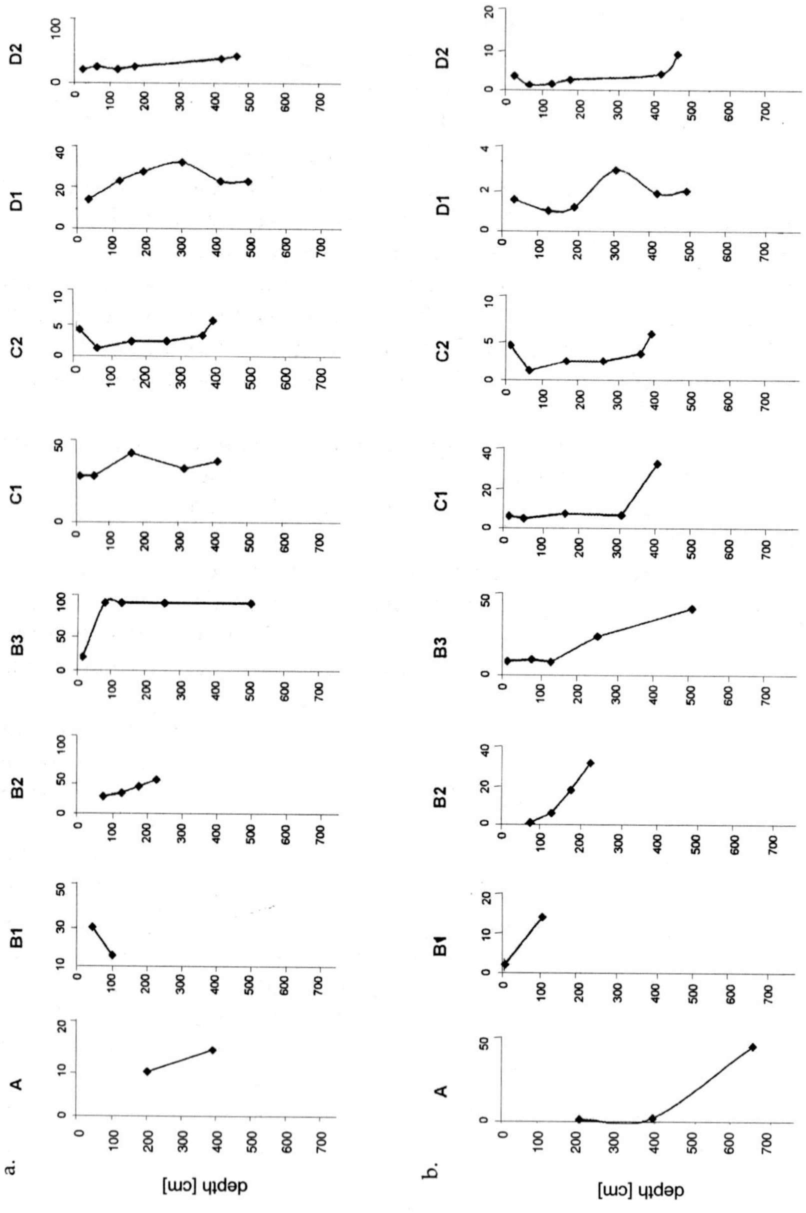

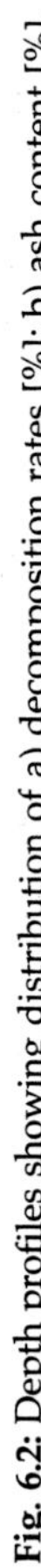

Fig. 6.2: Depth profiles showing distribution of a) decomposition rates [%]; b) ash content [%].

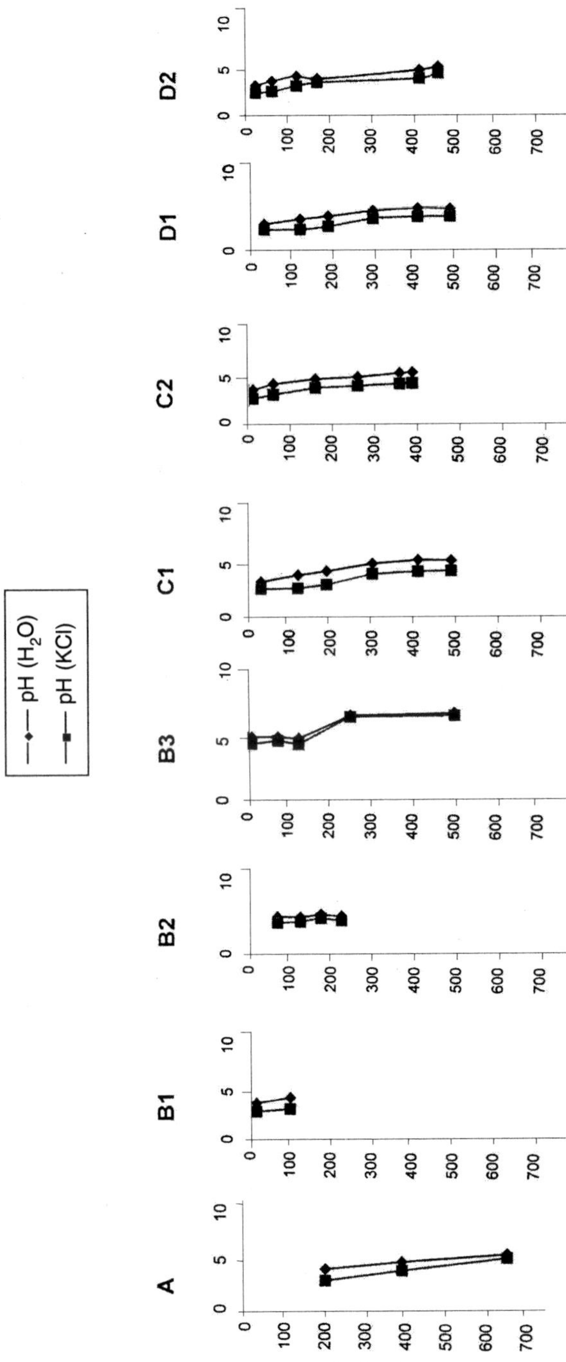

Fig. 6.3: Depth profiles showing pH distribution.

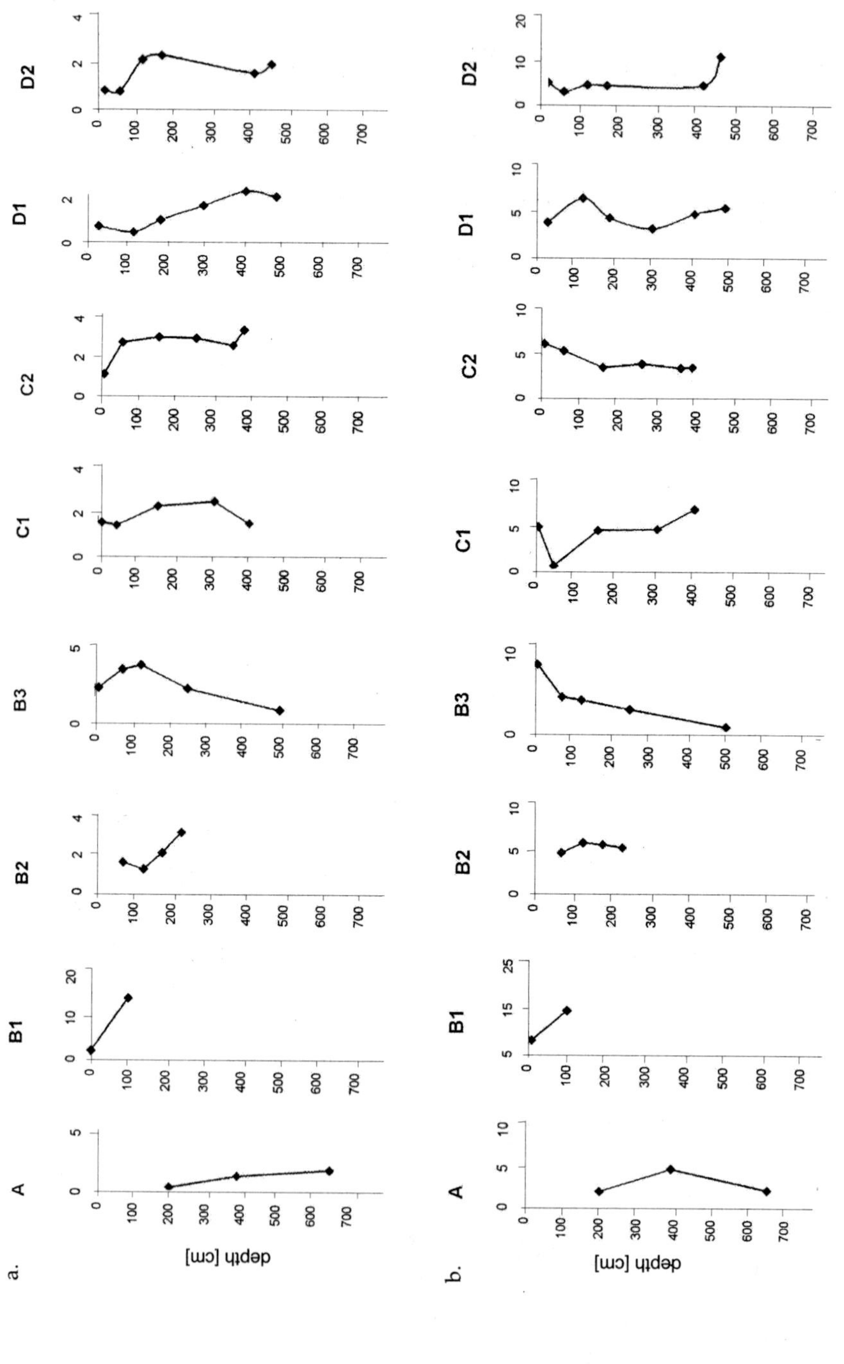

Fig. 6.4: Depth profiles showing distribution of a) C content [%]; b) N content [%].

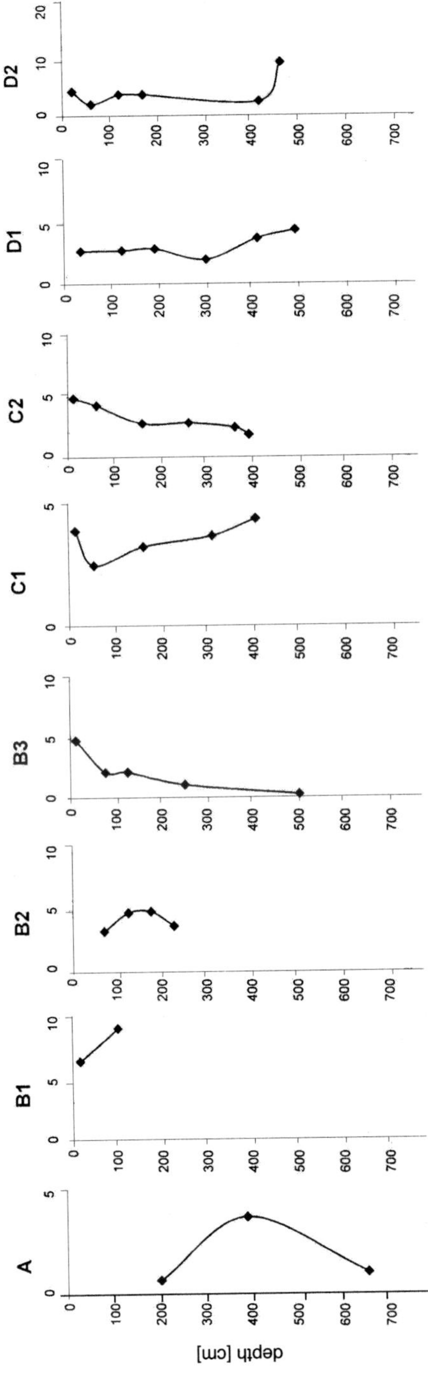

Fig. 6.5: Depth profiles showing distribution of C in humic substences [%].

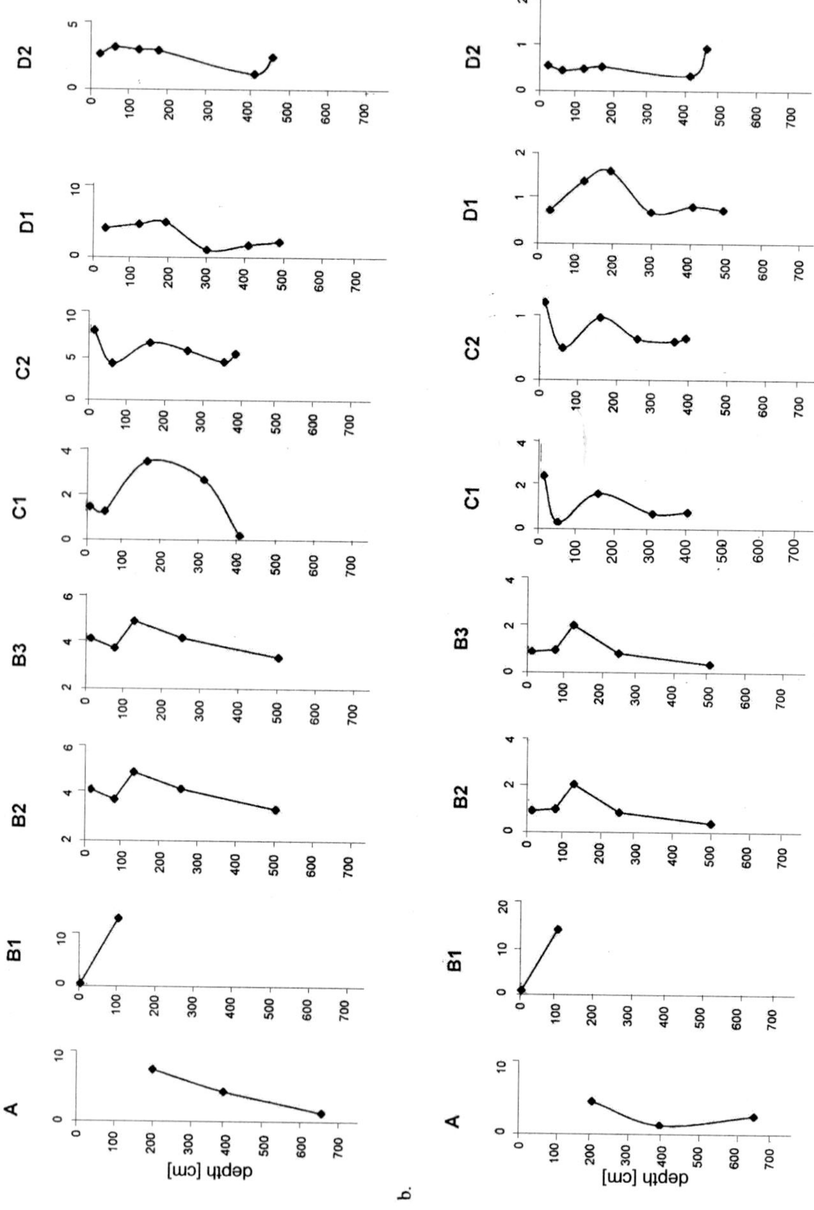

Fig. 6.6: Depth profiles showing distribution of a) Na content (mmol kg^{-1}); b) K content (mmol kg^{-1}).

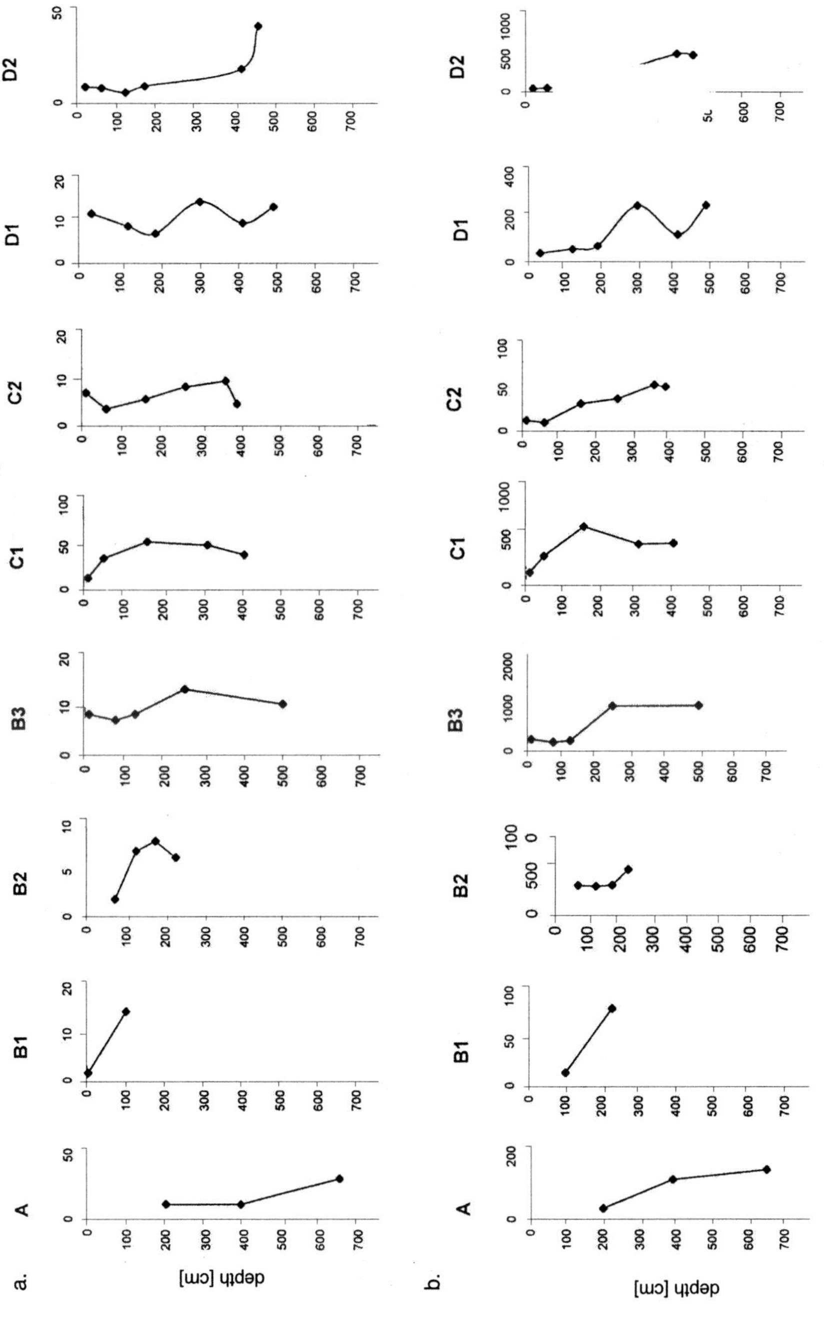

Fig. 6.7: Depth profiles showing distribution of: (a) Mg content [mmol kg^{-1}]; (b) Ca content [mmol kg^{-1}].

concentration in peat corresponds to the input of water (rain water or groundwater).

Table 6.2 presents the concentrations of 17 unsubstituted PAHs (16 anthropogenic PAHs and perylene) analyzed in peat samples from the four study sites. It was found that in all peat samples the level of 16 anthropogenic PAH concentrations was between 70 and 439 ng g^{-1}. Also, except the transitional bog Bagna Skieblewskie C2, the concentrations of the 16 PAHs in peat decreased with depth, reaching the level of 70–95 ng g^{-1} in the bottom layer. On site C2, the concentration of the 16 PAHs was high (173–327 ng/g^{-1}) throughout the profile and no decrease was found in depth gradient. Compared to peat samples, the levels of the 16 PAHs were much lower in gyttjas, especially in calcareous-clayey gyttja. The concentration of the 16 PAHs in algal gyttja from middle layers of the lowland bog Grądy Węgorzewskie B3 was about twice as high as that in calcareous-clayey gyttja, and varied between 190 and 205 ng g^{-1}.

The concentrations of PAHs with different number of rings are presented in Figure 6.8. Three-ring and 4-ring PAHs dominate in all peat and gyttja samples. Five-ring and 6-ring PAHs occurred only in top layers of peat bogs Grądy Węgorzewskie B2, B3, and Bagna Skieblewskie C1, C2 (immediately below the 5–20 cm surface layer) and at very low levels. Five-ring and 6-ring PAHs were not observed in the Żegary bogs (D1, D2).

Analysis involved 37 peat and gyttja samples. Of these, no perylene was found in top peat layers from sites A, C1, or C2, or in the calcareous-clayey gyttja of the lowland bog B3. In other samples, the perylene concentration varied between 6 and 957 ng g^{-1}, reaching the highest level in *Magnocaricioni* peat (CAP and CAR) at sites B2 and C1.

While studying an ombrotrophic peat bog in northwest England, Sanders et al. (1995) calculated PAH fluxes from air to peat and associated the high PAH concentration in peat at the level of about 30 cm with the industrial revolution in England. However, if we assume that the average growth rate of peat in a raised bog is 1 mm y^{-1}, then the deepest peat layers at site A (390–430 cm) are about 4,000–5,000 years old and the presence of PAHs in peat had ought to be attributed to local fires in peat bogs or surrounding forests or, perhaps, to chemical transformation of substances such as humus compounds. Wickstroem and Tolonen (1987) suspected that phenanthrene, methyl-phenanthrene, and methyl-anthracene may be produced biologically or come from the reduction of abietic and primaric acid, which occur, for example, in *Pinus* species.

Interestingly, the concentration of perylene in *Magnocaricioni* peat from sites B2 and C1 is very high. In their study of perylene concentration in lake sediments, Silliman et al. (1998) did not detect this compound in oxic surface sediments, finding it only in deeper sediments. The authors suggested that perylene is an indicator of diagenetic conditions and that its genesis may be process governed and independent of the type of organic matter within the sediment. Moreover, perylene may form from nonspecific precursor materials by microbial processes. It seems that the high perylene concentration in lower peat layers of the B2 and

Table 6.2: PAH content in various genera of peat [ng g^{-1}]

Site	Peat genus	Depth of peat [cm]	acenaphtylene	acenaphthene	fluorene	phenanthrene	anthracene	fluoranthene	pyrene	benzo[a] anthracene	chrysene	benzo[b] fluoranthene	benzo[k] fluoranthene	benzo[e] pyrene	benzo[a] pyrene	indeno[1,2,3-cd]pyrene	dibenzo[ah] anthracene	benzo[ghi] perylene	perylene	sum of 16 U.S.: EPAPAHs
A	cus	200-240	7	17	47	240	8	57	32	7	14	10	0	0	0	0	0	0	0	439
	ers	390-430	5	17	43	163	8	41	21	5	12	8	0	0	0	0	0	0	54	323
	detritial gyttja	650-740	0	7	15	37	0	10	5	0	0	0	0	0	0	0	0	0	125	74
B1	cab	015-080	3	8	18	74	3	26	14	0	10	8	0	0	0	0	0	0	76	164
	car	100-150	3	5	13	49	2	17	10	0	6	0	0	0	0	0	0	0	21	105
B2	cab	070-120	4	16	29	104	5	24	12	0	4	0	0	0	0	13	24	30	12	265
	car	125-170	4	9	21	68	2	22	13	0	5	0	0	0	0	0	0	0	140	144
	car	175-205	4	10	22	70	2	23	15	0	6	0	0	0	0	.0	0	0	957	152
	car	225-245	0	6	12	37	0	9	6	0	0	0	0	0	0	0	0	0	678	70
B3	cab	010—35	3	9	20	75	4	37	23	9	17	25	11	20	12	27	0	34	29	326
	algal gyttja	075-120	2	9	21	77	3	31	18	0	6	12	0	0	0	11	0	0	57	190
	algal gyttja	125-175	3	15	27	92	5	28	14	0	5	9	7	0	0	0	0	0	73	205
	calc.-clay.gyttja	250-275	0	5	13	48	0	13	6	0	0	0	0	0	0	0	0	0	0	85
	calc.-clay.gyttja	500-550	0	6	12	37	0	10	5	0	0	0	0	0	0	0	0	0	0	70
C1	spc	010-050	5	9	19	75	4	25	17	5	12	12	0	9	6	0	0	0	0	198
	cap	050-070	4	12	27	107	5	23	12	0	5	0	0	0	0	0	0	0	264	195
	cap	160-210	0	5	15	59	2	18	8	0	0	0	0	0	0	0	0	0	55	107

(Table 6.2 Contd.)

	car	310-360	2	5	17	67	4	21	11	0	0	0	0	0	0	0	0	0	6	127
	cap	405-430	0	6	15	55	0	12	7	0	0	0	0	0	0	0	0	0	662	95
C2	ers	010-050	0	8	19	82	0	39	25	10	25	27	13	14	9	16	0	15	0	302
	spc	060-110	0	18	27	104	3	43	28	11	17	13	8	9	9	0	0	0	11	290
	cab	160-200	0	22	34	122	3	27	14	0	0	0	0	0	0	0	0	0	0	234
	cab	260-300	0	14	24	95	2	22	11	0	0	0	0	0	0	0	0	0	50	173
	cab	360-390	0	17	35	147	3	34	17	4	9	9	0	6	9	16	0	21	18	327
	cab	390-420	0	13	28	107	3	32	17	4	7	9	7	0	0	0	0	0	9	227
D1	ers	030-070	6	14	35	134	6	33	17	0	8	0	0	0	0	0	0	0	0	253
	pis	120-160	4	9	28	101	4	21	12	0	5	0	0	0	0	0	0	0	18	184
	ers	190-220	26	23	55	179	8	44	23	0	7	0	0	0	0	0	0	0	38	365
	ers	300-350	9	16	32	111	4	27	14	0	7	0	0	0	0	0	0	0	506	220
	cab	410-460	7	16	42	157	6	40	20	0	21	0	0	0	0	0	0	0	140	309
	car	490-510	9	20	54	152	9	42	24	0	10	0	0	0	0	0	0	0	179	320
D2	ers	020-050	4	10	25	92	4	28	16	0	10	9	0	0	0	0	0	0	0	198
	ers	060-090	6	13	114	5	26	14	0	5	0	0	0	0	0	0	0	0	0	183
	cab	120-160	5	12	32	110	4	31	15	0	6	0	0	0	0	0	0	0	64	215
	cab	170-210	4	10	30	97	4	35	20	11	17	7	0	0	0	0	0	0	26	235
	aln	415-445	2	7	18	73	2	28	13	0	5	0	0	0	0	0	0	0	280	148
	aln	460-500	3	9	21	69	2	20	13	0	9	0	0	0	0	0	0	0	244	146

cus—Cuspidato-Sphagneti peat; cab—Carici-Bryaleti peat; ers—Eriophoro-Sphagneti peat; car—Cariceti peat;spc—Sphagno—Cariceti peat
Gyttja D—detritial gyttja; Gyttja A—algal gyttja; Gyttja CC—Calcareous-clayey Gyttja

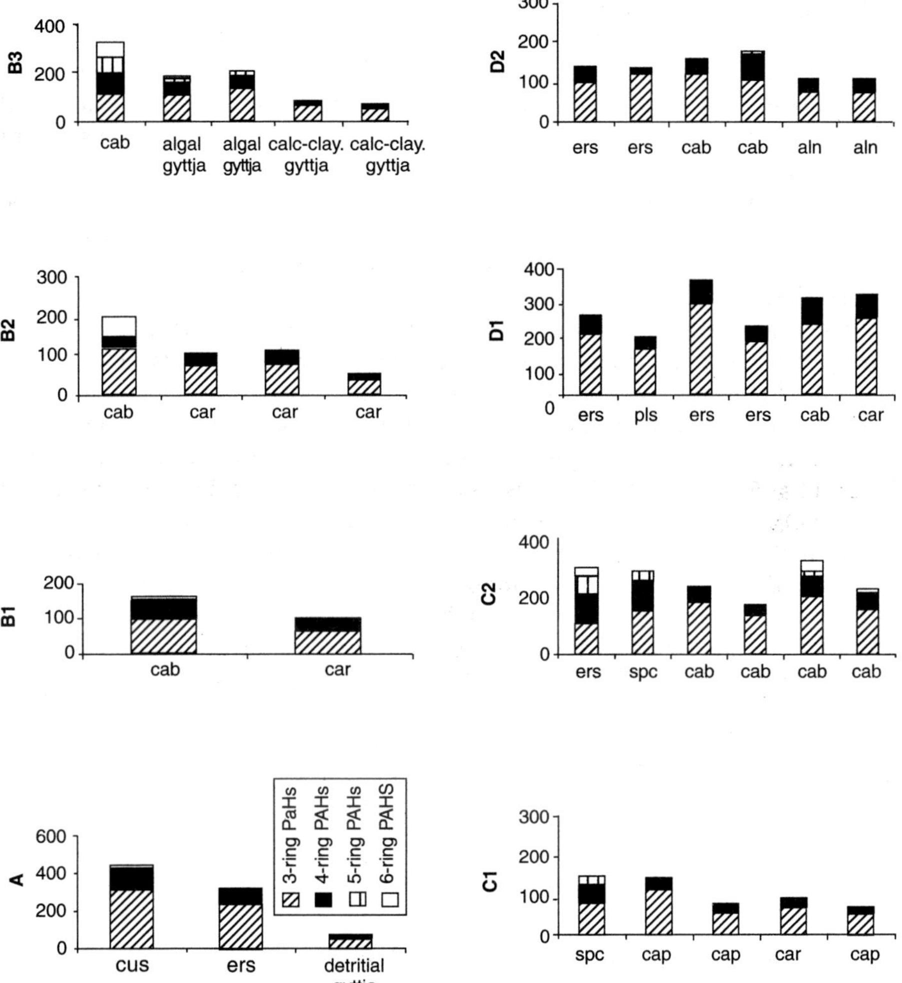

Fig. 6.8: Ratio of concentration of the 3, 4, 5 and 6-ring PAH ring number classes in the peat profiles

C1 bogs is due to their genesis and running water filtration processes during their early growth period.

Kulczyński (1939) claimed that in lowland bog profiles (the most common bog type on both sites) one could observe certain phenomena that indicated concurrence of peat bog growth with peat water swelling. Single plant associations have led to formation of thick layers of homogenous peat in lowland bogs (sites B2 and C2). The fact that a *Magnocaricioni* (B2) or a *Bryalo-Magnocaricioni* (C2) association "managed" to build up a 2-m-thick layer of homogenous peat could only have happened because the peat growth was accompanied by a gradual increase in groundwater level. Sedge peat and sedge-reed peat layers alternate at greater depths, while peat wood and transitional peat layers do so closer to the surface (C1). This indicates groundwater level changes in the initial phase of lowland bog plant succession and a subsequent change of the transitional bog into the forest stage (wood peat). The concentration of perylene in particular layers of peat at all study sites should, on the one hand, be associated with horizontal and vertical water level changes and, on the other, to the sorption of perylene created in river sediments by particular plant species.

River and lake sediments, as well as soils (also considered sediments), were dominated by 4-ring, 5-ring, and 6-ring compounds, mainly fluoranthene. Only in peat are high, dominant perylene levels detected.

No correlation was found between degree of peat decomposition or ash content and the PAH and perylene concentration in the peat samples analyzed. Chefetz et al. (2000) studied processes of perylene sorption by natural organic matter (NOM) in cuticule, humin, humic acid, degraded lignin, peat, and lignite. They found that the mechanism of perylene sorption by NOM depends on refractory aliphatic structures, aromatic components, and micropore structure. The amount of humic acids, organic carbon of humic substances, and nitrogen depends not only on the degree of decomposition, but also on the species of plants that undergo mineralization and humification processes in the top layers of the bog. It seems therefore that processes of perylene sorption from water during peat bog formation are more strongly influenced by hydrologic conditions, the type of plant association, and the sediments on which the peat bog grows than by the amount of compounds created during humification processes.

4 CONCLUSIONS

1. The total concentration of the 16 PAHs in all peat samples was between 70 and 439 ng g^{-1}.
2. Most of the PAHs were 3-ring compounds. Four-ring compounds were slightly less and 5-ring compounds were least common.
3. In some of the peat samples, the perylene concentration greatly exceeded the total concentration of the 16 PAHs.
4. The concentration of the 16 PAHs and perylene in peat was associated with hydrologic conditions, type of plant association, and sediment type during peat bog formation.

Acknowledgments

We acknowledge our indebtedness to the State Committee for Research for issuing grant No. 6 P04G 107 18, which made this work possible. We thank Geerd A. Smidt for his technical assistance.

References

Aamot E., Krane J., and Steinnes E. 1987. Determination of trace amounts of polycyclic aromatic hydrocarbons in soil. *Fresenius Z Anal. Chem.* 328: 569–571.

Boatman D. J. 1957. An ecological study of two areas of blanket bog on the Galway-Mayo peninsula, Ireland. *Proc. R. Ir. Acad.* Sect. B 59: 29–42.

Bracewell J.M., Hepburn A., and Thomson C. 1993. Levels and distribution of polychlorinated biphenyls on the Scottish land mass. *Chemosphere* 27: 1657–1667.

Bradley L., Magee B., and Allen S. 1994. Background levels of polycyclic aromatic hydrocarbons (PAH) and selected metals in New England urban soils. *J. Soil Contam.* 4: 1–13.

Chefetz B., Desmukh A.P., Hatcher P.G., and Guthrie E.A. 2000. Pyrene sorption by natural organic matter. *Environ. Sci. Tech.* 34: 2925–2930.

Coleman P., Lee R., Alcock R., and Jones K. 1997. Observation on PAH, PCB and PCDD/F trends in U.K. urban air, 1991–1995. *Environ. Sci. Tech.* 31: 2120–2124.

Damman A.W. H. 1978. Distribution and movement of elements in ombrotrophic peat bogs. *Oikos* 30: 480–495.

Gore A.J.P. and Allen S.E. 1956. Measurement of exchangeable and total cation content for H^+, Na^+, K^+, Mg^{2+}, Ca^{2+} and iron, in high level blanket peat. *Oikos* 7: 48–55.

Grosse-Bauckmann G. 1990. Ablagerungen der Moore. In: *Moor- und Torfkunde* K. Goettlich (ed.). E. Schweizerbart'sche Verlagsbuchhandlung. Berlin, Germany, pp. 175–236.

Harvey R.G. (ed.) 1985. *Polycyclic Hydrocarbons and Carcinogenesis.* Amer. Chem. Soc., Washington D.C.

Harvey R.G. 1998. Environmental chemistry of PAHs. In: *PAHs and Related Compounds*. A.H. Neilson (ed.) Springer-Verlag, Berlin, pp. 1–54.

Himberg K. K, and Pakarinen P. 1994. Atmospheric PCB deposition in Finland during the 1970s and 1980s on the concentration in ombrotrophic peat mosses (*Sphagnum*). *Chemosphere* 29: 431–440.

Howsam M. and Jones K. 1998. Sources of PAHs in the environment. In: *PAHs and Related Compounds*. A.H. Neilson (ed.). Springer-Verlag, Berlin, pp. 137–174.

Kononowa. M.M. 1968. *Substancje organiczne gleby, ich budowa, wlasciwosci i metody badan* PWRiL, Warszawa. (in Polish)

Kulczyński S. 1939. *Die Moore des Polesie-Gebietes*. Gebethner and Wolf, Kraków. (in Polish)

Lang G. 1994. *Quartaere Vegetationsgeschichte Europas. Methoden und Ergebnisse.* Gustav Fischer Verlag, Stuttgart.

Lichtfouse E., Budziñski H., Garrigues P., and Eglinton T. 1997. Ancient polycyclic aromatic hydrocarbons in modern soils:13C, 14C and biomarker evidence. *Org. Geochem.* 26, 353–359.

Maliszewska-Kordybach B. and Oleszek W. 1994. The use of high-performance liquid chromatography for determination of polycyclic aromatic hydrocarbons (PAH) in soil samples. *Acta Chromatogr.* 3: 84–93.

Neilson, A.H. (ed.) 1998. PAHs and related compounds. Springer-Verlag, Berlin, 137–174.

Ollivon D., Garbon B., and Chesterikoff A. 1995. Analyses of distribution of some polycyclic aromatic hydrocarbons in sediments and suspended matter in the river Seine (France). *Water, Air, Soil Pollut.* 81: 135–152.

Pereira W.E., Hostettler F.D., Luoma S.N., van Geen A., Fuller C.H., and Anima R. 1999. Sedimentary record of anthropogenic and biogenic polycyclic aromatic hydrocarbons in San Francisco Bay, California. *Marine Chem.* 64: 99–113.

Rapaport R.A. and Eisenreich S.J. 1986. Atmospheric deposition of toxaphene to eastern North America derived from peat accumulation. *Atmos. Environ.* 20: 2367–2379.

Sanders G., Jones K., Hamilton-Taylor J., and Doerr H. 1995. PCB and PAH fluxes to a dated UK peat core. *Environ. Pollut.* 89: 17–25.

Shotyk W. 1997. Atmospheric deposition and mass balance of major and trace elements in two oceanic peat bog profiles, northern Scotland and the Shetland Islands. *Chem. Geol.* 138: 55–72.

Silliman J.E., Meyers P.A., and Eadie J.M. 1998. Perylene: an indicator of alteration processes or precursor materials? *Org. Geochem.*, 29: 1737–1744.

Sumner M.E., and Miller W. P. 1996. Cation-exchange capacity and exchange coefficient. In: *Methods of Soil Analysis*. Part 3–*Chemical Methods*. D.L. Sporks (ed.), Soil Sci. Soc. Amer. Book Series 5, Madison, WI, USA, pp. 1201–1229.

Thiele S. and Brummer G.W. 2002. Bioformation of polycyclic aromatic hydrocarbons in soil under oxygen deficient conditions. *Soil Biol. Biochem.* 34: 733–735.

Tiurin I. W., and Tiurina E.I. 1940. Humus substances in peats. *Soil Sci.* no. 2 (in Russian).

Wickstroem K. and Tolonen K. 1987. The history of airborne polycyclic aromatic hydrocarbons (PAHs) and perylene as recorded in dated lake sediments. *Water, Air, Soil Pollut.* 32: 155–175.

Wilcke W., Muller S., Kanchanakool N., Niamskul C., and Zech W. 1999. Polycyclic aromatic hydrocarbons (PAHs) in hydromorphic soil of the tropical metropolis Bangkok. *Geoderma* 91: 297–309.

7

Polycyclic Aromatic Hydrocarbons Concentration in Peat Profiles and Plants Growing on Two Selected Raised Bogs in Poland

B. Wilkomirski*, *and* **M. Malawska**

Abstract

The geochemical parameters and the concentrations of polycyclic aromatic hydrocarbons (PAHs) were analyzed in samples of peat collected from two different peatlands (Zdory and Wolosate) located respectively in the lake district and the mountains of Poland. The concentration of PAHs was also analyzed in samples of four plant species growing on these peatlands. The total concentration of 16 anthropogenic PAHs in all peat samples was between 120 and 299 ng g^{-1}. A markedly high perylene concentration was found in wood peat (*Pineti*) from the Wolosate raised bog. The concentrations of PAHs in plants growing on this bog greatly exceeded the concentrations of PAHs in peat, suggesting anthropogenic pollution of this area.

1 INTRODUCTION

Polycyclic aromatic hydrocarbons (PAHs, polyarenes) are an important group of organic compounds that consist only of hydrogen atoms and sp^2 hybridized carbon atoms in two or more aromatic carbon rings, although some PAHs can also be methylated (Harvey, 1997). Polycyclic aromatic hydrocarbons play an important role not only in many areas of scientific and industrial activity, but also in environmental toxicology. Isolation of benzene and toluene, as well as some polyarenes, from coal tar in the early part of the 19th century began a

**Corresponding author:* Dr. B. Wilkomirski, Dept. Plant Systematics and Geography, Warsaw University, 00-478 Warszawa, Al, Ujazdowskie 4, Poland. E-mail: bowi@biol.uw.edu.pl period of great interest in this compound group. The number of papers dealing with PAHs rapidly grew when people realized that some polyarenes possessed highly toxic and carcinogenic activity. By the late 20th century it was clear that PAHs were present (sometimes in great abundance) in various ecosystems. In recent years many studies have been published concerning levels of PAHs in various biological materials, as well as their migration mechanisms (Gardner et al., 1995; Jones et al., 1995; Wilcke et al., 1996; Wilcke, 2000).

PAHs are important and well-known pollutants that have been identified in diverse environmental matrices worldwide. They are of environmental concern not only because of their toxicity and carcinogenicity, but also because they are very resistant to decomposition. These compounds are included in a group of derivatives termed persistent organic pollutants (POPs). They also have possible harmful effects on soil organisms and plants (Menzie et al., 1992).

PAHs can be produced by the incomplete combustion of compounds containing carbon and hydrogen or during some biotransformation processes. Natural hydrocarbons are generally encountered at trace levels, whereas anthropogenic hydrocarbons originating from various sources are found at higher levels, particularly in areas associated with industrial and transportation activities (Malawska and Wilkomirski, 2000; Niederer et al., 1995). The main anthropogenic sources of PAHs in the environment include oil refineries, the processing of mineral fuels in coke plants, and the transport and storage of liquid fuels, as well as combustion of coals and liquid fuels (Behymer and Hites, 1988; Harvey, 1998; Howsam and Jones, 1998).

PAHs do not originate only from anthropogenic sources. They can be formed during geochemical and biochemical processes or synthesized during volcanic eruptions and forest, bush, and peat fires (Evans et al., 1990; Capaccioni et al., 1995; Jiang et al., 1998; Koziński and Saade, 1998; Harvey, 1998; Howsam and Jones, 1998). A reasonable amount of polyarenes are present in environments with little anthropogenic pollution and in rocks and sediments. Many sediments of organic origin have been formed by the alteration of accumulated plant material, in which PAHs have formed. Their precursors were mainly terpenoids; for example, diterpenoids may change to pimanthene or retene, while triterpenoids may change to dimethylchrysene or trimethylpicene. Hopanoids (common prokaryotic triterpenoides) and steroids may also be the precursors of polyarenes (Smith et al., 1995; Neilson and Hynning, 1998; Simoneit, 1998). PAHs are present in small quantities in all subaqueous sediments, both recent and fossil (Pereira et al., 1998).

Polyarenes are also found in brown coal (Chaffee and Johns, 1983; Wang and Simoneit, 1991; Bojakowska and Sokolowska, 2001a), hard coals (Bojakowska and Sokolowska, 2001b), and crude oil coals (Bence et al., 1996; Bojakowska and Sokolowska, 2001c). However, the literature contains few papers concerning the presence of PAHs in peat (Sanders et al., 1995; Bojakowska et al., 2000; Berset et al., 2000; Malawska et al., 2002).

Peat is usually defined as a young Quaternary organogenic sedimentary rock in the first stage of coalification. The abundant plant residue present in peat is a site for biochemical transformations. Some of these transformations can be a source of polyarenes, e.g., biogeochemical conversion of steroids and triterpenes.

In Poland most peat bogs are localized in regions with little or very little anthropogenic pollution. One can suppose that the PAHs content in peat in these bogs derives mostly from natural sources. The aim of this study was to establish the concentrations of 16 main PAHs from the U.S. Environmental Protection Agency's (U.S. EPA) list (except for naphthalene), plus perylene, in peat from two selected peat bogs and plants overgrowing these peat bogs.

2 MATERIALS AND METHODS

2.1 Study Sites

Although peatlands and other wetland types are a common feature of the Polish landscape, their distribution across Poland is not regular. Generally, the number of peatlands decreases from north to south. In spite of this tendency there are some interesting peatlands in various areas of the mountain range located on the southern border of Poland. Overall, Poland has more than 50,000 peatlands covering a total surface area of more than 13,000 km^2.

Peat and plant samples were collected in two selected raised bogs, Zdory and Wolosate. The Zdory raised bog is situated in northeastern Poland (Masurian District) in areas created by Baltic glaciation and covered by many lakes and forests. Compared to the rest of Poland, the Masurian District is characterized by a considerable variety of site conditions and a high degree of largely undisturbed ecosystems. In contrast, the Wolosate raised bog is located in the Bieszczady Mountains (part of the Carpathian Mountains in the southeastern Poland) at an altitude of about 800 m. The Bieszczady mountain range is a national park with a very high percentage of forest land cover up to 90%. In spite of governmental protection, the Bieszczady Mountains experience considerable tourist activity and are the site for many small charcoal preparation plants.

The Zdory and Wolosate peat bogs are mapped in Fig. 7.1.

2.2 Soil and Plant Sampling

Peat and plant samples were collected in June 2001. Points for collection of biogenic sediment cores (with the Russian sampler, container length 50 cm) (Lang, 1994) were selected on the basis of geological documentation showing the geomorphology and hydrology of peat deposit. Peat samples were collected from various depths of the stratigraphic profile of the peat deposits. The peat samples were stored at –20°C until preparation for analysis was undertaken.

Plant samples (Scotch pine needles—*Pinus silvestris,* Marsh Labrador Tea leaves—*Ledum palustre,* swamp valerian leaves—*Vaccinium uliginosum,* and Cotton-grass leaves—*Eriophorum* sp.) were collected from each sampling station.

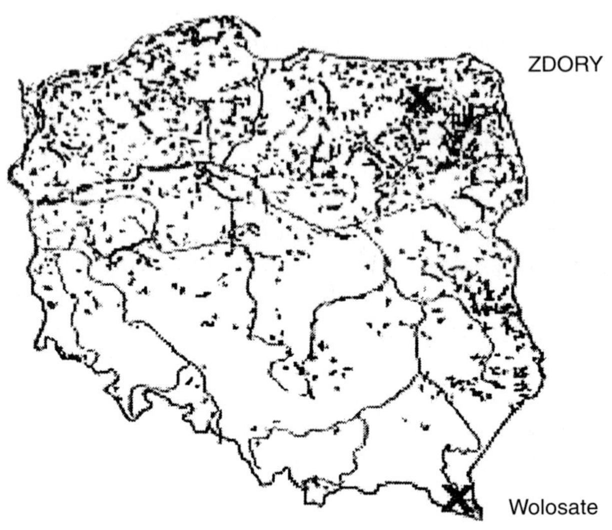

Fig. 7.1: Main wetlands in Poland and localization of the study area.

Each sampling station was represented by a pool of leaves (needles) collected from several trees or plants near each other. The plant samples (about 50 g for each station) were wrapped in aluminum foil.

2.3 Identification of Polycyclic Aromatic Hydrocarbons (PAHs) in Peat and Plants

The peat and plant samples were dried at room temperature for 2 days, then at 50°C for 2 to 3 hours, sieved through a 2-mm mesh to remove large particles and (in the case of peat) organic debris, and stored at 5°C prior to analysis.

Liquid extraction was performed with dichloromethane in a Soxtec apparatus. Further purification was carried out by column chromatography on Florisil. The quantitative analysis was performed using the external standard method in which the certified PM-612 standard (ULTRA Scientific Ltd.) was applied (Colmsjö, 1998).

PAH content analysis in plant and peat material was undertaken using a gas chromatograph (5890 II) equipped with a mass selective detector (GC/MSD Hewlett-Packard) and a nonpolar capillary column HP-5 (length 24 m, diameter 0.2 mm, 0.33 µm diphenyl—95 % dimethylpolysiloxane film). The following temperature programming was applied: 70°C at 10°C min^{-1} to 200°C, at 2.5°C min^{-1} to 300°C (7 min). The detector temperature was 280°C. The detection limit was ca. 0.1 ng g^{-1} dry weight, assuming 10 g of sample.

The following U.S. EPA PAHs were determined: acenaphthene (ACE), acenaphthylene (ACY), fluorene (FLU), phenanthrene (PHE), anthracene (ANT), fluoranthene (FLA), pyrene (PYR), benzo[a]athracene (BAA), chrysene (CHRY), benzo[b]fluoranthene (BBFLA), benzo[k]fluoranthene (BKFLA), benzo[a]pyrene (BAP), benzo[e]pyrene (BEP), indeno[1,2,3-cd]pyrene (IND), dibenzo[ah] anthracene (DBAH), and benzo[ghi]perylene (BGHI), as well as perylene (PER) (Malawska et al., 2002).

2.4 Quality Assurance and Quality Control (QA/QC)

The procedures described above were checked for recoveries and reproducibility. Extraction was tested prior to extracting by spiking Scots pine needles, swamp valerian, cotton-grass, and Marsh Labrador Tea leaves, as well as peat samples, with four increasing amounts of standards. For all the compounds analyzed, recovery results were in the range 81 to 96%. Ten percent of samples were extracted and analyzed in duplicate. After analysis of every 10 samples, standard samples with a known PAH content were analyzed. A solvent blank was analyzed after every 20 samples as a check on the response of gas chromatography.

3 RESULTS AND DISCUSSION

The type of peat deposit and selected parameters of stratigraphy and chemical characteristics are presented in Table 7.1.

Table 7.2 presents the content of 17 unsubstituted PAHs (16 anthropogenic PAHs and perylene) analyzed in peat samples from two study sites. It was found that in all Zdory peat samples the level of 16 EPA PAHs varied from 183 to 273 ng g^{-1}. At Wolosate, the level of these same 16 PAHs varied from 181 to 299 ng g^{-1}. At Zdory perylene was absent in high-moor (*Eusphagneti*) and (*Eriophoro-Sphagneti)* peat, whereas the perylene content in reed-sedge peat (*Carici-Phragmiteti*) was low (93 and 22 ng g^{-1}, respectively). At Wolosate perylene was absent in *Eusphagneti* and *Eriophoro-Sphagneti* peat. An unusually high level of perylene (2292 ng g^{-1}) was found in wood peat (*Pineti*) at Wolosate. Predominant among all peat samples were PAHs with 2 and 3 rings.

Table 7.3 presents the concentrations of 17 unsubstituted PAHs (16 anthropogenic PAHs and perylene) analyzed in plant samples from the two study sites. Three species of plants were collected at Zdory (*Ledum palustre, Eriophorum* sp., *Vaccinium uliginosum*), and four species of plants at Wolosate, (*Ledum palustre, Eriophorum* sp., *Vaccinium uliginosum*, and *Pinus sylvestris*). Results show that the content of PAHs in plants growing on the peat bog studied at Wolosate largely exceeds the content of PAHs in peat at that site. It is significant that benzo[a]pyrene (Bap), which is carcinogenic to humans, is absent in peat from both of the studied peat bogs as well as in plants growing on Zdory, but present in plants from Wolosate. This is probably due to combustion sources from tourist and sport centers located close to the peat bog. It is essential to know the number and intensity of sources of PAHs in determining absolute PAH concentration in the leaves of plants.

Table 7.1: Selected characteristics of peat in the study areas

Site	Depth of peat [cm]	Peat genus	pH [water]	pH [KCl]	Degree of decomposition [%] (Grosse-Baukmann,1990)	Ash content [%]	N content [%]	C content [%]	Mg content [g/100 g soil]	Ca content [g/100 g soil]
Zdory	050-100	eus	3.79	2.94	30	1.97	1.27	25.2	0.013	0.411
	150-200	cap	5.48	4.65	25	3.38	2.18	17.2	0.058	0.78
	250-300	gyttja	5.77	5.11	not detected	5.56	3.31	12.9	0.063	0.789
Wolosate	010-050	ers	4.73	3.62	35	7.62	1.64	24	0.022	0.107
	110-160	eus	4.75	3.28	25	2.06	0.98	17.7	0.017	0.103
	200-250	ers	4.78	3.1	40	2.15	1.09	18.2	0.012	0.132
	280-320	eus	4.5	2.99	55	2.74	1.77	24.6	0.01	0.094
	350-390	pin	4.59	3.09	65	14.24	1.78	25.6	0.014	0.151

ers—*Eriophoro-Sphagneti* peat; eus—*Eusphagneti* peat; cap—*Carici-Phragmiteti* peat; pin—*Pineti* peat

Table 7.2: Content of PAHs [ng g^{-1}] in peat from wetlands

Site	Peat genus	ACE	ACY	FLU	PHE	ANT	FLA	PYR	BAA	CHRY	BBFLA	BKFLA	BEP	BAP	IND	DBAH	BGHI	PER	Sum of 16 US EPA PAHs
Zdory	eus	7	17	45	142	6	32	15	0	9	0	0	0	0	0	0	0	0	273
	cap	4	9	24	113	5	36	17	0	5	0	0	0	0	0	0	0	93	213
	gyttja	2	6	18	92	6	35	18	0	6	0	0	0	0	0	0	0	22	183
Woolosate	ers	5	11	28	89	4	34	20	6	12	14	6	7	0	11	0	0	0	247
	eus	7	13	36	117	5	35	19	5	11	11	0	7	0	0	0	0	29	266
	ers	7	15	43	127	6	37	21	6	13	10	0	7	7	0	0	0	7	299
	eus	6	13	31	82	4	20	11	0	7	7	0	0	0	0	0	0	0	181
	pin	3	11	19	49	0	12	7	0	9	10	0	0	0	0	0	0	2292	120

eus—*Eusphagneti* peat; cap—*Carici-Pharagmiteti peat;* ers—*Eriophoro-Sphagneti* peat; pin—*Pineti* peat

Table 7.3: Content of PAHs [ng g^{-1}] in plant species growing at investigated peat-bogs

Site	Plant genus	ACE	ACY	FLU	PHE	ANT	FLA	PYR	BAA	CHRY	BBFLA	BKFLA	BEP	BAP	IND	DBAH	BGHI	PER	Sum of 16 US EPA PAHs
Zdory	*L.p.*	4	19	25	107	0	73	38	9	30	0	0	0	0	0	0	0	0	305
Wolosate	*L.p.*	4	31	786	380	17	599	473	200	357	371	203	259	212	186	0	265	0	4343
Zdory	*E.sp.*	0	9	11	46	0	11	6	0	0	89	0	0	0	0	0	0	0	83
Wolosate	*E.sp.*	0	15	23	156	5	95	75	39	72	89	52	64	68	48	0	50	0	851
Zdory	*V.ull.*	4	15	22	104	5	29	15	0	9	0	0	0	0	0	0	0	0	203
Wolosate	*V.ull.*	8	20	34	255	12	351	293	18	260	277	186	0	56	158	31	154	0	2113
Wolosate	*P.s.*	15	29	71	205	13	360	306	127	227	288	112	218	227	188	34	241	60	2661

L.p.—Ledum palustre; E.sp.—Eriophorum sp.; *V.ull.—Vaccinium ulliginosum; P.s.—Pinus silvestris*

A very high content of perylene was found in wooden peat (*Pineti*) collected from the bottom of the Wolosate peat bog. Perylene is a 5-ring PAH that is usually found in water sediments (Silliman et al., 1998). Wolosate peat bog was mainly supported by water from a small mountain river during formation; therefore, the concentration of perylene in the bottom layer of peat could be associated with water sediments.

4 CONCLUSIONS

In summarizing the above results, it should be stated that the content of U.S. EPA PAHs in different samples of peat qualifies them as unpolluted soil with increased content (second pollution class as designated by EPA). The PAHs level in plants growing on the studied peat bogs was generally much higher than in peat, suggesting anthropogenic pollution. This is much more significant at Wolosate, probably due to the neighboring tourist center. In wooden peat (*Pineti*) the perylene content largely exceeds the total content of EPA PAHs.

Acknowledgments

We are gratefully indebted to the State Commeette for Scientigic Research for grant No. P04G 107 18, which made this work possible.

References

Behymer T. and Hites R. 1988. Photolysis of polycyclic aromatic hydrocarbons adsorbed on fly ash. *Environ. Sci. Tech.* 22 (11): 1311–1319.

Bence A., Kvenvolden K., and Kennicutt M. 1996. Organic geochemistry applied to environmental assessment of Prince William Sound Alaska, after the Exxon Valdez oil spill—a review. *Org. Geochem.* 24 (1): 7–42.

Berset J.D., Kuehne P., and Shotyk W. 2000. Concentration and distribution of some polychlorinated biphenyls (PCBs) and polycyclic aromatic hydrocarbons (PAHs) in an ombrotrophic peat bog profile of Switzerland. *Sci. Tot. Environ.* 267: 67–85.

Bojakowska I., and Sokolowska G. 2001a. Polycyclic aromatic hydrocarbons in brown coals from Poland. *Geol. Quart.* 45(1): 93–98.

Bojakowska I. and Sokolowska G. 2001b. Polycyclic aromatic hydrocarbons in hard coals from Poland. *Geol. Quart.* 45 (1): 87–92.

Bojakowska I. and Sokolowska G. 2001c. Polycyclic aromatic hydrocarbons in crude oils from Poland. *Geol. Quart.* 45 (1): 81– 86.

Capaccioni B., Martini M., and Mangani F. 1995. Light hydrocarbons in hydrothermal and magmatic fumaroles; hints of catalytic and thermal reactions. *Bull. Volcanol.* 56(8): 593–600.

Chaffee A. and Johns R. 1983. Polycyclic aromatic hydrocarbons in Australian coals. I. Angularly fused pentacyclic tri- and tetraaromatic components of Victorian brown coal. *Geochim. Cosmochim. Acta* 47: 2141–2155.

Colmsjö A. Concentration and extraction of PAHs from environmental samples. In*: PAHs and Related Compounds*. A.H. Neilson (ed.). Springer-Verlag, Berlin, pp. 55–76.

Evans K.M., Gill R.A., and Robotham P.W. 1990 The source, composition and flux of polycyclic aromatic hydrocarbons in sediments of the River Derwent, Derbyshire, UK. *Water, Air, Soil Pollut.* 51: 1–12.

Gardner B., Hewitt C., and Jones K.C. 1995. PAHs in air adjacent to two inland water bodies. *Environ. Sci. Tech.* 29: 2405–2413.

Harvey R. 1997. *Polycyclic Hydrocarbons*. Wiley-VCH, New York, NY.

Harvey R. 1998. Environmental chemistry of PAHs. In*: PAHs and Related Compounds:* A.H. Neilson (ed.). Springer-Verlag, Berlin, pp. 1–54.

Howsam M. and Jones K. 1998. Sources of PAHs in the environment. In: *PAHs and Related Compounds*: A.H. Neilson (ed.). Springer-Verlag, Berlin. pp. 137–174.

Jiang C., Aleksander R., Kagi R., and Murray A. 1998. Polycyclic aromatic hydrocarbons in ancient sediments and their relationships to paleoclimate. *Org. Geochem.*, 29 (5–7): 1721–1735.

Jones K.C., Johnston A.E., and McGrath S.P. 1995. The importance of long- and short-term air-soil exchanges of organic contaminants. *Int. J. Environ. Anal. Chem.* 59: 167–178.

Koziński J. and Saade R. 1998. Effect of biomass burning on the formation of soot particles and heavy hydrocarbons. An experimental study. *Fuel* 77(4): 225–237.

Lang G. 1994. Quartäre Vegetationsgeschichte Europas. Methoden und Ergebnisse. Gustav Fischer Verlag, Jena Stuttgart, New York, NY.

Malawska M. and Witkomirski B. 2000. An analysis of soil and plant (*Taraxacum officinale)* contamination with heavy metals and polycyclic aromatic hydrocarbons (PAHs) in the area of the railway junction Ilawa Glówna, Poland. *Water, Air and Soil Pollut.* 127: 339–349.

Malawska M., Bojakowska I., and Witkomirski B. 2002. Polycyclic aromatic hydrocarbons (PAHs) in peat and plants from selected peat-bogs in the north-east of Poland. *J. Plant. Nutr. Soil Sci.* 165: 686–691.

Menzie C.A., Potocki B.B., and Santodano J. 1992. Exposure to carcinogenic PAHs in the environment. *Environ. Sci. Tech.* 26: 1278–1283.

Neilson A., Hynning P. 1998. PAHs: Products of chemical and biochemical transformation of alicyclic precursors. In*: PAHs and Related Compounds*.A.H. Neilson (ed.). Springer-Verlag, Berlin, pp. 223–269.

Niederer M., Maschka-Selig, A., and Hohl, Ch. 1995. Monitoring polycyclic aromatic hydrocarbons (PAHs) and heavy metals in urban soil, compost and vegetation. *Environ. Sci. Pollut. Res.* 2: 83–89.

Pereira W.E., Hostettler F.D., Luoma S.N., van Geen A., Fuller C.C., and Anima R.J. 1998. Sedimentary record of anthropogenic and biogenic polycyclic aromatic hydrocarbons in San Francisco Bay, California. *Marine Chem.* 64: 99–113.

Sanders G., Jones K., Hamilton-Taylor J. and Doerr H. 1995. PCB and PAH fluxes to a dated UK peat core. *Environ. Pollut.*, 89: 17–25.

Silliman J.E., Meyers P.A., and Eadie B.J. 1998. Perylene: an indicator of alteration processes or precursor materials? *Org. Geochem.*, 29: 1737–1744.

Simoneit B. 1998. Biomarker PAHs in the environment. In*: PAHs and Related Compounds*. A.H. Neilson (ed.). Springer-Verlag, Berlin, pp. 175–215.

Smith J., George S., and Batts B. 1995. The geosynthesis of alkylaromatics. *Org. Geochem.*, 23(1): 71–80.

Wang T., and Simoneit B. 1991. Organic geochemistry and coal petrology of Tertiary brown coal in the Zhoujing mine, Baise Basin, South China. *Fuel* 70(6): 819–827.

Wilcke W. 2000. Polycyclic aromatic hydrocarbons (PAHs) in soil—a review. *J. Plant Nutr. Soil Sci.* 163: 229–248.

Wilcke W., Baumler R., Deschauer H., Kaupenjohann M., and Zech W. 1996. Small scale distribution of Al, heavy metals and PAHs in an aggregated Alpine Podzol. *Geoderma* 71: 19–30.

Part III
Nitrogen, Phosphorus, Sulfur and Boron

8

Catalysis of the Maillard Reaction by δ-MnO_2: A Significant Abiotic Sorptive Condensation Pathway for the Formation of Refractory N-Containing Biogeomacromolecules in Nature

A. Jokic, H.-R. Schulten, J.N. Cutler, M. Schnitzer, *and* **P.M. Huang***

Abstract

Although mineral colloids are known to play a significant role in transforming organic matter in soils and sediments, there are still many gaps in our understanding of the mechanisms of organic-mineral interactions. The global nitrogen cycle, second in importance only to the carbon cycle, is of prime importance in natural ecosystems. However, the origin and nature of up to one-half total soil nitrogen remains either unknown or poorly understood despite all attempts to elucidate its nature. The Maillard reaction (condensation reaction between sugars and amino acids) is considered a major pathway in natural humification. This study investigated the role of a major oxide-mineral, δ-MnO_2 (a form of Mn(IV) oxide), in catalyzing the condensation reaction between sugars and amino acids for forming N-containing biogeomacromolecules. Using a suite of spectroscopic methods (including ^1H and ^{13}C NMR, Py-FIMS Py-GC/MS, and N-XANES) our data provide, for the first time, unequivocal evidence that the promoting action of Mn (IV) oxide on the Maillard reaction under ambient conditions results in the abiotic formation of two of the most important naturally occurring groups of organic nitrogen compounds, i.e., heterocyclic nitrogen compounds which are often referred to as *unknown* nitrogen, and amides which

**Corresponding author:* Dr. P.M. Huang, Dept. of Soil Science, University of Saskatchewan, 51 Campus Drive, Saskatoon SK S7N 5A8, Canada. E-mail: huangp@sask.usask.ca are apparently the dominant nitrogen moieties in nature. The information presented is of fundamental significance in understanding the role of mineral colloids in abiotic transformations of organic nitrogen moieties, the incorporation of nitrogen in the organic matrix of fossil fuels, and the global nitrogen cycle.

1 INTRODUCTION

Humification is one of the most important processes in the carbon cycle. It involves conversion of dead and decaying plant and animal remains into soil organic matter or humus (Wershaw, 1994). The dead biomass can be mineralized or transformed by enzymatic and/or nonenzymatic degradation reactions into a series of complex organic compounds (Bollag et al., 1998). The chemical composition of soil organic matter is estimated to consist of: carbohydrates 10%; N-components (including proteins, peptides, amino acids, purines, pyrimidines, and heterocyclics) 10%; lipids (alkanes, alkenes, fatty acids and esters) 10%, and humic substances approximately 70% (Schnitzer, 1991). These are average values for agricultural soils and may vary in specific environments.

Humic substances are a group of extremely complex, heterogeneous organic compounds formed from decomposing plant and animal detritus (Hatcher et al., 1980; Hedges 1988). They are important because they participate in numerous environmental interactions (Müller-Wegener 1988), form an integral part of the most important biogenic element cycles, and affect the toxicity and mobility of metals through complexation (Chapman and Kimstach, 1992). Abiotic processes are important in soil and related environments. These reactions are catalysed by various crystalline and noncrystalline soil minerals, metal ions, and sunlight (photocatalysis) and involve the formation of organomineral complexes, as well as the release of nutrients and toxic substances and their retention (Bartlett, 1986; Wang et al., 1986; McBride, 1994; Huang, 1995; Smolen and Stone, 1998).

Mayer (1994) examined the worldwide relationships between mineral specific surface area and organic carbon concentration for sediments and soil A horizons. He found that most of the mineral surface area lies present within small pores and that the organic coatings on the mineral grains likely reside on adsorption sites within these small pores. Small pores or micropores are defined by IUPAC as < 2 nm in diameter (Guggenberger and Haider, 2002). Low molecular mass organics can readily enter these small pores of matrix soils and sediments; contrarily microbes and enzymes cannot enter due to steric hindrance (Guggenberger and Haider, 2002). Van Veen and Kuikman (1990) calculated that 95% of the pore space in a silt loam soil is not accessible to bacteria. This restricted accessibility is extremely important for the protection of soil organic matter from decomposition (Elliot and Coleman, 1988), since adsorbed organic materials are thereby protected against microbial and enzymatic attack (Baldock and Skjemstad, 2000; Kaiser and Guggenberger, 2000). This sorption protection can thus account for the enigmatic preservation of intrinsically labile molecules such as amino acids and simple sugars in marine deposits (Keil et al., 1994). Collins et al. (1995) suggested that the work done by Mayer (1994) and Keil et al. (1994) indicated that, at least in the case of marine sediments, organic matter was sorbed to the surface of mineral grains and suggested that two tandem processes, i.e., adsorption and condensation, were involved. Adsorption on mineral surfaces of dissolved organic compounds also implies a mineral surface condensation model for the formation of amorphous kerogen, the insoluble macromolecular material that forms the bulk of organic matter in sedimentary rocks (Kennedy et al., 2002).

Of all the short-range ordered metal oxides, manganese oxides demonstrate by far the highest catalytic effect (Shindo and Huang, 1982, 1984; Huang 2000), being able to oxidize many environmentally important organic compounds and inorganic ions that include pollutants toxic to living organisms. Manganese oxide minerals are ubiquitous in terrestrial and aquatic environments (McKenzie, 1989) and are considered the most important redox-active minerals found in many soils and sediments (Risser and Bailey, 1992).

Enormous scientific effort has been committed to researching transformations of nitrogen in nature. The global nitrogen cycle is second in importance only to the carbon cycle and is of prime importance in terrestrial and aquatic ecosystems (Stankiewicz and van Bergen, 1998). Organic nitrogen compounds comprise a major reservoir of nitrogen on the Earth (Bada, 1998) and represent more than 90% of the nitrogen in soils and sediments (Stankiewicz and van Bergen, 1998; Stevenson, 1994). However, one of the most significant problems related to the nitrogen cycle is the source of the forms of organic nitrogen. As a result of a series of very complex reactions in the soil or related geochemical environments, the original biogenic, organic nitrogen compounds were transformed into heterogeneous macromolecules with obscure geochemical histories, ages and sources (Bada, 1998). Indeed, the origin and nature of approximately one-third to one-half of total soil nitrogen remains either *unknown* or poorly understood (Anderson et al., 1989; Schnitzer 2000; Schulten and Leinweber, 2000) despite all attempts to elucidate its nature. Detailed information on this *unknown* organic nitrogen is required in order to gain an understanding of nitrogen fluxes and the impact on ecosystem health. However, research on *unknown* nitrogen has been severely hampered by the absence of suitable methods that could provide definitive information on these nitrogen moieties (Stankiewicz and van Bergen, 1998).

Maillard (1912, 1913) first investigated the formation of melanoidins produced by refluxing solutions of glucose and lysine. The initial reaction in the Maillard reaction involves condensation between the carbonyl group of a reducing sugar and the amino group from an amino acid with formation of the Schiff base (Fig. 8.1). The Schiff base then undergoes the Amadori or Heyn's rearrangement producing reactive intermediates such as 1- or 3-deoxyglycosones. These compounds can undergo retroaldolization reactions (Ho, 1996) forming reactive α-dicarbonyl and α-hydroxyketone compounds such as pyruvaldehyde, diacetyl, and hydroxyacetone (Fig. 8.1). Amino acids react with α-dicarbonyl compounds, undergoing the Strecker degradation and

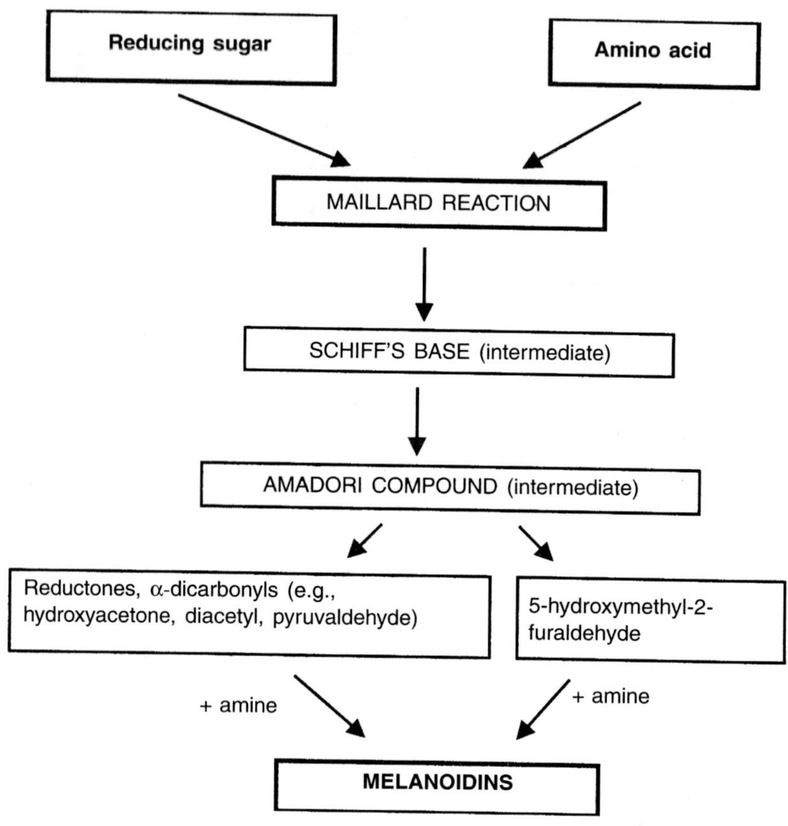

Fig. 8.1: The Maillard reaction summarized (adapted from Ikan et al., 1996).

forming α-amino ketones. The α-amino ketones may then condense together resulting in the formation of pyrazines, which are characteristic Maillard reaction products (Ho, 1996).

The Maillard reaction is believed to be a major abiotic humification pathway because of significant spectroscopic similarities between natural humic substances and melanoidins formed through this pathway (Ikan et al., 1996). The great appeal of the Maillard reaction in understanding humification processes lies in the two proposed precursors (sugars and amino acids) being among the most abundant constituents of terrestrial and aquatic environments (Anderson et al., 1989). Further support is lent by the presence of humic substances in marine environments where carbohydrates and proteins, because of their abundance, are more probable precursors of humic substances than are lignin or phenolic polymers (Nissenbaum and Kaplan, 1972; Hedges and Parker, 1976; Ikan et al., 1996). On the other hand, Flaig et al. (1975) have noted that phenols can react with ammonia, followed by autopolymerization under oxidative conditions, leading to the formation of nitrogenous polymers.

The Maillard reaction as a humification pathway in nature has been criticized because it is slow under ambient conditions (Hedges, 1978). Active microbial recycling of the precursor carbohydrates and proteins is also a problem. Evidence from ^{15}N NMR for the presence of heterocyclic N compounds in soils is also lacking, with amide appearing to be the major form of soil organic N (Burdon, 2001). However, the inability to detect heterocyclic nitrogen compounds is believed to be due to the inadequate sensitivity of ^{15}N NMR, given its low natural occurrence (Vairavamurthy and Wang, 2002). Knicker et al. (1993) and Knicker and Lüdemann (1995) have noted that "Solid-state ^{15}N NMR spectra of soils with natural ^{15}N with an acceptable signal-to-noise ratio abundance can only be obtained within reasonable measurement time if their organic N-content is higher than 1%." Such nitrogen concentrations, however, are rarely found in mineral soils and particle fractions, excluding them from ^{15}N NMR spectroscopic investigations (Knicker et al., 1999). It is not surprising therefore that heterocyclic N compounds are difficult to detect in soils using ^{15}N NMR spectroscopy. It is also likely that since more than a hundred different heterocyclic N compounds are formed microbially or abiotically in soils and humic substances ^{15}N NMR cannot, at present, resolve this complex mixture (Schulten and Schnitzer, 1988). Recently, however, ^{15}N NMR spectroscopy was used to detect the significant presence of pyrrolic (heterocyclic) N, ascribed to pyrolised plant material, in an Australian soil before and after UV-oxidation (Knicker and Skjemstaad, 2000). Mahieu et al. (2000), using solid-state ^{15}N NMR spectroscopy, noted that the heterocyclic N content ranged from 7 to 22% in humic substances with more heterocyclic N structures in more humified samples. Poirier et al. (2000, 2002) showed that the refractory organic matter isolated from two different soils consisted in part of cross-linked melanoidins poorly resolvable by ^{13}C NMR spectroscopy. The presence of amide structures in soil as elucidated by ^{15}N NMR is ascribed to either preserved proteinaceous structures or melanoidin type macromolecules (Derenne and Largeau, 2001).

Heterocyclic nitrogen compounds are significant N components of soils and humic substances (Schulten and Schnitzer, 1998). Sorge et al. (1993) and Schulten et al. (1995) using Curie-point pyrolysis-gas chromatography/mass spectrometry (Py-GC/MS) with N-selective detection analyzed a number of whole soils and humic substances and identified pyrroles, free and substituted imidazoles, and pyrazines as pyrolysis products. Evershed et al. (1997), using desorption headspace GC/MS, detected the presence of heterocyclic nitrogen compounds (alkyl pyrazines), believed to be characteristic Maillard reaction products, in archaeological plant remains from Egypt up to 1,500 years old. Also, using desorption headspace GC/MS, Poinar (2002) detected alkylpyrazines, furanones and furaldehydes, all volatile compounds derived from the Maillard reaction, in a coprolite originating from a 20,000-year-old ground sloth. In the first application of N K-edge XANES (x-ray absorption near edge structure) spectroscopy to environmental samples, Vairavamurthy and Wang (2002) discovered that pyrinidic structures occur widely in natural humic substances and sediments.

Very few studies have investigated the possible mechanisms of formation of melanoidins under naturally occurring conditions. On the basis of rate studies, Hedges (1978) suggested that melanoidins should form more readily in seawater or in marine sediments (pH ~ 8) than in terrestrial waters or soils which are generally more acidic (Becking et al., 1960). Hedges (1978) also noted that the relative affinities of kaolinite and montmorillonite for dissolved melanoidins varied with pH, and that montmorillonite adsorbed more melanoidins than did kaolinite. Taguchi and Sampei (1986) examined the roles of montmorillonite and calcium carbonate in accelerating the condensation of solutions of glucose (1.5M) and glycine (1.5M), and glucose (1.5M) and alanine (1.5M) at a temperature of 90°C for 131 h 36 min. They found that at these elevated temperatures, the rate of melanoidin formation was higher in the presence of montmorillonite. Collins et al. (1992) presented experimental evidence for condensation reactions between sugars and proteins in carbonate skeletons by heating mixtures of protein, polysaccharides and finely ground calcite crystals in sealed glass vials at 90°C. They postulated that fossil biomaterials such as shells protected proteins and polysaccharides from microbial degradation. Based on yield of melanoidins obtained, they estimated that at the global scale roughly 2.4×10^6 tons of calcified tissue matrix glycoproteins were processed annually through the melanoidin pathway.

Soil mineral colloids are significant in the turnover of organic matter (Huang, 1995; Torn et al., 1997) and manganese oxides are one of the most widespread mineral constituents of soils and sediments (McKenzie, 1989). Such mineral colloids could catalyze the Maillard reaction by adsorbing and promoting the condensation of sugars and amino acids, resulting in the accelerated formation of humic substances in nature. During the process the adsorbed substrates would be protected against microbial action. Recently, the catalytic role of δ-MnO_2 in the Maillard reaction, and the formation of humic substances was reported (Jokic et al., 2001). However, the role of mineral colloid surface-promoted acceleration of the Maillard reaction in the global nitrogen cycle, in the formation of heterocylic and other nitrogen compounds in nature, in the incorporation of nitrogen in the organic matrix of fossil fuels and the formation of kerogen is not known. The model reactants used, i.e., glucose and glycine, are important constituents of carbohydrates and proteins respectively, and are released from plant and animal materials by microbial decomposition (Cheshire, 1979; McKeague et al., 1986).

The objectives of the present study were (1) to further investigate the nature of the organic carbon structures produced by the catalytic action of δ-MnO_2 on the Maillard reaction, at ambient environmental temperatures, from glucose (carbohydrate) and glycine (amino acid), and (2) to investigate the nature of the nitrogen compounds formed as a means of illuminating the formation mechanisms of unknown nitrogen in soils and sediments and thereby the significance of the catalyzed Maillard reaction as an abiotic, sorptive condensation formation pathway for heterocyclic and other nitrogen compounds in nature. Analytical methods such as Py-FIMS (pyrolysis field ionization mass spectrometry), Curie-point Py-GC/MS (Curie-point pyrolysis-gas chromatography/mass spectrometry), ^{1}H and ^{13}C (liquid) NMR (nuclear magnetic resonance), and K-edge N-XANES (X-ray absorption near edge structure) spectroscopy were employed to obtain information.

2 MATERIALS AND METHODS

δ-MnO_2 was synthesized according to McKenzie's method (1971). It was characterized by X-ray diffractometry and Fourier transform infrared absorption (FTIR) spectroscopy. The X-ray diffractogram exhibited peaks at 0.720 nm, 0.360 nm, and 0.243 nm, which are characteristic for δ-MnO_2. The FTIR spectra had absorption bands at 3431, 1624, and 524 cm^{-1}, which are also diagnostic for δ-MnO_2 (Potter and Rossman, 1979). Glucose (ACS reagent grade) and glycine (Sigma Ultra grade purity) were obtained from Sigma Chemical Co. (St. Louis, MO). Two and one-half grams of δ-MnO_2 were suspended in 100 mL of an equimolar solution of glucose and glycine (0.05 mole each), and incubated in an oscillating water bath for 15 d at 45°C—the approximate temperature of the soil surface on a day when the air temperature is 25°C (Baver, 1956; Jury et al., 1991) and hence frequently encountered in many climate zones. Further, Russell (1983) stated that "July temperatures of surface soils at Rothamstead can reach 40–45°C on cloudless days." Rothamstead is located in England. Therefore, a temperature of 45°C is not unusual even in soils of the cool-temperate zone. Further, in tropical climates even higher soil surface temperatures are frequently encountered, e.g. in the tropical region of China maximum soil-surface temperatures are reported to range from 50°C to 80°C in direct sunlight [http://www.fas.org/china lands/intro.htm]. Therefore, 45°C is an environmentally relevant soil-surface temperature on a global scale.

The initial adjusted pH of all the systems was 7.0. Controls did not contain δ-MnO_2. The concentrations of glucose and glycine used were as recommended by Hedges (1978) in investigations of the Maillard reaction. Sterile, abiotic conditions were obtained by autoclaving all equipment and chemicals and adding an antibacterial agent (thimerosal, 0.02% w/v, final volume) to the reaction systems. Absence of microbial growth was verified by culturing aliquots from the reaction mixtures on Trypticase Soy Agar (TSA) plates (Dandurand and Knudsen, 1997). In the absence of δ-MnO_2 the reaction, as measured by the amount of browning by visible absorption spectroscopy, was much slower, and the yield of reaction products was negligible. Only in the presence of δ-MnO_2 was it possible to isolate reaction products (fulvic acid). Fulvic acid was isolated from the supernatant of the glucose-glycine-δ-MnO_2 system following the reaction period, according to the procedure described by Swift (1986). The solid residue remaining at the end of the reaction period in the system containing δ-MnO_2, was collected and after dialysis against distilled, deionized water was lyophilized.

Thimerosal was used at a concentration of 0.02% w/v (final volume), which corresponds to 20 milligrams in a volume of 100 mL. Since gram quantities of all the other reagents were used, the effect of thimerosal is negligible. This was confirmed in a control experiment run under the same conditions as the main experiments but in the absence of sugar and amino acid. At the end of the reaction period of 15 days at 45°C, and with an initial adjusted pH of 7.0, the supernatant was entirely clear (absorbance <0.01 over the range 400–700 nm) compared to the much darker supernatant in the systems which also contained sugar and amino acid, and no Mn(II) was detected in the supernatant. It was clear that thimerosal was not involved in the polymerization reaction. Further, it has been reported that thimerosal does not interfere with humification reactions (Wang et al., 1983). Moreover, thimerosal is often used as an antibacterial agent in humification studies (e.g., Jokic et al., 2001).

2.1 NMR Analyses

The FA was analyzed by ^{1}H and ^{13}C NMR (liquid) NMR spectroscopy. ^{1}H NMR spectroscopy was performed on a Bruker 500 spectrometer (Bruker Analytik, Rheinstetten/Karlsruhe Germany) operating at 500.13 MHz for ^{1}H. The sample was dissolved under nitrogen in 2 ml D_2O to which one drop of 30% NaOD solution was added. 256 scans were acquired. The sample was subsequently analyzed using ^{13}C NMR spectroscopy on the same spectrometer operating at 125 MHz for ^{13}C. Relaxation delay was 2 s. 40,000 FIDs (free induction decays) were accumulated.

2.2 Py-FIMS and Curie-point Py-GC/MS

Using Py-FIMS, about 100 μg fulvic acid or 5 mg residue from the glucose-glycine-δ-MnO_2 system were analyzed using a modified high performance mass spectrometer and, as complementary identification tool, Curie-point Py-GC/MS. For a detailed description of the methods see Schulten et al. (1998) and Schulten (2002).

2.3 Nitrogen K-edge XANES Spectroscopy

Nitrogen K-edge XANES spectroscopy, a sensitive, selective, and nondestructive method (Vairavamurthy and Wang, 2002), was utilized in a complementary manner to FIMS and Curie-point Py-GC/MS in order to unambiguously determine nitrogen speciation in the reaction products. The XANES spectra were collected at the Canadian Synchrotron Radiation Facility (CSRF) located on the 1 GeV storage ring, Aladdin, at the Synchrotron Radiation Center in Stoughton, WI. N K-edge (1s) XANES spectra were recorded using the spherical grating monochromator beamline (SGM) equipped with a 600 groove mm^{-1} ruled grating with a resolving power of ~8,000. The photon energy scale was calibrated using the 1s → Π* transition for nitrogen gas at 400.8 eV (Schwarzkopf et al., 1999). For calibration of the analysis, spectra were obtained for the reference compounds carbazole (a pyrrole), N-acetyl-D-glucosamine (an amide), glutamine (an amino acid containing an amide group), and 6-hydroxyquinoline (a pyridine). In addition, the literature, e.g. Mitra-Kirtley et al. (1993), Vairavamurthy and Wang (2002), and M. Kasrai et al. (unpubl. data, 2002) was consulted.

3 RESULTS AND DISCUSSION

3.1 ^{1}H NMR Spectroscopy

In the ^{1}H NMR spectrum (Fig. 8.2, peak assignations in Table 8.1) of the FA isolated from the supernatant of the glucose-glycine-δ-MnO_2 system, the small peak at 0.92 ppm is ascribed to terminal methyl groups on the end of methylene chains. Some sharp peaks observed at e.g. 1.15, 1.79, 2.27, 3.04, 3.81, 4.04, 7.36, and 8.30 ppm are distinct from the other broad ^{1}H resonances and their existence is ascribed to the presence of low molecular weight species originally covalently bonded to the fulvic acid macromolecules which later became separated by base hydrolysis and/or oxidation (Wilson et al., 1988). Peaks in the region of 1.10–1.15 ppm are assigned to protons on methyl groups of highly branched

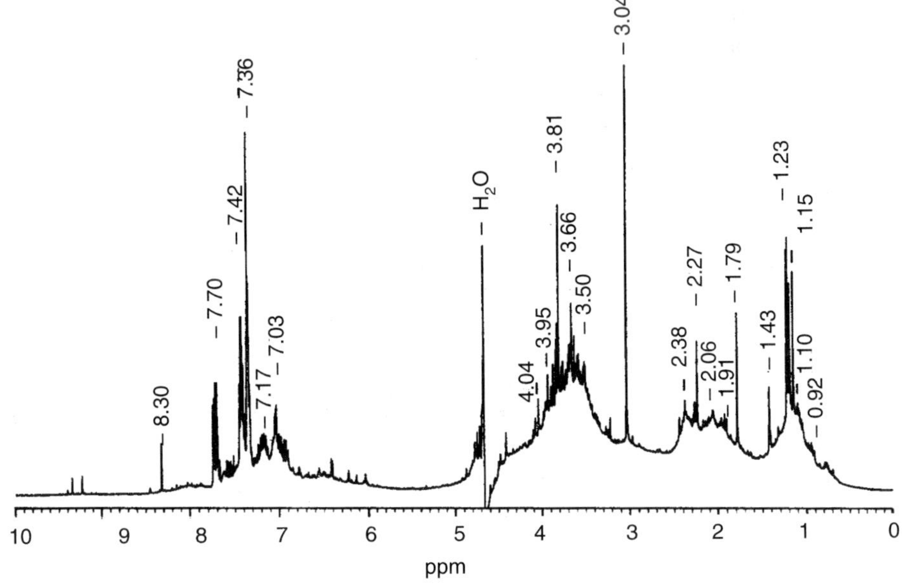

Fig. 8.2: ^{1}H (liquid) NMR spectra of fulvic acid isolated from the supernatant of the glucose-glycine-δ-MnO_2 system.

Table 8.1: Interpretation of the ^1H solution NMR signal assignations for the FA isolated from the supernatant of the glucose-glycine-δ-MnO_2 system.†

Assignation	ppm
Terminal methyl groups of methylene chains	0.92
Protons on methyl groups of highly branched aliphatic structures	1.10, 1.15
Protons on methyl or methylene groups of highly branched aliphatic structures and/or methyl in lactate	1.23
Protons on aliphatic carbons which are two or more carbons removed from aromatic rings or polar functional groups	1.43
Possibly from acetate	1.79
Protons attached to aliphatic carbons (methyl and methylene groups) which are alpha or attached to electronegative structures (e.g. a carboxyl group or an aromatic ring)	1.91, 2.06 2.27, 2.38
Protons on carbon attached to O or N heteroatoms, e.g. HCO of carbohydrates, methoxyl groups, aromatic amines	3.50, 3.66
Methylaryl ether or aromatic amines	3.81, 3.95
Methine in lactate	4.04
Amide, sterically unhindered aromatic protons, pyrrole, and indole NH groups	7.03, 7.17, 7.36, 7.42, 7.70
Probably from formate ion	8.30

†The FA was prepared from the system with an initially adjusted pH of 7.0 and subsequent reaction at 45°C for 15 d. Assignations are based on Wilson et al. (1988), Malcolm (1990), and Silverstein and Webster (1998).

aliphatic structures. The strong signals at around 1.23 ppm can be assigned to methyl or methylene groups of highly branched aliphatic structures and also to methyl in lactate. Signals in the region of 1.43 ppm arise from protons on aliphatic carbons that are two or more carbons removed from aromatic rings or polar functional groups. The sharp single peak at 1.79 ppm may arise from acetate. Peaks at 1.91, 2.06, 2.27, and 2.38 ppm are assigned to protons of methyl and methylene groups that are adjacent or attached to electronegative structures such as aromatic rings or carboxyl groups. The strong cluster of peaks in the region from approximately 3 to 4.4 ppm is assigned to protons on carbon attached to O or N heteroatoms. Signals at 3.50 and 3.66 ppm are assigned to protons on carbons attached to O or N heteroatoms, e.g. HCO of carbohydrates, also methoxyl groups and aromatic amines. Aromatic amines absorb in the general region from ~ δ 5.0 to 3.0 (Silverstein and Webster, 1998). Signals at 3.81 and 3.95 ppm are ascribed to methylaryl ethers or aromatic amines (Silverstein and Webster, 1998). The single peak at 4.04 ppm may arise from methine protons in structures such as lactate. Peaks at 7.03, 7.17, 7.36, 7.42, and 7.70 ppm are assigned to amide, pyrrole, and indole NH groups, which are known to absorb in this region (Silverstein and Webster, 1998). The presence of such structures is confirmed by the use of analytical pyrolysis and N-XANES, described later in the text. Sterically unhindered aromatic protons such as those of phenols and other oxygen containing aromatics, including also quinones (Choudhry and Webster, 1989), would likewise be expected to absorb in this region. The peak at 8.30 ppm is assigned to formate (Wilson et al., 1988), present as an artifact because of base hydrolysis of the FA biogeopolymer.

3.2 ^{13}C NMR Spectroscopy

The ^{13}C NMR spectrum (Fig. 8.3, peak assignations in Table 8.2) of the FA isolated from the supernatant of glucose-glycine-δ-MnO_2 system exhibits peaks at 20.2 and 22.7 ppm ascribed to terminal methyl groups and methylene carbon in alkyl chains. The peak at 31.8 ppm is assigned to methylene carbon α, β, δ, ε from a terminal methyl group. Peaks at 43.6 and 44.8 ppm are assigned to carbon in aliphatic amines. Signals at 52.9 and 55.7 ppm arise from aliphatic esters and ethers, methoxyl and ethoxyl groups, and from carbon bonded to amino groups. The signals at 61.3 and 62.9 ppm are assigned to CH_2OH groups, e.g. C_6 in carbohydrates and also to ether bonded aliphatic carbon. Peaks at 72.0 and 77.2 ppm are assigned to CHOH groups, e.g. ring carbon (C_2–C_5) of polysaccharides.

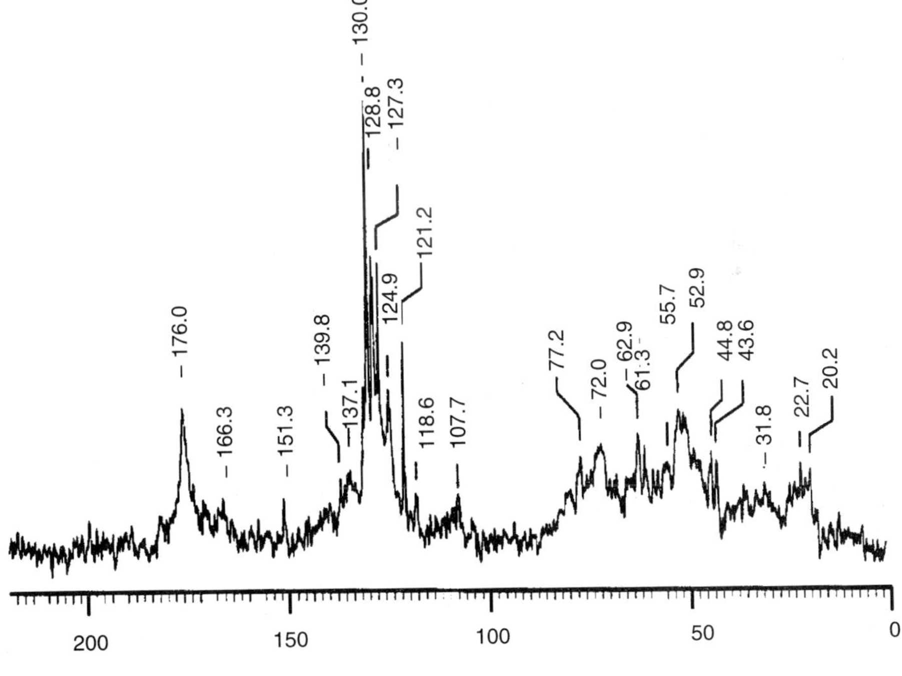

Fig. 8.3: ^{13}C (solution) NMR spectra of fulvic acid isolated from the supernatant of the glucose-glycine-δ-MnO_2 system.

Table 8.2: Interpretation of the ^{13}C (solution) NMR spectrum of the FA isolated from the supernatant of the glucose-glycine-δ-MnO_2 system†

Assignation	ppm
Terminal methyl groups and methylene carbon in alkyl chains	20.2, 22.7
Methylene carbon α, β, δ, ε from terminal methyl group	31.8
Methylene and methine carbon of branched alkyl chains	36.9, 37.9
Carbon in aliphatic amines	43.6, 44.8
Aliphatic esters and ethers, methoxyl and ethoxyl carbon, and carbon bonded to amino groups	52.9, 55.7
CH_2OH, e.g. C_6 in carbohydrates; ether bonded aliphatic carbon	61.3, 62.9
CHOH, e.g. ring carbon (C_2–C_5) of polysaccharides	72.0, 77.2
Carbon in furans or pyrroles; acetal or ketal carbon in carbohydrates	107.7
Aromatic carbon ortho to oxygen or nitrogen substituted aromatic carbon, e.g. in alkyl pyrroles	118.6, 121.2
Unsubstituted and alkyl substituted aromatic carbon, e.g. in pyrroles and pyridines	124.9, 127.3, 128.8 130.0, 137.1, 139.8
Carbon in alkyl pyrazines or phenols	151.3
Mostly carboxyl and amide carbon	166.3, 176.0

†The FA was prepared from the system with an initially adjusted pH of 7.00 and subsequent reaction at 45°C for 15 d. Assignations were based on Silverstein and Webster (1998).

Carbon in heterocylic compounds such as furans and pyrroles, or acetal or ketal carbon from carbohydrate structures, give rise to the signal at 107.7 ppm. Aromatic carbon ortho to oxygen or nitrogen substituted aromatic carbon, e.g. in alkyl pyrroles give rise to the signals at 118.6 and 121.2 ppm. The presence of pyrrolic structures is indirectly supported by ^1H NMR spectroscopy (Fig. 8.2 and Table 8.1) and directly by Py-FIMS and Py-GC/MS and N-XANES spectroscopy, the results of which are presented later. A cluster of peaks at 124.9, 127.3, 128.8, 130.0, 137.1, 139.8 ppm are ascribed, in part, to unsubstituted and alkyl substituted aromatic carbon, e.g. in pyridines and also pyrroles. The presence of such structures, which are characteristic Maillard reaction products (Ho, 1996), is unequivocally confirmed by Py-FIMS and Py-GC/MS and N-XANES spectroscopy. The signal at 151.3 ppm is assigned to carbon in phenols or in alkyl pyrazines, whose presence is confirmed by Py-GC/MS. Signals at 166.3 and 176.0 ppm arise from carboxyl and amide carbons. The presence of amide is unequivocally confirmed by N-XANES spectroscopy and is discussed in Section 3.4.

3.3 Py-FIMS and Curie Point Py-GC/MS

Table 8.3 summarizes the N-containing structures and low molecular mass (m/z 15–56) ion content obtained by Py-FIMS for the FA isolated from the

Table. 8.3: N-containing compounds and low molecular mass (m/z 15–56) ions content (as percentages of Total Ion Intensities) detected by Py-FIMS analysis of (1) fulvic acid isolated from the supernatant of the glucose-glycine-δ-MnO$_2$ system and (2) the lyophilized residue from the glucose-glycine-δ-MnO$_2$ system[†]

Sample	Total Ion Intensity	N-containing compounds	Peptides	m/z 15–56
		%		
(1) fulvic acid from glucose-glycine-MnO$_2$	100	7.3	0.7	0.9
(2) Residue from glucose-glycine-MnO$_2$	100	10.8	3.1	46.5

[†]The samples were obtained from systems with an initially adjusted pH of 7.0 following reaction at 45°C for 15 d.

supernatant of the glucose-glycine-δ-MnO$_2$ system (sample 1) and the lyophilized solid residue (sample 2) remaining after reaction was complete.

The Py-FIMS spectra and thermograms obtained are shown in Figures 8.4a and 8.4b. The Py-FIMS spectrum of the FA (Fig. 8.4a) is considerably more complicated than that of the residue (Fig. 8.4b). The residue contains a much higher proportion of low molecular mass ions, i.e. those of m/z 15–56 (46.5%) than the FA (0.9%). The thermogram (inset, Fig. 8.4a) of the FA shows that

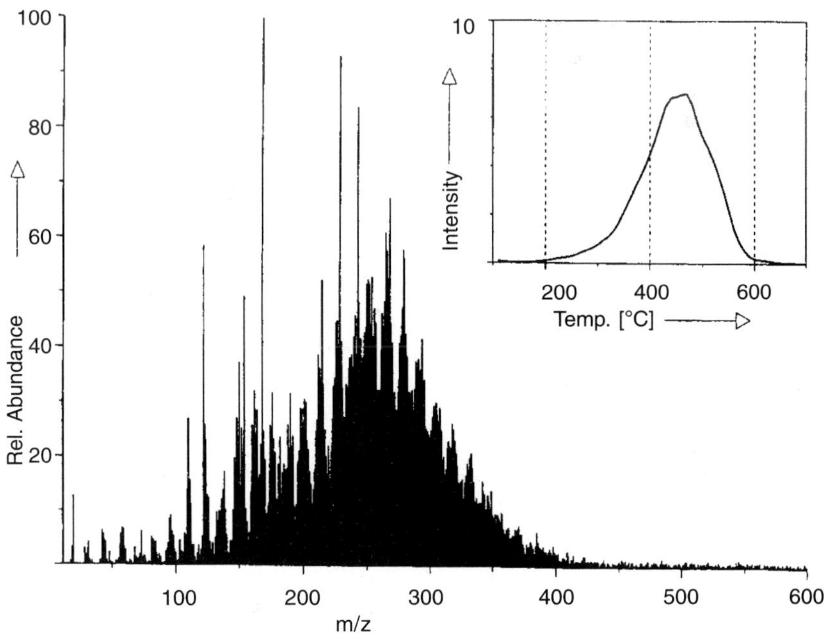

Fig. 8.4a: Py-FIMS spectra of fulvic acid isolated from the supernatant of the glucose-glycine-δ-MnO$_2$ system.

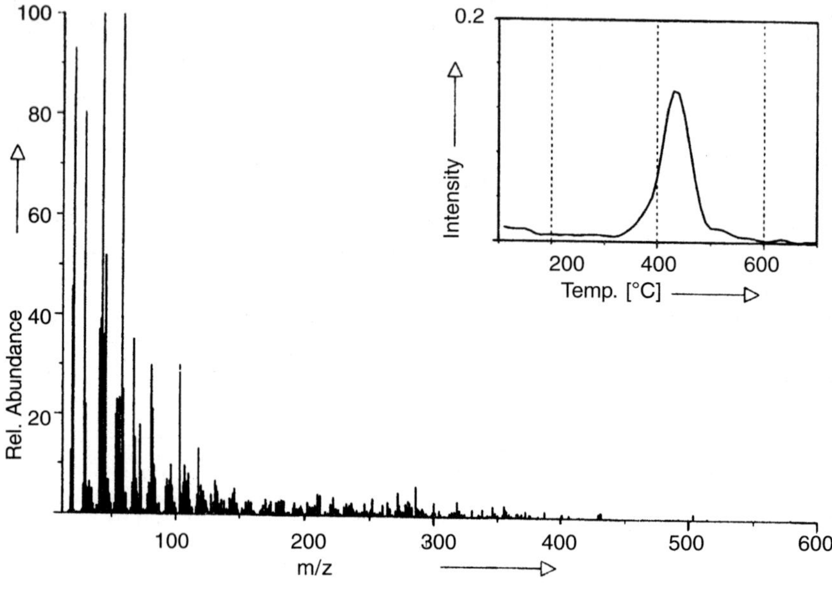

Fig. 8.4b: Py-FIMS spectra of the lyophilized solid phase of the glucose-glycine-δ-MnO_2 system.

maximum volatilization occurred at approximately 450–500°C and the organic compounds had volatilized by 600°C. The thermogram for the solid residue (insect, Fig. 8.4b) was considerably narrower, with a maximum at approximately 425–450°C, and most of the organic structures had volatilized by approximately 500°C. The thermograms indicate that, compared with the solid residue, the FA obtained from the supernatant contains organic structures with more thermally stable chemical bonds. Higher molecular weight products from the glucose-glycine-δ-MnO_2 system presumably are desorbed from the surface of the δ-MnO_2 and enter the supernatant.

The residue (solid phase) contains a much higher proportion of low molecular weight reaction products than the fulvic acid (supernatant). This is attributed to the more numerous functional groups and higher charge density of the lower molecular weight reaction products compared with the higher molecular weight reaction products. Therefore, compared with the latter the former have higher affinity to the solid phase and are thus more preferentially adsorbed. Consequently, higher molecular weight reaction products were observed in the supernatant solution than in the solid phase. In addition it is important to remember that the solid phase here consists of the mineral δ-MnO_2 coated with sorbed organic material, i.e., possibly unreacted substrates (glucose and glycine) as well as higher molecular weight reaction products which are clearly visible in Py-FIMS spectrum (Fig. 8.4b). Fulvic acid was extracted from the supernatant according to the method recommended by Swift (1996). The fraction of organic reaction products that are extracted in this manner are hydrophobic organic acids and some hydrophobic neutral compounds, whereas the remainder of the organic materials in the supernatant including unreacted sugar and amino acid, hydrophilic organic acids and some hydrophobic neutral compounds are not extracted and remain in the supernatant. Therefore the isolated fulvic acid obtained from the supernatant contains only some of the products of the Maillard reaction, with others remaining in the supernatant. However, isolation of the operationally defined "fulvic acid fraction" is important from the standpoint of environmental soil chemistry.

Using FIMS and Py-GC/MS analysis Figs. 8.4a and 8.4b, Table 8.4), the presence of heterocyclic N and O structures such as aldehydes, furans, furanones, alkylpyrroles, alkylpyridines, alkylpyrazines, and alkylpyrazoles as well as other N-containing compounds such as amides was detected. The alkyl substituents are mainly CH_3- and C_2H_5-groups in different positions. Some of the individual heterocyclic N and amide compounds identified by Py-FIMS and Py-GC/MS spectroscopy as components of the fulvic acid include acetamide (m/z 59), methylpyrrole (m/z 81), methylpyridine (m/z 93), dimethylpyrrole (m/z 95), trimethylpyrrole (m/z 109), trimethylpyrazine (m/z 122), tetramethylpyrrole

Table 8.4: N-containing structures identified by Py-GC/MS present in the lyophilized residue and the FA†

Freeze dried Area × 10^8	FA Area × 10^8	Mol. wt.	C	H	O	N	Heterocyclic N compounds
							Alkyl-Pyrroles
3.7	n.d.	67	4	5		1	1H-Pyrrole
0.90	0.65	81	5	7		1	1H-Pyrrole, 3-methyl-
0.59	n.d.	97	5	7	1	1	1H-Pyrrole-2-methanol
2.7	4.6	81	5	7		1	1H-Pyrrole, 2-methyl-
1.9	3.0	81	5	7		1	1H-Pyrrole, 1-methyl-
0.24	12	95	6	9		1	1H-Pyrrole, 2,5-dimethyl-
0.57	1.6	95	6	9		1	1H-Pyrrole, 2,3-dimethyl-
1.6	2.7	95	6	9		1	1H-Pyrrole, 2,4-dimethyl-
n.d.	3.0	109	7	11		1	Pyrrole, 2,3,4-trimethyl-
n.d.	6.2	109	7	11		1	Pyrrole, 1,2,5-trimethyl-
0.77	n.d.	109	7	11		1	1H-Pyrrole, 2,3,5-trimethyl-
n.d.	1.0	123	8	13		1	1H-Pyrrole, 2,3,4,5-tetramethyl-
0.22	1.8	123	7	9	1	1	Ethanone, 1-(1-methyl-1H-pyrrol-2-yl)-
13	**37**						**Sum**
							Alkyl-Pyridines
1.2	1.0	79	5	5		1	Pyridine
0.10	7.5	93	6	7		1	Pyridine, 2-methyl-
0.08	1.0	93	6	7		1	Pyridine, 3-methyl-
0.08	1.4	93	6	7		1	Pyri dine, 4-methyl-

(Table 8.4 contd.)

0.12	0.30	107	7	9		1	Pyridine, 2,6-dimethyl-
0.28	2.5	107	7	9		1	Pyridine, 2-ethyl-
n.d.	1.9	107	7	9		1	Pyridine, 2,5-dimethyl-
n.d.	1.9	109	7	11		1	Pyridine, 2,3-dihydro-2,3-dimethyl-, trans
n.d.	1.7	121	7	7	1	1	2-Acetylpyridine
n.d.	3.0	123	7	9	1	1	3-Pyridinemethanol, α-methyl-
n.d.	2.0	135	9	13		1	Pyridine, 3-butyl-
1.9	**24**						**Sum**
							Alkyl-Pyrazines
0.63	5.9	80	4	4		2	Pyrazine
0.88	n.d.	94	5	6		2	Pyrazine, methyl-
0.34	n.d.	108	6	8		2	Pyrazine, ethyl-
0.20	8.0	108	6	8		2	Pyrazine, 2,3-dimethyl-
0.44	0.92	122	7	10		2	Pyrazine, 2-ethyl-3-methyl-
0.52	n.d.	122	7	10		2	Pyrazine, propyl-
n.d.	12	122	7	10		2	Pyrazine, 2,3,5-trimethyl-
n.d.	1.4	136	8	12		2	Pyrazine, 3-ethyl-2,5-dimethyl-
n.d.	6.5	150	9	14		2	Pyrazine, 3,5-diethyl-2-methyl-
3.0	**35**						**Sum**
							Alkyl-Pyrazoles
n.d.	2.5	82	4	6		2	1H-pyrazole, 3-methylene-
0	**2.5**						**Sum**
0.05	n.d.	81	5	7		1	Butanenitrile, 2-methylene-
0.16	n.d.	94	5	6		2	Pyrimidine, 4-methyl-
0.09	0.80	94	5	6		2	4-Pyridinamine
0.29	n.d.	80	4	4		2	Pyridazine
0.05	n.d.	93	6	7		1	Benzenamine
0.11	2.4	108	6	8		2	Pyrimidine, 4,5-dimethyl-
0.54	3.1	103	7	5		1	Benzonitrile
n.d.	19	100	5	12		2	Piperazine, 1-methyl-
n.d.	0.58	131	9	9		1	1H-Indole, 1-methyl-
0.03	0.45	144	9	8		2	Quinoxaline, 5-methyl-
0.08	0.43	144	9	8		2	1,5-Naphthydrine, 4-methyl-
n.d.	0.51	166	7	6	3	2	Benzamide, 2-nitro-
1.4	**29**						**Sum**

†The samples were obtained from systems with an initially adjusted pH of 7.0 following reaction at 45°C for 15 d. The letters C, H, N, and O refer to the number of constituent atoms in the respective N-containing compound; n.d. = no data.

(m/z 123), ethyldimethylpyrazine (m/z 136), diethylmethylpyrazine (m/z 150), and nitrobenzamide (m/z 166). Compounds detected in the lyophilized solid residue include acetamide (m/z 59), methylpyrrole (m/z 81) methylpyridine (m/z 93), methylpyrazine (m/z 94), dimethylpyrrole (m/z 95) and trimethylpyrrole (m/z 109). Evershed et al. (1997) identified various alkyl pyrazines including methylpyrazine, ethyldimethylpyrazine, and trimethylpyrazine in their samples. Pyrazines, pyrroles and pyridines are considered characteristic products of the uncatalyzed Maillard reaction (Ho, 1996) and Schulten and Schnitzer (1998) have identified these structures in humic substances and whole soils by Py-FIMS and Py-GC/MS.

3.4 N XANES Spectroscopy

Figure 8.5 shows the N K-edge XANES spectra of (a) the fulvic acid produced from a δ-MnO_2-glucose-glycine mixture and (b) the resultant lyophilized solid phase. Examination of the spectrum of the fulvic acid (Fig. 8.5a) indicates the presence of pyridinic (398.5 eV), amide (401.2 eV), and pyrrolic (401.9) moieties. The broad peak, at ~401 to 402 eV, ascribed to amide and pyrrole moieties probably also arises from other heterocyclic N structures such as pyrazines whose presence was identified by FIMS and Py-GC/MS. The lyophilized solid residue (Fig. 8.5b) contained a substantial pyridinic (398.5 eV) component together with amide (401.2 eV) and pyrrolic (401.9 eV) content. Besides these moieties, the broad peak assigned to pyridinic compounds (Fig. 8.5b) may contain related N structures. The relative percent contributions of pyridinic, pyrrolic, and amide and other heterocyclic N moieties for the fulvic acid isolated from the glucose-glycine-δ-MnO_2 system (Fig. 8.5a) are pyridinic 19%, pyrrolic 23%, amide and other heterocyclic N 58% respectively, and for the lyophilized solid phase (Fig. 8.5b) pyridinic 56%, pyrrolic 15%, amide and other heterocyclic N 29%. Precision of the estimates lies in the range of 10%.

The energies corresponding to these components, indicated by the captions (Fig. 8.5), agree well with those obtained from several reference compounds. The results obtained unequivocally corroborate the formation of amide and heterocyclic N compounds by the Maillard reaction catalyzed by δ-MnO_2 at a temperature and pH frequently encountered under natural environmental conditions. The spectrum of the fulvic acid is similar to that of asphaltene (Mitra-Kirtley et al., 1993), which has implications with regard to the formation of sedimentary organic matter such as humic substances, fossil fuels, and kerogen. Pyrrolic N and pyridinic N are the dominant fractions in fossil fuels (Gorbaty and Keleman, 2001).

The action of δ-MnO_2 on glucose (carbohydrate) should generate dicarbonyl compounds (Fig. 8.6) (Wolff, 1996). These reactive dicarbonyl compounds would then react with ammonia (Fig. 8.7), formed by the known deamination of glycine catalyzed by δ-MnO_2 (Wang and Huang, 1987), as proposed by Vairavamurthy and Wang (2002), or with glycine (undergoing the Strecker degradation),

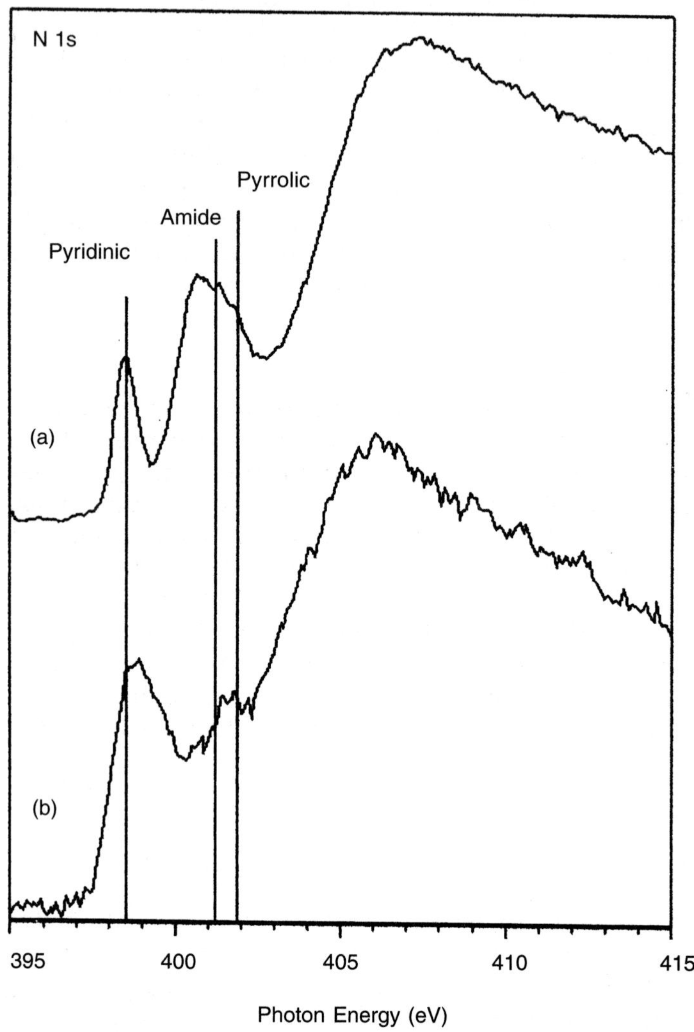

Fig. 8.5: Nitrogen 1s XANES spectra of (a) fulvic acid extracted from a glucose-glycine-δ-MnO_2 system and the resultant (b) lyophilized solid phase. Peaks are assigned to pyridinic (398.5 eV), amide (401.2 eV) and pyrrolic (401.9 eV) moieties. The relative percent contributions of pyridinic, pyrrolic and amide and other heterocyclic N moieties are for (a): pyridinic 19%, pyrrolic 23%, amide and other heterocyclic N 58%; (b): pyridinic 56%, pyrrolic 15%, amide and other heterocyclic N 29%. The precision of the estimates of the different constituents from XANES deconvolution is in the range of ±10%.

resulting in either case in the formation of heterocyclic nitrogen compounds (Wong and Shibamoto, 1996), as illustrated in Fig. 8.8. Ammonium ions are known to react with compounds containing reactive carbonyl groups to form pyridinic structures (Steelink, 1994). Carboxylic acids are produced during the

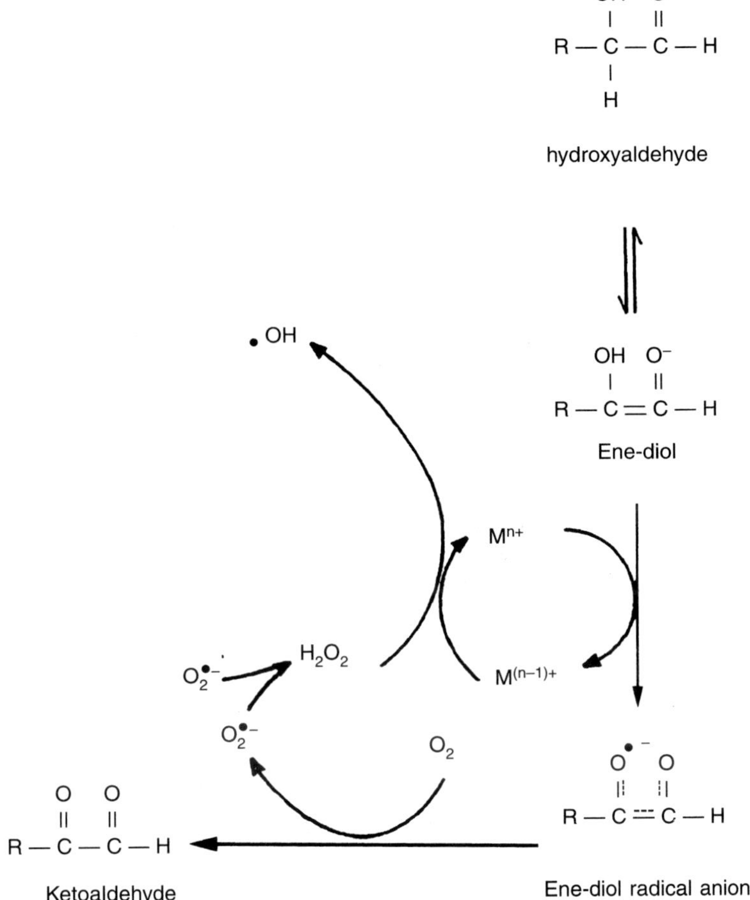

Fig. 8.6: Glucose can enolize and reduce transition metal thereby generating superoxide free radical ($O_2^{\bullet}$), hydrogen peroxide (H_2O_2), and the hydroxyl radical (.OH), forming dicarbonyl compounds (Wolff, 1996).

Strecker degradation of amino acids (Wong and Shibamoto, 1996), or by the action of manganese dioxide on simple carbohydrates (Bose et al., 1959). It is known that when heated, carboxylic acids react with ammonia to form amides (Smith and March, 2001). The scheme for possible amide formation is shown in Fig. 8.9. The present study indicates that the Maillard (sugar-amino acid condensation) reaction catalyzed by δ-MnO_2, is an abiotic pathway for the synthesis of biogeomacromolecules containing organic nitrogen originally derived from amino acids under ambient environmental conditions.

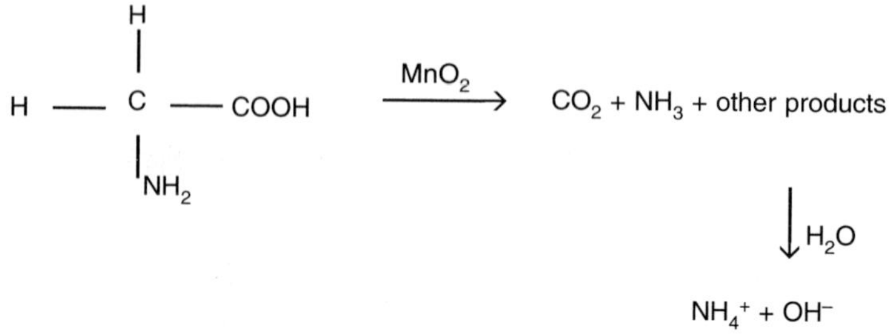

$$H_2N-CH_2-COOH \xrightarrow{MnO_2} CO_2 + NH_3 + \text{other products}$$

$$NH_3 \xrightarrow{H_2O} NH_4^+ + OH^-$$

Fig. 8.7: Decarboxylation and deamination of glycine catalyzed by δ-MnO_2 (Wang and Huang, 1987).

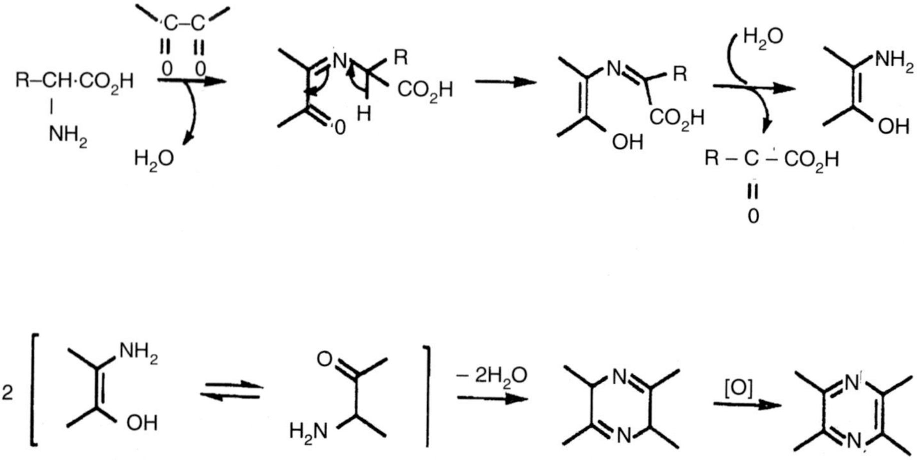

Fig. 8.8: Strecker degradation of amino acids and α-dicarbonyls to form heterocyclic compounds. For glycine, R = H (Wong and Shibamoto, 1996).

4 CONCLUSIONS

The data obtained in the present study provided for the first time unequivocal evidence that the Maillard reaction catalyzed by δ-MnO_2, which is common in soils and sediment environments, produces: (1) heterocyclic nitrogen compounds which are often referred to as *unknown* nitrogen and are the dominant organic nitrogen structures in fossil fuels, and (2) amides which are the dominant nitrogen types in humic substances, soils, and sediments. Therefore, the Maillard reaction (sugar-amino acid condensation) catalyzed by δ-MnO_2 provides an important, alternative, abiotic geochemical pathway for the synthesis of biogeomacromolecules containing organic nitrogen originally derived from

(1) Oxidation of carbohydrate or Strecker aldehyde by δ-MnO_2 to form carboxylic acid, or formation of carboxylic acid during Strecker degradation

(2) Deamination of amino acid by δ-MnO_2

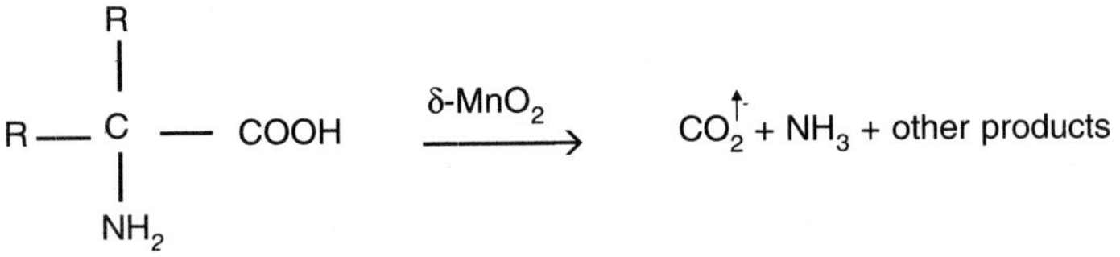

$$\mathrm{R{-}\underset{\underset{NH_2}{|}}{\overset{\overset{R}{|}}{C}}{-}COOH} \xrightarrow{\delta\text{-}MnO_2} CO_2\uparrow + NH_3 + \text{other products}$$

(3) Amide formation

$$\mathrm{R\overset{\displaystyle O}{\overset{\|}{C}}OH} + NH_3 \longrightarrow \mathrm{R\overset{\displaystyle O}{\overset{\|}{C}}NH_2} + H_2O$$

Fig. 8.9: Possible amide formation pathway including the key role of δ-MnO_2.

amino acids. Since carbon-nitrogen bonds formed in Maillard reaction products are known to be stable enough to last for thousands of years (Evershed et al., 1997; Poinar, 2002), the carbon-nitrogen bonds in biogeomacromolecules formed by the Maillard reaction catalyzed by δ-MnO_2 should also be equally refractory. This suggests that a significant fraction of the organic nitrogen component of surficial reservoirs may exist in the form of similar structures, i.e., Maillard reaction condensates formed by the abiotic, sorptive condensation of labile organic matter such as simple sugars and amino acids catalyzed by mineral colloids.

Moreover, the adsorption of low molecular weight organic substrates on mineral surfaces and their entry into the small pores (<2 nm in diameter) of the minerals provides the mechanism which protects them from microbial and enzymatic action. In these small pores, the low molecular weight organics would undergo the mineral catalyzed Maillard condensation reaction eventually resulting in the formation of refractory biogeomacromolecules. Our data show that this process of mineral catalyzed sorptive condensation occurs at an ambient temperature frequently encountered in many climate zones, and therefore is environmentally significant.

Since the two proposed precursors (sugars and amino acids) are extremely abundant constituents of terrestrial and aquatic environments (Anderson et al., 1989), and mineral colloids such as δ-MnO_2 are very reactive, as illustrated in this study, the data indicate that the mineral-catalyzed Maillard reaction is the source of at least some of the amide/peptide and heterocyclic N structures found in nature. Formation of amide and heterocyclic N compounds in the Maillard reaction catalyzed by mineral colloids such as δ-MnO_2 merit attention in our understanding of the transformation of organic moieties in the ecosystem and the global carbon and nitrogen cycles. Hence, the role of δ-MnO_2 catalysis in these processes is of crucial importance, which means the mineral colloid catalysis of the Maillard reaction in organic matter transformation in nature warrants further investigation.

Acknowledgments

This study was supported by the Natural Sciences and Engineering Research Council of Canada Discovery Grant GP 2383-Huang and the Deutsche Forschungsgemeinschaft (DFG, project Schu 416). We are grateful to Drs. Kim Tan and Yong-Feng Hu from CSRF and the staff of the Synchrotron Radiation Center (SRC), University of Wisconsin, Madison, for their technical support and the National Science Foundation (NSF) for supporting the SRC under Award no. DMR-0084402. The Py-GC/MS work by Dr. Gerald Jandl, Institute of Soil Science, University of Rostock, was invaluable.

References

Anderson H.A., Bick W., Hepburn A., and Stewart M. 1989. Nitrogen in humic substances. In: *Humic Substances, II. In Search of Structure*. M.H.B. Hayes, P. MacCarthy, R.L. Malcolm, and R.S. Swift (eds.). Wiley-Interscience, Chichester, UK, pp. 223–253.

Bada J.L. 1998. Biogeochemistry of organic nitrogen compounds. In: *Nitrogen-containing Macromolecules in the Bio- and Geosphere*. B.A. Stankiewicz and P.F. van Bergen (eds.). ACS Symp. Series 707. Amer. Chem. Soc., Washington DC, pp. 64–73.

Baldock J.A. and Skjemstad J.O. 2000. Role of the soil matrix and minerals in protecting natural organic materials against biological attack. *Org. Geochem*. 31: 699–710.

Bartlett J.R. 1986. Soil redox behavior. In: *Soil Physical Chemistry*. D.L. Sparks (ed.). CRC Press, Inc., Boca Raton, FL, pp. 179–207.

Baver L.D. 1956. *Soil Physics*. John Wiley and Sons, New York, NY (3rd ed.).

Becking L.G.M.B., Kaplan I.R., and Moore D. 1960. Limits of the natural environment in terms of pH and oxidation-reduction potentials. *J. Geol*. 68: 243–284.

Bollag J.-M., Dec J., and Huang P.M. 1998. Formation mechanisms of complex organic structures in soil habitats. *Adv. Agron*. 63: 237–266.

Bose J.L., Foster A.B., Stacey M., and Webber J.M. 1959. Action of manganese dioxide on simple carbohydrates. *Nature* 184: 1301–1302.

Bosetto M., Arfaioli P., Pantani O.L., and Ristori G.G. 1997. Study of the humic-like compounds formed from L-tyrosine on homoionic clays. *Clay Miner* 32: 341–349.

Burdon J. 2001. Are the traditional concepts of the structures of humic substances realistic? *Soil Sci*. 166: 752–769.

Chapman D. and Kimstach V. 1992. The selection of water quality variables. In: *Water Quality Assessments*. D. Chapman (ed.). Chapman and Hall, London, pp. 51–119.

Cheshire M.V. 1979. *Nature and Origin of Carbohydrates in Soils*. Acad. Press, New York, NY.

Choudhry G.G. and Webster G.R.B. 1989. Soil organic matter chemistry. Part 1. Characterization of several humic preparations by proton and carbon-13 nuclear magnetic resonance spectroscopy. *Toxic. Environ. Chem*. 23: 227–242.

Collins M.J., Bishop A.N., and Farrimond P. 1995. Sorption by mineral surfaces: Rebirth of the classical condensation pathway for kerogen formation? *Geochim. Cosmochim. Acta* 59: 2387–2391.

Collins M.J., Westbroek P., Muyzer G., and de Leeuw J.W. 1992. Experimental evidence for condensation reactions between sugars and proteins in carbonate skeletons. *Geochim. Cosmochim. Acta* 56: 1539–1544.

Dandurand L.-M.C. and Knudsen G.R. 1997. Sampling microbes from the rhizosphere and phyllosphere. In: *Manual of Environmental Microbiology*. C.J. Hurst (ed.-in-chief). Amer. Soc. Microbiol., Washington DC, pp. 391–399.

Derenne S. and Largeau C. 2001. A review of some important families of refractory macromolecules: Composition, origin, and fate in soils and sediments. *Soil Sci.* 166: 833–847.

Elliot E.T. and Coleman D.C. 1988. Let the soil work for us. *Ecol. Bull.* 39: 23–25.

Evershed R.P., Bland H.A., van Bergen P.F., Carter J.F., Horton M.C., and Rowley-Conwy. P.A. 1997. Volatile compounds in archaeolgoical plant remains and the Maillard reaction during decay of organic matter. *Science* 278: 432–433.

Flaig W., Beutelspacher H., and Rietz E. 1975. Chemical composition and physical properties of humic substances. In: *Soil Components*, Vol. 1. *Organic Components*. J.E. Gieseking (ed.). Springer, New York, NY, pp. 1–211.

Gorbaty M.L. and Kelemen S.R. 2001. Characterization and reactivity of organically bound sulfur and nitrogen fossil fuels. *Fuel Process. Tech.* 71: 71–78.

Guggenberger G. and Haider K.M. 2002. Effect of mineral colloids on biogeochemical cycling of C, N, P, and S in soil. In: *Interactions between Soil Particles and Microorganisms*, (eds.). P.M. Huang, J.-M. Bollag, and N, Senesi: John Wiley & Sons, Ltd., New York, NY, pp. 268–322.

Hatcher P.G., Breger I.A., and Mattingly M.A. 1980. Structural characteristics of fulvic acids from continental shelf sediments. *Nature* 285: 560–562.

Hedges J.I. 1978. The formation and clay mineral reactions of melanoidins. *Geochim. Cosmochim. Acta* 42: 69–76.

Hedges J.I. 1988. Polymerization of humic substances in natural environments. In: *Humic Substances and Their Role in the Environment*. F.H. Frimmel and R.F. Christman (eds.). John Wiley and Sons, Chichester, UK, pp. 45–58.

Hedges J.I. and Parker P.L. 1976. Land-derived organic matter in surface sediments from the Gulf of Mexico. *Geochim. Cosmochim. Acta* 40: 1019–1029.

Ho C.-T. 1996. Thermal generation of Maillard aromas. In: *The Maillard Reaction. Consequences for the Chemical and Life Sciences*. R. Ikan (ed.). John Wiley and Sons, Chickester, UK, pp. 27–53.

Huang P.M. 1995. The role of short-range ordered mineral colloids in abiotic transformation of organic components in the environment. In: *Environmental Impact of Soil Component Interactions: Vol. 1. Natural and Anthropogenic Organics*. P.M. Huang, J. Berthelin, J-M. Bollag, W.B. McGill, and A.L. Page (eds.). CRC/Lewis Publ., Boca Raton, FL, pp. 135–167.

Huang P.M. 2000. Abiotic catalysis. In: *Handbook of Soil Science*. M.E. Summer (ed.-in-chief). CRC Press, Boca Raton, Fl., pp. B303–B332.

Ikan R. Rubinsztain Y., Nissenbaum A., and Kaplan I.R. 1996. Geochemical aspects of the Maillard Reaction. In: *The Maillard Reaction. Consequences for the Chemical and Life Sciences*. R. Ikan (ed.). John Wiley and Sons, Chichester, UK, pp. 1–25.

Jokic A., Frenkel A.I., Vairavamurthy M.A., and Huang P.M. 2001. Birnessite catalysis of the Maillard reaction: its significance in natural humification. *Geophys. Res. Lett.* 28: 3899–3902.

Jury W.A., Gardner W.R., and Gardner W.H. 1991. *Soil Physics*. John Wiley and Sons, Inc. New York, NY, (5th ed.).

Kaiser K. and Guggenberger G. 2000. The role of DOM sorption to mineral surfaces in the preservation of organic matter in soils. *Org. Geochem*. 31: 711–725.

Keil R.G., Montiucon D.B., Prahl F.G., and Hedges J.I. 1994. Sorptive preservation of labile organic matter in marine sediments. *Nature* 370: 549–552.

Kennedy M.J., Pevear D.R., and Hill R.J. 2002. Mineral surface control of organic carbon in black shale. *Science* 295: 657–660.

Knicker H. and Lüdemann H.-D. 1995. N-15 and C-13 CPMAS and solution HR NMR studies on the chemical modifications of N-15 enriched plant materials during 600 days of microbial degradation. *Org. Geochem*. 23: 329–341.

Knicker H. and Skjemstad J.O. 2000. Nature of organic carbon and nitrogen in physically protected organic matter of some Australian soils as revealed by solid ^{13}C and ^{15}N NMR spectroscopy. *Austr. J. Soil Sci.* 38: 113–127.

Knicker H., Fründ R., and Lüdemann H.-D. 1993. The chemical nature of nitrogen in native soil organic matter. *Naturwissenschaften,* 80: 219–221.

Knicker H., Schmidt M.W.I., and Kögel-Knabner I. 1999. The structure of organic nitrogen in particle size fractions determined by ^{15}N CPMAS NMR. In: *Effect of Mineral-Organic-Microorganism Interactions on Soil and Freshwater Environments.* J. Berthelin, P.M. Huang, J.-M. Bollag, and F. Andreux (eds.). Kluwer Acad./Plenum Publ., New York, NY, pp. 143–149.

Mahieu, N., Olk, D.C., and Randall, E.W. 2000. Accumulation of heterocyclic nitrogen in humified organic matter: ^{15}N NMR study of lowland rice soils. *Eur. J. Soil Sci.* 51: 379–389.

Maillard L.C. 1912. Action des acides amines sur les sucres; formation des melanoidines par voie methodologique. *C.R. Acad. Sci.* 154: 66–68.

Maillard L.C. 1913. Formation de matieres humiques par action de polypeptides sur sucres. *C.R. Acad. Sci.* 156: 148–149.

Malcolm R.L. 1990. The uniqueness of humic substances in each of soil, stream and marine environments. *Anal. Chem. Acta* 232: 19–30.

Mayer L.M. 1994. Relationships between mineral surfaces and organic carbon concentrations in soils and sediments. *Chem. Geol.* 114: 347–383.

McBride M.B. 1994. *Environmental Chemistry of Soils.* Oxford University Press, Oxford, UK.

McKeague J.A., Cheshire M.V., Andreux F., and Berthelin J. 1986. Organo-mineral complexes in relation to pedogenesis. In: *Interactions of Soil Minerals with Natural Organics and Microbes.* P.M. Huang and M. Schnitzer (eds.). SSSA Spec. Publ. 17. SSSA, Madison, WI, pp. 549–592.

Mckenzie R.M. 1971. The synthesis of birnessite, cryptomelane, and some other oxides and hydroxides of manganese. *Mineral Mag.* 38: 493–502.

McKenzie R.M. 1989. Manganese oxides and hydroxides. In: *Minerals in Soil Environments.* J.B. Dixon and S.B. Weed (eds.). SSSA, Madison, WI, pp. 439–465 (2nd ed.).

Mitra-Kirtley S., Mullins O.C., Branthaver J.F., and Cramer S.P. 1993. Determination of the nitrogen chemical structures in petroleum asphaltenes using XANES spectroscopy. *J. Amer. Chem. Soc.* 115: 252–258.

Müller-Wegener U. 1988. Interaction of humic substances with biota. In: *Humic Substances and Their Role in the Environment.* F.H. Frimmel and R.F. Christman (eds.). John Wiley and Sons, Chichester, UK, pp. 179–192.

Nissenbaum A. and Kaplan J.R. 1972. Chemical and isotopic evidence for the *in situ* origin of marine humic substances. *Limnol. Oceanogr.* 17: 570–582.

Poinar N.H. 2002. The genetic secrets some fossils hold. *Acc. Chem. Res.* 35: 676–684.

Poirier N., Derenne S., Balesdent J., Rouzaud J.-N., Mariotti A., and Largeau C. 2000. Chemical structure and sources of the macromolecular, resistant, organic fraction isolated from a forest soil (Lacadee, Southwest France). *Org. Geochem.* 31: 813–827.

Poirier N., Derenne S., Balesdent J., Rouzaud J.-N., Mariotti A., and Largeau C. 2002. Abundance and composition of the refractory organic fraction of an ancient, tropical soil (Pointe Noire, Congo). *Org. Geochem.* 33: 383–391.

Potter R.M. and Rossman G.R. 1979. The tetravalent manganese oxides: identification, hydration, and structural relationships by infrared spectroscopy. *Amer. Mineral.* 64: 1199–1218.

Risser J.A. and Bailey G.W. 1992. Spectroscopic study of surface redox reactions with manganese oxides. *Soil. Sci. Soc. Amer. J.* 56: 82–88.

Russell E.W. 1983. *Soil Conditions and Plant Growth.* A Wild (ed.). Longman, London, UK, 285 pp. (11th ed.).

Schnitzer M. 1991. Soil organic matter: The next 75 years. *Soil Sci.* 151: 41–58.

Schnitzer M. 2000. A lifetime perspective on the chemistry of soil organic matter. *Adv. Agron.* 68: 1–58.

Schnitzer M. and Schulten H.-R. 1995. The analysis of organic matter in soil extracts and whole soils by pyrolysis-mass spectrometry. *Adv. Agron.* 55: 167–217.

Schulten H.R. 2002. New approaches to the molecular structure and properties of soil organic matter: humic-, xenobiotic-, biological-, and mineral-bonds. In: *Soil Mineral-Organic Matter-Microorganism Interactions and Ecosystem Health. Developments in Soil Science.* A. Violante, P.M. Huang, J.-M. Bollag, and L. Gianfreda (eds.). Elsevier, Amsterdam, vol. 28A, pp. 351–381.

Schulten H.-R. and Schnitzer M. 1989. The chemistry of soil organic nitrogen: A review. *Biol. Fertil. Soils* 26: 1–15.

Schulten H.-R and Leinweber P. 2000. New insights into organo-mineral particles: Composition, properties and models of molecular structure. *Biol. Fertil. Soils* 30: 399–432.

Schulten H.-R., Sorge C., and Schnitzer M. 1995. Structural studies on soil nitrogen by Curie-point pyrolysis-gas chromatography/mass spectrometry with nitrogen-selective detection. *Biol. Fertil. Soils* 20: 174–184.

Schulten H.-R., Leinweber P., and Schnitzer M. 1998. Analytical pyrolysis and computer modeling of humic and soil particles. In: *Structure and Surface Reactions of Soil Particles*. P.M. Huang, N. Senesi, and J. Buffle (eds.). John Wiley and Sons, Chichester, England, pp. 281–324.

Schwarzkopf O., Borchert M., Eggenstein F., Flechsig U. et al. 1999. The BESSY constant length Rowland circle monochromator. *J. Electron Spectrosc. Relat. Phenom.*, pp. 100–103, 997–1001.

Shindo H. and Huang P.M. 1982. Role of Mn (IV) oxide in abiotic formation of humic substances in the environment. *Nature* 298: 363–365.

Shindo H. and Huang P.M. 1984. Catalytic effects of manganese (IV, iron (III), aluminium and silicon oxides on the formation of phenolic polymers. *Soil Sci. Soc. Amer. J.* 48: 927–934.

Silverstein R.M. and Webster F.X. 1998. *Spectrometric Identification of Organic Compounds*. John Wiley & Sons, Inc. New York, NY.

Smith M.B. and March J. 2001. *March's Advanced Organic Chemistry Reactions, Mechanisms and Structure*. Wiley-Interscience, John Wiley & Sons, Inc., New York, NY (5th ed.).

Smolen J.M. and Stone A.T. 1998. Organophosphorus ester hydrolysis catalyzed by dissolved metals and metal-containing surfaces. In: *Soil Chemistry and Ecosystem Health*. P.M. Huang et al. (eds.). Soil Sci. Soc. Amer., Spec. Publ. 52, Madison, WI, pp. 157–171.

Sorge C., Schnitzer M., and Schulten H.-R. 1993. In-source pyrolysis-field ionization mass spectrometry and Curie-point pyrolysis-gas chromatography/mass spectrometry of amino acids in humic substances and soils. *Biol. Fertil. Soils* 16: 100–110.

Stankiewicz B.A., and van Bergen P.F. 1998. Nitrogen and N-containing macromolecules in the Bio- and Geosphere: An introduction. In: *Nitrogen-Containing Macromolecules in the Bio-and Geosphere.* B.A. Stankiewicz and P.F. van Bergen (eds.). ACS Symp. Series 707. Amer. Chem. Soc., Washington DC, pp. 1–12.

Steelink C. 1994. Application of N-15 NMR spectroscopy to the study of organic nitrogen and humic substances in the soil. In: *Humic Substances in the Global Environment*. N. Senesi and D. Miano (eds.). Elsevier Science, Amsterdam, pp. 405–427.

Stevenson F.J. 1994. *Humus Chemistry—Genesis, Composition, Reactions*. John Wiley and Sons, New York, NY (2nd ed.).

Swift R.S. 1996. Organic matter characterization In: *Methods of Soil Analysis*, Part 3. *Chemical Analysis*. D.L. Sparks et al. (eds.). Soil Sci. Soc. Amer., Madison, WI, pp. 1011–1069.

Taguchi K. and Sampei Y. 1986. The formation and clay mineral and $CaCO_3$ association reactions of melanoidins. *Org. Geochem.* 10: 1081–1089.

Torn M.S., Trumbore S.E., Chadwick O.A., Vitousek P.M., and Hendicks D.M. 1997. Mineral control of soil organic carbon storage and turnover. *Nature* 317: 613–616.

Vairavamurthy A. and Wang S. 2002. Organic nitrogen in geomacromolecules: Insights on speciation and transformation with K-edge XANES spectroscopy. *Environ. Sci. Tech.* 36: 3050–3057.

van Veen J.A. and Kuikman P.J. 1990. Soil structural aspects of decomposition of organic matter by micro-organisms. *Biogeochemistry* 11: 213–217.

Wang M.C. and Huang P.M. 1987. Polycondensation of pyrogallol and glycine and the associated reactions as catalysed by birnessite. *Sci. Tot. Environ.* 62: 435–442.

Wang T.S.C., Wang M.C., Ferng Y.L., and Huang P.M. 1983. Catalytic synthesis of humic substances by natural clays, silts and soils. *Soil Sci.* 135: 350–360.

Wang T.S.C., Huang P.M., Chou C-H., and Chen J-H. 1986. The role of soil minerals in the abiotic polymerization of phenolic compounds and formation of humic substances. In: *Interactions of Soil Minerals with Natural Organics and Microbes*. P.M. Huang and M. Schnitzer (eds.). SSSA Spec. Publ. no. 17. Soil Sci. Soc. Amer., Madison, WI, pp. 251–281.

Wershaw R.L. 1994. *Membrane-Micelle Model for Humus in Soils and Sediments and Its Relation to Humification*. U.S. Geol. Surv. Water-Supply Paper 2410. U.S. Govt. Printing Office, Wash. DC.

Wilson M.A., Collins P.J., Malcolm R.L., Perdue E.M., and Cresswell P. 1988. Low molecular weight species in humic and fulvic fractions. *Org. Geochem.* 12: 7–12.

Wolff S.P. 1996. Free radicals and glycation theory. In: *The Maillard Reaction. Consequences for the Chemical and Life Sciences*. R. Ikan (ed.). John Wiley and Sons, Chichester, UK, pp. 73–88.

Wong J.W. and Shibamoto T. 1996. Genotoxicity of Maillard reaction products. In: *The Maillard Reaction. Consequences for the Chemical and Life Sciences*. R. Ikan (ed.). John Wiley and Sons, Chichester, UK, pp. 129–159.

9

Humus and Nitrogen Distribution in Particle-size Fractions of Soils from the Chernozemic Zone of Kazakhstan

V. I. Kiryushin*

Abstract

The maximum humus content in automorphic soils of the Steppe zone (ordinary and southern Chernozems and the Solonetzes) was found to occur in the fine silt fraction. In these soils the humus content decreased both in coarser and finer fractions. To the contrary, in hydromorphic meadow soils, the humus content of particle-size fractions increased with decreases in the particle size fraction and reached maximum in the clay fraction.

1 INTRODUCTION

Investigations on the patterns of quantitative and qualitative distribution of humus by soil particle-size fractions are important for understanding the mechanisms of accumulation of humic substances and their interaction with the mineral part of the soil. There are several generalizing works on this problem (Hassink, 1997; Orlov, 1990; Stevenson, 1994). It should be noted that particular studies performed in Russia suggest that distribution patterns of humus in particle-size fractions may differ depending on the soil type. Thus, it was found that the humus content of particle-size fractions in Soddy-podzolic soils increased with a decrease in particle size (Kocherina, 1954; Lobitskaya, 1962). In steppe soils, the maximum humus content occurred in the fine silt fraction (Voronin, 1958; Trofimenko and Kizyakov, 1967). In forest-steppe soils and also in hydromorphic and semihydromorphic soils of the steppe zone, both variants of humus distribution by particle-size fractions were found (Trofimenko and

Correspondence author: Dr. V.I. Kiryushin Timiryazev Agricultural Acedemy, Russian Academy of Sciences, ul. Timiryazeva 49, Moscow 127550, Russia. E-mail: akiruishin@mail.ru

Kizyakov, 1967; Grati et al., 1965; Aidinyan, 1947). In this study the humus distribution in particle-size fractions is considered with respect to different soils in the steppe zone of Kazakhstan.

2 MATERIALS AND METHODS

The following soils were studied: (1) Ordinary Chernozem (pit 5-76, Kustanai oblast, Uritsk district), (2) Steppe Solonetz (pit 3-76, the same area), (3) Southern Chernozem (pit E-6, Turgai oblast, Esil'skii district), (4) Meadow-Chernozemic soil (pit E-8, Turgai oblast, Esil'skii district), and (5) Meadow soil (pit K-81, Kustanai oblast, Komsomol'skii district). All these soils develop from clayey mantle loams and are characterized by a heavy texture containing 50–60% particle size <0.01 and 30–38% of <0.001 mm. Mixed soil samples characterizing a given genetic horizon were collected from different parts of the pit walls; thus, spatial variability in the soil properties was partly "averaged" through the sampling procedures.

The following particle-size fractions were separated from the air-dried soil samples using elutriation of soil suspensions: 1–0.25, 0.25–0.01, 0.01–0.005, 0.005–0.001, and <0.001 mm. Soil suspensions were first saturated with 1 N NaCl as a dispersing agent to ensure complete peptization of soil microaggregates. It is known that soil rubbing in the paste state is insufficient to achieve complete peptization of soil microaggregates (Titova, 1970). Moreover, some authors believe that a high humus content of the fine silt fraction (0.005–0.001 mm, according to the scale adopted in Russia) is explained by the fact that this fraction partly consists of clay particles glued together by humic substances.

The clay fractions (<0.001 mm) from the Ordinary Chernozem and the Steppe Solonetz were further subdivided into the colloidal (finest clay) fraction (<0.0002 mm) and two near-colloidal fractions (0.0002–0.0003 and 0.0003–0.001 mm) using an FS-45 centrifuge (20,000 rpm, suspension feeding at the rate of 200 ml min^{-1}). The colloidal fraction was obtained after the evaporation of the centrifuged colloidal solution at 50°C. Near-colloidal fractions (0.0002–0.0003 and 0.0003–0.001 mm) were scraped from the upper and lower portions of the clayey film on the walls of the centrifuge respectively.

The contents of organic carbon (C_{org}, %) and total nitrogen were determined in the particle-size fractions obtained by standard methods (the wet combustion method according to Tyurin and the Kjeldahl procedure respectively); the humus content was calculated as $C_{org}(\%) \times 1.724$. We also analyzed the group composition of humic substances according to the pyrophosphate method suggested by Kononova and Bel'chikova (1961): the extraction of mobile humic and fulvic acids with a 0.1 M $Na_4P_2O_7$ + 0.1 N NaOH solution with further separation of humic acids; the residual fraction is conventionally referred to as the humin fraction. Finally, the contents of total and oxalate-extractable (amorphous) sesquioxides were determined in both the bulk soil mass and separated particle-size fractions.

3 RESULTS AND DISCUSSION

3.1 Humus Distribution in Various Particle-size Fractions

Results showed that in the automorphic soils, i.e., not affected by groundwater including the Chernozems and the Steppe Solonetz, the maximum humus content occurred in the fine silt fraction (0.005–0.001 mm), decreasing in coarser and finer fractions (Tables 9.1 and 9.2). It is noteworthy that the humus content in the fine silt fraction was considerably higher than that in the clay (<0.001 mm) fraction (Table 9.1).

Table 9.1: Humus content in particle-size fractions

Depth, (cm)	Humus content in bulk soil (%)	Humus content (%) Particle-size fraction (mm)			
		<0.001	0.001–0.005	0.005–0.01	0.01–0.25
		Southern Chernozem, pit E-6			
0–10	5.5	7.1	11.8	5.1	2.2
50–60	1.7	3.1	3.7	1.4	0.4
		Meadow-Chernozemic soil, pit E-8			
0–12	6.2	10.4	11.2	6.3	2.6
30–40	3.8	7.0	5.2	2.0	0.8
		Meadow soil, pit K-81			
0–21	7.7	13.8	11.1	7.1	3.0
70–80	1.9	4.8	2.9	1.2	0.5

In the Meadow-Chernozemic soil, the humus contents in the fine silt and clay fractions were approximately similar in the humus (A1) horizon, but in the B horizon, the humus content of the clay fraction was higher than that of the fine silt fraction. In the Meadow soil, the humus content of the clay fraction was higher than that of the fine silt fraction in all soil horizons (Table 9.1).

In the automorphic soils, the humus content of near-colloidal fractions decreased with a decrease in size of particles (Table 9.2). The humus content in the fraction 0.0003–0.001 mm is 1.5–2 times higher than that in the 0.0002–0.0003 mm fraction.

A considerable rise in humus content of the colloidal fraction (<0.0002 mm) compared with the next larger size fraction (0.0002–0.0003 mm) was also observed (Table 9.3). This could be due to the transition of a part of mobile humic substances from coarser fractions into the colloidal solution upon repeated elutriations of soil suspensions saturated with 1 N NaCl. This is indirectly supported by a very high amount of pyrophosphate-extractable humic

Table 9.2: Humus content of particle-size fractions, including the colloidal and near-colloidal fractions, in automorphic soils

Depth, (cm)	Humus content in bulk soil (%)	Humus content (%) Particle-size fractions (mm)					
		<0.0002	0.0002–0.0003	0.0003–0.001	0.001–0.005	0.005–0.01	0.01–0.25
		Ordinary Chernozem, pit 5–76					
0–13	5.8	11.1	5.2	8.6	12.3	12.5	1.8
13–27	3.6	7.7	2.9	5.3	7.6	6.5	1.2
27–37	3.4	7.4	3.0	5.7	5.6	6.7	0.7
100–115	0.6	0.9	0.8	0.7	1.0	0.8	—
		Steppe Solonetz, pit 3-76					
0–11	4.7	12.5	5.3	7.7	11.5	7.8	1.2
11–25	4.1	4.8	1.8	3.3	5.2	3.8	0.6
25–35	1.9	2.1	1.8	2.5	3.5	3.3	1.0
85–95	0.6	0.9	0.5	0.7	0.8	0.6	0.4

substances present in the colloidal fraction and a relatively low amount of nonextractable humic substances (the humin fraction) (Table 9.3).

In general, the distribution of humic substances in particle-size fractions of automorphic soils (the Ordinary Chernozem and the Steppe Solonetz) was characterized by two maximums, one in the colloidal and the other in the fine silt fractions. In the Chernozem, the maximum in the colloidal (<0.0002 mm) fraction was somewhat lower than the maximum in the fine silt (0.001–0.005 mm) fraction.

3.2 Group Composition of Humic Substances in Particle-size Fractions

The group composition of humus in different particle-size fractions of Ordinary Chernozem and Steppe Solonetz is shown in Table 9.3. The highest C_{ha}/C_{fa} ratio was observed in the fine silt fraction. The portion of fulvic acids increased with both a decrease and an increase in size of soil particles. In the topsoils, the C_{ha}/C_{fa} ratios in the colloidal fractions were two to three times lower than those in the fine silt fractions. In deeper horizons, pyrophosphate-extractable humic substances were mainly represented by fulvic acids.

The nonextractable residue (humin) content increased with an increase in particle size fraction, especially in the humus (A1) horizon (Table 9.3). In colloidal and near-colloidal fractions, the content of nonextractable residue increased with increasing soil depth, while in the silt fractions, it remained approximately constant in all soil horizons.

Table 9.3: Group composition of humus in bulk soil and in particle-size fractions

Particle-size	Ordinary Chernozem, pit 5–76						Steppe Solonetz, pit 3–76					
fractions (mm)	Depth	C_{org}	Group composition (% of C_{org})				Depth (%)	$C_{org\ (\%)}$	Group composition (% C_{org})			
	(cm)	(%)	C_{ha}	C_{fa}	C_{ha}/C_{fa}	C_{humin}			C_{ha}	C_{fa}	C_{ha}/C_{fa}	C_{humin}
Bulk soil (<1 mm)	0–13	3.36	31.2	18.4	1.69	50.3	0–11	2.70	32.6	20.4	1.60	47.0
	13–27	2.11	26.5	18.0	1.47	55.5	11–25	1.81	22.1	18.8	1.18	59.1
	27–37	1.95	19.4	15.4	1.27	65.1	25–37	1.12	16.0	17.0	0.95	67.0
	100–115	0.34	8.8	20.6	0.43	70.6	83–95	0.36	5.6	19.4	0.25	75.0
<0.0002	0–13	6.42	36.4	34.4	1.06	29.2	0–11	7.25	30.6	41.8	0.73	27.6
	13–27	4.44	35.6	35.1	1.01	29.3	11–25	2.78	22.3	45.0	0.50	32.7
	27–37	4.30	31.7	33.8	0.94	34.5	25–37	1.24	12.1	51.6	0.23	36.3
	100–115	0.51	3.9	31.4	0.12	64.7	83–95	0.55	1.8	27.3	0.06	70.9
0.0002–0.0003	0–13	2.99	30.4	23.8	1.28	45.8	0–11	3.05	28.9	27.2	1.06	43.9
	13–27	1.70	24.7	21.2	1.17	54.1	11–25	1.06	20.7	27.4	0.76	51.2
	27–37	1.77	14.7	14.7	4.00	70.6	25–37	1.07	11.2	15.0	0.75	73.8
	100–115	0.46	4.4	13.0	0.33	82.6	83–95	0.30	6.7	16.6	0.40	76.7
0.0003–0.001	0–13	4.96	17.5	12.1	1.44	70.4	0–11	4.45	24.5	14.2	1.73	61.3
	13–27	3.08	15.6	15.9	0.98	68.5	11–25	1.92	17.7	14.6	1.21	67.7
	27–37	3.31	13.3	12.4	1.07	74.3	25–37	1.47	12.2	11.6	1.06	76.2
	100–115	0.39	7.7	17.9	0.43	74.4	83–95	0.39	5.1	18.0	0.28	76.9
0.001–0.005	0–13	7.14	16.9	8.0	2.12	75.1	0–11	6.68	18.0	8.4	2.14	73.6
	13–27	4.38	15.7	8.4	1.86	75.8	11–25	3.00	14.3	8.7	1.65	77.0
	27–37	3.24	14.2	8.3	1.70	77.5	25–37	2.04	18.1	11.8	1.54	70.1
	100–115	0.56	7.1	14.3	0.50	78.6	83–95	0.44	6.8	15.9	0.43	77.3
0.005–0.01	0–13	7.26	10.2	7.7	1.32	82.1	0–11	4.53	12.8	7.9	1.61	79.3
	13–27	3.96	12.2	9.6	1.17	78.2	11–25	2.21	12.2	7.7	1.59	80.1
	27–37	3.92	9.7	8.2	1.19	82.1	25–37	1.90	11.0	7.9	1.40	81.1
	100–115	0.47	6.4	8.5	0.57	85.1	83–95	0.36	8.3	13.9	0.60	77.8

These data pose a problem for explaining the existence of a lower content of humic substances in the clay (and colloidal) fractions vis-à-vis the fine silt fraction. We suggest the following two hypotheses: (1) clayey (clay and colloidal) fractions have a lower adsorption capacity for humic substances compared to the fine silt fraction and / or (2) humification products of plant residues tend to accumulate in the fine silt fraction.

The first hypothesis is commonly used to explain the weak bonding of humic substances with clay minerals in the illuvial horizons of the Solonetzes; the weakness of adsorption bonds is attributed to a low content of iron hydroxides in the clay fraction of illuvial horizons of the Solonetzes (Kozyreva and Rubilin, 1981).

In order to test the validity of this hypothesis, we determined the contents of total and oxalate-soluble (amorphous) sesquioxides in the bulk soil mass and separated particle-size fractions (Table 9.4). The maximum amount of oxalate-extractable sesquioxides was found in the colloidal and near-colloidal fractions (the difference between these fractions with respect to the contents of amorphous R_2O_3 was low). With increase in particle size, the contents of both total and oxalate-extractable sesquioxides decreased. Thus, the fine silt fraction contained 1.5–2.0 times less oxalate-extractable sesquioxides than the clay fraction. It is noteworthy that the distribution of sesquioxides by separate particle-size fractions in the Chernozem was the same as that in the Solonetz soil.

Considering these data, we can state that the enrichment of the fine silt fraction with humic substances (compared with the clay fraction) cannot be explained by the stronger binding of these substances by amorphous sesquioxides in the fine silt fraction.

Thus, the second hypothesis suggesting the direct accumulation of humic substances in the fine silt fraction seems to be more probable. Microscopic studies showed that the coagulates of humic substances, iron–humus microaggregates, and humified remains of finest roots tend to concentrate in the fine silt fraction of skeletal soil particles (Lobitskaya, 1962; Trofimenko and Kizyakov, 1967). Humic substances also form films on the surfaces of mineral skeletal grains. In the clay fraction, humic substances are only represented by films on mineral surfaces. It can be assumed that in the course of decomposition of plant and animal remains, humification products tend to accumulate in the fine silt fraction. A highly dispersed part of humic substances is transferred into soil solution, from which these substances are adsorbed on the surface of fine soil particles. This process may take place in the course of both humification proper and secondary dissolution of humic substances. Thus, a redistribution of humic substances between particle-size fractions takes place and is accompanied by a firmer attachment of humic substances on the surface of fine silt particles. This process is especially intense in the Solonetz soil due to the high mobility of humic substances in conditions of the alkaline reaction and the presence of exchangeable sodium. This is also the reason for the predominance of fulvic

Table 9.4: Contents of total and oxalate-extractable sesquioxides in particle-size fractions from Ordinary Chernozem (pit 5-76) and Steppe Solonetz (pit 3-76).

Depth cm	Soil		Particle-size fractions (mm)											
	1	2	<0.0002		0.0002–0.0003		0.0003–0.001		0.001–0.005		0.005–0.01		0.01–0.25	
			1	2	1	2	1	2	1	2	1	2	1	2
% of air-dried sample														
Ordinary Chernozem. Bulk content of sesquioxides														
0–13	4.1	11.3	7.2	20.3	9.2	22.7	8.0	20.2	5.5	13.5	3.3	11.1	1.0	5.0
43–27	4.4	10.8	8.2	23.4	9.2	21.4	8.4	21.2	6.1	11.6	3.4	11.2	0.9	5.0
27–37	4.5	10.9	8.7	24.7	8.7	20.2	8.2	21.7	5.6	12.9	3.7	12.1	1.0	5.0
100–115	4.6	10.8	9.2	25.1	8.4	19.9	7.0	18.1	6.4	10.6	2.7	10.2	1.0	4.0
Ordinary Chernozem. Oxalate-extractable (by Tamm) sesquioxides														
0–13	0.49	0.53	0.75	0.87	nd	nd	0.87	0.91	0.42	0.54	0.25	0.36	0.03	0.04
13–27	0.54	0.73	0.72	0.85	»	»	0.77	0.83	0.32	0.42	0.28	0.41	0.04	0.06
27–37	0.47	0.64	0.85	0.72	»	»	0.75	0.68	0.33	0.40	0.28	0.34	0.03	0.07
100–115	0.46	0.54	0.93	0.70	»	»	0.50	0.58	0.22	0.37	0.22	0.31	0.03	0.05
Steppe Solonetz. Bulk content of sesquioxides														
0–11	3.1	9.3	7.0	20.1	9.2	21.4	8.1	20.2	5.9	12.5	2.5	9.0	0.8	4.2
11–25	5.3	13.3	8.5	24.0	9.4	22.4	8.5	21.9	5.6	12.8	3.2	11.9	1.1	5.5
25–35	4.6	11.7	8.8	24.6	8.5	20.1	8.2	20.7	5.5	13.4	3.8	12.3	1.2	5.5
83–95	4.3	10.2	9.7	25.3	8.2	19.3	6.9	17.6	6.3	10.9	2.7	9.9	1.1	4.3
Steppe Solonetz. Oxalate-extractable sesquioxides														
0–11	0.48	0.42	0.53	0.86	nd	nd	1.17	0.84	0.52	0.43	0.20	0.31	0.04	0.03
11–25	0.79	0.86	0.88	1.08	»	»	1.04	0.77	0.46	0.41	0.24	0.36	0.03	0.05
25–35	0.58	0.64	1.12	0.74	»	»	0.80	0.62	0.37	0.34	0.21	0.29	0.03	0.03
83–95	0.46	0.43	1.01	0.72	»	»	0.54	0.40	0.20	0.28	0.19	0.26	0.02	0.03

Note: Columns (1) contain data on Fe_2O_3, columns (2), on Al_2O_3

nd = not determined.

acids in the pyrophosphate extract from the colloidal fraction of the Solonetz soil, whereas humic acids predominated in the pyrophosphate extract from the fine silt fraction of this soil (Table 9.3). Predominance of fulvates in the colloidal fractions of B horizons of the Solonetzes is related to the illuvial origin of fulvic acids in these horizons.

Redistribution of humic substances between particle-size fractions (with the predominant accumulation in the fine silt fraction) is accompanied by a relative increase in the content of not only humic acids, but also the nonextractable residue of humic substances in the fine silt fraction. The latter phenomenon is also explained by the accumulation of partly decomposed (not fully humified) products of the decomposition of plant and animal tissues in this fraction. The portion of incompletely humified products in the organic matter of silty fractions seems to be rather high. This assumption is supported by data on the radiocarbon age of separate fractions of humic substances. Thus, Kozyreva and Rubilin (1981) reported that the radiocarbon age of humic substances from the Chernozemic and the Chestnut soils was minimal in free (0.1 N NaOH-extractable) humic acids on the one hand, and nonextractable residue on the other; these fractions of humic substances were especially youthful within the root zone. Goh et al. (1976, 1977) also found that the humin fraction was more enriched with bomb ^{14}C than the humic acid and the fulvic acid in New Zealand soils. These suggested that the humin fraction represented a part of the most recently formed organic matter that, being strongly bound to the mineral components of the soil, became nonextractable.

Taking into account these considerations, the observed phenomenon of an increase in the humus content of the clay fraction with an increasing degree of soil hydromorphism may be explained by a more active redistribution of humic substances due to periodic waterlogging and the development of reducing processes. Under these conditions, activation of sesquioxides in the clay fraction favors additional adsorption of humic substances from the soil solution.

3.3 C/N Ratio in Different Particle-size Fractions

The C/N ratio decreased with increases in the relative content of fulvic acids and decreases in the relative content of the nonextractable residue (Table 9.5). It also decreased with decreases in the particle size (from the fine silt to the colloidal fraction). This may be explained by a parallel increase in the relative content of fulvic acids among humic substances. As can be seen from Table 9.6, the C/N ratio in fulvic acids was lower than in humic acids. Thus, the distribution of total nitrogen by particle-size fractions is related to the distribution of organic matter corrected for its qualitative group composition.

A decrease in C/N ratio with a decrease in the size of soil particles can also be attributed to a corresponding increase in the content of adsorbed ammonium ions as shown by Mogilevkina (1965). The same phenomenon can explain a lower C/N ratio in lower soil horizons than in upper horizons. In the coarser

Table 9.5: Nitrogen content and C/N ratio in bulk soil mass and particle-size fractions

Depth cm	Bulk soil		Particle-size fractions (mm)									
			0.0002–0.0003		0.0003–0.001		0.001–0.005		0.005–0.01		0.01–0.25	
	N (%)	C/N	N (%)	C/N	N (%)	C/N	N (%)	C/N	N (%)	C/N	N (%)	C/N
Ordinary Chernozem, pit 5-76												
0–13	0.317	10.6	0.336	8.9	0.498	10.0	0.618	11.6	0.573	12.7	0.066	16.1
13–27	0.195	10.8	0.198	8.6	0.336	9.2	0.421	10.4	0.324	11.6	0.047	15.3
27–37	0.186	10.5	0.224	7.9	0.330	10.0	0.309	10.5	0.344	11.4	0.029	14.5
100–115	0.049	6.9	0.074	6.2	0.056	7.0	0.084	6.7	0.054	8.7	0.022	–
Steppe Solonetz, pit 3-76												
0–11	0.267	10.1	0.389	7.8	0.489	9.1	0.548	12.2	0.348	13.0	0.046	15.7
11–25	0.185	9.8	0.139	7.6	0.227	8.5	0.263	11.4	0.186	11.9	0.024	14.2
25–35	0.122	9.2	0.149	7.2	0.181	7.2	0.181	8.1	0.192	10.6	0.184	14.3
85–95	0.058	6.2	0.068	4.4	0.066	5.9	0.065	6.8	0.048	7.5	0.021	10.5

Table 9.6: Carbon and nitrogen contents in humic and fulvic acids of the Steppe soils of Kazakhstan

		Humic acids			Fulvic acids		
Soil	Depth (cm)	C	N (mg/100 g soil)	C/N	C	N (mg/100 g soil)	C/N
Ordinary Chernozem, pit 300	0–10	1740	140	12.4	760	79	9.6
	80–90	260	22	11.8	150	17	8.8
Shallow Steppe Solonetz, pit 68	0–8	540	45	12.0	420	51	8.2
	19–30	60	5.5	10.9	120	16	7.5

particle-size fractions, the C/N ratio was much wider (Table 9.5), which might be explained by the accumulation of slightly humified detritus of plant remains (see discussion on the humus distribution) with a low content of nitrogen.

4 CONCLUSIONS

Results of the present study showed that the distribution of humus in particle-size fractions of soils is governed by the process of predominant accumulation of the products of decomposition and humification of plant residues in the fine silt fraction and by the adsorption of humic substances from the soil solution on the surface of clay minerals in finer fractions. The intensity of redistribution of humic substances increased with an increase in degree of soil hydromorphism (from the automorphic Chernozems to the Meadow soils) with a corresponding rise in the relative content of pyrophosphate-extractable organic substances and oxalate-extractable sesquioxides that tend to concentrate in the finest particle-size fractions. As a result, the maximum content of humus in the automorphic Chernozemic soils was registered in the fine silt fraction, while in the hydromorphic Meadow-Solonetzic and Meadow soils, the maximum humus content occurred in the clay fraction.

The regular pattern of changes in the group composition of humic substances within different particle-size fractions confirms our hypothesis on humus distribution processes in the Steppe soils studied. The highest ratio of humic acids to fulvic acids was found in the fine silt fraction, whereas the minimal ratio was observed in the colloidal fraction. The relative content of nonextractable residue (humin) in different particle-size fractions within the humus horizon increased with an increase in size of these fractions, probably due to the predominant accumulation of half-humified plant residues in the fine silt and coarser fractions.

The distribution of total nitrogen in different particle-size fractions was strongly related to the distribution of humus in these fractions and humus qualitative composition. C/N ratio decreased with a decrease in soil particle size due to increases in nitrogen-rich fulvic acids and a decreasing portion of nitrogen-poor non-extractable residue.

References

Aidinyan R.Kh. 1947. Vydelenie pochvennykh kolloidov bez khimicheskoi Obrabotki [Separation of soil colloids without chemical treatment]. *Kolloidnyi Zh.* 9(1): 3–12. (in Russian)

Goh K.M., Stout J.D., and Rafter T.A. 1977. Radiocarbon enrichment of soil organic matter fractions in New Zealand soils. *Soil Sci.* 123: 385–391.

Goh K.M., Rafter T.A., Stout J.D., and Walker T.W. 1976. The accumulation of soil organic matter and its carbon isotope content in a chronosequence of soils developed on aeolian sand in New Zealand. *J. Soil Sci.* 27: 89–100.

Grati V.P., Sikevich Z.A., and Kleshch F.I. 1965. Soderzhanie i sostav gumusa otdel'nykh mekhanicheskikh fraktsii v pochvakh Moldavii. [Humus contents and composition in particle-size fractions from Moldavian soils.] *Pochvovedenie,* 10: 72–81 (in Russian with English abstract)

Hassink J. 1997. The capacity of soils to preserve organic C and N by their association with clay and silt particles. *Soil Sci. Soc. Amer. J.* 61: 131–139.

Kocherina Ye.I. 1954. Nekotorye khimicheskie i fizicheskie svoistva otdel'nykh mekhanicheskikh fraktsii dernovo-podzolistoi pochvy. (Some chemical and physical properties of separate particle-size fractions from a soddy-podzolic soil.) *Pochvovedenie,* 12: 56 (in Russian with English abstract)

Kononova M.M. and Bel'chikova N.P. 1961. Uskorennye metody opredeleniya sostava gumusa mineral'nykh pochv [Rapid [methods to study the humus composition in mineral soils] *Pochvovedenie,* 10: 75–87 (in Russian with English abstract)

Kozyreva, M.G. and Rubilin Ye.V. 1981. O radiouglerodnom opredelenii vozrasta sovremennykh pochv i nachale ikh formirovaniya. [Determination of radiocarbon ages of surface soils and the beginning of their formation] In: *Preobrazovanie pochv Nechernozem'ya pri sel'skokhozyaistvennom osvoenii [Transformation of Soils in the Non-chernozemic Region of Russia Consequent Agricultural Development].* Moscow, p. 101. (in Russian)

Lobitskaya L.V. 1962. Raspredelenie gumusa v granulomtricheskikh fraktsiyakh chernozema, krasnozema i dernovo-podzolistoi pochvy [Humus distribution in particle-size fractions of chernozems, red ferrallitic, and soddy-podzolic soils]. *Zap. Leningr. Skhi,* 90(1): 20–25 (in Russian)

Mogilevkina I.A. 1965. Soderzhanie fiksirovannogo ammoniya v nekotorykh tipakh pochv SSSR i ikh NH_4^+-fiksiruyushchaya sposobnost [Contents of fixed ammonium in some soil types of the USSR and soil capacity for fixing NH_4^+.] *Agrokhimiya,* 7: 26–36 (in Russian)

Orlov D.S. 1990. *Gumusovye kisloty pochv i obshchaya teoriya gumifikatsii [Soil humus acids and the general theory of humification].* Mosk. Gos. Univ., Moscow, 326 pp. (in Russian)

Stevenson F.J. 1994. *Humus Chemistry: Genesis, Composition, Reactions.* John Wiley & Sons, New York. 496 pp.

Titova N.A. 1970. *Priroda gumusa i khrakter ego svyazi s mineral'noi chast'yu v pochvakh sukhostepnogo ryada. [The nature of humus and the character of its binding with the mineral soil matrix in dry-steppe soils],* PhD thesis, Dokuchaev Soil Science Institute, Moscow, p. 17 [in Russian]

Trofimenko K.I. and Kizyakov Yu.E. 1967. Organicheskoe veshchestvo otdel'nykh granulometricheskikh fraktsii osnovnykh tipov pochv Predkavkaz'ya. [Organic matter in particle-size fractions from cultivated soils of North Caucasus]. *Pochvovedenie* 2: 82–90. [in Russian with English abstract]

Vinokurov M.A. and Kornilova, A.M. 1972. Khimicheskii sostav mekhanicheskikh fraktsii svetlo-seroi lesnoi pochvy i vyshchelochennogo chernozema i izmenenie ego pod vliyaniem okul'turivaniya [Chemical composition of particle-size fractions from a light gray forest soil and its transformation upon soil cultivation] *Pochvovedenie,* 8: 30–37 [in Russian with English abstract]

Voronin A.D. 1958. Nekotorye svoistva fraktsii mekhanicheskikh elementov kompleksa pochv svetlo-kashtanovoi podzony [Some properties of particle-size fractions in soils of the light chestnut subzone]. *Vestnik Mosk. Gos. Univ. Ser. Biol., pochvoved., geol., geogr.* 4: 93–102 (in Russian).

10

Effects of Organic Acids on Desorption of Phosphate from the Surfaces of Aluminum Hydroxide and Complexes

H. Hu*, X. Li, *and* J. He

Abstract

The desorption of phosphate from aluminum hydroxides (Al(OH)x) by 0.01 mol L^{-1} KCl in the presence of different organic acids and from $Al(OH)_x$, Al-P, Al-oxalate, and Al-citrate complexes by organic acids was studied. When phosphate, adsorbed on $Al(OH)_x$ in the presence or absence of oxalate, was desorbed with 0.01 mol L^{-1} KCl (two times in sequence), total desorption rates of adsorbed phosphate in the presence of 0.5–2.0 mmol L^{-1} oxalate were 4.1%–4.8% of the amount of P adsorbed. However, in the absence of the organic acid, the desorption rate decreased to 2.1%. The amount of P desorbed by 0.01 mol L^{-1} KCl increased with increasing concentration of organic acids in the adsorption solution. Some phosphate, which could not be desorbed by oxalate, was desorbed by citrate at the same molar concentration. When phosphate was desorbed by a series of oxalate and citrate, the total desorption rate of phosphate from Al-oxalate complex was lower than that from Al hydroxide, while the rate from Al-citrate complex was the highest. Citrate could release the phosphate from the Al-P complex, and the amount of phosphate desorbed increased with increment of citrate concentration. These results seemed to indicate that the mechanism of phosphate desorption involved ligand exchange and dissolution by organic acids. Organic acids increase the desorption of phosphate and improve plant availability of phosphate in acid soils.

1 INTRODUCTION

Desorption of phosphorus in the soil is the reverse of the adsorption and fixation (Bache and Christina, 1980; He et al., 1992a). It influences the form,

**Corresponding author:* Dr. Hongqing Hu College of Resource and Environment, Huazhong Agricultural University, Wuhan 430070, P.R. China. E-mail: hqhu@mail.hzau.edu.cn.

transformation, availability, and fate of P in the environment (Barrow, 1987; Sharpley and Tunney, 2000; Wang and Chen, 1991). The impact of phosphorus on plant nutrition and water environments has been the cause of much concern. Therefore, adsorption and desorption of phosphorus have been studied extensively (Bhatti and Comerford, 2002; Hu et al., 2001, Hu et al., 2002; Hue, 1991). It has been recognized that the desorption mechanism of phosphate involves ligand exchange, diffusion and dissolution (He et al., 1992a; Nagarajah, 1968), and the relative importance of these reactions depends on the surface properties of soil colloids (He et al., 1992b; Zhao, 1988). In the field, outside the rhizosphere or organic fertilization sites the dissolution of phosphate in soils is not predominant and diffusion is more important than the exchange mechanism (Barrow, 1987; He et al., 1992b).

Organic acids are abundant in the rhizospheric and organic fertilization sites (Gerke, 1994; He et al., 1992b; Jones, 1998; Stevenson, 1967, 1986). Their role in phosphate adsorption has been widely studied (Bhatti et al., 1998; He et al., 1999; Hu, et al., 2001; Hu et al., 2002; Hue, 1991). Generally, organic acids can reduce P adsorption by soils and oxides through competition for common sites. However, the reports on phosphate desorption by organic acids are rather scarce. The role of organic acids, in P desorption may vary depending on the kind, concentration of organic acids, and pH of reaction systems (Bhatti et al., 1998, Bhatti and Comerford, 2002; Gerke, 1994; Nagarajah et al., 1968). Oxalic and citric acids are common in plant root exudates (Jones, 1998; Stevenson, 1967, 1986). They were employed to desorb phosphate from Al oxides to simulate the rhizosphere of plants in this study.

Aluminum oxides and hydroxides are common constituents in acid soils of southern and central China (Hu et al., 2001, Hu et al., 2002; Wang and Chen, 1991). Some of them exist either as noncrystalline compounds or complexes with organic/inorganic anions, or crystalline compounds such as gibbsite, or both. The adsorption/desorption of phosphate by Al oxides are very important for P fate in the environment but the surface behaviors of aluminum (hydr-) oxides are unclear up to date.

Our objective was to investigate the mechanism by which different types of organic (anions) acids desorb phosphate adsorbed on synthetic systems representing variable charged mineralogy including Al oxides, Al-P and Al-organic acid complexes. The information obtained in this study will provide a scientific basis for determining the supply of phosphate in the environment and promoting P efficiency in the soil rhizosphere zone.

2 MATERIALS AND METHODS

2.1 Preparation of Al Hydroxide Complexes

Five solutions were prepared by adding the following into 0.2 mol L^{-1} $AlCl_3$ solution: (1) H_2O; (2) 0.2 mol L^{-1} oxalic acid (OX); (3) 0.2 mol L^{-1} citric acid (CT); (4) 0.2 mol L^{-1} KH_2PO_4; (5) 0.1 mol L^{-1} KH_2PO_4–0.1 mol L^{-1} oxalic acid (OXP), to make the initial anion ligand/Al ratios of 0 or 0.1. Then 0.5 mol L^{-1} NaOH was slowly titrated (about 5 ml min^{-1}) into the solutions to increase the pH to 5.50. The suspensions were aged for 2 days, then centrifuged for 10 min at 5,000 g. The precipitates were washed with H_2O and dialyzed until Cl^- free. They were then air-dried and ground to pass through a 0.25-mm mesh sieve. The samples were (in the same order of the 5 solutions used for their preparation just described): Al(OH)x, Al-OX, Al-CT, Al-P, and Al-OXP. They were noncrystalline substances as determined by the X-ray diffraction technique (Fig. 10.1) except for Al(OH)x, which contains some crystalline compounds. The chemical constituents and the point of zero charge (PZC) of these compounds are shown in Table 10.1.

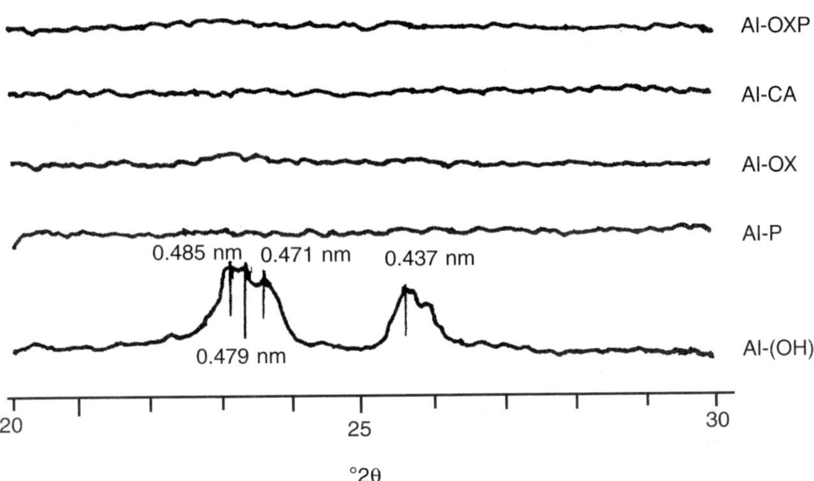

Fig. 10.1: X-ray diffraction patterns of Al hydroxide samples synthesized. Diffraction of the samples was done for Fe K_α radiation at 36 KV, 20 mA. Al(OH)x was synthesized in the absence of any interfering ligands, while Al-P, Al-OX, Al-CT and Al-OXP were synthesized in the presence of various ligands, i.e. phosphate, oxalate, citrate and oxalate + phosphate, respectively.

2.2 Experimental Methods

Phosphate adsorption

Adsorption of phosphate on the Al hydroxide complexes was carried out in four series (Table 10.2). All P solutions were made from KH_2PO_4 and the ionic strength was kept at 0.01. Samples of 50 mg Al(OH)x (Al hydroxide) were added to 25 ml of the solutions of the above series. The combinations were shaken for 2 h at 25°C, equilibrated for 24 h, and centrifuged. Concentrations of P in equilibrating solutions were determined by the molybdenum-blue colorimetric

Table 10.1: Chemical constituents (dissolved in 1 mol L^{-1} HCl) and point of zero charge (PZC) of the samples synthesized

Samples	Chemical constituents (mg/g)		PZC
	Al	ligands	
$Al(OH)_x$	281	—	6.78
Al-OX	220	202	7.42
Al-CT	230	588	7.14
Al-P	246	146	7.31
Al-OXP	249	78 PO_4, 189 OX	7.18

Table 10.2: Treatments of phosphate adsorption on $Al(OH)_x$ surface

Series	pH	Cp (mg L^{-1})	Cox (mmol L^{-1})
I	5.50	20, 40, 80, 160, 320	0
II	3–8	40	0
III	5.50	40	0.5, 1, 2
IV	3–8	40	0.8 or 1.0

method. The amount of P adsorbed was calculated by the difference of initial and final concentrations of P in solution. The remaining Al hydroxide solid was weighed to determine the amount of the remaining solution and the amount of the remaining P was deducted for the desorption experiments.

Phosphate desorption

The desorbents used were 0.01 mol L^{-1} KCl, 0.1–50 mmol L^{-1} citric acid and 0.1–10 mmol L^{-1} oxalic acid. To compare the desorption of phosphorus in two situations, the desorbents were individually or sequentially added into the samples of the preceding adsorption experiment at a solid:solution ratio of 1:500 (w/w). The combinations were shaken for 2 h at 25°C, equilibrated for 24 h, and centrifuged. Concentrations of P of the equilibrium solutions were determined by the molybdenum-blue colorimetric method. The rate of desorption at time n (Rn) was calculated using the following equation:

$$R_n = A_n/(Q - A_{n-1})$$

where A_n is the amount of P desorbed at time n; Q is the amount of P adsorbed; A_{n-1} is the sum of P desorbed from time 1 to time n–1.

The total rate of P desorption was calculated as the sum of P desorbed divided by the amount of P adsorbed.

The desorption experiments were replicated twice.

3 RESULTS

3.1 Desorption of Phosphate Adsorbed on $Al(OH)_x$ in the Presence and Absence of Oxalate by KCl

In absence of oxalate

Desorption rate of phosphate adsorbed on $Al(OH)_x$ surface at various pH values by 0.01 mol L^{-1} KCl in the absence of oxalate (series II in Table 10.2) was affected by pH (Fig. 10.2). It was lowest at pH 4 and slightly higher at pH 3 than pH 4. This might be caused by the different adsorption states of P at the two pH values (Liu et al., 1995a). Increasing pH from 4 to 8 increased desorption rate of P for the first desorption and for total desorption. It was reported that the amount of P adsorbed decreased with increasing pH of equilibrium solution due to the competition of OH^- (Liu et al., 1995a; Nagarajah et al., 1968). The greater the pH of the initial solution, the higher the pH of the equilibrating solution.

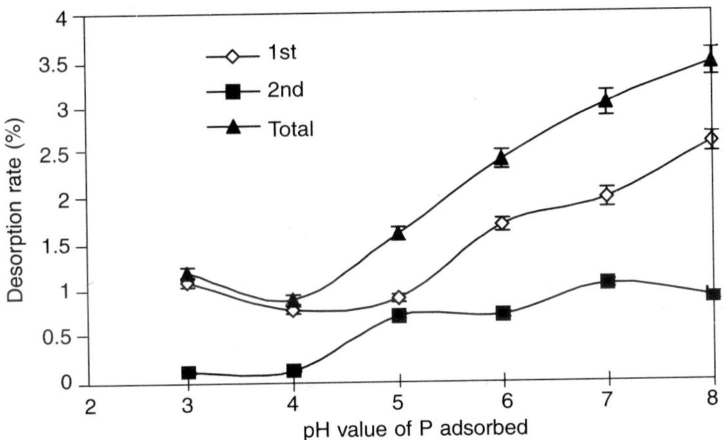

Fig. 10.2: Desorption rate of phosphate adsorbed on $Al(OH)_x$ surface at various pH by 0.01 mol L^{-1} KCl. (Vertical bars represent standard deviation.)

In Presence of oxalate

After P was adsorbed from the combined solution of 40 mg L^{-1} P and oxalate at pH 5.5 (series III in Table 10.2), the desorption of P by 0.01 mol L^{-1} KCl increased with increasing concentrations of oxalate (Table 10.3). This showed that the desorption of P was much easier when oxalate was present because of the competition between P and OX. It was probably related to the desorption of OX by KCl and the replacement of P with OX at the adsorption sites.

The desorption rate of P adsorbed in the presence of 1.0 mmol L^{-1} oxalate (sample series IV) by KCl increased from pH 3 to 6 and decreased from pH 6 to 8 (Fig. 10.3). This was likely due to valence change of oxalate (pK_1 =1.25, pK_2 = 4.27) (Lide, 1994). When the pH increased from 3 to 6, oxalate changed from

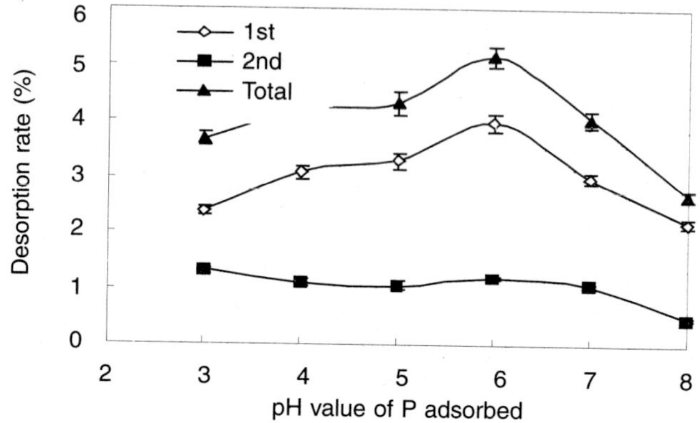

Fig. 10.3: Desorption rate of phosphate adsorbed on Al(OH)x in the presence of 1.0 mmol L^{-1} oxalate at various pH by 0.01 mol L^{-1} KCl. (Vertical bars represent standard deviation.)

Table 10.3: Desorption rate of phosphate adsorbed on Al(OH)x by 0.01 mol L^{-1} KCl in the presence of oxalate (pH 5.5).

Sequence of desorption*	Concentration of oxalate (mmol L^{-1})			
	0	0.5	1	2
		%		
1st	1.62	2.26	3.56	3.68
2nd	0.43	1.88	1.09	1.11
Total	2.05	4.10	4.49	4.75

*Desorption was conducted twice in sequence using 0.01 mol L^{-1} KCl.

monovalence to divalence while phosphate remained monovalent. Adsorption affinity generally increased as valence increased; hence competition of oxalate became stronger as pH increased. However, as pH increased further from 6 to 8, oxalate competition stabilized and desorption of P decreased.

3.2 Desorption of P Adsorbed on $Al(OH)_x$ and Al-organic Acid Complexes by Organic Acid Anions

When 1.0 mmol L^{-1} oxalate was used to desorb P adsorbed on $Al(OH)_x$ surfaces (series I in Table 10.2), desorption rates were 2.1% for 20 mg P L^{-1} and 8.6%–11.3% for 40–320 mg P L^{-1} (Table 10.4). When 1 mmol L^{-1} citrate was used to desorb P after oxalate desorption, the desorption rate of adsorbed P was low at 20 mg P L^{-1}, but relatively high without substantial difference at 40–320 mg P L^{-1}. Total desorption rate was similiar to the first desorption: 8.9% at 20 mg L^{-1} and 19.2%–20.9% at 40–320 mg L^{-1}.

Table 10.4: Desorption rate (%) of different concentration of phosphate adsorbed on several Al oxides by two organic acids

Adsorbent	Desorption reagent used in sequence	Concentration of phosphate in solution (mg P L^{-1})				
		20	40	80	160	320
$Al(OH)_x$	1 mmol L^{-1} OX	2.1	8.6	9.8	10.5	11.3
	1 mmol L^{-1} CT	6.9	11.6	10.6	10.2	11.1
	Total[a]	8.9	19.2	19.3	19.7	20.9
Al-OX	0.1 mmol L^{-1} OX	1.5	1.7	2.2	3.4	4.7
	0.1 mmol L^{-1} CT	1.2	1.6	2.1	2.4	3.1
	Total[a]	2.7	3.2	4.3	5.7	7.7
Al-CT	0.1 mmol L^{-1} CT	16.0	10.4	9.7	7.8	5.3
	1 mmol L^{-1} CT	15.6	14.2	9.7	7.3	9.8
	Total[a]	29.2	23.1	18.4	14.6	14.5

[a]The total rate of P desorption was calculated as the sum of P desorbed divided by the amount of P adsorbed.

When oxalate and citrate at 0.1 mmol L^{-1} were used sequentially to desorb phosphate adsorbed on the Al-OX complex, individual and total desorption rates increased with added concentration of P (Table 10.4). Due to the presence of oxalate on the Al-OX complex, increments in P were adsorbed through the exchange with easily desorbed OX.

When 0.1 mmol L^{-1} and 1.0 mmol L^{-1} citrate were used sequentially to desorb the P adsorbed on the Al-CT complex, the desorption rate of P decreased with increments in P concentration. The reason could well be a very high affinity between citrate and Al for the Al-CT complex; hence phosphate could not replace the citrate from the surface of the Al-CT complex, but bonded instead with the exposed aluminum ion to form Al-H_2PO_4. Furthermore, it would be difficult for P to be desorbed by citrate at low concentrations of the organic acid.

Meanwhile, some phosphate that could not be desorbed by a low concentration of organic acids could be desorbed with a higher concentration of an organic acid. For example, 1 mmol L^{-1} citrate could desorb P after it had been desorbed by 0.1 mmol L^{-1} citrate on the Al-CT complex (Table 10.4). At the same concentration of organic ligand, citrate could desorb phosphate better than oxalate (samples of Al(OH)x and Al-OX in Table 10.4). This was due to their different affinity to Al.

Desorption of P on Al(OH)x and Al-OX by the organic acid varied with the pH of the adsorption system (Table 10.5). When the pH of the P adsorption system increased from 3 to 8, the desorption rate of phosphate from Al(OH)x and Al-OX complex decreased from 46.6% to 32.5% and from 37.5% to 33.4% respectively. Under the same desorption conditions, the desorption rate of P from $Al(OH)_x$ was greater than that from Al-OX complex.

Table 10.5: Desorption rate (%) of phosphate adsorbed on Al oxides at different pH by citrate at various concentrations

Adsorbents	Desorp. order	Concn. of citrate (mol L^{-1})	pH of phosphate solution					
			3	4	5	6	7	8
Al(OH)x	1st	0.01	24.4	22.2	21.3	19.5	16.4	15.1
	2nd	0.01	15.8	12.8	12.6	12.7	10.6	9.6
	3rd	0.05	15.4	13.0	14.3	13.6	11.8	11.9
	4th	0.05	1.8	0.6	0.1	0.7	0.2	0.1
	Total[a]		46.6	42.1	41.1	39.6	34.1	32.5
Al-OX	1st	0.01	13.1	12.6	11.8	12.1	11.0	11.3
	2nd	0.01	12.4	12.1	11.8	10.7	11.3	10.1
	3rd	0.05	16.3	14.8	15.1	15.9	16.3	15.2
	4th	0.05	2.0	0.8	1.6	0.9	0.7	1.6
	Total[a]		37.5	35.1	35.0	34.8	34.4	33.4

[a]The total rate of P desorption was calculated as the of P desorbed divided by the amount of P adsorbed.

Desorption rate of phosphate from $Al(OH)_x$ by 0.05 mol L^{-1} citrate in the fourth run was as low as 0.1%–1.8% (Table 10.5). Thus, it can be deduced that the total desorption of four runs would approach the whole desorbable P. This shows that 53.4%–67.5% of P adsorbed on Al(OH)x and 62.5%–66.6% on Al-OX could not be desorbed by 0.05 mol L^{-1} citrate. The results demonstrated that Al oxides displayed very strong adsorption of phosphate.

A comparison of P desorption rates in Table 10.5 and Figure 10.2 showed that the desorption rates by organic acids were much greater than by KCl at the same concentration of P. The desorption rate of P by 0.01 mol L^{-1} KCl was lowest at pH 4 while that by citrate declined steadily with increaments in pH. This tends to illuminate different desorption mechanisms involved in the two desorption reactions. For KCl, the desorption of P would deplete protons as shown by the pH increase, while for citric acid the main mechanism would be ligand exchange (Nagarajah et al., 1968).

3.3 Desorption of P Adsorbed on $Al(OH)_x$ and Al-organic Acid Complexes by Organic Acids as Affected by pH

Phosphate adsorbed on various Al oxide complexes at different pH levels was desorbed sequentially 8 times with oxalate and citrate of various concentrations. As for the Al(OH)x, the desorption rate was 31.4%–43.1% (Table 10.6), i.e., more than half of phosphate adsorbed in the presence of oxalate could not be desorbed by the two organic acids. Desorption rate of P adsorbed on Al-OX complex was 25.3%–35.9%. Obviously the P adsorbed on the Al-OX complex was more difficult to desorb than that on $Al(OH)_x$ in the presence of oxalate. These implied

Table 10.6: Desorption rate (%) of phosphate adsorbed in the mixture of 0.8 mmol L^{-1} oxalate and 40 mg L^{-1} P* on Al oxides by various organic acids

Adsorbents	pH of adsorp. solution	Sequence of P desorption								Total
		1	2	3	4	5	6	7	8	
		Desorption reagent and concentration (mmol/L^{-1})								
		0.1 OX	1.0 OX	10 OX	10 OX	10 CT	10 CT	50 CT	50 CT	
Al(OH)x	3	1.6	5.2	12.1	10.1	14.5	9.0	0.8	0	43.1
	4	1.6	3.9	9.8	8.2	12.9	7.9	0.6	0	37.5
	5	1.3	3.1	9.1	8.3	11.4	7.8	0.7	0	35.3
	6	1.1	3.4	9.0	8.1	11.7	7.9	0.5	0	35.2
	7	1.5	3.8	8.5	7.3	11.6	7.4	0.8	0	34.7
	8	0.8	2.3	7.8	6.4	10.3	7.5	1.0	0	31.4
Al-OX	3	1.1	2.9	8.7	5.1	13.9	10.1	0.6	0	35.9
	4	1.0	2.7	7.1	3.6	10.5	9.2	0.6	0	30.2
	5	0.3	2.0	5.8	3.5	10.6	9.9	0.9	0	29.1
	6	0.4	0.8	6.0	3.9	10.1	8.8	0.6	0	28.3
	7	0.8	1.9	5.5	3.6	9.9	9.8	0.6	0	28.2
	8	0.2	1.3	5.1	3.3	8.5	9.0	0.5	0	25.3
Al-CT	3	9.0	8.1	22.0	9.5	14.3	13.9	nd*	nd	56.4
	4	7.8	14.7	21.3	6.5	12.0	8.5	nd	nd	53.8
	5	9.3	14.4	25.6	9.2	18.7	20.0	nd	nd	65.9
	6	7.0	9.5	24.8	24.0	21.4	16.1	nd	nd	68.3
	7	7.9	17.9	27.3	9.8	21.3	14.7	nd	nd	66.7
	8	2.2	6.8	21.5	8.4	16.6	10.2	nd	nd	50.9

Notes: nd = not determined.

*See Seires IV in Table 10.2.

that the presence of oxalate in Al oxide complexes would result in more difficult desorption of P from Al oxides. Furthermore, the more the P the stronger the bond in the presence than in the absence of an organic acid.

In the presence of oxalate, the P adsorbed on the Al-CT complex was desorbed by organic acids in larger quantities than on the Al-OX complex. Desorption rate (6 times in sequence) by the organic acids ranged from 50.9% to 68.3% and reached the highest value around pH 6. This might indicate that P occupies lower energy sites in the Al-CT complex in the presence of oxalate than in the Al-OX complex.

Some other important observations that may be drawn from Table 10.6 are as follows: The desorption rate of phosphate on Al oxide complexes increased with increment in desorption reagent concentration. At the same concentration of desorption reagent, citrate desorbed more P than oxalate. For the same organic acid, the desorption of P at the second run was usually lower than the first run.

The cumulative desorption rate of P was lower at high pH of the adsorption system.

3.4 Desorption of Phosphate from P-containing Al Oxides by Citric Acid

When 0.1–2.0 mmol L^{-1} citric acid was added to Al-P and Al-OXP complexes, the amounts of P released were 1.20–3.11 cmol kg^{-1} for Al-P and 4.68–11.4 cmol kg^{-1} for Al-OXP (Table 10.7). The amounts accounted for 0.78%–2.03% and 5.67%–13.8% of total P in the samples respectively. Desorption of P from Al-OXP complex was greater than from the Al-P complex. Results suggested that the availability of P in the rhizosphere would be enhanced when Al and P precipitated in the presence of oxalate.

Table 10.7: Amount of P desorbed from Al-P and Al-OXP complexes by citrate

Phosphorus compounds	Concentration of citrate (mmol L^{-1})				
	0.1	0.2	0.5	1.0	2.0
	cmol kg^{-1}				
Al-P	1.20	1.60	2.05	2.64	3.11
Al-OXP	4.68	6.08	9.95	11.10	11.40

4 DISCUSSION

4.1 Desorption of Phosphate and Organic Acid Adsorbed by Indifferent Ions

When 0.01 mol L^{-1} KCl was used to desorb adsorbed phosphate, there was no doubt that some organic acids would be desorbed as well. The following changes would take place given the introduction of organic acids. First, the concentration of organic acid in the solution would increase, resulting in increasing exchange of adsorbed phosphate. Second, the ionic strength of solution would increase, bringing about disturbance of the equilibrium of ion concentration within the diffused double layer; consequently phosphate would be desorbed through the diffusion effect. The finding that phosphate adsorbed in the presence of the organic acid was easily desorbed might be related to the exchange of the organic acid and P. The mixture of organic matter and P could promote the availability of P in the soils with variable charges. One reason is that P has higher desorption rate when adsorbed in the presence of organic acids.

The highest desorption rate of phosphate adsorbed on the $Al(OH)_x$ was found at pH 6 in the presence of oxalate (Fig 10.3); it decreased sharply when pH increased to 8. The change in pH corresponded to P states at different pHs and their competition power with organic acids on the Al oxide surface. Results suggest that soil pH will be raised when organic substances and P are applied in combination to acid soils.

4.2 Desorption of Phosphate by Organic Anions

Desorption rate of P by organic acids (organic anions) gradually increased when P concentration rose from 20 to 320 mg L^{-1}. It decreased with increasing pH of P solution. These were related to the coordination of P at different concentration and pH ranges. At low P concentrations, P formed binuclear or polynuclear coordination with Fe and Al so P was difficult to desorb; at high pH, P adsorption was the same as at low concentration (Liu et al., 1995b). Although these results were reported with inorganic desorption reagent, similar effects were expected for organic acids.

Desorption of P on Al oxides by organic acids increased with increasing concentration of organic acids. At the same concentration, citrate could desorb more P than oxalate. This was related to the principle of ligand exchange and the affinity between Al oxides and ligand anions. Citrate had stronger affinity to Al than oxalate (Hu et al., 2001).

The P-contained in Al oxides is one of the main P pools in acid soils but is not available to plants. Some of the P could be desorbed by organic acids, thereby improving its availability. Our results showed that the P in the Al-OXP complex was more readily replaced by citrate than Al-P. This implies that oxalate and citrate will increase the availability of P in the rhizosphere.

4.3 Availability of P Adsorbed on Al Complexes

Oxalate and citrate were employed to desorb phosphate from Al oxides to simulate the rhizosphere of plants. The accumulative desorption rates were less than 44% for the $Al(OH)_x$ and Al-OX complex (both were desorbed for 8 times in sequence) and 50–68% for Al-CT complex (desorbed for 6 times in sequence). Some phosphate could not be desorbed even after 50 mmol L^{-1} citric acid was used for the 8th time. However, more than half of the P adsorbed on the Al-CT complex could be desorbed. This showed that the organic acids are more effective than KCl in releasing P from the Al-CT complex.

Acknowledgements

This study was financed by the Natural Science Foundation of China (Grant No. 40371065). We thank Dr C. Tang for invaluable comments on the manuscript.

References

Bache B.W. and Christina I. 1980. Desorption of phosphate from soil using anion exchange resins. *J. Soil Sci.* 31: 297–306.

Barrow N.J. 1987. Reactions with variable charge soils. *Fertil. Res.* 14(1): (special issue).

Bhatti J.S. and Comerford N.B. 2002. Measurement of P desorption from a spodic horizon using different methods and pH control. *Commun. in Soil Sci. Plant Anal.* 33: 845–853.

Bhatti J.S., Comerford N.B., and Johnston C.T. 1998. Influence of oxalate and soil organic matter on sorption and desorption of phosphate onto a Spodic horizon. *Soil Sci. Soc. Amer. J.* 62: 1089–1095.

Gerke J. 1994. Kinetics of soil phosphate desorption as affected by citric acid. *Z. Pflanzenernahr Bodenk* 157: 17–22.

He J. Z., De Cristofaro A., and Violante A. 1999. Comparison of adsorption of phosphate, tartrate and oxalate on hydroxyl aluminum montmorillonite complexes. *Clays, Clay Miner.* 47: 226–233.

He Z.L., Yuan K.N., and Zhu Z.X. 1992a. Studies on desorption mechanism of phosphate influenced by electrolyte kinds and concentration. *Acta. Pedological* 29: 26–33. (in Chinese).

He Z.L., Yuan K.N., and Zhu Z.X. 1992b. Effects of organic anions on phosphate adsorption and desorption from variable charge clay minerals and soils. *Pedosphere* 2(1):1–11.

Hu H.Q., He J.Z., and Li X.Y. 2002. Effects of organic ligands on adsorption of phosphate on a non-crystalline Al hydroxide. In: *Developments in Soil Science*, A. Violante et al. (eds.). Elsevier, Amsterdam, The Netherlands, vol. 28A, pp. 311–317.

Hu H.Q., He J.Z., Li X.Y., and Liu F. 2001. Effect of organic acids on phosphate adsorption by variable charge soils of central China. *Environ. Int.* 26(5 & 6): 353–358.

Hue N.V. 1991. Effects of organic acids/anions on P sorption and phytoavailability in soils with different mineralogies. *Soil Sci.* 152(6): 463–471.

Jones D.L. 1998. Organic acids in the rhizosphere—a critical review. *Plant Soil* 205: 25–44.

Lide D.R. 1994. *Handbook of Chemistry and Physics*. CRC Press, Boca Raton, FL, (74th ed.).

Liu F., Jie X.L., Zhou D.H., and Li X.Y. 1995a. Effect of pH on chemical forms of phosphate on surface of goethite. *Pedosphere* 5: 229–235.

Liu F., He J.Z., Li X.Y., and Wang D.F. 1995b. Chemical states of phosphorus adsorbed on goethite surfaces at various phosphate concentrations. *Chinese Sci. Bull.* 40: 506–511.

Nagarajah, S., Posner, A.M., and Quirk, J.P. 1968. Desorption of phosphate from kaolinite by citrate and bicarbonate. *Soil Sci. Soc. Amer. Proc.* 32: 507–510.

Sharpley A. and Tunney H. 2000. Phosphorus research strategies to meet agricultural and environmental challenges of the 21st century. *J. Environ. Qual.* 29: 176–181.

Stevenson F.J. 1967. Organic acids in soil. In: *Soil Biochemistry.* A.D. McLaren and J.H. Peterson (eds). Marcel Dekker, New York, NY, pp. 119–146.

Stevenson F.J. 1986. *Cycles of Soil Carbon, Nitrogen, Phosphorus, Sulfur, Micronutrients*. John Wiley & Sons, New York, NY, 380 pp.

Wang J.L., and Chen J.F. 1991. Desorption characteristics of P on the surface of variable charge soils. *Acta Pedol.* 28: 14–23 (in Chinese).

Zhao M.Z. 1988. Desorption of P from several soils and clay minerals. *Acta Pedol.* 25: 156–161. (in Chinese).

11

Pedotransfer Function for Estimation of Phosphate Adsorption Capacity on a Wide Range of Soils

O.K. Borggaard*, L.H. Rasmussen, A.L. Gimsing, ***and*** **C. Szilas**

Abstract

To make plant nutritional and environmental needs compatible, a soil must contain sufficient available phosphate in the root zone to cover plant need, but to protect surface waters against phosphate pollution, leaching losses must be minimized. As the phosphate adsorption capacity (PAC) together with phosphate content determines the magnitude of available and fixed phosphate in soils, PAC is an indispensable key parameter in handling these different needs. Accordingly, PAC has been found important in estimating the plant available phosphate pool in soils and in predicting the risk of phosphate pollution of the aquatic environment due to overfertilization and to re-establishment of formerly drained and cultivated wetlands enriched in phosphate. Furthermore, PAC may be an important factor in assessing the risk of drainage and groundwater contamination with the widely used glyphosate herbicide because phosphate already bound on soil adsorption sites may inhibit glyphosate adsorption. PAC can therefore be considered an important soil quality indicator in relation to ecosystem functioning and human welfare.

As aluminium and iron oxides are the main phosphate adsorbents in many soils, PAC can be predicted by pedotransfer functions (PTFs) based on various aluminium and iron oxide fractions such as oxalate-extractable aluminum and iron (Al_{Ox}, Fe_{Ox}) and dithionite-citrate-bicarbonate-extractable iron (Fe_{DCB}). The experimentally determined PAC taken as the sum of oxalate-extractable phosphate (P_{Ox}) and the Langmuir maximum of the phosphate adsorption isotherm, were found to be well predicted by the PTF: $P_{calc} = 0.207 \times Al_{Ox} +$

**Corresponding author*: Dr. O.K. Borggaard, Department of Natural Sciences, Royal Veterinary and Agricultural University, 40 Thorvaldsensvej, DK-1871 Frederiksberg C, Denmark.
E-mail: okb@kvl.dk.

$0.150 \times Fe_{Ox} + 0.016 \times (Fe_{DCB} - Fe_{Ox})$ for a wide range of soil samples. The soil samples were from Denmark, Canada, Ghana, and Tanzania representing noncalcareous Alfisols, Entisols, Histosols, Inceptisols, Mollisols, Oxisols, Spodosols, and Ultisols. The samples included in the test totaled 95. Even though individual groups of soils such as some Spodosols were rather well described by a simpler PTF (e.g. $P_{calc} = \alpha \times (Al_{Ox} + Fe_{Ox})$), the proposed PTF is to be preferred given its considerably wider application range, since it takes into account the various contributions from amorphous and poorly crystalline aluminum and iron oxides (Al_{Ox}, Fe_{Ox}) and includes the effect of crystalline iron oxides ($Fe_{DCB} - Fe_{Ox}$). The latter term is especially important in old, strongly weathered soils in tropical and subtropical regions enriched in well-crystallized iron oxides that are less reactive than amorphous and poorly crystalline oxides. However, while close or fairly close relationships between experimentally determined and predicted PACs were obtained with these 95 soil samples, the PTF failed to predict PAC of two Tanzanian Andisols. Future studies are needed to more precisely delineate groups of soils falling within the range of application of the proposed PTF from those falling outside its application range.

1 INTRODUCTION

To make plant nutritional and environmental needs compatible, modern agriculture faces two problems with phosphate, which may have serious impacts on ecosystem functioning and human welfare. The soil must contain sufficient available phosphate in the rhizosphere to cover plant needs but to protect surface waters against phosphate pollution, leaching losses must be minimised (Brady and Weil, 1999). To ensure adequate supply of plants with this essential element, the soil solution in the root zone must, depending on the crop, contain 1–10 μM phosphate corresponding to 0.03–0.3 mg P L^{-1} (Fox, 1981; Marschner, 1998). On the other hand, phosphate concentrations above 1–3 μM (0.03–0.1 mg P L^{-1}) may trigger eutrofication in open waters (Sharpley and Menzel, 1987; Taylor and Kilmer, 1980). Lack of available phosphate is a serious problem in many lesser developed countries of tropical and subtropical regions, while phosphate pollution of the aquatic environment due mainly to long-term overfertilization of cultivated soils, is a serious and widespread problem in many industrialized countries in Europe and North America (Brady and Weil, 1999; Del Campillo et al., 1999; Kang, 1989; Leinweber et al., 1999; Lookman et al., 1995; Maguire and Sims, 2002; Menon et al., 1991; Novak et al., 2000; Szilas, 2002). Furthermore, high phosphate concentrations are suspected to lead to pollution of the aquatic environment given the wide use of glyphosate herbicide, because phosphate and glyphosate compete for adsorption sites (Gimsing and Borggaard, 2001, 2002a, b; De Jonge et al., 2001; Dion et al., 2001).

Phosphate is strongly adsorbed by most soils as determined by the number of adsorption sites, which, vary greatly among soils (Borggaard, 1990; Borggaard et al., 1990; Freese et al., 1992; Singh and Gilkes, 1991; Yuan and Lavkulich, 1994). Therefore, the availability of soil phosphate as well as the soil solution concentration of phosphate will depend on the degree of phosphate saturation, rather than on the total phosphate content (Del Campillo et al., 1999; Kleinman et al., 1999; Owusu-Bennoah et al., 1997; Szilas, 2002). Phosphate saturation is the proportion of adsorption sites occupied by phosphate, which is normally taken as the ratio between adsorbed phosphate and the phosphate adsorption capacity (PAC) of the soil (Borggaard and Møberg, 1991; Freese et al., 1992; Hooda et al., 2000; Leinweber et al., 1999; Lookman et al., 1995; Maguire and Sims, 2002). From plant growth and environmental points of view, phosphate saturation and PAC can be considered soil quality indicators.

Close correlations have been found between a soil's capacity to adsorb phosphate and its content of aluminum and iron oxides, suggesting these oxides to be the main phosphate adsorbents in soils (Borggaard, 1990, 2002; Freese et al., 1992; Torrent, 1997; van der Zee and van Riemsdijk, 1986). In contrast, soil organic matter (SOM) and 2:1 layer silicates, which play a decisive role for many other important soil properties, have limited capacity to adsorb phosphate (Borggaard, 2002; Brady and Weil, 1999; Gimsing and Borggaard, 2002a, b). However, SOM may at least transiently block adsorption sites and since PAC of aluminum and iron oxides is closely related to their specific surface area, SOM can indirectly affect soil PAC by retarding crystal growth of amorphous and poorly crystalline aluminum and iron oxides, which because of high specific surface areas have very high PACs (Afif et al., 1995; Borggaard, 1983a; Borggaard et al., 1990; Huang and Schnitzer, 1986; Huang and Wang, 1997; Torrent, 1997).

Experimentally, PAC can be determined by applying the Langmuir equation to phosphate adsorption data (phosphate adsorption isotherm) obtained by measuring the amounts of phosphate adsorbed at different solution concentrations (Borggaard et al., 1990; Yuan and Lavkulich, 1994). Since aluminium and iron oxides are the main or only phosphate adsorbents in many soils, PAC can also be estimated by means of a pedotransfer function (PTF) based on various aluminium and iron oxides pools (Borggaard, 1986, 2002; Borggaard et al., 1990; Freese et al., 1992; Yuan and Lavkulich, 1994). To use a PTF is clearly an advantage as it is much cheaper and less time consuming than to determine the phosphate adsorption capacity by means of phosphate adsorption experiments (Kleinman et al., 1999; McBratney et al., 2002). Reliable prediction of PAC is a prerequisite for using a PTF. Unfortunately, fair predictability of PAC by existing PTFs is limited to a rather narrow soils range. Thus, various PTFs were needed for PAC prediction of aluminum oxide-dominated soils in south-western Australia (Singh and Gilkes, 1991), of some iron oxide-dominated soils from Denmark and Tanzania (Borggaard, 1983b, 1986), of Mediterranean soils (Pena and Torrent, 1984), of heavily P fertilized-soils in northwestern Europe (Freese et al., 1992) and of Northeuropean and Canadian sandy soils (Borggaard et al., 1990; van der Zee and van Riemsdijk, 1986; Yuan and Lavkulich, 1994). Considering the plant-nutritional and environmental importance of soil PAC, a reliable PTF that can be used on a wider range of soils is greatly needed and would be very much appreciated.

This chapter evaluates the prediction of soil PAC by commonly used PTFs and proposes a PTF that can be used on a wider range of soils than presently possible. Accordingly, data from a large number of contrasting soil samples from various parts of the world were tested on different PTFs by multiple linear regression analysis to ascertain the best function. In addition, potential PTF application areas are pinpointed with emphasis on plant growth and environmental issues. These issues include PAC, and hence PTFs, in relation to availability of phosphate both as a plant nutrient and as a water pollutant. An additional, indirect application area comprises water pollution with glyphosate because high phosphate contents compared to PAC can cause waters to be contaminated by glyphosate, which competes with phosphate for adsorption sites on soil solids.

2 SOILS AND ANALYTICAL METHODS

In the evaluation of existing PTFs as well as in the development of a more widely applicable PTF, a rather high number of contrasting soil samples from different parts of the world were used. The soils were characterized by different analytical methods and both the methods and soils are presented below.

2.1 Soils

A number of soil samples from Denmark, Ghana, and Tanzania comprising Alfisols, Entisols, Histosols, Inceptisols, Mollisols, Oxisols, Spodosols, and Ultisols according to the Soil Taxonomy system (USDA, Soil Survey Staff, 1999) were used in the investigation. Most of the samples are from soils previously described (Borggaard, 1983b, 1986; Borggaard et al., 1990; Borggaard and Møberg, 1991; Møberg, 1973; Owusu-Bennoah et al., 1997; Szilas, 2002; Szilas et al., 1998). Names and classification of the soils according to the WRB system (Belgium ISSS Working Group RB, 1998) are shown in Table 11.1 together with other relevant data of the various soil samples. These soil characteristics were obtained by the analytical methods described below.

2.2 Analytical Methods

The pH was measured potentiometrically in a 1:2.5 soil to water suspension or in a 1:2.5 soil to 0.01 M calcium chloride ($CaCl_2$) suspension with a pH-meter equipped with glass and calomel electrodes, either as two separate electrodes or in a combination electrode. Soil-water suspensions were used with all tropical and some Danish soil samples, while pH in the remaining Danish samples was determined in 0.01 M $CaCl_2$. To make the two kinds of pHs comparable, pHs in 0.01 M $CaCl_2$ were added 0.5 and reported as pH_{H2O} ($pH_{H_2O} = pH_{CaCl_2} + 0.5$) (Borggaard, 2002). The organic carbon content was measured by dry combustion in an oxygen atmosphere and either detection of evolved CO_2 by an infrared detector in a computer-controlled Eltra CS 500 analyser or by gravimetric determination of the evolved CO_2 in a LECO apparatus (or LECO-like

Table 11.1: Characteristics of the Danish, Ghanaian and Tanzanian soils used.

Name	Horizon/ Depth	WRB classification	pH_{H2O}	Org. C (%)	Clay (%)	Al_{Ox}	Fe_{Ox}	Fe_{DCB}	P_{max}
						(mmol kg^{-1})			
Danish soil samples									
Asnæs	Ap (0–20)	Eutric Cambisol	5.9	1.5	10	10.5	50.1	91.7	10.3*
	Bw (30–50)		5.5	0.5	9	12.0	21.5	88.9	7.1*
Dannemare	Cg (90–110)	Mollic Fluvisol	8.0	0.1	29	19.9	67.8	159.0	23.3*
Farris	Cg1 (30–50)	Mollic Gleysol	7.5	0.8	24	52.0	181.0	263	27.0*
Gengård 1	Ap (0–20)	Gleyic Arenosol	6.1	1.6	1	14.0	170.0	291	23.0
Gengård 2	Bhs (29–47)	Gleyic Arenosol	5.1	0.6	2	9.0	344	512	57.0
Hem	Ap (0–25)	Eutric Cambisol	6.8	1.0	19	25.9	51.7	150.0	10.7*
Holmsgård	Bt (45–70)	Haplic Luvisol	5.2	0.2	46	66.2	85.9	551	41.1*
Højbakke	Ap (0–25)	Luvic Phaeozem	7.3	1.5	16	29.1	46.6	109.2	16.0
	E (25–40)		7.2	0.4	20	33.3	35.8	134.3	15.2
	Bt1 (40–80)		7.3	0.3	26	33.3	32.2	159.4	18.9
	Bt2 (80–120)		7.9	0.1	22	25.9	26.9	139.7	19.1
Knudshoved	Bt (74–130)	Umbric Alisol	4.6	0.2	30	33.3	48.1	207	22.6*
Kærenge	Ap (0–25)	Histic Fluvisol	4.8	13.9	—	102.0	505	657	102.0
Rørskifte	Ap (0–29)	Thionic Histosol	5.4	17.5	—	118	253	291	76.0
Galtlund	A (0–5)	Umbric Podzol	3.8	3.1	1	9.4	3.0	8.3	3.7
	E (5–20)		4.0	0.6	1	2.6	1.4	1.8	1.9
	Bh (20–25)		4.3	3.4	8	66.4	165.0	200	36.3
	Bhs (25–45)		4.6	0.9	6	155.0	38.0	118.0	46.7
	Bs1 (45–90)		4.7	0.3	6	82.0	13.0	69.0	26.5
Lundgård	Ap (0–35)	Fragic Podzol	4.9	1.9	2	24.3	13.4	20.6	7.0
	Bh (35–50)		6.0	1.7	1	54.4	16.9	20.2	12.4
	Bs (50–90)		5.9	0.3	1	34.2	12.4	18.1	10.1
Fire Huse	Bhs (28–50)	Umbric Cambisol	5.7	0.9	4	110.7	16.9	101.1	25.0
	C (60–125)		5.4	0.2	1	29.8	19.5	67.6	11.0
Lundbæk	Bhs (35–43)	Humic Podzol	4.3	1.4	2	65.2	1.3	2.7	17.2
Stråsø	Bhb (100–110)	Spodic Arenosol	4.3	4.3	2	125.6	5.5	8.1	27.3
	Bhsb (110–120)		4.8	1.1	1	98.6	8.4	13.2	23.3
Ghanaian soil samples									
Abenia	A (0–15)	Haplic Ferralsol	4.0	2.6	27	28	29	145	17.0
Ankasa	A (0–15)	Haplic Ferralsol	4.4	2.2	20	27	13	88	13.1
Aiyinasi	A (0–15)	Haplic Ferralsol	4.3	2.2	18	29	16	40	13.8
Boi	A (0–15)	Ferric Plinthosol	4.1	1.7	40	30	28	660	17.2
Kwaben	A (0–15)	Haplic Ferralsol	4.9	1.4	17	15	14	55	9.3
Tiboko	A (0–15)	Haplic Ferralsol	5.0	1.0	18	14	10	168	10.3

Contd...

(Table 11.1 Contd.)

Tanzanian soil samples									
Igabiro	Ap (0–18)	Umbric Ferralsol	5.6	1.5	26	59.6	18.9	207	16.8
Ukiriguru	Ap (0–28)	Haplic Acrisol	5.3	0.3	9	20.4	8.3	50.7	10.0
Mlingano	Ap (0–25)	Rhodic Ferralsol	5.3	1.2	54	48.5	14.4	433	15.4
Magadu	Ap (0–10)	Chromic Acrisol	4.9	1.2	55	53.2	16.3	327	20.8
Msimba	Ap (0–18)	Haplic Luvisol	6.5	2.3	27	35.4	17.6	234	11.3
Suluti	Ap (0–15)	Rhodic Acrisol	6.2	1.0	24	30.4	17.5	132.4	11.4
Lubonde	Ap (0–25)	Humic Ferralsol	5.7	2.6	43	157.4	45.5	257.2	36.4
Nkundi	Ap (0–8)	Chromic Acrisol	5.9	2.0	18	23.9	10.5	109.8	6.8
A6	BA (15–30)	Ferric Acrisol	4.8	2.4	20	156	215	343	34.0*
	Bto (90–150)		4.5	0.8	30	107	100	400	51.0*
BuI2A	Bo (90–150)	Ferralic Arenosol	4.0	0.3	9	18.4	104	300	20.9*
BuII2A	Bto (60–90)	Ferric Acrisol	4.1	0.7	54	110	64.5	1220	55.6*
BuII2B	Bo (90–150)	Acric Ferralsol	4.2	0.5	46	87	71.6	1100	47.4*
BuIV3A	Bto (30–60)	Ferric Acrisol	4.3	1.4	47	189	35.8	816	47.7*
F2	Bo (90–150)	Rhodic Ferralsol	4.2	—	40	112	12.5	732	38.8*
G1A	Bto (30–60)	Ferric Acrisol	3.9	—	37	136	397	954	56.6*
Morogoro	Ap (0–20)	Rhodic Ferralsol	5.0	1.3	47	66.4	21.5	963	38.8*

*P_{max} for these samples is the sum of the Langmuir maximum (*b*) and sulphuric acid-extractable phosphate instead of oxalate-extractable phosphate.

apparatus) as described by Tabatabai and Bremner (1970). Particle size-distribution was determined by sieving in combination with either the hydrometer method or the pipette method after dispersion in 0.002 M sodium pyrophosphate ($Na_4P_2O_7 \cdot 10H_2O$) (Day, 1965).

Oxalate-extractable aluminum, iron, and phosphate (Al_{Ox}, Fe_{Ox}, P_{Ox}) were determined after extraction for 2 h with 0.2 M ammonium oxalate at pH 3 in darkness (Schwertmann, 1964). Dithionite-citrate-bicarbonate-extractable iron (Fe_{DCB}) was determined by two extractions for 15 min at 70°C as described by Mehra and Jackson (1960). For some Danish and Tanzanian samples, oxalate-extractable phosphate was replaced by 0.1 M sulfuric acid-extractable phosphate (Borggaard, 1983b). Concentrations of aluminium and iron in the clear extracts were determined by flame atomic absorption spectroscopy. The phosphate concentration was determined by the molybdophosphate blue-method as described below.

Phosphate adsorption isotherms for each sample were based on phosphate adsorption data obtained by shaking six to ten 1-g samples of soil for 7 days at room temperature and pH 7 with 50 mL 0.01 M calcium acetate or 0.02 M sodium chloride containing between 0 and 1.5 mmol L^{-1} of KH_2PO_4. After shaking, the suspensions were centrifuged and the phosphate concentrations in the clear supernatants determined by the molybdophosphate-blue method using flow injection analysis (FIA) (Janse et al., 1983; Ruzicka and Hansen, 1988). The difference between phosphate concentrations before and after shaking with the soil samples was used to calculate adsorbed phosphate. The adsorption data were fitted to the Langmuir equation and the adsorption maximum (*b*) calculated for each sample:

$$Q = \frac{bKC_{eq}}{1 + KC_{eq}} \quad \text{... (1)}$$

where Q is adsorbed phosphate (mmol kg^{-1}) at an equilibrium concentration of C_{eq} (mmol L^{-1}), K a binding constant, and b the phosphate adsorption maximum under the given experimental conditions including phosphate concentrations, time of equilibration, pH, and temperature (Borggaard, 1983b, 2002).

All determinations were carried out in duplicate or triplicate and the coefficient of variation for most samples and analyses was approximately 5% or lower. The correlation coefficients of the Langmuir linear or nonlinear regressions were ≥ 0.92 with most coefficients being larger than 0.96.

In addition, data for 43 Canadian Spodosol samples presented by Yuan and Lavkulich (1994) were included in the investigation of testing existing PTFs as well as in the development of a PTF with broader range of application. Compared to the phosphate adsorption method described above, Yuan and Lavkulich (1994) did adsorption at soil pH (pH range 3–5) and used only 24 hours for equilibration. Lower pH tends to increase phosphate adsorption while shorter equilibration time may decrease adsorption (Afif et al., 1995; Borggaard, 1983b, 1990; Violante et al., 1991) indicating comparable data by the two methods. Accordingly, use of the Yuan and Lavkulich (1994) data as test data seems justified.

2.3 Statistical Analyses

In order to unveil correlations between the phosphate adsorption capacity of the soils and their content of oxalate- and DCB-extractable iron and aluminum, multivariate statistical analysis was performed in the Unscrambler ver. 7.6 SR-1 software package (Camo ASA software, Oslo, Norway). Five datasets (soil sample groups) were used in the multivariate analyses: 1) Danish samples (n = 29), 2) African samples including Ghanaian and Tanzanian soils (n = 23), 3) Canadian samples (n = 43), 4) all samples (n = 95) and 5) Danish + African samples (n = 52). Multivariate analysis (multiple linear regression, MLR) was performed using the following variables: Al_{Ox}, Fe_{Ox}, $Al_{Ox} + Fe_{Ox}$, and $Fe_{DCB}-Fe_{Ox}$. The models (PTFs) applied to the data are shown in Tables 11.2 to 11.4 together with the test results. Model investigations were done by means of jackknifing techniques and Martens' uncertainty test (removal of outliers and nonimportant variables) (Esbensen, 2000). The models were validated and

Table 11.2: Multiple linear regression models of the form proposed by van der Zee and van Riemsdijk (1986): $P_{calc} = \alpha \times (Al_{Ox} + Fe_{Ox}) + \gamma$, where Al_{Ox} and Fe_{Ox} denote oxalate-extractable aluminum and iron

Soil groups	$Al_{Ox} + Fe_{Ox}$ (α)	Intercept (γ)	[a]2SEP	[b]Slope	[b]Offset	[c]Q^2
Danish	0.165 ± 0.009	4.10 ± 1.66	13.13	0.920	2.04	0.910
African	0.103 ± 0.019	13.39 ± 3.32	25.57	0.622	10.65	0.468
Canadian	0.169 ± 0.013	18.28 ± 6.55	35.36	0.814	18.62	0.788
All (95)	0.185 ± 0.006	5.82 ± 2.33	29.94	0.905	5.63	0.899

[a]Standard error of performance, 2SEP corresponds to 95% confidence interval.
[b]Parameters obtained from plot of measured PAC against calculated PAC (slope and intercept of the straight).
[c]R^2 for validated correlation coefficient.

Table 11.3: Multiple linear regression models of the form: $P_{calc} = \alpha \times Al_{Ox} + \beta \times Fe_{Ox} + \gamma$, where Al_{Ox} and Fe_{Ox} denote oxalate-extractable aluminum and iron

Soil groups	Al_{Ox} (α)	Fe_{Ox} (β)	Intercept (γ)	[a]2SEP	[b]Slope	[b]Offset	[c]Q^2
Danish	0.240 ± 0.025	0.150 ± 0.009	1.15 ± 1.71	11.36	0.930	1.73	0.933
African	0.227 ± 0.043	0.040 ± 0.026	8.54 ± 3.21	20.56	0.693	8.24	0.630
Canadian	0.171 ± 0.014	0.144 ± 0.058	20.32 ± 8.13	36.45	0.809	19.06	0.776
All (95)	0.200 ± 0.008	0.132 ± 0.017	7.78 ± 2.29	29.25	0.917	5.04	0.903

[a]Standard error of performance, 2SEP corresponds to 95% confidence interval.
[b]Parameters obtained from plot of measured against calculated PAC.
[c]R^2 for validated correlation coefficient.

Table 11.4: Multiple linear regression models of the form proposed by Borggaard et al. (1990): $P_{calc} = \alpha \times Al_{Ox} + \beta \times Fe_{Ox} + \delta \times (Fe_{DCB} - Fe_{Ox}) + \gamma$, where Al_{Ox} and Fe_{Ox} denote oxalate-extractable aluminum and iron and Fe_{DCB} is dithionite-citrate-bicarbonate-extractable iron

Soil groups	Al_{Ox} (α)	Fe_{Ox} (β)	$Fe_{DCB} - Fe_{Ox}$ (δ)	Intercept (γ)	[a]2SEP	[b]Slope	[b]Offset	[c]Q^2
Danish	0.244 ± 0.023	0.143 ± 0.008	0.028 ± 0.011	–0.68 ± 1.68	10.47	0.939	1.53	0.943
African	0.146 ± 0.031	0.052 ± 0.017	0.023 ± 0.004	4.76 ± 2.12	13.10	0.911	2.66[a]	0.851
Canadian	0.168 ± 0.015	0.170 ± 0.068	–0.045 ± 0.062	23.45 ± 9.22	38.18	0.804	19.36	0.756
All (95)	0.202 ± 0.008	0.130 ± 0.017	0.009 ± 0.007	6.12 ± 2.58	29.23	0.918	4.92	0.904
DanAfr[d]	0.207 ± 0.017	0.150 ± 0.008	0.016 ± 0.003	1.52 ± 1.17	10.58	0.962	1.75	0.926

[a]Standard error of performance, 2SEP corresponds to 95% confidence interval.
[b]Parameters obtained from plot of measured against calculated PAC.
[c]R^2 for validated correlation coefficient.
[d]Danish and African soils.

evaluated using full cross-validation based on the following parameters:

(i) Q^2, variance in calculated phosphate adsorption capacity according to cross-validation.
(ii) 2SEP, 2 × the standard error of performance corresponding to the standard deviation of the prediction residuals, comparable to a 95% confidence interval for the calculated PACs.
(iii) Slope and offset of the regression line between measured and calculated PACs (a slope of 1 and a offset of 0 indicates a perfect model).
(iv) Standard error of regression coefficients.

The final model (Table 11.4) was developed using the combined Danish and African dataset, tested on the Canadian dataset.

3 RESULTS

The soils and soil samples investigated and the statistical tests of various PTFs are presented in this section.

3.1 Soil Characteristics

The investigation comprises a broad range of noncalcareous soils and soil samples (Table 11.1). The soils were from different parts of the world including Denmark, Ghana and Tanzania with very different geological conditions, climates, topographies, soil uses and soil development times. Nearly half of all the reference soil groups of the WRB system (Belgium ISSS Working Group RB, 1998) were covered by the soils investigated. Most of the soils were well drained or rather well-drained upland soils but groundwater-affected lowland soils such as Fluvisols, Gleysols and Histosols were also represented.

Soil samples covered surface soils, subsurface soils, and subsoils. They differed in pH from very acidic to neutral and weakly alkaline (Table 11.1). C contents exhibited a wide range of 0.1% to 17.5% but most surface soils contained 1–3% C, while the carbon contents are negligible in some subsoils. The texture of the soil samples differed markedly, with clay contents from 1% to 55%. Tropical soils had the highest clay contents but a few Danish soil samples were also rather rich in clay. Al_{Ox} and Fe_{Ox} corresponding to amorphous to poorly crystalline aluminum and iron oxides ranged from 3 to 189 mmol kg^{-1} and from 1 to almost 400 mmol kg^{-1}, respectively. Fe_{DCB} corresponding to so-called free iron (Borggaard, 1990, 2002; Mehra and Jackson, 1960) also exhibited a wide range of 2 to 1,220 mmol kg^{-1} with the highest contents in some of the Tanzanian soils. ($Fe_{DCB} - Fe_{Ox}$) corresponding to well-crystallized iron oxides (Borggaard, 1990, 2002; Torrent, 1997) varied from less than 10 mmol kg^{-1} in some Danish samples to about 1000 mmol kg^{-1} in some iron-rich Tanzanian samples.

Since the aluminum and iron oxides, the main phosphate adsorbents, show wide ranges in contents, PAC exhibits great variability among the soil samples. PAC is taken as the sum of the Langmuir maximum (*b*) and oxalate-extractable or sulfuric acid-extractable phosphate (P_{Ox}):

$$P_{max} = b + P_{Ox} \qquad \text{... (2)}$$

P_{Ox} is normally considered to represent adsorbed phosphate (Freese et al., 1992; Hooda et al., 2000; Leinweber et al., 1999; Lookman et al., 1995) and a close relationship exists between oxalate-extractable and sulfuric acid-extractable phosphate (Borggaard, unpubl. results, 1990). Adsorbed phosphate occupied adsorption sites not available for added phosphate during the phosphate adsorption measurement and hence for determination of the Langmuir maximum (b). Therefore $b + P_{Ox}$ (P_{max}) must be the true PAC (Freese et al., 1992; Szilas et al., 1998). In Table 11.1, P_{max} is seen to vary from less than 2 mmol kg^{-1} to more than 50 mmol kg^{-1}. By means of the soil samples used by Yuan and Lavkulich (1994), the range is further extended from 20 mmol kg^{-1} to 170 mmol kg^{-1} corresponding to higher contents of, in particular, Al_{Ox} ranging from 20.6 mmol kg^{-1} to 169.5 mmol kg^{-1}.

3.2 PTF Testing

While the Danish soils covered several soil groups, the Ghanaian and Tanzanian soils were restricted to strongly weathered soils, mainly Acrisols and Ferralsols (Table 11.1) and the Canadian soils were Podzols. Therefore the soil samples were grouped according to country/geography as shown in Tables 11.2 to 11.4 containing the results of statistical analyses. The PTF of van der Zee and van Riemsdijk (1986) is seen to give fine prediction of PAC of the mixed group of Danish soil samples as shown by a small intercept and offset and a high correlation coefficient and slope (Table 11.2). PAC of the Canadian samples as well as all samples was also fairly well predicted with this PTF, which failed with the African soils. The fitting constant α is 0.103–0.185, which is in the lowest end of an earlier reported range 0.14–0.48 (Freese et al., 1992; Reddy et al., 1998; Yuan and Lavkulich, 1994). For estimation of PAC in phosphate pollution assessments this PTF is commonly used with $\alpha = 0.5$ (Del Campillo et al., 1999; Leinweber et al., 1999; Lookman et al., 1995). With this value, PAC of the soil samples in Table 11.1 would be seriously overestimated, which of course would be even worser using $\alpha = 1$, as proposed by Hooda et al. (2000).

In an attempt to improve the PTF of van der Zee and van Riemsdijk (1986) different coefficients to Al_{Ox} and Fe_{Ox} were used in the Table 11.3 model. As expected, good and rather good PAC estimates were obtained for the Danish and Canadian soil samples, while PACs of the African samples were still poorly predicted. The PTF of Borggaard et al. (1990) gave good estimates of Danish soil sample PACs (Table 11.4) as well as of the African soils and all soils, whereas the prediction potential for the Canadian group is comparable to those of the PTFs in Tables 11.2 and 11.3.

To test the Borggaard et al. (1990) model on a dataset reflecting a wide range of soils not biased by a high number of samples of a single soil type, the statistical analysis was performed by means of the combined Danish, Ghanaian, and Tanzanian data (omitting the 43 Canadian Podzol samples). In the resultant

PTF the coefficients to all three independent variables (Al_{Ox}, Fe_{Ox}, Fe_{DCB}–Fe_{Ox}) are highly significant, the slope and Q^2 are close to 1 and the intercept is insignificant (Table 11.4). Accordingly, the PTF: $P_{calc} = 0.207 \times Al_{Ox} + 0.150 \times Fe_{Ox} + 0.016 \times (Fe_{DCB} - Fe_{Ox})$, where the intercept has been removed, gives good or fair estimates of all 95 soils as shown in Figure 11.1.

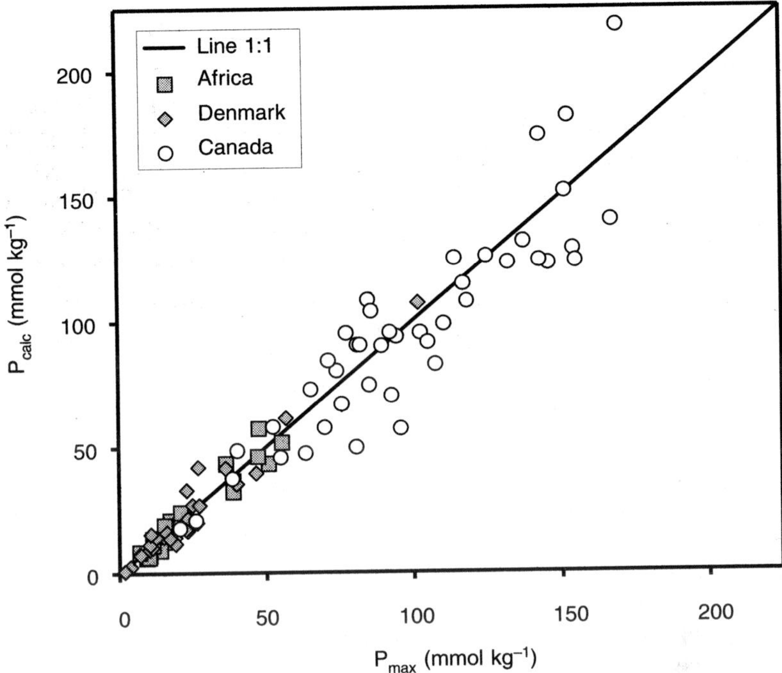

Fig. 11.1. Plot of experimentally determined and calculated phosphate adsorption capacities (PACs) for 95 soil samples from Denmark, Canada, Ghana, and Tanzania. Calculated PAC corresponds to the pedotransfer function: $P_{calc} = 0.207 \times Al_{Ox} + 0.150 \times Fe_{Ox} + 0.016 \times (Fe_{DCB} - Fe_{Ox})$.

4 DISCUSSION

Results of the statistical tests of the various PTFs are discussed below with emphasis on their potential and limitation for good prediction of PAC on as many and as different soil samples as possible. In addition, the plant nutritional and environmental importance and suitability of a widely applicable PTF for PAC estimation is discussed briefly in relation to impacts of available phosphate on the ecosystem and human welfare.

4.1 Prediction by PTFs

Despite the great variability of the soil samples (Table 11.1), the investigation clearly demonstrated the very close relationship between PAC and contents of aluminum and iron oxides, and hence the fairly good predictability of PAC by means of aluminum and iron oxide-based PTFs (Tables 11.2 to 11.4). This is in general agreement with results of previous studies carried out, however, on fewer and considerably less different soil samples (Borggaard, 1983b, 1986; Borggaard et al., 1990; Freese et al., 1992; Pena and Torrent, 1984; Singh and Gilkes, 1991; van der Zee and van Riemsdijk, 1986; Yuan and Lavkulich, 1994). The importance of the great variability among the soil samples of the present investigation is clearly demonstrated by the results of the statistical testing of the PTFs shown in Tables 11.2 to 11.4. The PAC of soil samples belonging to groups of similar soils such as the Canadian Podzol samples of Yuan and Lavkulich (1994) is almost equally well predicted by all tested PTFs. Results also show that a PTF predicting well for one group of soil samples may seriously fail with another group. This is clearly demonstrated by the PTF of van der Zee and van Riemsdijk (1986), which gives fairly good prediction of PAC of the samples in the Podzol group but exhibits serious limitations when it comes to soil samples of different groups, as shown by samples from different countries in Table 11.2. Accordingly, it has been recommended to use different PTFs for different groups of soils (Freese et al., 1992; Yuan and Lavkulich, 1994). On the other hand, a PTF with a wider application range than the PTFs currently known would definitely be very useful provided its range of application is clear or fairly easy to determine. Furthermore, delineation of soils that fall within as well as those falling outside its application range must be simple and cheap.

According to the results in Table 11.4, the following PTF could be suggested for PAC estimation of noncalcareous Alfisols, Entisols, Histosols, Inceptisols, Mollisols, Oxisols, Spodosols, and Ultisols:

$$P_{calc} = 0.207 \times Al_{Ox} + 0.150 \times Fe_{Ox} + 0.016 \times (Fe_{DCB} - Fe_{Ox}) \qquad \text{... (3)}$$

In Figure 11.1 PAC calculated by this equation has been plotted against experimentally determined PAC (P_{max}). For some soil samples substantial differences between calculated and experimentally determined PACs can be seen but the overall impression of Figure 11.1 is that the proposed PTF gives a good to fair estimate of PAC for a wide range of soils.

Further testing of this PTF was performed by data from two Tanzanian Andisols. The results shown in Table 11.5 demonstrate, however, that PAC of these soils is substantially overestimated by the function. The poor predictability of PAC of soils developed on volcanic materials compared to that of the other soils (Fig. 11.1) is probably due to differences of the main phosphate adsorbents. While poorly crystalline aluminum silicates (allophane and imogolite) can dominate phosphate adsorbents in Andisols, the other soils adsorb most phosphate onto more or less crystallized aluminum and iron oxides (Borggaard, 1990, 2002; USDA, Soil Survey Staff, 1999). In fact, the clay fractions from the Mpangala and Sasanda soils (Table 11.5) contained allophane/imogolite according to mineralogical analysis (Szilas, 2002).

Table 11.5: Oxalate-extractable aluminum and iron (Al_{Ox}, Fe_{Ox}), dithionite-citrate-bicarbonate-extractable iron (Fe_{DCB}) in two Tanzanian Andisols together with experimentally determined PAC (P_{max}) and the PAC estimated by the PTF: $P_{calc} = 0.207 \times Al_{Ox} + 0.150 \times Fe_{Ox} + 0.016 \times (Fe_{DCB} - Fe_{Ox})$. $P_{max} = b + P_{Ox}$, where β is the Langmuir maximum of phosphate adsorption isotherms and P_{Ox} is oxalate-extractable phosphate

Soil	mmol kg^{-1}				
	Al_{Ox}	Fe_{Ox}	Fe_{DCB}	P_{max}	P_{calc}
Mpangala	234	91	137	39	63
Sasanda	841	79	203	82	188

Whether or not the PTF in eqn (3) fails to give fair estimates of PAC of Andisols in general as well as of other groups of soils can only be decided by means of further testing. Hopefully, future testing will more precisely delineate the kinds of soils in which the suggested PTF can be used as well as those in which it would give unacceptable PAC estimates.

4.2 PTF Application Areas

Agriculture has two problems with phosphate, i.e., the soil must contain sufficient available phosphate in the root zone to cover the need of plants, but concomitantly leaching must be minimised in order to avoid pollution of surface waters (Brady and Weil, 1999). PAC and phosphate saturation (P_{Ox}/PAC) seem to be the most useful indicators (soil quality indicators) to address both needs (Del Campillo et al., 1999; Hooda et al., 2000; Leinweber et al., 1999; Lookman et al., 1995; Szilas, 2002). Depending on the crop, soil solution of the root zone must contain 1–10 μM phosphate to ensure adequate supply of plants with this element, while phosphate concentrations above 1–3 μM may lead to eutrophication in open waters (Fox, 1981; Marschner, 1998; Sharpley and Menzel, 1987; Taylor and Kilmer, 1980).

As phosphate leaching increases continuously at increasing phosphate saturation it is difficult to determine a threshold for phosphate saturation separating nonpolluting soils from soils that release phosphate in concentrations above what can be tolerated by the aquatic environment. Using the PTF of Borggaard et al. (1990), studies of cultivated Danish sandy soils indicated that soil layers with a phosphate saturation less than about 0.3 reduced the concentration of phosphate in the soil solution to nonpolluting levels, while phosphate leaching occurred in soil layers with a phosphate saturation percentage above 30% (Borggaard and Møberg, 1991). The latter value is in good agreement with the threshold value of 25% found in the Netherlands and that of 32% found in New York State (Kleinman et al., 1999). In apparent contrast, Hooda et al. (2000) found a phosphate saturation of 8% as the threshold between polluting and nonpolluting soils. The discrepancy between these thresholds is mainly due to the use of different PTFs for the calculation of PAC (Borggaard and Møberg, 1991; Hooda, 2000; Kleiman et al., 1999).

Streams draining intensively fertilised farmlands are often strongly polluted by nitrate and other pollutants but elevated nitrate loadings can be reduced through denitrification by passing the nitrate-polluted water through wetlands (Brady and Weil, 1999). Therefore, construction and reestablishment of wetlands on former agricultural areas could be of interest (Kovacic et al., 2000; Szilas et al., 1998). However, when formerly well-aerated soils become saturated with water, reduction and dissolution of the iron oxides will lead to mobilization of phosphate adsorbed on the iron oxides (Reddy et al., 1998; Ruiz et al., 1997). Flooding of cultivated and phosphate-enriched soils in connection with reestablishment of wetlands on former reclaimed agricultural areas might therefore result in severe phosphate pollution. That this may be a serious environmental problem is substantiated by results from both laboratory and field studies (Newman and Pietro, 2001; Reddy et al., 1998; Szilas, 1998).

Apart from these direct effects, adsorbed phosphate in relation to PAC may also have indirect effects on the environment because adsorbed phosphate blocks surface sites that could bind compounds that otherwise could threaten ground and drainage water quality. This is well illustrated by glyphosate. The popular herbicide glyphosate (N-(phosphonomethyl)glycin) is a phosphonic acid that resembles phosphate. Accordingly, adsorption of glyphosate and phosphate by soils and selected soil components seems to occur in very much the same way and on the same adsorbents (aluminum and iron oxides, in particular), suggesting competition between the two compounds for adsorption sites (De Jonge et al., 2001; Dion et al., 2001; Gimsing and Borggaard, 2001, 2002; Hance, 1976; Sheals et al., 2002). Although adsorption behavior of glyphosate and phosphate appear similar, differences are expected due to differences between the two compounds, such as larger molecular weight and additional functional groups of glyphosate compared to phosphate (Sheals et al., 2002). Gimsing and Borggaard (2001) found that both glyphosate and phosphate were strongly adsorbed by synthetic goethite but while adsorbed glyphosate could be almost completely desorbed by phosphate, glyphosate could only desorb small amounts of adsorbed phosphate, indicating stronger bonding of phosphate than of glyphosate. Figure 11.2 shows the results of similarly conducted competitive adsorption studies using synthetic gibbsite instead of goethite (Gimsing and Borggaard, 2002a). Glyphosate and phosphate are both effectively adsorbed by gibbsite and both adsorptions are fast as was also shown for adsorption onto goethite (Gimsing and Borggaard, 2001). However, phosphate can replace most glyphosate, while little adsorbed phosphate is desorbed by glyphosate. Accordingly, phosphate is more strongly adsorbed by gibbsite than is glyphosate, as was also found for goethite (Gimsing and Borggaard, 2001). If extrapolated to soils, addition or presence of phosphate will reduce adsorption and increase mobilization of glyphosate, which, in turn, may lead to pollution of the aquatic environment by this xenobioticum in soils with little available PAC. A very

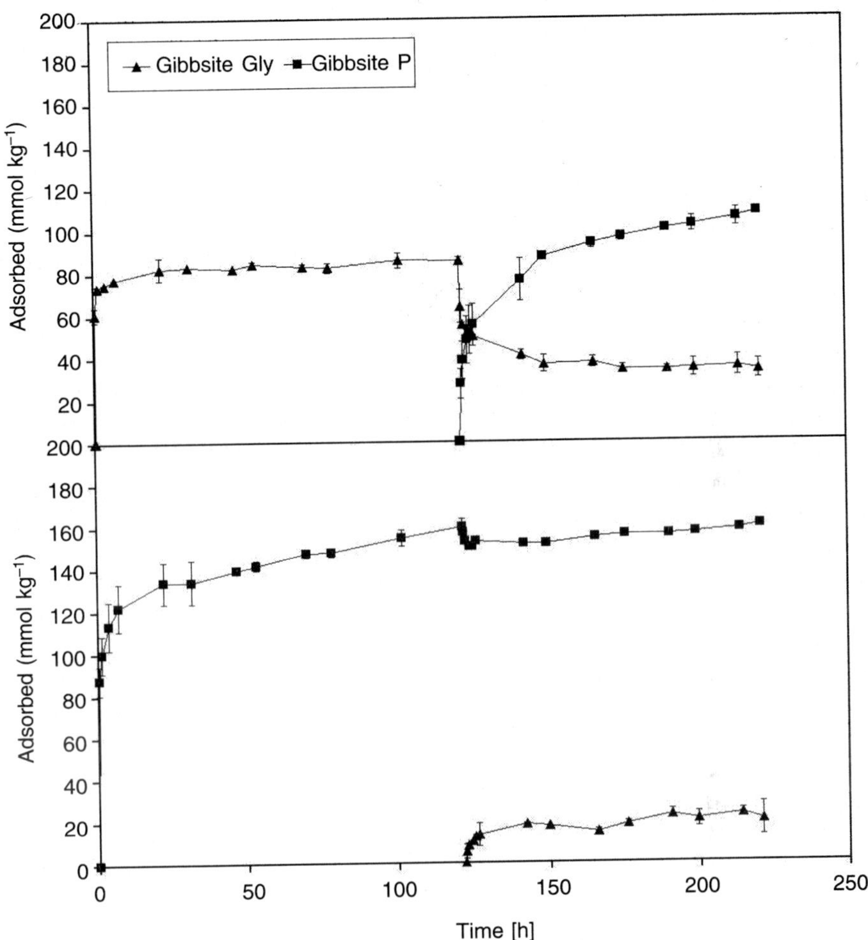

Fig. 11.2: Adsorption of glyphosate and phosphate onto synthetic gibbsite in relation to time. Upper figure shows results of applying glyphosate (▲) at time 0 and of phosphate (■) after 120 hours, while lower figure shows application of phosphate (■) at time 0 and of glyphosate (▲) after 120 hours. (From Gimsing and Borggaard, 2002a).

recent study of five soil samples indicated, however, limited competition for adsorption sites between glyphosate and phosphate (Gimsing and Borggaard, 2002b; Gimsing et al., 2004). Further studies with a larger number of contrasting soil samples are therefore needed to determine more precisely the influence of phosphate on soil glyphosate adsorption.

References

Afif E., Barrón V., and Torrent J. 1995. Organic matter delays but does not prevent phosphate adsorption by Cerrado soils from Brazil. *Soil Sci.* 159: 207–211.

Belgium ISSS Working Group RB. 1998. *World Reference Base for Soil Resources*. E.M. Bridges, N.H. Batjes, and F.O. Nachtergaele (eds.). ISRIC-FAO-ISSS-Acco. Leuven, Belgium.

Borggaard O.K. 1983a. Effect of surface area and mineralogy of iron oxides on their surface charge and anion-adsorption properties. *Clays Clay Miner.* 31: 230–232.

Borggaard O.K. 1983b. The influence of iron oxides on phosphate adsorption by soil. *J. Soil Sci.* 34: 333–341.

Borggaard O.K. 1986. Iron oxides in relation to phosphate adsorption by soils. *Acta Agric. Scand.* 36: 107–118.

Borggaard O.K. 1990. Dissolution and adsorption properties of soil iron oxides. DSc thesis, Royal Veterinary and Agricultural University, Copenhagen, Denmark.

Borggaard O.K. 2002. *Soil Chemistry in a Pedological Context*. 6th Edition, DSR Forlag, Frederiksberg, Denmark (6th ed.).

Borggaard O.K. and Møberg P.M. 1991. Estimation of critical phosphate loads in Danish sandy soils. In: *Soil Vulnerability to Pollution in Europe*. N.H. Batjes and E.M. Bridges (eds.). ISRIC, Wageningen, The Netherlands, pp. 71–75.

Borggaard O.K., Jørgensen S.S., Møberg J.P., and Raben-Lange, B. 1990. Influence of organic matter on phosphate adsorption by aluminium and iron oxides in sandy soils. *J. Soil Sci.* 41: 443–449.

Brady N.C. and Weil R.R. 1999. *The Nature and Properties of Soils*. Prentice-Hall, Upper Saddle River, NJ, USA (12th ed.).

Day P.R. 1965. Particle fractionation and particle-size analysis. In: *Methods of Soil Analysis*. C.A. Black, D.D. Evans, J.L. White, L.E. Ensminger, and F.E. Clark (eds.). Agronomy no. 9. Amer. Soc. Agron., Madison, WI, USA, pp. 545–567.

De Jonge H., de Jonge L.W., Jacobsen O.H., Yamaguchi T., and Moldrup P. 2001. Glyphosate sorption in soils of different pH and phosphorus content. *Soil Sci.* 166: 230–238.

Del Campillo M.C., van der Zee S.E.A.T.M., and Torrent J. 1999. Modelling long-term phosphorus leaching and changes in phosphorus fertility in excessively fertilized acid sandy soils. *Eur. J. Soil Sci.* 50: 391–399.

Dion H.M., Harsh J.B., and Hill Jr. H.H. 2001. Competitive sorption between glyphosate and inorganic phosphate on clay minerals and low organic matter soils. *J. Radioanal. Nucl. Chem.* 249: 385–390.

Esbensen K.H. 2000. *Multivariate Data Analysis—in Practice*. Camo ASA, Oslo, Norway (4th ed.).

Fox R.L. 1981. External phosphorus requirements of crops. In: *Chemistry in Soil Environment*. M. Stelly (ed.). Amer. Soc. Agron., Madison, WI, USA, pp. 223–239.

Freese D., van der Zee S.E.A.T.M., and van Riemsdijk W.H. 1992. Comparison of different models for phosphate sorption as a function of the iron and aluminum oxides of soils. *J. Soil Sci.* 43: 729–738.

Gimsing A.L. and Borggaard O.K. 2001. Effect of KCl and $CaCl_2$ as background electrolytes on the competitive adsorption of glyphosate and phosphate on goethite. *Clays Clay Miner.* 49: 270–275.

Gimsing A.L. and Borggaard O.K. 2002a. Competitive adsorption and desorption of glyphosate and phosphate on clay silicates and oxides. *Clay Miner.* 37: 509–515.

Gimsing A.L. and Borggaard O.K. 2002b. Effect of phosphate on the adsorption of glyphosate on soils, clay minerals and oxides. *Intern. J. Environ. Anal. Chem.* 82: 545–552.

Gimsing A.L., Borggaard O.K., and Bang M. 2004. Influence of soil composition on adsorption of glyphosate and phosphate by contrasting Danish surface soils. *Eur. J. Soil Sci.* 55: 183–191.

Hance R.J. 1976. Adsorption of glyphosate by soils. *Pestic. Sci.* 7: 363–366.

Hooda P.S., Rendell A.R., Edwards A.C., Withers P.J.A., Aitken M.N., and Truesdale V.W. 2000. Relating soil phosphorus indices to potential phosphorus release to water. *J. Environ. Qual.* 29: 1166–1171.

Huang P.M. and Schnitzer M. 1986. *Interactions of Soil Minerals with Natural Organics and Microbes*. SSSA Spec. Publ. 17. Soil Sci. Soc. Amer., Madison, WI, USA.

Huang P.M. and Wang M.K. 1997. Formation chemistry and selected surface properties of iron oxides. In: *Soil and Environment—Soil Processes from Mineral to Landscape Scale*. K. Auerswald,

H. Stanjek, and J.M. Bigham (eds.). Advances in GeoEcology 30. Catena Verlag, Reiskirchen, Germany, pp. 241–270.

Janse T.A.H., van der Wiel P.F.A., and Kateman G. 1983. Experimental optimization procedures in the determination of phosphate by flow injection analysis. *Anal. Chim. Acta* 155: 89–102.

Kang B.T. 1989. Nutrient management for sustained crop production in the humid and subhumid tropics. In: *Nutrient Management for Food Crop Production in Tropical Farming Systems*. J. Van der Heide (ed.). Inst. Soil Fertility, Haren, The Netherlands, pp. 3–28.

Kleinman P.J.A., Bryant R.B., and Reid W.S. 1999. Development of pedotransfer functions to quantify phosphorus saturation of agricultural soils. *J. Environ. Qual.* 28: 2026–2030.

Kovacic D.A., David M.B., Gentry L.E., Starks K.M., and Cooke R.A. 2000. Effectiveness of constructed wetlands in reducing nitrogen and phosphorus export from agricultural tile drainage. *J. Environ. Qual.* 29: 1262–1274.

Leinweber P., Meissner R., Eckhardt K.-U., and Seeger J. 1999. Management effects on forms of phosphorus in soil and leaching losses. *Eur. J. Soil Sci.* 50: 413–424.

Lookman R., Vandeweert N., Merckx R., and Vlassak K. 1995. Geostatistical assessment of the regional distribution of phosphate sorption capacity parameters (Fe_{ox} and Al_{ox}) in northern Belgium. *Geoderma* 66: 285–296.

Maguire R.O. and Sims J.T. 2002. Soil testing to predict phosphorus leaching. *J. Environ. Qual.* 31: 1601–1609.

Marschner H. 1998. *Mineral Nutrition of Higher Plants*. Acad. Press, London, UK (2nd ed).

McBratney A.B., Minasny B., Cattle S.R., and Vervoort R.W. 2002. From pedotransfer functions to soil inference systems. *Geoderma* 109: 41–73.

Mehra O.P. and Jackson M.L. 1960. Iron oxide removal from soils and clays by a dithionite-citrate system buffered with sodium bicarbonate. *Proc. 7th Nat. Conf. Clays Clay Miner. 1958*, 317–327.

Menon R.G., Chien S.H., and Gadalla A.N. 1991. Phosphate rocks compacted with superphosphates vs partially acidulated rocks for beans and rice. *Soil Sci. Soc. Amer. J.* 55: 1480–1484.

Møberg J.M. 1973. *An Edaphological and Pedological Investigation of the Soils in the West Lake Region of Tanzania*. Danida Report 104. Dan. 8/110.

Newman S. and Pietro K. 2001. Phosphorus storage and release in response to flooding: implications for Everglades stormwater treatment areas. *Ecol. Engin.* 18: 23–38.

Novak J.M., Watts D.W., Hunt P.G., and Stone K.C. 2000. Phosphorus movement through a Coastal Plain soil after a decade of intensive swine manure application. *J. Environ. Qual.* 29: 1310–1315.

Owusu-Bennoah E., Szilas C., Hansen H.C.B., and Borggaard O.K. 1997. Phosphate sorption in relation to aluminum and iron oxides of Oxisols from Ghana. *Commun. Soil Sci. Plant Anal.* 28: 685–697.

Pena F. and Torrent J. 1984. Relationships between phosphate sorption and iron oxides in Alfisols from a river terrace sequence of Mediterranean Spain. *Geoderma* 33: 283–296.

Reddy K.R., Oconnor G.A., and Gale P.M. 1998. Phosphorus sorption capacities of wetland soils and stream sediments impacted by dairy effluent. *J. Environ. Qual.* 27: 438–447.

Ruiz, J.M., Delgado, A. and Torrent, J. 1997. Iron-related phosphorus in overfertilized European soils. *J. Environ. Qual.* 26: 1548–1554.

Ruzicka J. and Hansen E.H. 1988. *Flow Injection Analysis*. Wiley & Sons, New York, NY (2nd ed.).

Schwertmann U. 1964. Differenzierung der Eisenoxide des Bodens durch Extraktion mit Ammoniumoxalat-Lösung. *Z. Pflanzenernähr. Düng. Bodenk.* 105: 194–202.

Sharpley A.N. and Menzel R.G. 1987. The impact of soil and fertiliser phosphorus on the environment. *Adv. Agron.* 41: 297–324.

Sheals J., Sjöberg S., and Persson P. 2002. Adsorption of glyphosate on goethite: Molecular characterization of surface complexes. *Environ. Sci. Tech.* 36: 3090–3095.

Singh B. and Gilkes R.I. 1991. Phosphorus sorption in relation to soil properties for the major soil types of south-western Australia. *Aust. J. Soil Res.* 29: 603–618.

Szilas C. 2002. The Tanzanian Minjingu phosphate rock—possibilities and limitations for direct application. PhD Thesis, Royal Veterinary and Agricultural University, Copenhagen, Denmark.

Szilas C.P., Borggaard O.K., Hansen H.C.B., and Rauer J. 1998. Potential iron and phosphate mobilization during flooding of soil material. *Water Air Soil Poll.* 106: 97–109.

Tabatabai M.A. and Bremner J.M. 1970. Use of the Leco automatic carbon analyser for total carbon analysis of soils. *Soil Sci. Soc. Amer. Proc.* 34: 608–610.

Taylor A.W. and Kilmer V.J. 1980. Agricultural phosphorus in the environment. In: *The Role of Phosphorus in Agriculture.* F.E. Khasawneh, E.C. Sample, and E.J. Kamprath (eds.). Amer. Soc. Agron., Madison, WI, USA, pp. 545–557.

Torrent J. 1997. Interactions between phosphate and iron oxide. In: *Soil and Environment—Soil Processes from Mineral to Landscape Scale.* K. Auerswald, H. Stanjek, and J.M. Bigham (eds.). Advances in GeoEcology 30. Catena Verlag, Reiskirchen, Germany, pp. 321–344.

USDA, Soil Survey Staff, 1999. *Keys to Soil Taxonomy.* US Dept. Agriculture, Soil Conservation Service. Pocahontas Press, Blacksburg, VI (8th ed.).

Van der Zee S.E.A.T.M. and van Riemsdijk W.H. 1986. Sorption kinetics and transport of phosphate in sandy soil. *Geoderma* 38: 293–309.

Violante A., Colombo C., and Buondonno A. 1991. Competitive adsorption of phosphate and oxalate by aluminum oxides. *Soil Sci. Soc. Amer. J.* 55: 65–70.

Yuan G. and Lavkulich L.M. 1994. Phosphate sorption in relation to extractable iron and aluminium in Spodosols. *Soil Sci. Soc. Amer. J.* 58: 343–346.

12

Sorption/Desorption of Sulfate on/from Variable Charge Minerals and Soils: Competitive Effects of Organic and Inorganic Ligands

A. Violante[*], M. Pigna, *and* F. Liu

Abstract

Many factors (pH, nature of sorbents, presence, concentration and nature of inorganic and organic anions) influence the mobility of sulfate in soil environments. Ligands with a high affinity for Al or Fe, such as phosphate and low molecular mass organic ligands (LMMOLs), effectively prevent sulfate sorption and readily desorb sulfate on/from metal oxides and variable charge soils. However, at pH $\leq$ 4.0 large amounts of LMMOLs were not able to completely replace sulfate. Macroscopic studies of sulfate sorption show evidence that in strongly acidic environments sulfate may form, at least in part, inner-sphere complexes. Consequently, the capacity of sulfate to compete with phosphate and LMMOLs for sorption sites of variable charge minerals and soils is inversely related to pH.

LMMOLs, coprecipitated with Al (and Fe) promote formation of active sites for sorption of ligands by distorting the structure of precipitation products of aluminum (and iron) and enhancing their specific surface. Maintenance of short-range structure of the precipitates with a large surface area by the presence of critical concentrations of chelating ligands helps to promote a relatively high anion (including sulfate) retention capacity of organomineral complexes.

Finally, we have also demonstrated that sulfate coprecipitated with aluminum that is part of the structural network, may be only partly removed by large amounts of phosphate or oxalate or by repeated washing with these ligands.

**Corresponding author*: Dr. A. Violante, Dipt. Sciene del Suolo, della Pianta e dell'Ambiente, Università di Napoli Federico II, 80055 Portici (Napoli), Italy. E-mail: violante@unina.it

1 INTRODUCTION

Sulfate is an essential mineral element for plant growth. Plants adsorb sulfate from the soil solution, therefore replenishment from organic and adsorbed sources is important in maintaining sulfate supply to the plant (Sumner, 2000).

Hydrous oxides of aluminum and iron and short-range ordered aluminosilicate materials can retain sulfate in soils against leaching. Ultisols, Alfisols, and Oxisols have a high sorption capacity due to an abundance of Fe and Al oxides. Adsorption of sulfate on variable charge minerals and soils has been studied but the mechanisms for its adsorption and the structures of its surface complexes are still not fully known. The mechanisms proposed for sulfate retention in soils are: i) electrostatic attraction by positively charged sites; ii) specific adsorption through ligand exchange; and iii) formation of aluminum hydroxysulfate precipitates (Chao et al., 1962; Tabatabai, 1982; Curtin and Syers, 1990; Pasricha and Fox, 1993). Macroscopic studies of sulfate adsorption have suggested that sulfate adsorbs via an outer-sphere (electrostatic) adsorption mechanism, but there is microscopic and spectroscopic evidence of sulfate inner-sphere surface complexation (Peak et al. 1999; 2001). According to Peak et al. (1999; 2001) sulfate forms only outer-sphere surface complexes above pH 6.0 and a mixture of outer- and inner-sphere surface complexes at pH less than 6.0. It is well known that the sulfate adsorption capacity of variable charge minerals and soils is decreased by increased phosphate content but the effect of phosphate on the sorption/desorption of sulfate on/from soil components at low pH values is not completely known. Furthermore, the effect of low molecular mass organic ligands (LMMOLs) on the mobility of sulfate in soil environments has received little attention (Chao et al., 1964; Inskeep, 1989; Liu et al., 1999). LMMOLs are common in rhizosphere as they are secreted by plant roots and produced by bacteria and fungi which are particularly abundant at plant-soil interface.

Because sulfate in aluminum precipitates in the subsoil appears to be available to plant roots, the effect of LMMOLs and phosphate on the removal of sulfate from aluminum hydroxysulfate precipitate is particularly important.

We carried out studies on the influence of phosphate and strongly chelating organic ligands (e.g. oxalate and malate) on the sorption of sulfate on variable charge minerals and soils (Andisols), and on an organomineral complex as influenced by (i) the initial phosphate or organic ligand/sulfate molar ratio, (ii) the order of anion addition, and (iii) pH (4.0–8.0). We also have studied the sorption of oxalate and phosphate on a noncrystalline aluminum hydroxysulfate precipitate and the associated release of sulfate as a function of pH (4.0–9.0).

2 MATERIAL AND METHODS

2.1 Materials

Some metal oxides (goethite, ferrihydrite, and gibbsite) and hydroxy-Al-oxalate or sulfate complexes were obtained as follows:

Goethite: Goethite was prepared following the method described by Atkinson et al. (1967). This procedure consisted of the slow addition, under vigorous stirring, of 200 mL of 2.5 mol L^{-1} NaOH to a solution containing 50 g of $Fe(NO_3)_3 \cdot 9\ H_2O$ in 825 mL of deionized water. The final suspensions was aged 6 days at 60°C and then washed free of salts by dialysis, freeze-dried, and ground slightly to pass through a 0.16 mm sieve. The X-ray diffraction (XRD) pattern of a randomly oriented sample obtained with a Rigaku diffractometer with Fe-filtered CoK_α radiation generated at 40 kV and 30 mA (Rigaku Co., Tokyo), showed XRD peaks at 0.418, 0.269, and 0.245 nm, which are characteristic of goethite.

The specific surface area was determined gravimetrically using the retention method of ethylene glycol monoethyl ether (EGME) (Eltanawy and Arnold, 1973). Point of zero salt effect (PZSE) was determined by the method of Sakuray et al. (1988). The goethite sample had a surface area of 82 $m^2\ g^{-1}$ and a PZSE of 7.8.

Gibbsite: The gibbsite used in our study was obtained from Aldrich Chemical Company (Milwaukee, WI, USA). The surface area determined by EGME was 135 $m^2\ g^{-1}$ and the PZSE was of 8.9.

Ferrihydrite: A stock solution of 0.01 mol L^{-1} $Fe(NO_3)_3$ was potentiometrically titrated to pH 5.5 by adding 1 mol L^{-1} NaOH at a feed rate of 0.5 ml min^{-1}. A Potentiograph E 536 Metrom Herisau automatic titrator in conjunction with an automatic syringe burette 655 Dosimat was used. The final volume of the sample was adjusted to 2 L and the final Fe concentration was 0.05 mol L^{-1}. The suspension was kept in a polypropylene container and aged at 20°C for one week. After ageing, a sample was collected and dialyzed [Molecular Weight (M. W.) cutoff 15,000] in deionized water, freeze dried, and lightly ground to pass through a 100 μm mesh sieve. Ferrihydrite showed a specific surface area (EGME) of 198 $m^2\ g^{-1}$.

Hydroxy-Al-sulfate complex: A stock solution of 0.01 mol L^{-1} $AlCl_3$ was titrated to pH 7.5 with 0.1 mol L^{-1} KOH (at a rate of 2 $cm^3\ min^{-1}$), in the presence of K_2SO_4 (SO_4 / Al molar ratio of 20) and under vigorous stirring. The suspension was aged for 24 h and the precipitate was collected and washed four times with deionized distilled water, followed by centrifugation at 10,000 g for 20 min. The solid phase was then dialyzed against deionized water until free of sulfate, freeze dried, and lightly ground to pass through a 0.1 mm sieve. The XRD pattern of a randomly oriented sample was obtained with a Rigaku diffractometer (Rigaku Co., Tokyo) with Ni-filtered CoK_α radiation generated at 40 kV and 30 mA. The infrared spectrum of the sample prepared by the KBr disk technique was recorded in the 400–4000 cm^{-1} range on a Perkin Elmer 1720X-FT-IR instrument (Perkin Elmer Ltd, Beaconsfield, UK). The complex was noncrystalline to both XRD and FT-IR analyses (not shown).

Aliquots (20–50 mg) of the complex were dissolved in 0.5 mol L^{-1} NaOH, and Al and SO_4 were determined as described below. The aluminum hydroxysulfate complex contained 1.35 mol SO_4 kg^{-1}, 11.9 mol Al kg^{-1} and only 21 mmol K kg^{-1}.

Hydroxy-Al-oxalate complex: A hydroxy-Al-oxalate complex [$Al(OH)_x$-oxalate] was prepared by titrating a stirred solution of 0.1 mol L^{-1} $AlCl_3$ with oxalic acid (oxalic acid / Al molar ratio of 0.5) with 0.1 mol L^{-1} NaOH at a rate of 2 mL min^{-1} to pH 7.0. The final suspension volume was then adjusted with deionized water to bring the Al to 0.03 mol L^{-1}. The suspension was aged for 24 h, centrifuged at 10,000 g, washed twice with deionized water, dialyzed until Cl^- free, and freeze-dried. The Al precipitate was lightly ground to pass a 0.1 mm sieve.

The $Al(OH)_x$–OX complex showed a surface area (EGME) of 698 m^2 g^{-1}, a PZSE of 5.14 and an oxalate content of 2.7 mol kg^{-1}.

The oxalate present in the organomineral precipitate was determined by ion chromatography (described below) after dissolution of the precipitate in 0.5 mol L^{-1} NaOH (Jackson, 1969).

Soil samples: For this study we selected two soil samples (Andisols) derived from volcanic materials present in the caldera of Roccamonfina volcano (south-central Italy). The classification and chemical and mineralogical properties of which were described in a previous work (Vacca et al., 2003).

Sample 1 was collected from an A1 horizon of a Eutric Fulvudand (pedon 1); sample 2 was collected from a 2C horizon of a Typic Hapludand (pedon 4). Selected chemical properties of these soils are reported in Table 12.1.

Table 12.1: pH in H_2O and NaF and organic carbon, ECEC, and allophane contents of selected soil samples (Andisols)

Sample	Horizons	pH (H_2O)	pH (NaF)	Organic C g kg^{-1}	ECEC** cmol kg^{-1}	Allophane %
Andisol 1 *	A1	6.7	10.9	184	30.56	17
Andisol 2 *	2C	5.5	11.0	2	1.95	42

*see Vacca et al. (2003).
**Effective cation-exchange capacity.

2.2 Phosphate, Sulfate, Oxalate, and Malate Sorption Isotherms

Fifty mg of goethite, gibbsite, or ferrihydrite were equilibrated with 45 mL of 0.05 mol L^{-1} KCl. Predetermined quantities of freshly prepared sulfate (as K_2SO_4), oxalate (as oxalic acid), malate (as malic acid), or phosphate (as KH_2PO_4) solutions, previously adjusted to pH 4.5, were added to give an initial concentration ranging from 0.04 to 0.4 mmol L^{-1}.

Soil samples (300 mg) were equilibrated with 20 mL of 0.05 mol L^{-1} KCl at pH 4.5. Suitable amounts of 0.1 mol L^{-1} solutions containing phosphate or sulfate were then added to produce initial phosphate, sulfate, or malate concentrations in the range of 5×10^{-4} to 10^{-2} mol L^{-1}. The pH of each suspension (20 mL) was kept at the initial value ± 0.01 by adding 0.1 or 0.01 mol L^{-1} HCl or KOH. Suspensions were shaken for 24 h, centrifuged at 10,000 *g* for 20 min, and filtered through a 0.22-μm membrane filter.

2.3 Sorption of Sulfate and Phosphate on Variable Charge Soils as a Function of pH

At constant pH ranging from 4.0 to 8.0, 1.4 mL of 0.1 mol L^{-1} solutions containing sulfate or phosphate were added to 300 mg soil samples in reaction flasks. As mentioned above, the pH of each suspension (20 mL) was kept at the initial ± 0.01 value by adding 0.1 or 0.01 mol L^{-1} HCl or KOH.

Determination of competitive sorption of sulfate and phosphate, added as a mixture, was carried out by adding suitable amounts of sulfate in the presence of increasing quantities of phosphate to achieve initial phosphate / sulfate molar ratios ranging from 0 to 2.0. The pH of each suspension was kept constant at 2.5, 3.5, or 4.5 for 24 h by adding 0.1 or 0.01 mol L^{-1} HCl or KOH. Some experiments were carried out at pH 4.5 by adding sulfate 24 h before phosphate (*SO_4 before PO_4* systems).

To study the kinetics of sorption of sulfate at pH 4.5 in the presence of phosphate on Andisol 2, suspensions containing 150 mmol kg^{-1} of sulfate (initial SO_4/PO_4 molar ratio = 2) were shaken from 0.08 to 168 h. The final suspensions (20 mL) were centrifuged at 10,000 g for 20 min, and filtered through a 0.22-μm filter.

2.4 Sorption of Sulfate, Phosphate, and Oxalate on Goethite as a Function of pH

Forty-five mL of 0.1 mol L^{-1} KCl were added to 50 mg of goethite in reaction flasks. Predetermined quantities of 0.02 mol L^{-1} solutions containing sulfate, phosphate, oxalate or a mixture of sulfate and oxalate, the pH values of which were previously adjusted (pH 3.0–8.0), were pipetted into the flask to give 300 mmol ligand per kg sample. Competition in sorption between sulfate and oxalate was also determined at pH 4.0 by adding 300 mmol kg^{-1} sulfate (or oxalate) in the presence of increasing quantities of oxalate (or sulfate) in order to have SO_4/OX (or OX/SO_4) molar ratios ranging from 0.17 to 3.33.

The suspensions (47 mL) were shaken for 24 h at 20°C and pH values were then adjusted to the original values, prior to centrifugation.

2.5 Effect of Increasing Concentrations of Malate on Sorption of Sulfate

Sorption of sulfate in the presence of increasing concentrations of malate on gibbsite, ferrihydrite, and Andisol 2 was determined by adding suitable amounts of sulfate (100, 300, and 150 mmol kg^{-1}) at pH 4.5. After 4 h malate was introduced into the system in order to obtain a malate/sulfate molar ratio ranging from 0 to 6.

To study the kinetics of sorption of sulfate at pH 4.5 on Andisol 2 in the presence of malate suspensions containing 150 mmol kg^{-1} of sulfate (initial SO_4/MAL molar ratio = 2) were shaken for 0.08–168 h. Sulfate was determined in the supernatant as described below.

2.6 Sulfate and Phosphate Sorption on $Al(OH)_x$-Oxalate Complex

Twenty-three milliliters of 0.02 mol L^{-1} KCl were added to 25 mg of the complex in reaction flasks. The pH values (4.0–9.0) of the mixtures were adjusted with 0.1 or 0.01 mol L^{-1} HCl or KOH. Suitable volumes of 0.05 mol L^{-1} solutions of phosphate or sulfate adjusted to pH 4.0–9.0, were pipetted into each flask to obtain 1,500 mmol of ligand per kilogram of complex. Competitive sorption experiments were carried out at pH 5.0 by adding 2,000 mmol sulfate per kg of complex and different amounts of phosphate (200–1,800 mmol kg^{-1}).

A few drops of toluene were initially added to each flask to inhibit microbial activity. The suspensions were shaken for 24 h at 20°C. The pH of the suspensions was periodically adjusted with 0.1 or 0.01 mol L^{-1} HCl or KOH by a Radiometer automatic titrator (Copenhagen, Denmark). The final suspensions (25 mL) were centrifuged and filtered through 0.2 μm Millipore M.F. filters (Bedford, MA), and phosphate, sulfate, and oxalate concentrations in the solution were then determined.

2.7 Sorption of Phosphate and Oxalate on Al-Hydroxy-Sulfate Complex

Twenty-three milliliters of 0.1 mol L^{-1} KCl was added to 25 mg of the complex in reaction flasks. Suitable quantities of 0.05 mol L^{-1} solutions containing phosphate or oxalate, adjusted to pH 4.0–9.0, were pipetted into the flasks to give 1,000–2,000 mmol of each ligand per kg of complex. The suspensions were shaken for 24 h at 20°C and their pH values were periodically adjusted to the original value. Samples were also kept at constant pH (4.0–9.0) for 24 h without addition of phosphate or oxalate. The final suspensions (25 ml) were centrifuged and ultrafiltered, as described before. No crystalline minerals were detected by XRD analyses at the end of the experiments.

2.8 Phosphate, Sulfate and Oxalate Determination

Sulfate, phosphate, and oxalate were determined in the final solutions by ion chromatography using a Dionex DX-300 ion chromatograph (Dionex Co, Sunnyvale, CA), an IonPac AS4A column (4.0 mm), an eluent of Na_2CO_3 and $NaHCO_3$ at a flow rate of 2 mL min^{-1}, and a CD20 conductivity detector combined with autosuppression. Average sulfate, phosphate, and oxalate retention times were 4.0, 6.0, and 12 min respectively. The standard concentrations were 0.2 to 2 mmol L^{-1} for each ligand. The amount of ligands adsorbed was determined by the difference between the initial and final concentrations. The data are the mean of either two or three determinations. Coefficients of variation ranged from 1.5 to 5%.

Aluminum and iron in the final solutions were determined by atomic adsorption spectrophotometry.

3 RESULTS AND DISCUSSION

3.1 Adsorption of Sulfate, Phosphate and LMMOLs on Variable Charge Minerals and Soils

Figure 12.1 shows the adsorption of sulfate (SO_4), phosphate (PO_4), oxalate (OX), and/or malate (MAL) on ferrihydrite, gibbsite, goethite, and Andisol 2 at a constant electrolyte (KCl) concentration and at pH 4.5. Adsorption data conformed to the Langmuir equation in the following form:

$$S = Sm\,K\,c/(1 + K\,c)$$

where S is the amount of adsorbate taken up per unit mass of adsorbent (mmol kg^{-1}), Sm is the maximum amount of adsorbate that may be bound, c the equilibrium solution concentration (mmol L^{-1}), and K a constant related to the binding energy (Giles et al., 1974). The sulfate adsorption capacity was usually lower than the oxalate, malate, and phosphate adsorption capacity.

Sulfate was adsorbed in lower amounts than phosphate, oxalate, and malate on most of the sorbents, clearly because phosphate and strongly chelating LMMOLs bind more strongly to aluminum or iron than sulfate does (Parfitt and Smart, 1978; Parfitt, 1980; Tabatabai, 1982; Lopez-Hernandez et al., 1996) and because of differences in the coordination forms of each ligand on the surfaces of the sorbents (binuclear and mononuclear, mono or bidentate) and their relative distribution (Sparks, 1999). Many authors have suggested that the mechanisms of sulfate and phosphate adsorption are similar and that both ions compete for the same sorption sites, although adsorbed sulfate does not compete strongly with phosphate. However, more recently some studies have demonstrated that sulfate may be adsorbed mainly by an outer-sphere complexation mechanism, that is, sulfate is adsorbed on the positive sites by electrostatic attraction (Curtin and Syers, 1990; Zhang and Sparks, 1990; Turner and Kramer, 1991), in spite of the formation of inner-sphere complexation by sulfate, which may be possible under certain conditions (Peak et al., 2001; Pigna and Violante, 2003). Bhatti et al. (1998) found that oxalate adsorbed by a soil clay fraction formed monodentate and binuclear surface complexes at pH 3.5, but binuclear surface complexes were formed at pH 4.5 and 5.5.

Figure 12.2A and B show the amounts of phosphate, sulfate (and oxalate) sorbed on goethite and the two Andisols at various pH values. Sorption of both phosphate and sulfate mainly decreased with increasing pH. However,

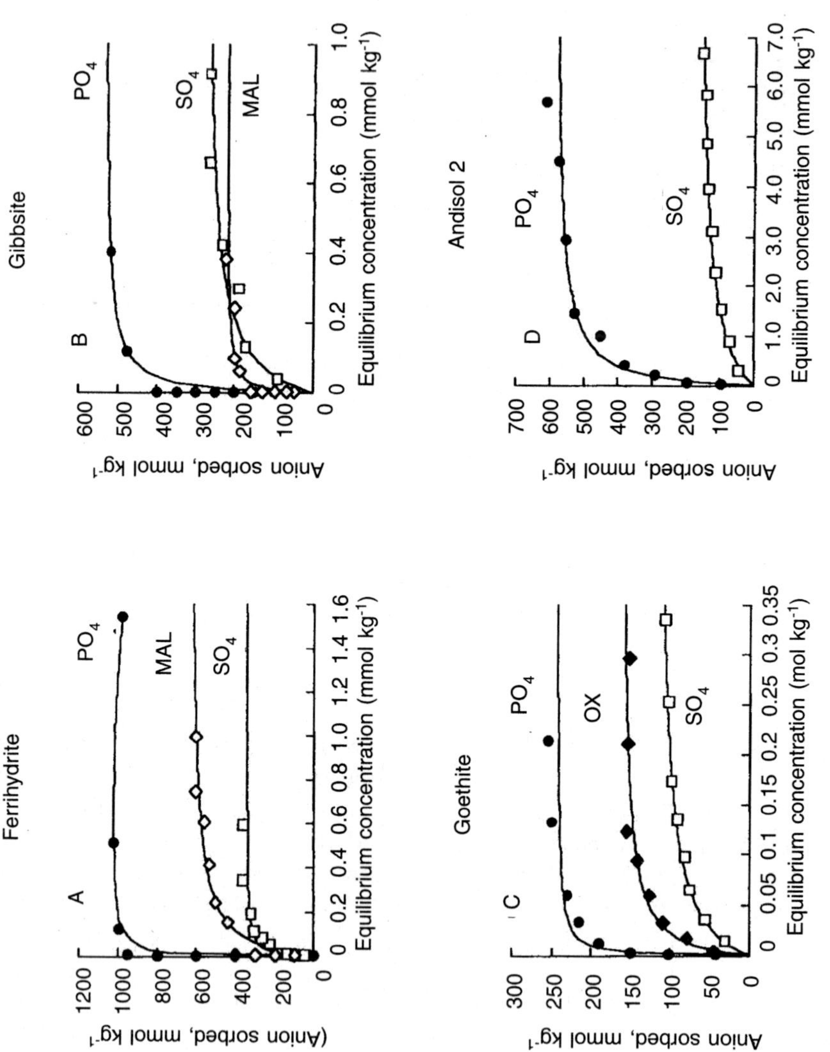

Fig. 12.1: Sorption isotherms of sulfate (SO_4), phosphate (PO_4), oxalate (OX), and/or malate (MAL) on ferrihydrite, gibbsite, goethite, and Andisol 2 at pH 4.5.

A

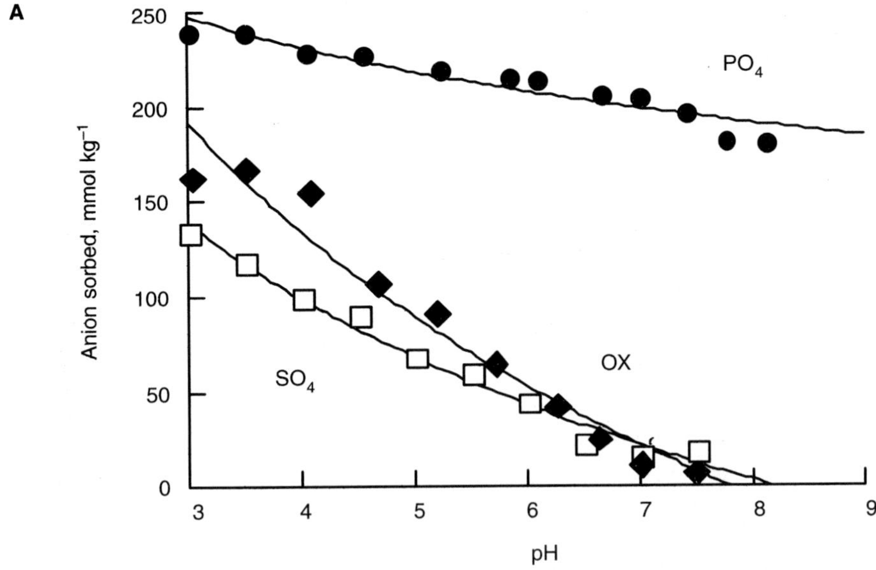

B

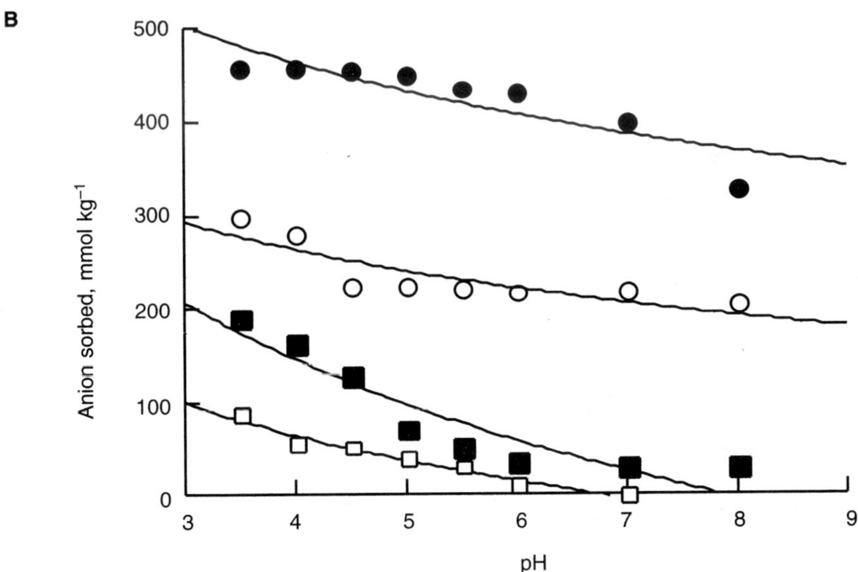

Fig. 12.2: Sorption of sulfate (SO_4), phosphate (PO_4), and oxalate (OX) at different pH values: (A) on goethite (300 mmol SO_4, PO_4 or OX kg^{-1}) and (B) on Andisols 1 and 2 (470 mmol SO_4 or PO_4 kg^{-1}); dark symbols (Andisol 2), white symbols (Andisol 1).

phosphate sorption decreased slightly by increasing the pH from 3.0 to 8.0, whereas sulfate sorption was negligible at pH > 6.0. Usually, no sorption of sulfate is seen above the point of zero charge of the sorbents (Kamprath et al., 1956; Chao et al., 1962; Peak et al., 1999; Pigna and Violante, 2003). Higher amounts of phosphate and sulfate were sorbed on Andisol 2 than on Andisol 1 because the quantities of allophanic materials in the former were greater (42%) than in the latter (17%; Table 12.1).

3.2 Adsorption of Sulfate on Andisols in the Presence of Phosphate

Experiments were carried out on the sorption of sulfate in the presence of phosphate at pH 4.5 on Andisols 1 and 2. The anions were added as a mixture at initial phosphate/sulfate (PO_4/SO_4) molar ratios ranging from 0 to 1 (Fig. 12.3); sulfate was added at its maximum surface coverage on each sample and phosphate added was relatively low but was completely adsorbed even in the presence of sulfate (Fig. 12.1). Sulfate sorption was strongly inhibited by the presence of phosphate, even at low PO_4/SO_4 molar ratios. In fact, at PO_4/SO_4 = 0.1, sulfate sorption decreased by 35% on Andisol 2 and by 67% on Andisol 1 compared to sulfate sorbed in the absence of phosphate. At PO_4/SO_4 = 1, sulfate adsorption was completely inhibited on Andisol 1 and was negligible on Andisol 2 (Fig. 12.3).

The decrease of sulfate sorption on soil samples cannot be attributed to just direct competition between phosphate and sulfate because i) the sulfate decrease

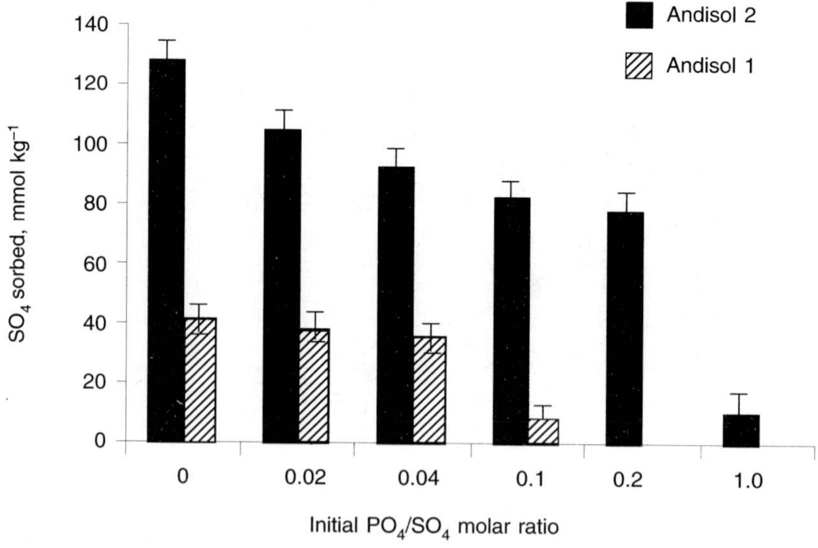

Fig. 12.3: Sorption of sulfate (SO_4) on the Andisols 1 and 2 at pH 4.5 and in the presence of increasing phosphate (PO_4) concentrations (470 mmol kg^{-1} sulfate added) (redrawn from Pigna and Violante, 2003).

was much greater than the amounts of phosphate sorbed and ii) many sites for phosphate sorption were available on both samples (Fig. 12.1d). A possible explanation of these findings is that phosphate sorption on soil samples decreased their surface charge and point of zero charge, preventing the fixation of sulfate ions on more negative surfaces (Parfitt, 1990; Barrow, 1992). According to Barrow (1992), the competition of anions for sorption sites of variable charge minerals was largely due to changes in the electric potential of the surface. We recently demonstrated that competition for sorption sites is a very important mechanism (Violante and Pigna, 2002) but without doubt reduction in the surface charge of the sorbents is also very significant. This is particularly true when a ligand weakly or moderately (like sulfate) sorbed on a sorbent competes with a ligand strongly adsorbed (like phosphate).

When sulfate was added 24 h before phosphate (*SO_4 before PO_4*) (Fig. 12.4), greater quantities of sulfate were sorbed on Andisol 2 then when phosphate and sulfate were added as a mixture (*SO_4 + PO_4* systems). At $PO_4/SO_4 = 0.1$, 0.2, and 1.0, the sulfate sorption decreased by 10, 22, and 70%, respectively in *SO_4* before *PO_4* systems versus 34, 45, and 94%, respectively, in the *PO_4 + SO_4* systems. Different processes may occur simultaneously in the sorption reactions of different ligands on soil components, as i) the kind of surface complexes formed by the anions when added as a mixture in different amounts, ii) the

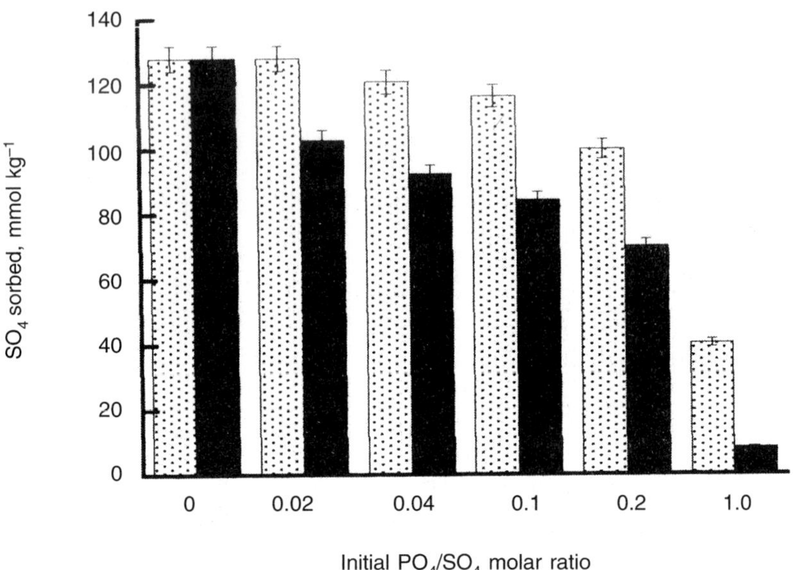

Fig. 12.4: Sorption of sulfate (SO_4) on Andisol 2 in the presence of increasing quantities of phosphate (PO_4), when phosphate and sulfate were added as a mixture (*SO_4 + PO_4* system ■) or when sulfate was added 24 h before phosphate (*SO_4 before PO_4* system ▤) (redrawn from Pigna and Violante, 2003).

change in surface charge after anion sorption, and iii) the effect of time on competition. We have demonstrated that the extent of competition between phosphate and sulfate is also related to sorption kinetics (see below).

3.3 Sorption of Sulfate in Presence of Phosphate as Affected by pH

Experiments on sorption of sulfate on Andisol 2 in the presence of increasing amounts of phosphate (PO_4/SO_4 ranging from 0 to 1) were carried out at acidic pH values (pH 2.5, 3.5, and 4.5) (Pigna and Violante, 2003). It was found that decreasing pH enhanced sulfate competition with phosphate (Fig. 12.5). For example, at PO_4/SO_4 = 0.2, phosphate inhibited sulfate sorption by 7% at pH 2.5 versus 11% and 46% at pH 3.5 and 4.5, respectively.

In our experiments it is possible that the solubilization of Al at low pH values (mainly at pH < 4.5; data not shown) promoted the formation of sparingly soluble Al-phosphates containing sulfate ions coprecipitated after oxyanion addition to soil samples. Therefore, other experiments were carried out by adding phosphate 24 h before sulfate so that sulfate ions could not be coprecipitated into the network of Al-phosphates that eventually formed. It is evident that even in the presence of large amounts of sulfate, phosphate was not desorbed by sulfate at pH 4.5, whereas 15% of phosphate initially sorbed was replaced by sulfate (SO_4/PO_4 molar ratio of 50) at pH 2.5 (Fig. 12.6). Removal of phosphate previously adsorbed by sulfate is surprising and explainable only

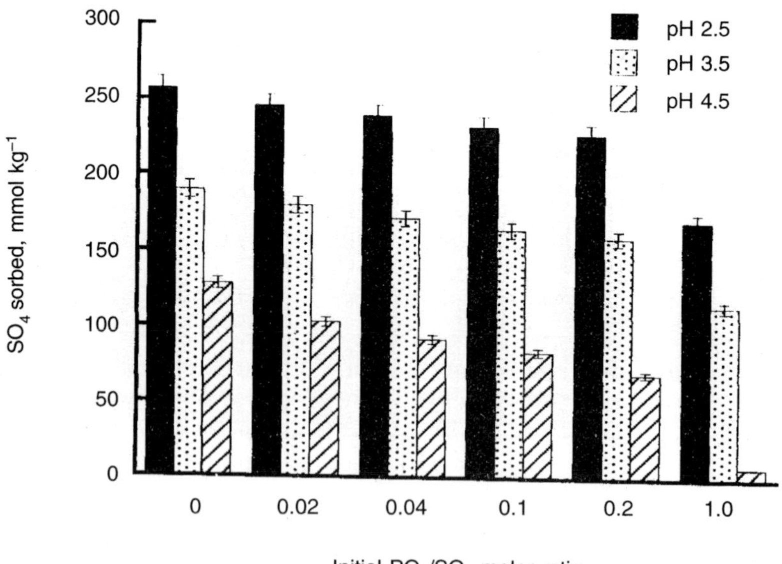

Fig. 12.5: Sorption of sulfate (SO_4; 470 mmol SO_4 kg^{-1}) on Andisol 2 at pH 4.5, 3.5, and 2.5 in the presence of increasing concentration of phosphate (PO_4) (modified from Pigna and Violante, 2003).

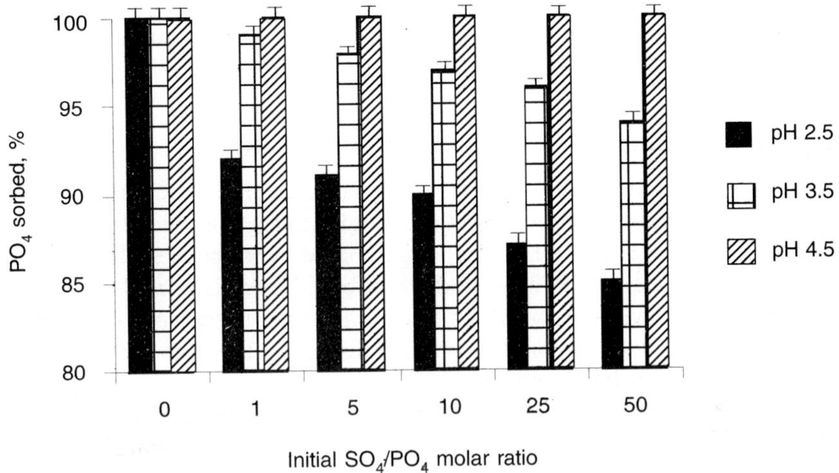

Fig. 12.6: Phosphate (PO_4) sorbed (%) on Andisol 2 in the presence of increasing concentrations of sulfate (SO_4) at pH 4.5, 3.5, and 2.5 (470 mmol PO_4 kg^{-1}). Phosphate was added 24 h before sulfate (redrawn from Pigna and Violante, 2003).

through the formation of some strong inner-sphere complexes formed by sulfate at very low pH values on the surfaces of allophanic materials.

According to Peak et al. (2001), previous macroscopic studies of sulfate sorption have generally suggested that sulfate adsorbs via an outer-sphere (electrostatic) adsorption mechanism on both soils and reference minerals. However, our findings show clear macroscopic evidence that sulfate may form strong complexes on the surfaces of variable charge sorbents at very low pH values. In other words, it seems evident that sulfate ions can form strong surface complexes at low pH. These strong surface complexes enhance the ability of sulfate to compete with phosphate (Figs. 12.5 and 12.6).

Indeed, sulfate sorption on variable charge minerals is quite complex. Sposito (1984) suggested that sulfate sorption might be of an intermediate nature, sorbing under different conditions as an outer-sphere complex versus an inner-sphere complex. According to some authors (Turner and Kramer, 1991; Eggleston et al., 1998; Rietra et al., 1999; Sparks, 1999), there is spectroscopic evidence that as pH is lowered and the concentration of sulfate increased, higher a percentage of inner-sphere complexes are formed by sulfate. Peak et al. (2001) claimed that it is important to consider not only the effects of pH, ionic strength, and reaction concentration on sulfate sorption, but also the nature of the sorbent under study. These authors found that sulfate forms inner-sphere monodentate surface complexes on hematite from pH 8.0 to 3.5 and across a wide range of surface loadings, whereas on goethite, sulfate forms only outer-sphere surface complexes at pH $\geq$ 6.0 and forms a mixture of outer-sphere and inner-sphere complexes at pH < 6.0. Finally, sulfate forms predominantly outer-sphere surface complexes on ferrihydrite.

3.4 Sorption of Sulfate in Presence of Organic Ligands

Low molecular mass organic ligands, such as oxalate, malate, citrate, succinate, and tartrate, are commonly present in soil environments, especially at the soil-root interface, and can influence the sorption/desorption of sulfate on/from soil components (Inskeep, 1989; Ali and Dzombak, 1996; Liu et al., 1999). Studies of competitive sorption of LMMOLs and sulfate are extremely important in understanding the role of biomolecules present in the rhizosphere on the availability of this nutrient for plants.

Liu et al. (1999) provided useful information on the competitive sorption of sulfate and oxalate on goethite. When equimolar amounts of sulfate and oxalate were added to goethite, the quantities of sorbed sulfate decreased over a wide range of pH (3.0 – 7.0) compared with those sorbed when only sulfate was added (Figs. 12.2 and 12.7). Sorbed oxalate significantly decreased in the presence of sulfate only at pH < 5.0 (Fig. 12.8). Furthermore, the order of anion addition influenced the adsorption of both sulfate and oxalate only at pH $\leq$ 6.0 (Table 12.2). Slightly more sulfate was adsorbed on goethite surfaces when added concurrently or before oxalate (*SO_4 + OX* or *SO_4 before OX* systems) than when

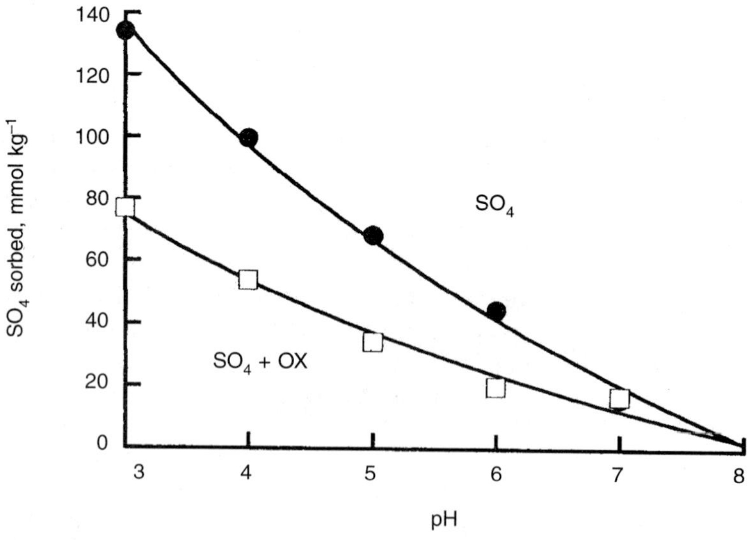

Fig. 12.7: Sorption of sulfate (SO_4) on goethite at different pH values: sulfate added alone (300 mmol kg^{-1}) or as a mixture with oxalate (OX) (*SO_4 + OX* system) (redrawn from Liu et al., 1999).

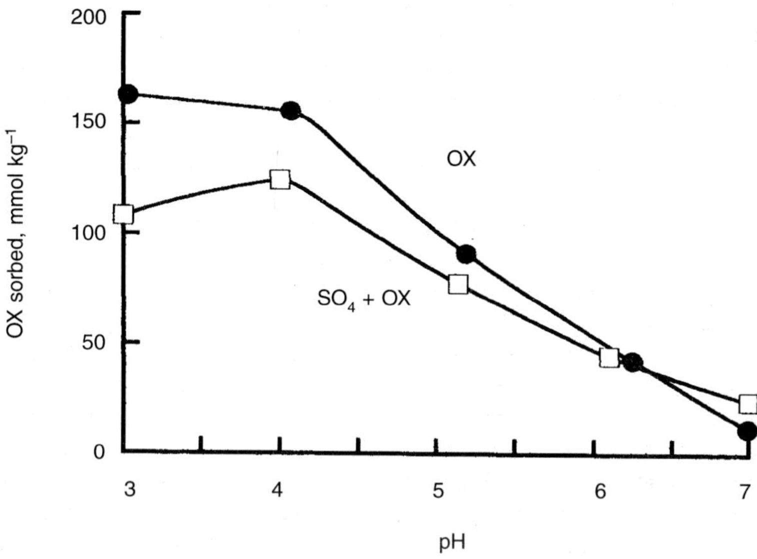

Fig. 12.8: Sorption of oxalate (OX) on goethite at different pH values: oxalate added alone (300 mmol kg^{-1}) or as a mixture with sulfate (*SO_4 + OX* systems) (redrawn from Liu et al., 1999).

Table 12.2: Amounts of sulfate (SO_4) and oxalate (OX) sorbed on goethite at different pH values when added alone or as a mixture (*SO_4 + OX* systems) or by adding oxalate before sulfate (*OX before SO_4* systems) at initial SO_4/OX molar ratio = 1 (300 mmol SO_4 and/or OX added kg^{-1})

			SO_4 + OX		*OX before SO_4*	
pH	SO_4 alone	OX alone	SO_4	OX	SO_4	OX
			mmol kg^{-1}			
3.0	134	163	77	108	68	152
3.5	118	167	60	126	53	153
4.0	100	156	54	124	36	135
4.5	91	110	37	106	31	116
5.0	68	92	34	78	27	100
5.5	60	65	24	57	22	62
6.0	44	43	20	45	22	45
6.5	22	26	19	25	16	28
7.0	15	12	17	24	14	24

added after oxalate (*OX before SO_4* systems). The sorption of sulfate in the *SO_4 + OX* systems did not differ significantly from that of *SO_4 before OX* systems (data not shown). Similarly, more oxalate was adsorbed when it was added before sulfate than when added in a mixture with or after sulfate (Table 12.2).

More oxalate than sulfate was always adsorbed at comparable pH values. The molar ratio of sorbed oxalate to sorbed sulfate (R) was particularly high at pH 4.0–5.0 both in SO_4 + *OX* and *OX before* SO_4 systems, showing that oxalate inhibited sulfate adsorption to a greater extent in slightly acidic environments (Fig. 12.9).

In SO_4 + *OX* systems the efficiency of sulfate in inhibiting oxalate sorption was particularly high at pH ≤ 4.0 ranging from 21 to 34%, whereas at pH > 4.5 it decreased (< 15%). Furthermore, in all the systems oxalate prevented sulfate sorption more at pH 4.5–6.0 than at pH ≤ 4.0. The efficiency of sulfate (or oxalate) to depress oxalate (or sulfate) sorption was calculated according to the expression of Deb and Datta (1967):

Efficiency of SO_4 (or OX) (%) = 1 – [OX (or SO_4) sorbed in the presence of SO_4 (or OX) / OX (or SO_4) sorbed when applied alone] × 100

Usually at pH ≥ 6.0 the competition in sorption between oxalate and sulfate was negligible (Fig. 12.7 and Table 12.2).

A possible explanation for these findings is that oxalate forming inner-sphere complexes on the surfaces of goethite easily replaced large amounts of sulfate from sites where this anion was not strongly held (forming outer-sphere complexes), especially at a pH range close to the pK_2 of the oxalic acid (pK_2 = 4.27). Furthermore, sulfate competed with oxalate particularly at pH < 4.0 (Fig. 12.9) because, as discussed above, at very low pH values sulfate may be strongly adsorbed on the goethite surfaces (Turner and Kramer, 1991) forming, at least in part, inner-sphere complexes that may compete with oxalate much more than in mildly acidic environments.

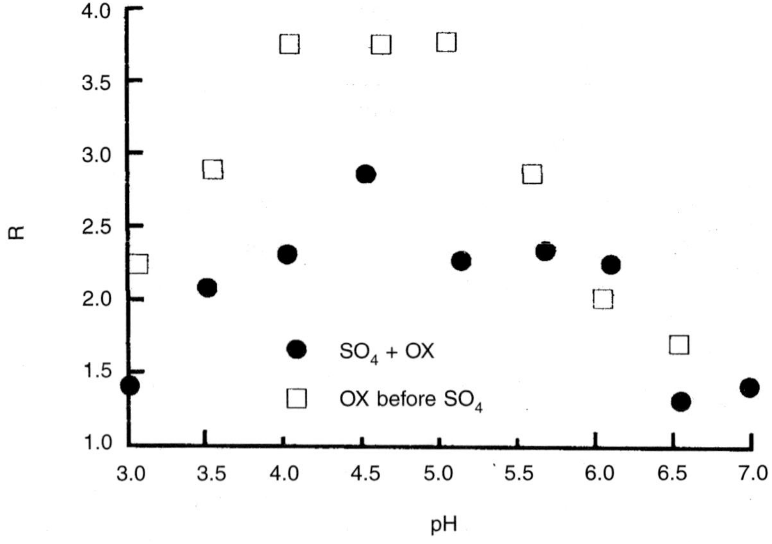

Fig. 12.9: Sorbed oxalate/sorbed sulfate molar ratio (R) versus pH in *OX before* SO_4 and SO_4 + *OX* systems (modified from Liu et al., 1999).

3.5 Effect of Increasing Concentrations of LMMOLs on Sulfate Sorption

Sorption of sulfate in the presence of increasing concentrations of oxalate or malate was likewise studied. Figure 12.10 shows sulfate and oxalate sorbed at pH 4.0 (300 mmol sulfate or oxalate kg^{-1}) in the presence of increasing amounts of oxalate or sulfate, respectively. The ligands were added as a mixture and the initial OX/SO_4 (r_i) or SO_4/OX (r_i') molar ratios ranged from 0.17 to 3.33.

Sulfate sorption decreased by 46% at $r_i = 1$ and by 72% at $r_i = 3.33$. It is evident that oxalate up to $r_i = 2$ strongly inhibited the sorption of sulfate while the quantities of sulfate adsorbed at $r_i > 2$ remained practically constant. These results indicate that 25–30% of SO_4 was sorbed so strongly on the oxide that even large amounts of oxalate were not able to replace it. Vice versa, the decrease in sorption of oxalate in the presence of increasing sulfate concentrations was 20.1% at $r_i' = 1$ and 38.7% at $r_i' = 3.33$.

Figure 12.11 shows sulfate sorption at pH 4.5 on gibbsite, ferrihydrite, and Andisol 2 in the presence of increasing malate concentrations (MAL/SO_4 molar ratio ranging from 0 to 4.0). Sulfate sorption decreased tremendously with increasing malate concentration; it was completely inhibited at a MAL/SO_4 molar ratio of 2 on the Andisol and of 4 on gibbsite and ferrihydrite.

Considerable evidence shows that the inhibition of sulfate sorption as well as of other oxyanions on soil components is affected by pH, and the nature and

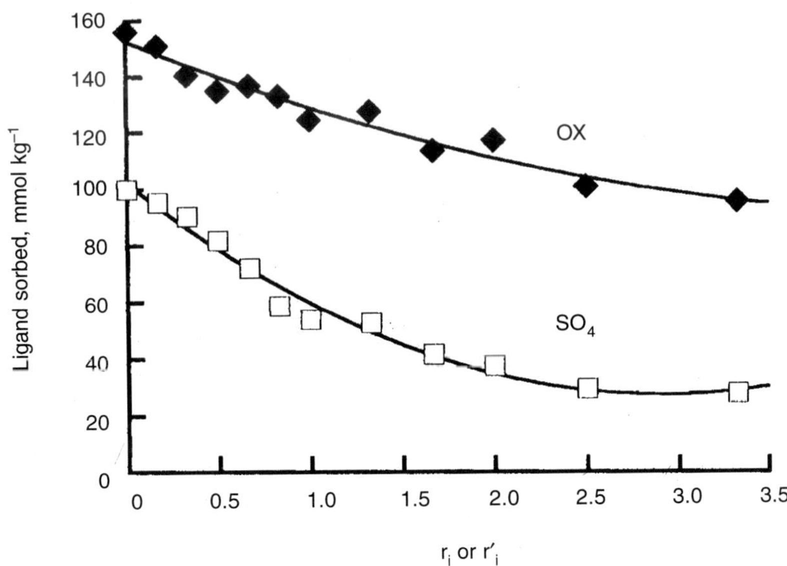

Fig. 12.10: Sulfate (SO_4) and oxalate (OX) sorbed at pH 4.0 when sulfate (300 mmol kg^{-1}) was added to goethite in the presence of increasing oxalate concentrations (initial OX/SO_4 molar ratio $r_i = 0.17 - 3.33$) and when oxalate (300 mmol kg^{-1}) was added in the presence of increasing sulfate concentrations (initial SO_4/OX molar ratio $r' = 0.17 - 3.33$).

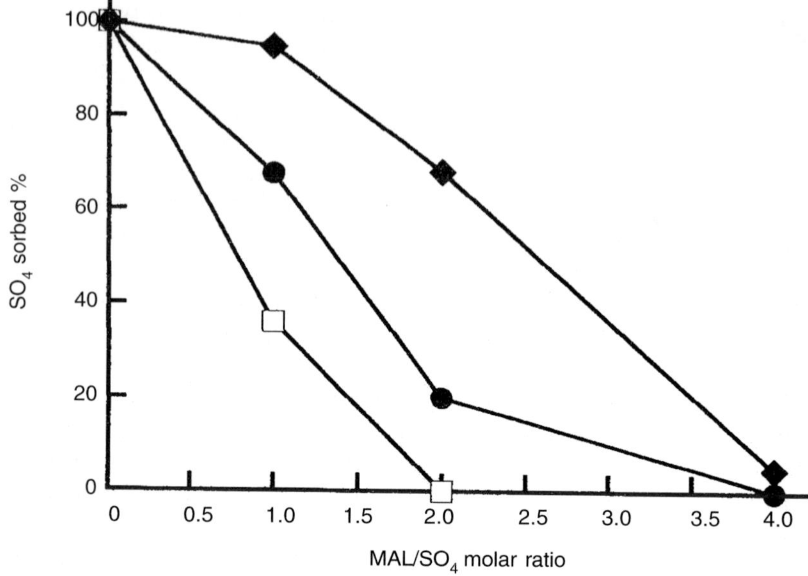

Fig. 12.11: Adsorption of sulfate (SO_4) expressed as percentage of sulfate initially adsorbed on gibbsite ●, ferrihydrite ■, and Andisol 2 □ at pH 4.5 in the presence of increasing malate/sulphate molar ratio. gibbsite, ferrihydrite, and Andisol 2 adsorbed respectively 100, 290, and 67 mmol kg^{-1} in the absence of malate at pH 4.5.

Table 12.3: Effect of time on the sorption of sulfate (SO_4) on Andisol 2 at pH 4.5 when added alone (150 mmol kg^{-1}) and in the presence of phosphate (PO_4) or malate (MAL). Initial SO_4/PO_4 or MAL molar ratio of 2.

Time (h)	SO_4	$SO_4 + PO_4$ mmol kg^{-1}	SO_4 + MAL
0.08	54	23	16
0.25	68	30	32
0.50	69	36	39
1	72	40	42
3	75	48	54
24	83	60	59
48	86	63	75
168	88	63	76

concentration of competitive organic and inorganic ligands and surface properties of sorbents (Peak et al., 2001; Violante and Pigna, 2002; Pigna and Violante, 2003).

3.6 Kinetics of Sorption of Sulfate in Presence of Foreign Ligands

We studied the effect of time on the sorption of sulfate on Andisol 2 at pH 4.5, when added alone (150 mmol kg^{-1}) or in the presence of phosphate or malate (SO_4/PO_4 or MAL molar ratio of 2). Sorption of sulfate on Andisol 2 increased with time, reaching maximum (83–88 mmol kg^{-1}) after 24–168 h (Table 12.3). No further increase in sulfate sorption was found after 7–15 days of reaction (data not shown). The presence of low amounts of phosphate or malate strongly prevented sulfate sorption even though many sites were still available for sulfate sorption (Fig. 12.1d).

In the presence of phosphate or malate the maximum amount of sulfate sorbed was reached after 24 and 48 h, no further increase was observed even after 15 days (data not shown).

Compared to the sulfate-only experiments about 70% of sulfate was sorbed in the presence of phosphate and 86% in the presence of malate after 168 h. However, after 7–15 days of reaction the amounts of sulfate sorbed in the presence of phosphate or malate did not increase further (data not shown).

3.7 Sorption of Sulfate on Organomineral Complex

Organomineral complexes have important effects on the physical and chemical properties and reactivity of soils. Sorption of nutrients and pollutants on organomineral complexes is of great importance in soil environments but surprisingly has received scarce attention (Violante and Gianfreda, 2000).

De Cristofaro et al. (2000) studied the sorption of sulfate and phosphate on an $Al(OH)_x$-oxalate complex [$Al(OH)_x$-OX] obtained by coprecipitating Al and oxalate at pH 7.0 and initial OX/Al molar ratio of 0.1.

Table 12.4 shows the influence of pH on the sorption of sulfate and phosphate when 1,500 mmol kg^{-1} of either ligand was added to the Al(OH)x-

Table 12.4: Sorption of sulfate (SO_4) and phosphate (PO_4) (1,500 mmol kg^{-1}) and desorption of oxalate (OX) at different pH values on/from an $Al(OH)_x$-OX complex

pH	OX desorbed*	PO_4 sorbed	OX desorbed	r**	SO_4 sorbed	OX desorbed	r′**
	mmol kg^{-1}				mmol kg^{-1}		
4.0	1283	1500	742	0.49	n.d.	n.d.	n.d.
5.0	450	1277	618	0.48	618	765	1.2
6.0	490	1175	625	0.53	455	607	1.3
7.0	775	1216	957	0.79	194	850	4.4
8.0	1300	1180	1200	1.02	185	1211	14.1

*OX desorbed in absence of phosphate or sulfate.
**r and r′ = OX desorbed/SO_4 (or PO_4) sorbed molar ratio.
n.d. = not determined.

OX complex. Table 12.4 reports the amounts of oxalate removed in the range of pH 4.0 – 8.0, both in the absence or presence of the inorganic ligands and the *OX* removed/SO_4 (or PO_4) sorbed molar ratios (*r* and *r'*).

In the absence of the inorganic ligands, high quantities of oxalate were removed from the complex at pH < 5.0 and > 7.0. At acidic pH, large quantities of Al were solubilized (2.78 mol kg^{-1} at pH 4.0 and 0.45 mol kg^{-1} at pH 5.0; data not shown), and proportionally high amounts of oxalate were found in solution (1,283 mmol kg^{-1} at pH 4.0 and 450 mmol kg^{-1} at pH 5.0). In contrast, at pH ≥ 6.0, although relatively low amounts of Al were solubilized, the quantities of oxalate removed from the complex increased tremendously with increaments in pH (up to 1,300 mmol kg^{-1} at pH 8.0). The removal of oxalate must be attributed mainly to the replacement of the organic ligands by OH^- ions (Violante et al., 1996).

The amounts of phosphate sorbed decreased from 1,500 mmol kg^{-1} at pH 4.0 to 1,180 mmol kg^{-1} at pH 8.0. The *OX* removed/PO_4 sorbed molar ratio was lower than 1 at pH < 7.0, indicating that phosphate was also held on sites free of oxalate. The quantities of sulfate sorbed on the Al(OH)x-oxalate complex were lower than those of phosphate but relatively high, 618 mmol kg^{-1} at pH 5.0. At pH > 6.0 the amounts of sulfate retained by the complex did not exceed 200 mmol kg^{-1} but on some natural (Andisol) or synthetic sorbents, negligible quantities of sulfate were fixed at pH > 6.0 (Fig. 12.2). The *OX* removed/SO_4 sorbed molar ratio values were 1.2–1.3 in the pH range 5.0–6.0 and then increased dramatically to 4.4 at pH 7.0 and 14.1 at pH 8.0. These data indicate that sulfate was able to partially remove oxalate from the complex at pH < 6.0, but in alkaline environments much more oxalate was replaced by OH^- ions than by sulfate.

Unfortunately, the influence of organic molecules on cation and anion adsorption on Al- or Fe-organic matter associations has not been thoroughly studied and is not clearly understood. Organic matter has often been correlated positively with phosphate sorption. Gunjigake and Wada (1981) found that phosphate sorption correlated highly with the pyrophosphate extractable Fe and Al content of soils and showed that Al-organic substances associations have significant phosphate sorption. Furthermore, phosphate sorption by Al precipitation products formed in the presence of LMMOLs has been reported to be much greater than sorption on Al and Fe oxides (De Cristofaro et al., 2000). Sulfate sorption on organomineral complexes may have similar behavior. However, reduction of sulfate sorption by LMMOLs has been demonstrated (as discussed above; Figs. 12.7 and 12.11; Table 12.2) due to competition of sulfate and organic ligands for sorption sites of variable charge mineral and soils. In fact, it is evident that many factors influence the sorption of sulfate (as well as other nutrients and pollutants) on organomineral complexes.

Organic ligands complex Al and Fe in aqueous solution and, subsequently, hamper the crystallization of Al and Fe oxides. The effectiveness of a biomolecule in perturbing hydrolytic reactions of Al and Fe is related to its chemical composition, molecular structure, size, functional groups, and subsequent affinity for Al and Fe. As a consequence, crystallinity, specific surface, PZC, excess surface charge, and reactivity of precipitation products of Al and Fe formed in the presence of organic ligands vary with kinds and amounts of ligands coprecipitated and aging of the samples (Huang and Violante, 1986).

Many authors (Wada, 1977; Huang and Violante, 1986; Violante et al., 1996) have demonstrated that organic acids such as malic, citric, aspartic, oxalic, and tannic acid, promote formation of active sites for the sorption of phosphate by distorting the structure of precipitation products of aluminum and enhancing their specific surface. Maintenance of the short-range structure of the precipitates with a large specific surface area by the presence of critical concentrations of some biomolecules helps to promote a high sulfate retention capacity of organomineral complexes.

Table 12.5 shows the amount of sulfate sorbed at pH 5.0 on the organomineral complex in the presence of phosphate when sulfate (2,000 mmol kg^{-1}) and phosphate were added together at different PO_4/SO_4 molar ratios ranging from 0 to 0.9. The presence of phosphate strongly prevented the sorption of sulfate. Sulfate sorption decreased by 40% in the presence of only 200 mmol of phosphate kg^{-1}. These results are similar to those found for the competitive sorption of phosphate and sulfate on Andisols (Fig. 12.3).

Table 12.5: Competitive sorption of phosphate (PO_4) and sulfate (SO_4) at pH 5.0 on the $Al(OH)_x$-oxalate complex

PO_4 added	SO_4 added	Ri*	PO_4 sorbed	SO_4 sorbed	Rf**
mmol kg^{-1}			mmol kg^{-1}		
0	2000	0	0	618	0
200	2000	0.1	196	380	0.5
400	2000	0.2	398	227	1.75
600	2000	0.3	595	190	3.13
1000	2000	0.5	912	148	6.16
1400	2000	0.7	1162	67	17.3
1800	2000	0.9	1275	80	15.9

*Ri = initial phosphate/sulfate molar ratio.
**Rf = final phosphate/sulfate molar ratio.

3.8 Desorption of Sulfate from an $Al(OH)_x$-Sulfate Precipitate by Phosphate and Oxalate

Formation of aluminum hydroxysulfate precipitates, including alunite $[KAl_3(OH)_6(SO_4)_2]$ and basaluminite $[Al_4(OH)_{10}SO_4]$, in acid soils containing sulfate or receiving sulfate has been widely reported (Adams and Rawajfih, 1977; Evans, 1991; Wolt et al., 1992; Sumner, 1993). Courchesne and Hendershot (1990) demonstrated that formation of basic aluminum sulfate minerals in the mineral horizons of two Spodosols is involved in the retention of sulfate.

Precipitation of hydroxy Al sulfates can be promoted by applying gypsum as an ameliorant for acid subsoils (Sumner, 1993). Although some researchers (Wolt and Adams, 1979; Evans, 1991; Wolt et al., 1992) claim that sorption is not the only retention mechanism in sulfur-retentive soils, with sulfate sorption and desorption also due to the possible presence and solubility behavior of aluminum and iron hydrous sulfates, information about the removal of sulfate from such compounds is scant.

Violante et al. (1996) studied the sorption of phosphate and oxalate on a synthetic aluminum hydroxysulfate complex containing 1.35 mol sulfate kg^{-1} and the associated release of sulfate from this complex (Table 12.6). These authors found that the amount of sulfate removed from the complex by 0.1 mol L^{-1} KCl solution increased only slightly (from 110 to 170 mmol kg^{-1}) on increasing pH from 4.0 to 6.0, but a much greater release of sulfate (up to 945 mmol kg^{-1}) was seen on raising the pH from 6.5 to pH 9.0 (Table 12.6). At pH 7.0–9.0, 25–70% of the sulfate initially present in the complex was brought into solution. These data support the observation of many authors that sulfate ions are easily released from soils by alkaline solutions (Kamprath et al., 1956; Chao et al., 1962; Tabatabai, 1982).

Table 12.6 shows the quantities of sorbed phosphate and oxalate and removed sulfate when 1,500 mmol $kg^{-1}$of each ligand was added. Phosphate retention by the complex decreased with increase in pH, whereas oxalate retention was greatest at pH 5.5 and decreased at both higher and lower pH. Relatively large amounts of oxalate promoted partial dissolution of the complex

Table 12.6: Amounts of phosphate (PO_4) and oxalate (OX) sorbed on, and of sulfate (SO_4) removed from, the aluminum hydroxy sulfate complex at different pH values (modified from Violante et al., 1996)

		P added*		OX added*	
pH	SO_4 removed**	PO_4 removed	SO_4 removed	OX sorbed	SO_4 removed
		mmol kg^{-1}			
4.0	110	1,130	220	340	330
4.5	116	1,065	250	474	442
5.0	157	950	260	570	420
5.5	162	990	280	697	470
6.0	170	800	300	583	552
6.5	240	780	308	591	532
7.0	336	547	371	306	596
8.0	678	480	401	250	686
9.0	945	470	586	220	966

*1,500 mmol kg^{-1} of P and OX added.
**SO_4 removed in the absence of P and OX.

in the entire pH range studied, but particularly at pH < 6.0. The decrease in sorption of oxalate at pH < 5.0 may be attributed to dissolution of the complex.

More sulfate was replaced in the presence of oxalate than in the presence of phosphate, despite the fact that between 1.3 and 3.3 times more phosphate than oxalate was retained by the complex (Table 12.6). At pH > 8.0 in the presence of phosphate and at pH > 6.5 in the presence of oxalate, more sulfate was desorbed from the complex than the quantities of phosphate or oxalate retained by it.

At pH < 6.0 and in the presence of fairly large amounts of oxalate or phosphate, the larger quantities of sulfate in solution in oxalate systems must be attributed mainly to greater solubilization of the complex in the presence of the organic ligand (1,832–630 mmol Al released kg^{-1}). Conversely, in acid systems and in the presence of small amounts of phosphate or oxalate (200–500 PO_4 or OX mmol kg^{-1}), usually slightly more sulfate was desorbed by phosphate than by oxalate, clearly because the solubilization of the complex in oxalate systems was minimal (data not shown). However, at pH ≥ 6.0, the fairly weak dissolution of the complex (429–74 mmol Al released kg^{-1}), even in the presence of 1,500 mmol oxalate kg^{-1}, cannot explain the much greater removal of sulfate in the presence of oxalate than in the presence of phosphate.

Apparently, in neutral and alkaline environments, hydroxyl ions, which have a strong affinity for Al and are powerful in removing sulfate from clay minerals (Chao, 1964; Pasricha and Fox, 1993), compete with phosphate or oxalate for sorption sites and sulfate removal. On increasing the pH, sulfate ions were replaced by OH^- more easily in the presence of oxalate than in the presence of phosphate, possibly because oxalate competed poorly with OH^- ions. The reason why much more sulfate was removed in alkaline systems from the complex in the presence of oxalate than phosphate may also be partially attributed to differences in aggregate stability. Usually, at pH > 7.0 the aggregates in short-range ordered Al precipitation products are unstable with a consequent partial dispersion of the particles and increase in the specific surface area (Goh et al., 1986). Differences in the stability of aggregated samples are attributed to the crystallinity of the sample and the nature and concentration of foreign ligands present in the precipitates (Goh et al., 1986; Violante and Huang, 1992). Long-term laboratory studies have demonstrated that phosphate promotes aggregation of Al precipitation product much more than oxalate does (Goh et al., 1986; Huang and Violante, 1986; Violante and Huang, 1992). As a consequence, in alkaline systems the greater particle dispersion of the aluminum hydroxysulfate complex in the absence or presence of oxalate than in the presence of phosphate could explain the larger increase in the removal of sulfate from the much more exposed surfaces of the complex. Curtin and Syers (1990), who compared the release of sulfate in 1 mol L^{-1} NaCl and KH_2PO_4, found that six extractions with NaCl removed slightly more sulfate than did six extractions with KH_2PO_4 from the Palace Leas soil.

Sulfate was only partly removed from the complex even after repeated washings with phosphate or oxalate at pH 6.0. Data reported in the literature on the removal of sulfate from clay minerals or soils is conflicting indeed. Some scientists reported replacing very large percentages or all of the sulfate sorbed in their sample soils or clay minerals with phosphate solutions (Chao et al., 1962; Bornemisza and Llanos, 1967; Pasricha and Fox, 1993; Figs. 12.3–12.5). Conversely, others found that extractions with KH_2PO_4 did not remove all the sulfate present in Latosols (Bornemisza and Llanos, 1967) and Brown Forest soils (Haque and Walmsley, 1973). According to Haque and Walmsley (1973), only sulfate sorbed on surface groups of hydrous oxides was desorbed, whereas sulfate that penetrated into some amorphous region of the crystal surface was retained. Our research seems to strengthen this view. In fact, according to Huang and Violante (1986), when sulfate ions (or other inorganic or organic ligands) are coprecipitated with aluminum, they are either sorbed on the external surfaces or present in the network of the initially formed short-range ordered aluminum precipitation products. However, it is not possible to determine how much of the sulfate is surface-adsorbed and how much is part of the structure of the aluminum precipitate. In our synthetic aluminum hydroxysulfate complex, the sulfate not removed at high pH by ligands with a strong affinity for aluminum may have been present in the structural network of the precipitate.

4 CONCLUSIONS

Mobility of sulfate in soil environments is affected by many factors, such as pH, nature and crystallinity of the sorbents, presence, concentration and nature of inorganic and organic anions, and possible formation of Al- and Fe-hydroxysulfate precipitates of different crystallinity. Sulfate sorption appears negligible at pH > 6.0 and no sorption of sulfate usually occurs above the PZC of the sorbents. In the presence of inorganic and organic ligands, which have a high affinity for aluminum or iron, sulfate is easily and completely desorbed at pH ≥ 4.5, but in strongly acidic environments the capacity of sulfate to compete with phosphate or oxalate increases. Our macroscopic studies of sulfate sorption strengthen the observation that sulfate may be adsorbed mainly by outer-sphere complexation mechanisms, but at low pH values it may also form inner-sphere complexes via a ligand exchange mechanism. As a consequence, at pH < 4.5, sulfate may not only compete with phosphate and LMMOLs for sorption sites of variable charge minerals and soils, but also desorbs these ligands, at least partially, from the surfaces of the sorbents.

Organomineral complexes adsorb relatively high amounts of sulfate. Maintenance of the short-range structure of the precipitates with a large specific surface area by the presence of critical concentrations of some organic ligands helps to promote a high sulfate retention capacity of organomineral complexes. Finally, we demonstrated that noncrystalline Al-hydroxysulfate compounds, likely to be present in soil, react differently at different pHs with phosphate or oxalate. Sulfate coprecipitated with aluminum that is part of the structural network may be only partly removed by large amounts of phosphate or oxalate or after repeated washing with these ligands. In acidic environments, especially at the soil-plant interface, the presence of strongly chelating anions may promote solubilization of Al- (or Fe^{-}) sulfate precipitates, favoring the release of sulfate.

Acknowledgments

This study was supported by the Italian Research Program of National Interest (PRIN), year 2002. DISSPA no. 0046.

References

Adams F. and Rawajfih Z. 1977. Basaluminate and alunite: a possible cause of sulfate retention by acid soils. *Soil Sci. Soc. Amer. Proc.* 41: 686–692.

Ali M.A. and Dzombak D.A. 1996. Competitive sorption of simple organic acids and sulfate on goethite. *Environ. Sci. Technol.* 30 (4): 1061–1071.

Atkinson R.J., Posner A.M., and Quirk J.P. 1967. Adsorption of potential determining ions on the ferric oxide-aqueous electrolyte interface. *J. Phys. Chem.* 71: 550–558.

Barrow N.J. 1992. The effect of time on the competition between anions for sorption. *Soil Sci. J.* 43: 424–428.

Bhatti J.S., Comerford N.B., and Johnston C.T. 1998. Influence of soil organic matter removal and pH on oxalate sorption onto spodic horizon. *Soil Sci. Soc. Amer. J.* 62: 152–158.

Bornemisza E. and Llanos R. 1967. Sulfate movement, adsorption, and desorption in three Costa Rica soils. *Soil Sci. Soc. Amer. Proc.* 31: 356–360.

Chao T.T. 1964. Anionic effects on sulfate adsorption by soils. *Soil Sci. Soc. Amer. Proc.* 31: 581–583.

Chao T.T., Harward M.E., and Fang S.C. 1962. Adsorption and desorption phenomena of sulfate ions. *Soil. Sci. Soc. Amer. J.* 26: 234–237.

Courchesne F. and Hendershot W.H. 1990. The role of basic aluminum sulfate minerals in controlling sulfate retention in the mineral horizons of two Spodosols. *Soil Sci.* 150: 571–578.

Curtin D. and Syers J.K. 1990. Mechanisms of sulfate adsorption by two tropical soils. *J. Soil Sci.* 41: 295–304.

De Cristofaro A., He J.Z., Zhou D.H., and Violante A. 2000. Adsorption of phosphate and tartrate on hydroxy-aluminum-oxalate precipitates. *Soil Sci. Soc. Amer. J.* 64: 1347–1355.

Deb D.L. and Datta N.P. 1967. Effect of associating anions on phosphorus retention in soils. I. Under variable phosphorous concentration. *Plant Soil* 26: 303–316.

Eggleston C.M., Hug S., Stumm W., Sulzberg B., and Dos Santos Alfonso M. 1998. Surface complexation of sulfate by hematite surfaces: FTIR and STM observation. *Geochim. Cosmochim. Acta* 62: 585–593.

Eltanawy I.M. and Arnold P.W. 1973. Reappraisal of ethylene glycol monoethyl ether (EGME) method for surface area estimation of clays. *J. Soil Sci.* 24: 232–238.

Evans Jr. A. 1991. The interactions of aliphatic acids with basic aluminum sulfates in a forested Ultisol. *Soil Sci.* 152: 53–60.

Geelhaed J.S., Hiemstra T., and Van Riemsdijk W.H. 1997. Phosphate and sulfate adsorption on goethite: single anion and competitive adsorption. *Geochim. Cosmochim. Acta* 61: 2389–2396.

Giles C.H., Smith D., and Huitson A. 1974. A general treatment and classification of the solute adsorption isotherms. *I. Theoret. J. Colloid Interface Sci.* 47: 755–765.

Goh T.B., Violante A., and Huang P.M. 1986. Influence of tannic acid on retention of copper and zinc by aluminum precipitation products. *Soil Sci. Soc. Amer. J.* 50: 820–825.

Gunjigake N. and Wada K. 1981. Effects of phosphorus concentration and pH on phosphate retention by active aluminium and iron of Andosoils. *Soil Sci.* 132: 347–353.

Haque I. and Walmsley D. 1973. Adsorption and desorption of sulfate in some soils of the West Indies. *Geoderma* 9: 269–278.

Huang P.M. and Violante A. 1986. Influence of organic acids on crystallization and surface properties of precipitation products of aluminum. In: *Interactions of Soil Minerals with Natural Organics and Microbes*. P.M. Huang and M. Schnitzer (eds). Spec. Publ. no. 17, Soil Sci. Soc. Amer., Madison WI, pp. 159–221.

Inskeep W.P. 1989. Adsorption of sulphate by kaolinite and amorphous iron oxide in the presence of organic ligands. *J. Envir. Qual*. 18: 379–385.

Jackson M.L. 1969. *Soil Chemical Analysis—Advanced Course*. Published by the author, Madison, WI, 895 pp.

Kamprath E.J., Nelson W.L., and Fitts J.W. 1956. The effect of pH, sulfate and phosphate concentration on the adsorption of sulfate by soils. *Soil Sci. Soc. Amer. Proc*. 28: 463–466.

Liu F., He Z., Colombo C., and Violante A. 1999. Competitive adsorption of sulfate and oxalate on goethite in the absence or presence of phosphate. *Soil Sci*. 164: 180–189.

Lopez-Hernandez, D., Siegert G., and Rodriguez J.V. 1986. Competitive adsorption of phosphate with malate and oxalate by tropical soils. *Soil Sci. Soc. Am. J*. 50: 1460–1462.

Parfitt R.L. 1980. Chemical properties of variable charge soil. In: *Soils with Variable Charge*. B.K.G. Theng (ed.). New Zeland Soc. Soil Science, Lower Hutt, NZ, pp. 167–194.

Parfitt R.L. 1990. Allophane in New Zealand—A review. *Aust. J. Soil Res*. 28: 343–360.

Parfitt R.L. and Smart S.C. 1978. The mechanism of sulfate adsorption on iron oxides. *Soil Sci. Soc. Amer. J*. 42: 48–50.

Pasricha N.S. and Fox R.L. 1993. Plant nutrient sulfur in tropics and subtropics. *Adv. Agron*. 50: 209–269.

Peak D., Ford R.G., and Sparks D.L. 1999. An in situ ATR-FTR investigation of sulfate bonding mechanisms on goethite. *J. Colloid Interface Sci*. 218: 289–299.

Peak D., Elzinga E.J., and Sparks D.L. 2001. Understanding sulfate adsorption mechanisms of iron(III) oxides and hydroxides: Results from ATR-FTIR spectroscopy. In: *Heavy Metals Release in Soil*. H.M. Selim and D.L. Sparks (eds.). Lewis Publ., Boca Raton, FL, pp. 167–190.

Pigna M. and Violante A. 2003. Adsorption of sulfate and phosphate on Andisols. *Comm. Soil Sci. Plant Anal*. 34: 2099–2113.

Rietra, R.P.J.J., Hiemstra T., and Van Riemsdijik W.H. 1999. Sulfate adsorption on goethite. *J. Colloid Interface Sci*. 218: 511–521.

Sakuray K., Ohdate Y., and Kyuma K. 1988. Comparison of salt titration and potentiometric titration methods for the determination of zero point of charge. *Soil Sci. Plant Nutr*. 34: 171–182.

Sparks D.L. 1999. Kinetics and mechanisms of chemical reactions at the soil mineral / water interface. In: *Soil Physical Chemistry*. D.L. Sparks (ed.). CRC Press: Boca Raton, FL, pp.135–191 (2nd ed.).

Sposito G. 1984. *The Surface Chemistry of Soils*. Oxford University Press New York, NY.

Sumner M.E. 1993. Gypsum and acid soils: the world scene. *Adv. Agron*. 51: 1–32.

Sumner M.E. 2000. *Handbook of Soil Science*. CRC Press Boca Raton, FL.

Tabatabai M.A. 1982. Sulfur. In: *Methods of Soil Analysis, Part 2*. A.L. Page and R.H. Miller (eds.). Agron. no. 9. Amer. Soc. Agron., Madison, WI, pp.501–583.

Turner L.J. and Kramer J.R. 1991. Sulfate ion binding on goethite and hematite. *Soil Sci*. 152: 226–230.

Vacca A., Adamo P., Pigna M., and Violante P. 2003. Properties and classification of selected soils from the Roccamonfina volcano, Central-Southern Italy. *Soil Sci. Soc. Amer. J*. 67: 198–207.

Violante A. and Huang P.M. 1992. Effect of tartaric acid and pH on the nature and physicochemical properties of short-range ordered aluminum precipitation products. *Clays Clay Miner*. 40: 462–469.

Violante A. and Gianfreda L. 2000. Role of bio-molecules in the formation and reactivity toward nutrients and organics of variable charge minerals and organomineral complexes in soil enviroment. In: *Soil Biochemistry*. J.M. Bollag and G. Stotsky (eds.). Marcell Dekker, New York, NY, vol 10, pp. 207–270.

Violante A. and Pigna M. 2002. Competitive sorption of arsenate and phosphate on different clay minerals and soils. *Soil Sci. Soc. Amer. J*. 66: 1788–1796.

Violante A., Rao M.A., De Chiara A., and Gianfreda L. 1996. Sorption of phosphate and oxalate by a synthetic aluminum hydroxysulfate complex. *Eur. J. Soil Sci*. 47: 241–247.

Wada K. 1977. Allophane and imogolite. In: *Minerals in Soil Environments*. J.B. Dixon and S.B. Weed (eds.). Soil Sci. Soc. Amer., Madison, WI, pp. 603–638.

Wolt J.D. and Adams F. 1979. The release of sulfate from soil-applied basaluminate and alunite. *Soil Sci. Soc. Amer. J.* 43: 118–121.

Wolt J.D., Hue N.V., and Fox R.L. 1992. Solution sulfate chemistry in three sulfur-retentive Hydrandepts. *Soil Sci. Soc. Amer. J.* 43: 118–121.

Zhang P.C. and Sparks D.L. 1990. Kinetics and mechanisms of sulfate adsorption/desorption on goethite using pressure-jump relaxation. *Soil Sci. Soc. Amer. J.* 54: 1266–1273.

13

The Effect of Organic Matter and Soil Chemical Properties on Sulfate Sorption in Chilean Volcanic Soils

M.L. Mora*, C. Shene, A. Violante, R. Demanet, *and* **N.S. Bolan**

Abstract

Almost 70% of the volcanic ash-derived acid soils in Chile are deficient in plant available sulphur (S). Furthermore, the low organic matter mineralization rates in these soils necessitate regular application of S fertilizers to achieve sustainable production. It is therefore important to examine the effect of various factors that regulate the behavior of this nutrient anion in these soils. The difference in sulfate sorption capacity between Chilean Ultisol and Andisol, which vary in organic matter content and natural acidity levels (pH, aluminum saturation), and the effects of different metal ions and phosphate on sulfate sorption were determined.

Results indicated that sulfate sorption capacity for both soil types was strongly pH-dependent, with sorption decreasing with increase pH. Sulfate sorption was highest at pH 3.5 and decreased markedly as pH increased. However, the initial natural soil acidity influenced sulfate sorption, which was higher in highly acid soils than in low acidity soils; this was attributed to the higher contents of exchangeable Al and acidity of reactive sites in the former soils. Sulfate sorption was concentration dependent showing a significant amount of sorption even at pH 8.0 at high sulfate concentration (20 mM) in solution, while at low sulfate concentration (6 mM) in solution, sorption was almost zero at pH > 6.5.

Removal of fulvic acids (FA) and humic acids (HA) resulted in an increase in sulfate sorption in both soil types. Increase in sorption was more pronounced with the removal of humin fraction, whereby the PZC changed from 4.1 to 7.6

**Corresponding author*: Dr. M.L. Mora, Dept. Ciencias Quimicas, Universidad de La Frontera, PO Box 54-D, Temuco, Chile. E-mail: mariluz@ufro.cl in the Andisol and from 4.0 to 8.3 in the Ultisol, indicating a large increase in positive charge at low pH. Sorption maximum ranged from 20 to 350 mmol kg^{-1} in the Andisol and 15 to 180 mmol kg^{-1} in the Ultisol, which was consistent with the greater increase in the number of active sites (Ns) in the former soil (from 80 to 600 cmol kg^{-1} in the Andisol and from 100 to 300 cmol kg^{-1} in the Ultisol). These results indicated a strong reaction between sulfate ions and active surface sites on inorganic minerals, such as allophane and ferrihydrite in Andisol.

There was a significant effect of metal cations on sulfate sorption; the latter followed the order Ca > Mg > K, which was consistent with the hydrated ionic radius of these metal cations. Results from the competitive sorption study indicated that while sorption of phosphate was not affected by the presence of sulfate ions, sorption of sulfate was greatly inhibited by the presence of phosphate ions.

1 INTRODUCTION

The predominant soil types in southern Chile, Ultisols and Andisols, derived from volcanic ashes, contain high amounts of extractable and exchangeable aluminum (Al) and have acidic pH values ranging from 4.5 to 5.5. With intensive farming, these soils increasingly become acidified, due partly to large-scale continued use of acid forming nitrogenous fertilizers, and partly to the biological nitrogen fixation process by legumes. Therefore, liming is commonly practiced in these soils in order to overcome the problems associated with acidification, especially to reduce the toxic effects of Al on plant growth (Bolan et al., 2003).

Dairy and meat production in southern Chile is based on the use of highly productive forage species, especially ryegrass (*Lolium perenne),* alone or mixed with clover *(Trifolium pratense and Trifolium repens).* The most important cereal-producing region of the country is also found in the south. The cereal crops grown in this region (wheat, oats, rape and barley) show significant differences in their tolerance to Al, both within the species and within the cultivars (Gallardo et al., 1999; Gallardo and Borie, 1999).

About 50% of the Chilean Andisols present a high soil acidity level, one of the main factors limiting agricultural production (Mora, 1993; Mora et al., 1999a). Acidification, strongly affected by the rate of nitrogen acquisition, increases Al concentration in the root environment, but decreases calcium and magnesium uptake by roots (Calba et al., 1999, Mora and Demanet, 1999). High Al content in soil solution restricts plant growth mainly due to malnutrition and Al toxicity (Peryea and Burrows, 1999). Figure 13.1 illustrates the relationship between pH and Al saturation in Chilean volcanic ash-derived soils (Mora and Demanet, 1999). Previous studies indicated that the Al concentration in the leaves of ryegrass plants growing in soils with an Al saturation of 40% was > 3,000 mg kg^{-1} (Mora, 1993), which is much higher than the maximum phytotoxic threshold concentration in plants (200 mg kg^{-1}) (Benton et al., 1991).

Monoammonium phosphate (MAP) and diammonium phosphate (DAP) fertilizers are commonly used in band applications because Chilean volcanic

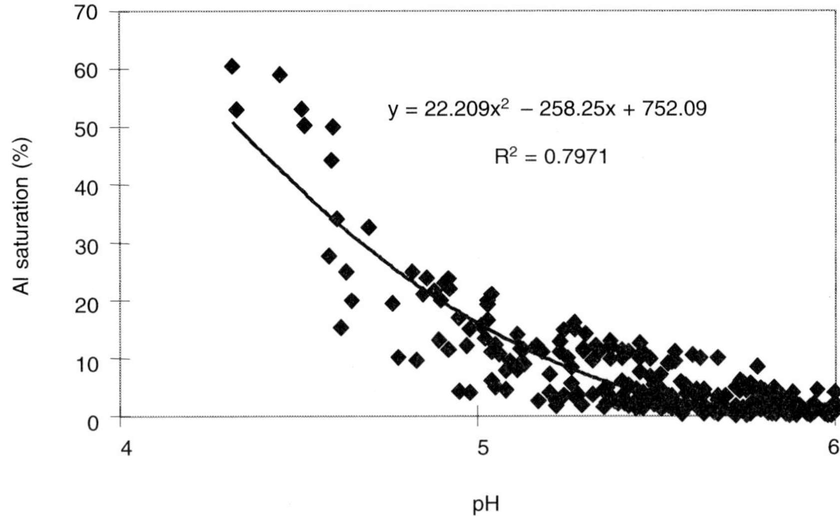

Fig. 13.1: Relationship between pH and Al saturation (%) in volcanic ash soils of southern Chile (adapted from Mora and Demanet, 1999)

soils present a high P-fixation capacity. These fertilizers generate acidity in soils, thereby resulting in a decrease in soil pH and an increase in Al saturation. It has been observed that continuous application of MAP or DAP for a wheat-oat-wheat rotation, increases Al saturation of soils, thereby decreasing the yield (Fig. 13.2) (Mora et al., 1999a). The concentration of exchangeable aluminum in the soil solution of many Chilean acidic soils was found to increase below pH 5.5, reaching values as high as 1.5–3.5 cmol kg^{-1} (Mora et al., 1999b); similar values were reported for New Zealand soils (Haynes and Naidu, 1998). The high soluble Al concentration in these soils is likely to have a toxic effect on plant growth, especially on root growth (Marschner, 1991) as reported in Figure 13.3 for wheat (Puken, a Chilean cultivar), grown in solution culture with 200 μM Al concentration (Gallardo and Borie, 1999).

It has often been shown that the addition of dolomite and gypsum to acidic pasture soils causes significant increases in dry matter yield during the active growing season in spring (Fig. 13.4). Furthermore, these soil amendments increase plant availability of a number of major nutrients (Mora et al., 1999c) and the uptake of nutrients by the pasture, thereby improving pasture quality (Mora et al., 2002). Also Mora et al. (1999b) in a greenhouse experiment showed that although the addition of gypsum to an acid soil caused only a small change in soil pH (0.1–0.4 units), it increased calcium and sulfur concentration in the soil, thereby resulting in the reduction of Al saturation. The dry matter response to gypsum application was very similar to that obtained with lime application, confirming the theory that aluminum sulfate complexes formed through gypsum application are not toxic to plants (Kinraide and Parker, 1987).

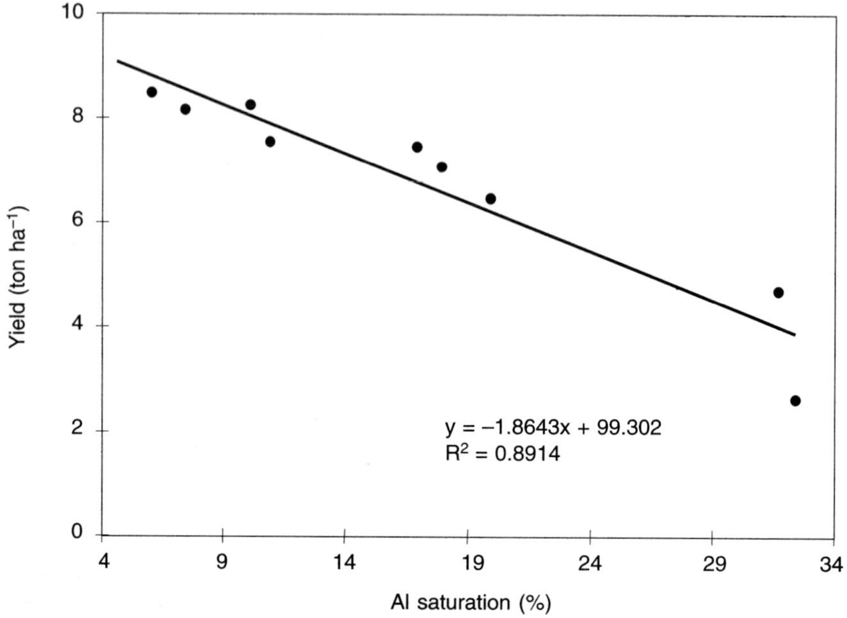

Fig. 13.2: Effect of aluminum saturation on wheat production in Chilean Andisol after ammonium phosphate (DAP and MAP) application for three years (adapted from Mora et al., 1999a).

Fig. 13.3: Effect of different Al concentration on roots of wheat (Puken an Chilean cultivar) growing in nutritive solution (adapted from Gallardo et al., 1999).

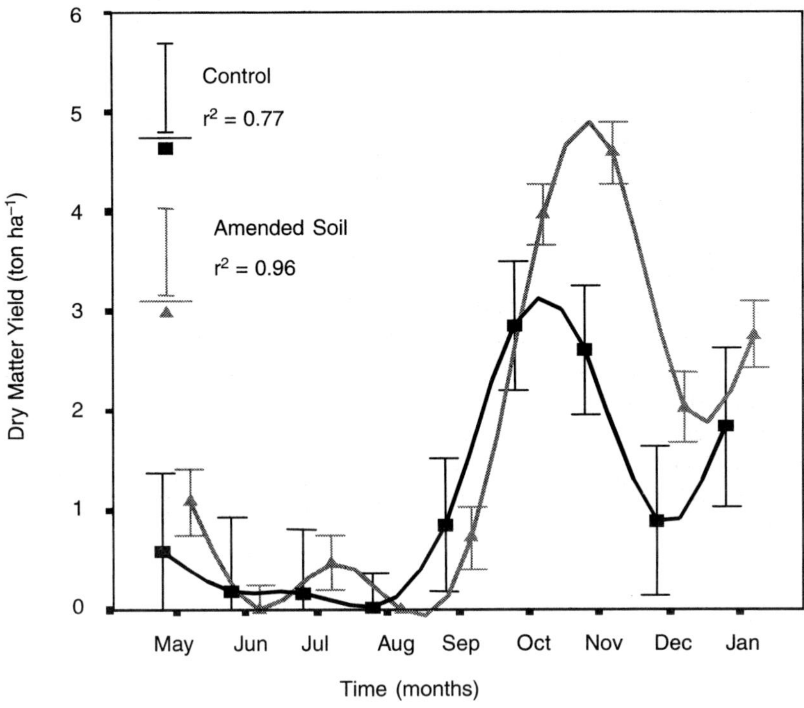

Fig. 13.4: Effect of lime and gypsum on pasture production on an acid Andisol in Southern Chile (adapted from Mora et al., 2002).

The application of high rates of phosphate fertilizers, natural acidity conditions, low temperature, and high rainfall are the main factors controlling the dynamics of S in soils of southern Chile. Liming materials are often applied several weeks before seedling development mainly to ensure their dissolution in soils. The increase in pH due to liming is likely to decrease sulfate sorption, thereby increasing the risk of sulphur leaching losses. However the combined use of lime and gypsum has been shown to overcome this problem because while lime helps in the precipitation of Al, gypsum helps in the formation of $AlSO_4^+$ ion–pair or solid $Al(OH)SO_4$ compounds (Pavan et al., 1982; Weaver et al., 1985). According to Mora et al. (1999b) and other studies, sulfate as gypsum applied to acid Andisols with high Al content in solution improved ryegrass production primarily because $AlHSO_4^+$ complexes developed. Increase in pH resulting from lime applications is also likely to increase the negative charge in variable charge soils (Mora and Barrow, 1996), thereby controlling sulfate availability for plants.

The Soil Service Laboratory of La Frontera University database indicates that nearly 70% of the soils in southern Chile contain less than 10 mg kg^{-1} plant available sulphur, which is likely to be a limiting factor for crop and pasture production in these soils (Fig. 13.5). Although the total sulfur concentration in Chilean volcanic soils is relatively high (0.04 to 0.17%) compared to agricultural soils in other areas of the globe (0.006–0.06%), almost 100% of the sulfur is found bonded to organic molecules (Aguilera et al., 2002). Organic matter is the main source of sulfur, but the mineralization process is strongly affected by high acidity in soil and climatic conditions. Therefore it is necessary to apply S fertilizer to achieve optimum pasture and crop production. However, the sulfur available for plant nutrition is regulated by the sorption mechanism onto soil surfaces. Although the nature of the sorption mechanism is not clear, it has been proposed that the sulfate anions are adsorbed onto hydrous oxides of Al and Fe in clay soils by ligand exchange with surface hydroxyl groups (Hingston et al., 1972; Bowden et al., 1973; Barrow, 2000). However spectroscopic studies have provided evidence for different types of surface complexes of sulfate on metal (hydroxy) oxides. While outer-sphere complexes were found by Hayes et al. (1987) and Persson and Lövgren (1996), inner-sphere complexes were suggested by Hu et al. (1997). Turner and Kramer (1991) presented evidence confirming that sulfate sorption is partly an outer-sphere phenomenon of

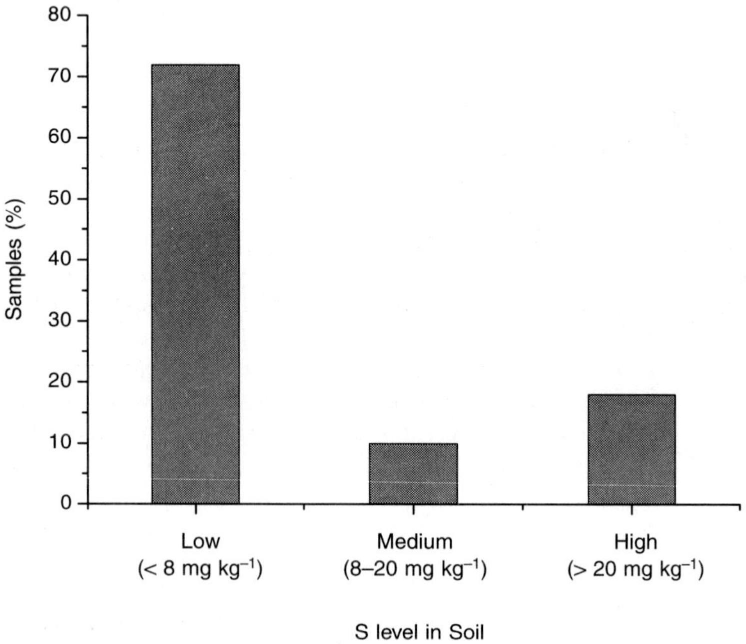

Fig. 13.5: Frequency distribution (%) of different levels of sulfur (low = < 8 mg kg^{-1}; medium = 8–20 mg kg^{-1}; high = > 20 mg kg^{-1}) in a number (2,800) of Chilean volcanic soils measured by Service Laboratory of Soil and Plant Analysis, Universidad de La Frontera-Chile.

electrostatic attraction and partly an inner-sphere one whereby one or two sulfate ion ligands exchange with potential determining hydroxyl groups to form mononuclear and binuclear bridging complexes. Eggleston et al. (1998) suggested that an intermediate behavior is most likely for sulfate sorption and adsorbates classified as outer-sphere may have small subpopulations of inner-sphere complexes at a given point and time.

Organic anions are adsorbed onto soil mineral surfaces, thereby affecting the sorption of other ions such as phosphate and sulfate (Violante et al., 1991; Violante and Gianfreda, 1993). For example, humic and fulvic acids present on the surface of clay minerals in Chilean Andisols have been shown to decrease the phosphate sorption capacity (Mora and Canales, 1995). Also competitive sorption experiments in Andisols and allophanic synthetic compounds showed that phosphate ions inhibited sulfate sorption (Pigna and Violante, 2003). Furthermore, sulfate sorption may be influenced by the simultaneous sorption of cations. For example, Marcano-Martinez and McBride (1989) and Bolan et al. (1993) showed that sulfate sorption was higher in the presence of di- and trivalent cations than monovalent ions, especially in variable charge soils.

The objective of this research was to evaluate the effects of various factors that regulate sulfate sorption in Chilean Ultisols and Andisols. The factors evaluated included: (i) natural acidity conditions of the soil; (ii) pH of the equilibrium solution; (iii) organic matter content; and (iv) metal cations and phosphate anion in soil solution.

2 MATERIALS AND METHODS

2.1 Characterization of Soil Sample

Soil samples from the Metrenco Series (Ultisol) and the Pemehue Series (Andisol) (southern Chile, 37° 45′ S, 73°00′ W and 39°30′ S, 72°20′ W, respectively) were collected at a depth of 0–20 cm. For each soil series, samples with two different natural acidity conditions (low acidity and high acidity) were taken. The chemical composition of the samples was determined according to the methodology described by Sadzawka et al. (2000). Soil pH was measured by potentiometry in a soil: solution suspension of 1:2.5 H_2O. Organic Matter (OM) was estimated by wet digestion with a modified Walkley-Black procedure. Exchangeable-cations (Ca, Mg, Na and K) were extracted with 1 M NH_4Ac at pH 7.0 and analyzed using atomic absorption spectrometry (AAS). Exchangeable aluminum was extracted with 1 M KCl and analyzed using ASS. The forms of Fe and Al in the soils were determined by a selective dissolution analysis (Blakemore et al., 1987). Iron, Al, and Si were extracted by sodium dithionite citrate (Fe_d, Al_d, Si_d) and by ammonium oxalate (Fe_o, Al_o, Si_o); Fe and Al were also extracted by sodium pyrophosphate (Fe_p, Al_p). Silica, Al and Fe concentrations in the extractants were determined by AAS. The allophane content of the soil samples was estimated by multiplying Si_o by a factor depending on the Al/Si molar ratio as proposed by Parfitt and Wilson (1985), and ferrihydrite concentration was estimated according to Parfitt and Childs (1988).

The point of zero charge (PZC) was determined by potentiometric titration at three ionic concentrations of KCl background electrolyte at 298 K. The acidity constants were calculated from the potentiometric titration data using the Constant Capacitance model (Stumm et al., 1980). The number of active sites was calculated from potentiometric titration at pH 10.

Organic matter was extracted by 0.5 M NaOH solution according to Schnitzer (1978) and was fractionated into humic (HA) and fulvic (FA) acids and purified according to the technique adapted by Aguilera et al., (1997) for this type of soil. Chemical Characterization was obtained by techniques described by Schnitzer (1978). After, humic and fulvic acid removal (i.e. soil without FA and HA), the residual fraction was exhaustively washed with double distilled water and redispersed by ultrasonification under stirring. Part of the remaining fraction was treated with 30% hydrogen peroxide until no dark residues were observed in order to remove the humin fraction (i.e. soil without organic matter).

2.2 Sorption Studies

Experiment 1. Sulfate sorption at different pH in soils with different natural acidity

Sulfate sorption studies in soils with different natural acidity (pH and Al saturation, see Table 13.1) were carried out in a batch system to determine sorption as a function of pH. To measure sulfate sorption, 3 g air-dried soil samples were mixed with 30 mL of K_2SO_4 solution containing 0–20 mM of sulfate (equivalent to 0–200 mmol kg^{-1} soil) in 0.1 M KCl background electrolyte at 298 K for 24 h. The initial pH of the soil suspensions was adjusted to 3.5–8.0 by adding HCl 0.1N or KOH 0.1N solution. The samples were then centrifuged and the sulfate concentration in the solution determined turbidimetrically (Tabatabai, 1982).

The sulfate sorption isotherm data at pH 4.5, 5.5 and 6.5 were fitted to the Freundlich equation in the form:

$$X = k_f C^n$$

where X is the amount adsorbed, C = the equilibrium solution concentration, and k_f and n are constants ($n < 1$).

Experiment 2. Effect of organic matter on sulfate sorption

Pemehue and Metrenco soil samples with low natural acidity were selected to examine the effect of the removal of organic matter fractions on sulfate sorption. Sulfate sorption isotherms were carried out for the whole soil and its fractions, soil FA, HA, and OM. In this study 0.30 g of the soil material was equilibrated at 298 K with 30 mL of K_2SO_4 solution with sulfate concentration ranging from 0 to 20 mmol L^{-1} in 0.1 M KCl. The pH of the soil suspensions was adjusted to 5.0 by adding required amounts of 0.1 M HCl or 0.1 M KOH. After this the samples were centrifuged and the sulfate concentration in solution was determined turbidimetrically (Tabatabai, 1982).

Experiment 3. Effect of cations on sulfate sorption
To examine the effect of cations on sulfate sorption, the soil samples were incubated with 0–20 mmol kg^{-1} sulfate as $CaSO_4$, $MgSO_4$, and K_2SO_4 for 24 h at 333 K and pH 5.0 (Mora and Barrow, 1996). Then samples were then extracted with $Ca(H_2PO_4)_2$ (Blakemore, 1987) and sulfate concentration in solution was determined turbidimetrically (Tabatabai, 1982). The amount of sulfate adsorbed was calculated from the difference between the amount added and the amount extracted by the phosphate solution.

Experiment 4. Effect of phosphate on sulfate sorption
Competitive sorption studies between sulfate and phosphate were carried out at pH 5.0 and 298 K. Samples of 3 g of dry soil were mixed with 30 ml of 0.1 M KCl containing various concentrations of phosphate and sulfate (0 and 20 mM) for 24 h. In the first part of the experiment, the sulfate concentration was 3.0 mM and the phosphate concentration varied between 0 and 20 mM (0–200 mmol kg^{-1}). In the second part of the experiment, the phosphate concentration was 2.0 mM and the sulfate concentration varied between 0 and 20 mM (0–200 mmol kg^{-1}). Final sulfate and phosphate concentration in solution were measured as already described.

3 RESULTS AND DISCUSSION

3.1 Soil Characteristics

The chemical characteristics of Pemehue and Metrenco soils with two soil acidity levels are typical of Andisols and Ultisols, respectively, in southern Chile (Table 13.1). The low pH and high Al saturation in Pemehue soil is a consequence of the low basic cation content (1.32 cmol kg^{-1}) caused by leaching losses and hydric erosion induced by the high rainfall in this area (1,800–2,000 mm per annum). The relationship between pH and % Al saturation was similar to the pattern shown previously for the ash-derived soils in Fig. 13.1 (Mora and Demanet, 1999) and was regulated by the initial soil pH in both soil types. In both soils, phosphate and sulfate contents were very low (Table 13.1) due mainly to high fixation capacity for anions at low pH and the rapid conversion to nonlabile organic forms of S (Aguilera et al., 2002) and P (Borie and Zunino, 1983; Escudey et al., 2001). The sulfate content of the soils measured in this study was close to the values reported by Aguilera et al. (2002) for Chilean ash soils, indicating the need for application of S fertilizer to maintain production. The chemical composition of these soils indicates that Al and Fe are present as short-range ordered compounds and Al-humus complexes are the most important components in Pemehue soil and iron oxides in Metrenco soil (Table 13.2). Organic matter content was very high in Pemehue soil, resulting in high Al-humus complexes in this Andisol in the first 20 cm of depth (Besoain, 1985). In Metrenco soil, however the organic matter accumulation is always less than

Table 13.1: Chemical characteristics of soils

Soil series	Pemehue low acidity	Pemehue high acidity	Metrenco low acidity	Metrenco high acidity
pH	5.65	5.08	5.67	5.16
S (mg kg^{-1})	2	2	9	7
P (mg kg^{-1})	10	6	6	10
	Exchangeable cations (cmol kg^{-1})			
Ca	2.77	0.76	6.83	2.68
Mg	0.55	0.29	0.13	0.92
K	0.27	0.16	0.79	1.73
Al	0.12	1.74	0.08	0.61
Σ (K Ca Mg)	3.36	1.36	7.85	5.41
Al saturation (%)	3.17	56.13	1.01	10.13

Table 13.2: Organic matter content, fractions of Al, Fe, and Si, and allophane and ferrihydrite contents of soils

Soil series	Pemehue	Metrenco
Class	Andisol	Ultisol
Organic matter (g kg^{-1})	180	90
Al_o[1] (g kg^{-1})	47.0	5.9
Si_o (g kg^{-1})	15.8	1.7
Fe_o (g kg^{-1})	17.0	10.7
Al_{pp}[2] (g kg^{-1})	9.2	3.3
Fe_{pp} (g kg^{-1})	1.4	1.6
Al_d[3] (g kg^{-1})	18.8	5.7
Si_d (g kg^{-1})	2.6	1.2
Fe (g kg^{-1})	49.1	84.4
Al:Si* (g kg^{-1})	2.39	1.52
Allophane[4] (g kg^{-1})	110	—
Ferrihydrite[5] (g kg^{-1})	29	18

Except where otherwise indicated, analyses were done according to the methods of Blakemore et al. (1987).

[1] Extracted by acid ammonium oxalate.

[2] Extracted by sodium pyrophosphate.

[3] Extracted by citrate-dithionite-bicarbonate.

[4] Si_o × 7 (Parfitt and Wilson, 1985).

[5] Fe_o × 1.7 (Parfitt and Childs, 1988).

* (Al_o-Alp)/Si_o (Parfitt and Wilson, 1985).

that of Andisols which may be attributable to the difference in the mineralogy between these two soil types (Table 13.2).

3.2 Effect of pH on Sulfate Sorption

Results presented here clearly show that sulfate sorption in both soils types, Ultisols and Andisols decreases with an increase in soil pH (Fig. 13.6). Similar results were reported in various studies carried out earlier using different soil types and synthetic Al and Fe oxides (Barrow, 1970; Parfitt, 1978). With an

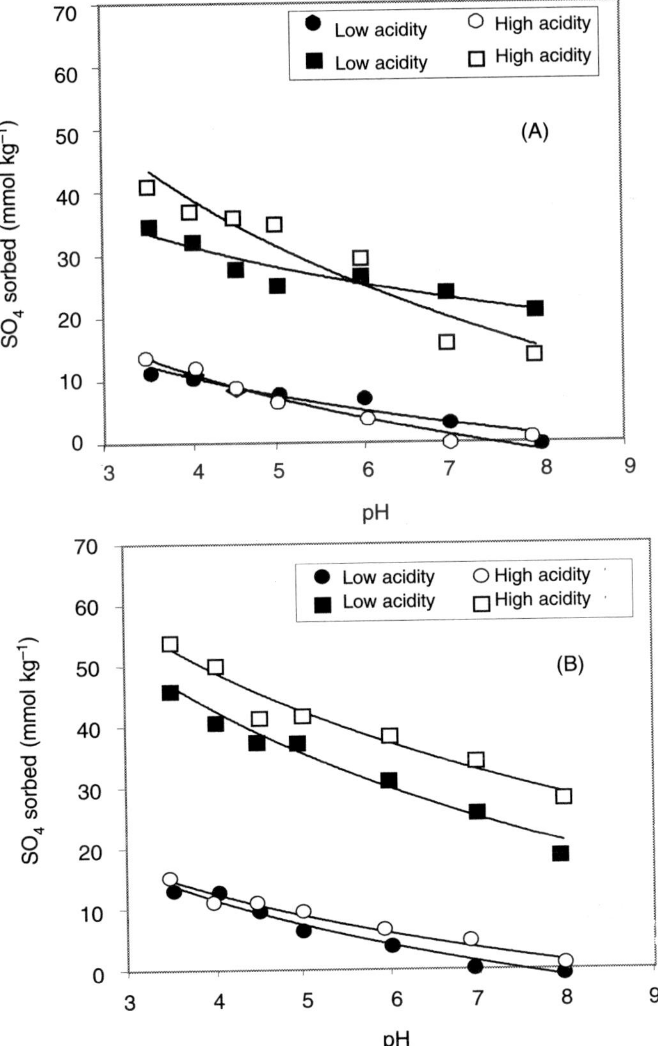

Fig. 13.6: Effect of pH on sulfate sorption isotherm for Pemehue (A) and Metrenco (B) soils with two different natural acidities. (●, ○ = low sulfate level (6 mM); ■, □ = high sulfate level (20 mM)).

increase in soil pH, the positive charge decreases in these variable charge soils (Mora and Barrow, 1996) resulting thereby in a decrease in sulfate sorption (Ajwa and Tabatabai, 1995; Pigna and Violante, 2003). Therefore, the natural acidity condition in soil determines the acid force of reactive sites, as shown by the pKa_1 and pKa_2 values (Table 13.3), indicating that reactive sites of Al or Fe are more acidic as a consequence of irreversible acid hydrolysis produced by low pH. Sulfate sorption was highest at pH 3.5, and sorption decreased with an increase in pH. Sorption capacity was higher in soils with high acidity condition than in soils with low acidity, which is attributable to the high exchangeable Al and high positive charge as determined by potentiometric titration.

Table 13.3: Acidity constant* of Pemehue and Metrenco soils with and without sulfate in solution

Electrolyte	Metrenco low acidity		Metrenco high acidity		Pemehue low acidity		Pemehue high acidity	
	pKa_1	pKa_2	pKa_1	pKa_2	pKa_1	pKa_2	pKa_1	pKa_2
KCl 10^{-1} M	3.10	8.80	2.92	8.37	3.10	7.80	2.80	6.80
KCl 10^{-1} M + SO_4^{-2} 10^{-3} M	2.90	9.20	2.96	9.80	3.56	9.41	2.99	6.66

*From potentiometric titration data and calculated by Capacitance Constant Model (Stumm et al., 1980).

The effect of pH on sulfate sorption was more pronounced in Andisol soil (Pemehue soil) than in Ultisol (Metrenco soil) and attributable to the higher level of variable charge components, such as amorphous aluminum and iron oxides, and allophane Al-humus complexes in the former soil. The decrease in sulfate sorption with increasing soil pH is likely to have an important effect on sulfur supply for plants, indicating that sulfate sorption by both types of soils are controlled by pH and liming. Almost 70% of acid soils in southern Chile have been shown to be deficient in plant-available sulfur. This has been attributed to regular lime application, resulting in low sulfate retention capacity and/or formation of $Al(OH)SO_4$ (Weaver et al., 1985). Hence regular sulfate application is mandatory to maintain optimal plant production in these soils. Sulfate retention in soils can be explained by dynamic interactions among SO_4^{-2}, OH^-, and Al^{+3} ions in solution and solid phases.

Sulfate sorption was found to be concentration dependent; while sorption was low at pH > 6.0 when the concentration used for the sorption isotherm measurement was very low (60 mmol kg^{-1}), significant amount of sulfate was sorbed even at pH 8.0 when sulfate concentration in the equilibrium solution was high (200 mmol kg^{-1}), probably due to formation of aluminum hydroxysulfate precipitates such as alunite and basaluminalite (Wolt et al., 1992; Sumner, 1993). Sulfate sorption reaction is promoted by two factors: (i) reaction of the sulfate ions with Al in solution favouring the ion-pair formation such as

$AlHSO_4^+$ and (ii) increase in the cationic exchange by H^+ generates more $AlOH_2^+$ groups that react with the sulfate ion by charge attraction forming outer-sphere complexes. The isotherms measured at selected pH values clearly show this effect (Fig. 13.7) and the sorption behavior was adequately described by the Freundlich model, in which the parameters, k_f and n decrease with increasing pH (Table 13.4). The substitution of the pH-dependent equation for k_f and n was

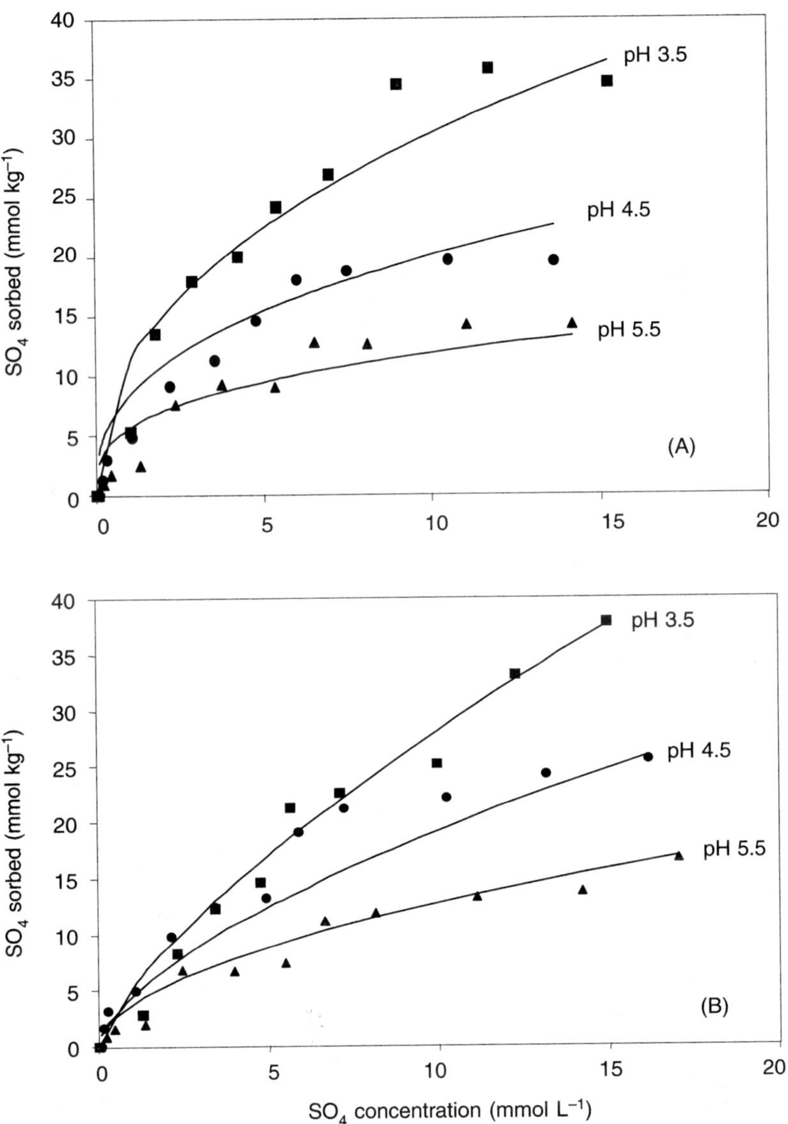

Fig. 13.7: Sulfate sorption isotherm at three different pHs as described by the Freundlich model. Metrenco (A) and Pemehue (B) soils. Temperature 298 K and 0.1 M KCl background electrolyte.

determined by regression (Table 13.5) and showed a greater pH dependence of sulfate sorption in Andisol than Ultisol. It is known that negative charge sites are activated when soils are limed, and consequently the capacity for sulfate sorption decreases (Bolan et al., 1986).

Table 13.4: Freundlich isotherm constants for sulfate sorption at 3 pH values

	Metrenco Soil		
pH	k_f	n	R^2
3.5	11.48	0.4361	0.9341
4.5	7.65	0.3784	0.8705
5.5	5.99	0.3246	0.8768
	Pemehue Soil		
pH	k_f	n	R^2
3.5	5.35	0.6935	0.9636
4.5	4.32	0.6966	0.9058
5.5	3.86	0.4935	0.9216

Table 13.5: Regression equations describing the relationship between pH and Freundlich isotherm constants

Metrenco soil		R^2
	$k_f = 20.721 - 2.7444$ pH	0.9510
	$n = 0.6306 - 0.0558$ pH	0.9996
Pemehue soil		R^2
	$k_f = 7.8549 - 0.7435$ pH	0.9556
	$n = 1.0779 - 0.1000$ pH	0.7384

3.3 Effect of Organic Matter on Sulfate Sorption

The organic matter content and reactivity was higher in Pemehue soil than in Metrenco soil, which was attributed to high concentration of the carboxylic groups of fulvic acids in the former soil (Table 13.6). Results in Table 13.2 show that the concentration of humus-Al complexes is higher in Pemehue soil (9.2 g kg^{-1}) than in Metrenco soil (3.3 g kg^{-1}) indicating that these complexes are preferentially formed between the carboxylates of fulvic acids and the Al^{+3} ions in soil solution. On the other hand, the higher level of humus-Al complexes could have a greater blocking effect of the active sites in the mineral matrix, resulting in higher sorption with the removal of organic matter by the Metrenco soil than Pemehue soil at low pH (Fig. 13.6). In both soils PZC is close to pH 4.0

Table 13.6: Amount and characterization of humic acid and fulvic acid in Chilean soils

	Amount (%)	Total acidity	Carboxylic acidity	Fenolic acidity	E_4/E_6
			(mmol g^{-1})		
			Pemehue		
HA	1.33	5.4	3.8	1.6	4.5
FA	0.54	10.3	9.4	8.0	8.4
			Metrenco		
HA	0.50	4.2	3.5	0.7	0.46
FA	0.35	11.9	3.1	8.2	5.4

Table 13.7: PZC and active number of surface sites (Ns) in whole soils and in soil from which humic acid (HA), fulvic acid (FA), and organic matter has been removed

Soil	PZC	Ns (mmol kg^{-1})
Pemehue	4.10	80
Pemehue without HA and FA	6.20	200
Pemehue without organic matter	7.60	600
Metrenco	4.01	100
Metrenco without HA and FA	5.90	200
Metrenco without organic Matter	8.30	300

(Table 13.7), a typical value of the organic matter coating of the soil colloids as demonstrated by Mora et al. (1994) in allophanic soil synthetic compounds. This indicates that the surface is mainly negative at pH > 4.0 resulting in low affinity for sulfate ions above this pH. Removal of fulvic and humic acids increased the PZC values from 4.1 to 6.2 in Pemehue and from 4.0 to 5.9 in Metrenco soils, respectively. As a consequence of this release of the surface active sites sulfate sorption increases (Fig. 13.8). On the other hand, removal of all OM including the more recalcitrant humin fraction results in the uncoating of clay surface and the oxides of Fe and Al in the soil. The number of surface active sites increases from 80 to 600 mmole kg^{-1} in the Andisol, in which allophane and ferrihydrite are predominant, and from 100 to 300 mmol kg^{-1} in the Ultisol (Table 13.7), in which the crystalline iron oxide and clay materials dominated (Besoain, 1985). Complete removal of OM increased sulfate fixation by 3.75 and 7.5 times in the Andisol and Ultisol, respectively (Fig. 13.8). This was consistent with the results obtained for phosphate sorption in Chilean ash-derived soils (Mora and Canales, 1995). Our results indicate that the mechanism of sulfate sorption in the mineral fraction differs in soils with high organic matter content. The OM fractions in these soils block surface active sites of aluminum

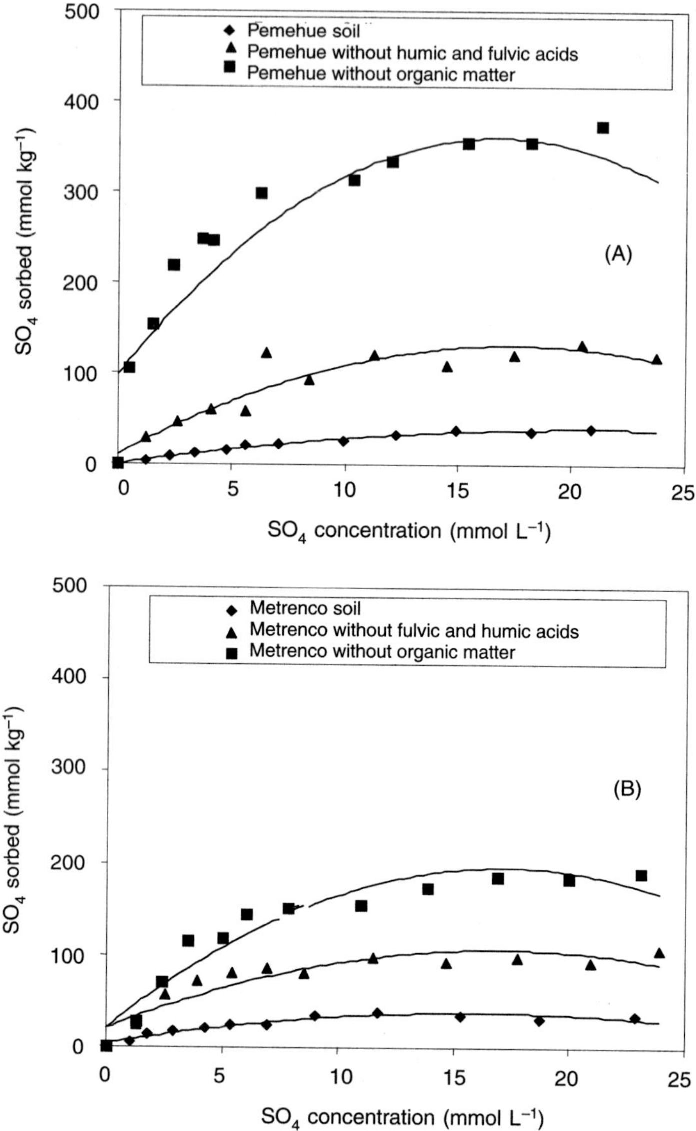

Fig. 13.8: Effect of removal of fulvic plus humic acids and organic matter on sulfate sorption in Pemehue (A) and Metrenco (B) soils. Temperature 298 K, pH 5.0, and 0.1 M KCl background electrolyte.

and iron, thereby inhibiting the sorption of sulfate by the mineral fraction. Although some studies involving anion sorption by goethite have indicated that similar to phosphate, sulfate is adsorbed through ligand exchange (Turner and Kramer, 1991), this mechanism is not a major process of sulfate sorption by these ash-derived acid soils.

3.4 Effect of Metal Cations on Sulfate Sorption

The amount of adsorbed sulfate was influenced by the nature of cation saturation (Fig. 13.9). In both soils, sulfate sorption increased in the order: $CaSO_4 > MgSO_4 > K_2SO_4$. Similar results were reported by Marcano and McBride (1989) and Bolan et al. (1993). In their studies, the effect of cation on sulfate sorption was more pronounced in soils dominated by variable charge components. Ajwa and Tabatabai (1995) indicated that calcium as well as magnesium form ionic pairs

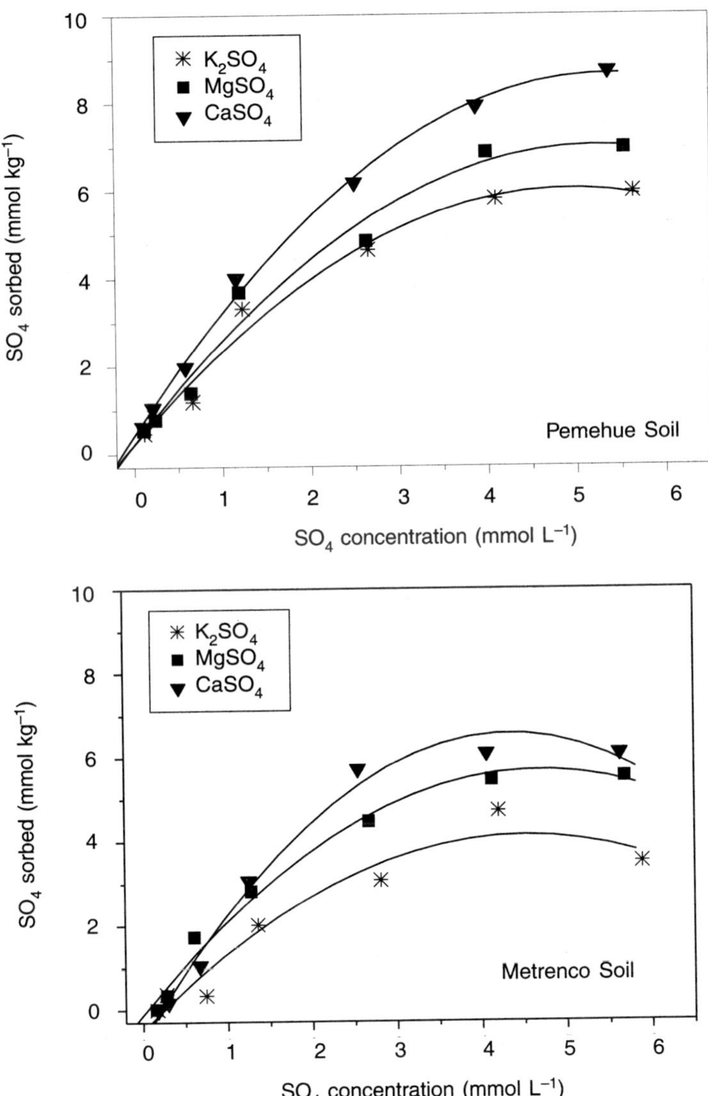

Fig. 13.9: Effect of cation saturation on sulfate sorption isotherm. Metrenco and Pemehue soils at pH 5.5. Temperature 298 K and 0.1 M KCl background electrolyte.

at high concentrations of sulfate and metal cations in the solution as a result of the simultaneous sorption of the metal and sulfate (Fig. 13.6). However, this mechanism is almost absent in the presence of potassium. These findings are very important for the Chilean acid soils, which are normally amended with lime or gypsum. Bolan et al. (1993) have reported similar results for variable charge soils. These authors postulated that the increase in sulfate sorption in the presence of calcium is firstly due to the surface complex formation between the sulfate ion and calcium. This involves coordination of one Ca to two adsorbed anion groups, reducing the repulsive force between adjacent anion groups, thereby enhancing further sorption. Furthermore, specific sorption of calcium increases positive charge, thereby enhancing anion sorption. Therefore, liming of soils is likely to influence sulfate sorption through two processes: (i) the pH-induced increase in surface negative charge decreases the affinity for sulfate sorption and (ii) the Ca-induced increase in sulfate sorption. The latter mechanism could explain the high sulfate retention in this study even at pH 8 in the presence of high concentration of sulfate (20 mM) in solution (Fig. 13.6). However, Marsh et al. (1987) observed a steady decrease in sulfate sorption with increasing levels of lime, which they attributed to a decrease in the positive charge. It is important to emphasize that, unlike the present study, Marsh et al. (1987) measured sorption at a low equilibrium solution concentration (20 mg L^{-1}) of sulfate.

3.5 Effect of Phosphate Anion on Sulfate Sorption

The significant difference in the amount of phosphate sorption by the soils is attributed to the difference in the nature and amount of clay mineral in them (Fig. 13.10). The high reactivity of Andisols attributed to the presence of allophane, ferrihydrite, and Al-humus complexes is one of the main reasons why Pemehue soil adsorbed a higher level of phosphate. However, this reactivity depends on the type of anion because sulfate sorption was less than phosphate sorption and the maximum sorption in both soils was around 40 and 50 mmol kg^{-1}. Although the pattern of sorption curves varied slightly between the anions, Metrenco soil adsorbed more sulfate at low sulfate concentration in solution. It is well known that the mechanism of phosphate sorption is through ligand exchange and coordinate covalent bonding between the metal ions and phosphate. At equilibrium, the maximum sulfate sorption was approximately 30% and 50% that of phosphate adsorbed in Pemehue and Metrenco soils, respectively, at 10 mM solution concentration.

The strong competitive effect of phosphate on sulfate sorption observed in this experiment indicates that the formation of inner-sphere complexes is not the most important mechanism for the sulfate sorption process, although there was a slight increase in pKa_2 values in the presence of sulfate (Table 13.3), indicating some chemical interaction with surface complexes. On the contrary, the competitive effect of sulfate on phosphate sorption was very poor in these

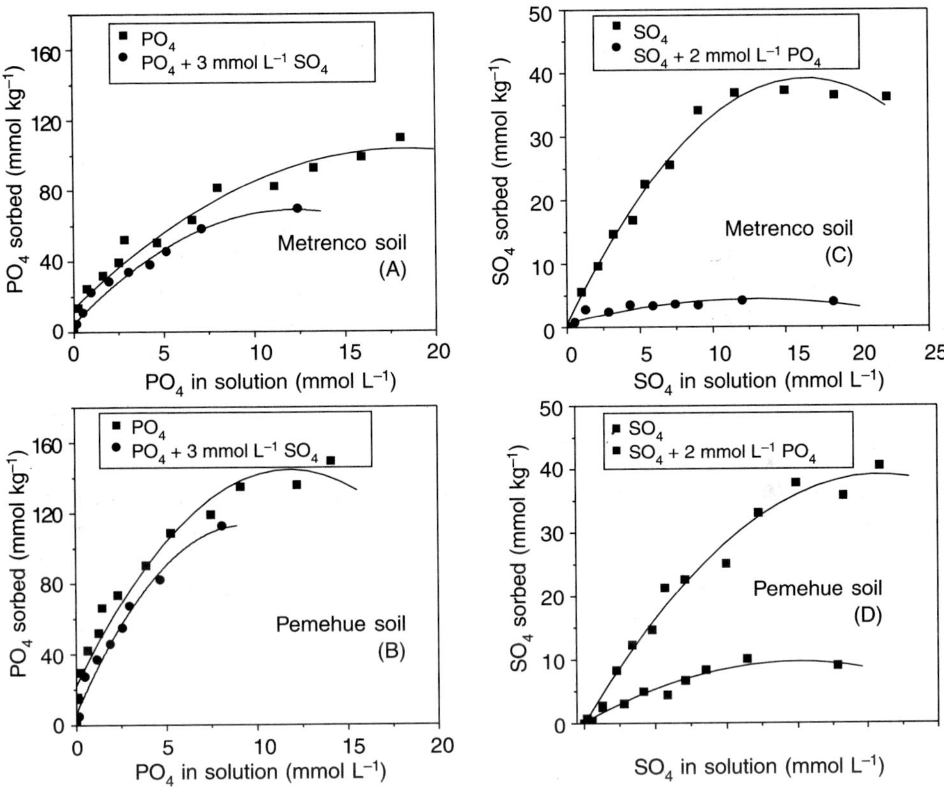

Fig. 13.10: Effect of sulfate ion on phosphate sorption in (A) Metrenco soil and (B) Pemehue soil. Effect of phosphate ion on sulfate sorption in (C) Metrenco soil and (D) Pemehue soil. At 298 K temperature, pH 4.5 and 0.1 M KCl background electrolyte.

ash-derived soils, indicating that the bonding energy between sulfate and the surface active sites of the colloids is lower than that of phosphate. For example, Huang and Violante (1986) showed a strong affinity between phosphate and organic aluminum complexes which dominate in these acid soils.

These observations are of great agronomic implication since in these Chilean acid soils phosphorous fertilizers are applied at heavy rates. This is particularly true in Andisols because these soils posses high phosphate fixation capacity and also are liable for sulfate leaching, especially after liming as shown by Baeza (2002) for Chilean soils and Bolan et al. (1988) for New Zealand soils. Nowadays, farmers are applying potassium sulfate during the seedling time because they normally use between 100–120 kg P per hectare for crops and pasture and one or two tons of lime per hectare. From the economic and environmental point of view it is appropriate to use gypsum as a source of sulphur in these acid soils. Previous studies in glasshouse and field conditions showed that application of a gypsum and dolomite mixture is very effective in overcoming the problems associated with acid conditions including the supply of S, thereby enhancing dry matter production (Mora et al., 1999b; Mora et al., 2002).

4 CONCLUSIONS

Results from the present study on sulfate sorption in Chilean ash-derived soils (Andisols and Ultisols) indicated the following:

(1) The sulfate sorption process was strongly controlled by the initial acidity conditions and by the pH of the equilibrium sulfate solution.
(2) Organic matter blocked the sites of Al-OH and Fe-OH present on soil surfaces, thereby decreasing the sulfate sorption capacity.
(3) Sulfate sorption was strongly regulated by the charge of the metal ions in the solution and followed the order: Ca> Mg > K.
(4) Phosphate ions strongly competed with sulfate for sorption by the active sites of Al and Fe in soil surface and decreased sulfate sorption by 70–80%.

In summary, soil pH, organic matter content, Al and Fe humus complexes, and charge of cations had a significant effect on sulfate sorption. Sulfate apparently form outer-sphere complexes on ash-derived Chilean soils through electrostatic forces. From an economic and an environmental point of view it is important to include gypsum application in the agronomic management of sulphur fertilizer in these acid soils.

Acknowledgment

This research was supported by the National Research Council of Chile by Fondecyt 1990-873 and Fondecyt 1020934, International Cooperation Fondecyt Grant 7020934, and Andes Foundation by C Grant 13755-28.

References

Aguilera M., Borie G., Peirano P., and Galindo G. 1997. Organic matter in volcanic soils in Chile. Chemical and biochemical characterization. *Commun. Soil Sci. Plant Anal.* 28: 899–912.

Aguilera M., Mora M.L., Borie G., Peirano P., and Zunino H. 2002. Cycling and balance of sulphur in volcanic ash-derived soils in Chile. *Soil Biol. Biochem.* 34: 1355–1361.

Ajwa Ha and Tabatabai Ma. 1995. Metal-induced sulfate adsorption by soils .1. Effect of pH and ionic-strength. *Soil Sci.* 159(1): 32–42.

Baeza G. 2002. Dinámica del reciclaje de N, P y S proveniente de excretas animales en un Andisol acidificado. Tesis de Doctorado, Universidad de Santiago de Chile.

Barrow N.J. 1970. Comparison of the adsorption of molibdate, sulphate, and phosphate by soils. *Soil Sci.* 109: 282–288.

Barrow N.J. 2000. Towards a single-point method for measuring phosphate sorption by soils. *Austr. J. Soil Res.* 38: 1099–1113.

Benton J., Wolf B., and Mills H.A. 1991. *Plant Analysis Handbook*. Micro-Macro Publ., Athens, GA.

Besoain E. 1985. Los Suelos. In: *Suelos volcánicos de Chile.* Tosso (ed). INIA, Chile. pp. 25–95.

Blakemore L., Searle P., and Daley B. 1987. Methods for chemical analysis of soils. Dept. Scientific and Industrial Research, Lower Hutt, New Zealand.

Bolan N.S., Syers J.K., and Tillman R.W. 1986. Ionic strength effects on surface charge and adsorption of phosphate and sulfate by soils. *J. Soil Sci.* 37: 379–388.

Bolan N.S., Syers J.K., and Sumner M.E. 1993. Calcium induced sulfate adsorption by soil. *Soil Sci. Soc. Amer. J.* 57 : 691–696.

Bolan N.S., Adriano D.C., and Curtin D. 2003. Soil acidification and liming interactions with nutrient and heavy metal transformation and bioavailability. *Adv. Agron.* 78: 216–272.

Bolan N.S., Syers J.K., Tillman R.W., and Scotter D.R. 1988. Effect of liming and phosphate application on sulphate leaching. *J. Soil Sci.* 39: 493-504.

Borie F. and Zunino H. 1983. Organic matter-phosphorus associations as a sink in P-fixation processes on allophanic soils of Chile. *Soil Biol. Biochem.* 15: 599–603.

Bowden J.W., Bolland M.D., Posner A.M., and Quirk J.P. 1973. Generalized model for anion and cation adsorption at oxide surfaces. *Nature Phys. Sci.* 245: 81–83.

Calba H., Cazevieille P., and Jaillard B. 1999. Modelling of the dynamics of Al and protons in the rhizosphere of maize cultivated in acid substrate. *Plant Soil.* 209(1): 57–69.

Eggleston C.M., Hug S., Stumm W., Sulzberger B., and Dos Santos Afonso M. 1998. Surface complexation of sulfate by hematite surfaces: FTIR and STM observations. *Geochim. Cosmochim. Acta.* 62: 585–593.

Escudey M., Galindo G., Förster J.E., Briceño M., Díaz P., and Chang A. 2001. Chemical forms phosphorus of volcanic ash-derived soils in Chile. *Commun. Soil Sci. Plant Anal.* 32: 601–616.

Gallardo F., and Borie F. 1999. Sensibilidad de especies y cultivares a condiciones de acidez. Tests rápidos de diagnóstico. *Frontera Agrícola* 5: 3–18.

Gallardo F., Borie F., Alvear M., and von Baer E. 1999. Evaluation of aluminum tolerance of three barley cultivars by two short-term screening methods and field experiments. *Soil Sci. Plant Nutr.* 45(3): 719–719.

Hayes K.F., Roe A.L., Brown Jr. G.E., Hodgson K.O., Leckie J.O., and Parks G.A. 1987. In situ X-ray absorption study of surface complexes: Selenium oxyanions on α-FeOOH. *Science* (Washington, DC). 238: 783–786.

Haynes R.J. and Naidu R. 1998. Influence of lime, fertilizer and manure applications on soil organic matter content and soil physical conditions: a review. *Nutr. Cycl. Agroecosys.* 51(2): 123–137.

Hingston F.J., Posner A.M., and Quirk J.P. 1972. Anion adsorption by goethite and gibbsite. I. The role of the proton in determining adsorption envelopes. *J. Soil Sci.* 23: 177–192.

Hu L.M., Zelazny L.W., Baligar V.C., Ritchey K.D., and Martens D.C. 1997. Ionic strength effects on sulfate and phosphate adsorption on γ-alumina and kaolinite: Triple-layer model. *Soil Sci. Soc. Amer. J.* 61: 784–793.

Huang P.M. and Violante A. 1986. Influence of organic acids on crystallization and surface properties of precipitation products of aluminum. In: *Interactions of Soil Minerals with Natural Organics and Microbes.* P.M. Huang and M. Schnitzer (eds.). Spec. Publ. 17. *Soil Sci. Amer.*, Madison WI, pp. 159–221.

Kinraide T.B. and Parker D.R. 1987. Cation amelioration of aluminum toxicity in wheat. *Plant Physiol* 83: 546–551.

Marcano-Martinez E. and McBride M.B. 1989. Calcium and sulphate retention by two Oxisols of the Brazilian Cerrado. *Soil Sci. Soc. Amer. J.* 53: 63–69.

Marschner H. 1991. Mechanisms of adaptation of plants to acid soils. *Plant Soil* 134(1):1–20.

Marsh K.B., Syers J.K., and Tillman R.W. 1987. Charge relationships of sulfate adsorption by soils. *Soil Sci. Soc. Amer. J.* 51: 318–323.

Mora M.L. 1993. Nivel de fertilidad de los Suelos de la IX Región y su relación con la acidificación. *Frontera Agrícola* 1: 5–12.

Mora M.L. and Canales J. 1995. Humin-clay interactions on surface reactivity in Chilean Andisols. *Commun. Soil Sci. Plant Anal.* 26: 2819–2828.

Mora M.L. and Barrow N.J. 1996. The effects of time of incubation on the relation between charge and pH of soil. *Eur. J. Soil Sci.* 47: 131–136.

Mora M.L. and Demanet R. 1999. Uso de enmiendas calcáreas en suelos acidificados. *Frontera Agrícola.* 5: 43–58.

Mora M.L., Escudey M., and Galindo G. 1994. Synthesis and characterization of allophanic soils. *Bol. Soc. Chil. Quim.* 39(3): 237–243.

Mora M.L., García J.C., Santander J., and Demanet R. 1999a. Rol de los fertilizantes nitrogenados y fosfatados en los procesos de acidificación de los suelos. *Frontera Agrícola* 5: 59–81.

Mora M.L., Schnettler B., and Demanet R. 1999b. Effect of liming and gypsum on soil chemistry, yield and mineral composition of ryegrass grown in an acidic Andisol. *Commun. Soil Sci. Plant Anal.* 30: 1251–1266.

Mora M.L., Baeza G., Pizarro C., and Demanet R. 1999c. Effect of calcitic and dolomitic lime on physico-chemical properties of a Chilean Andisol. *Commun. Soil Sci. Plant Anal.* 30: 427–439.

Mora M.L., Cartes P., Demanet R., and Cornforth I.S. 2002. The effects of lime and gypsum on pasture growth and composition on an acid Andosol in Chile, South America. *Commun. Soil Sci. Plant Anal.* 33: 2069–2081.

Parfitt R.L. 1978. Anion adsorption by soils and soil materials. *Adv. Agron.* 30: 1–50.

Parfitt R.L. and Wilson A.D. 1985. Estimation of allophane and halloysite in three sequences of volcanic soils, New Zealand. In: *Volcanic Soils. Weathering and Landscape Relationships of Soils on Tephra and Basalt.* E.F. Caldas and D.H. Yaalon (eds.). Catena-Verlag, Germany, Supplement 7, pp. 1–8.

Parfitt R.L. and Childs C.W. 1988. Estimation of forms of Fe and Al: a review and analysis of contrasting soils by dissolution and Moessbauer methods, *Aust. J. Soil Res.*, 26: 121–144.

Pavan M.A., Bingham F.T., and Pratt P.T. 1982. Toxicity of aluminum to coffee in Ultisols and Oxisols amended with $CaCO_3$, $MgCO_3$, and $CaSO_4{\cdot}2H_2O$. *Soil Sci. Soc Amer. J.* 46: 1201–1207.

Persson P. and Lovgren L. 1996. Potentiometric and spectroscopic studies of sulfate complexation at the goethite-water interface. *Geochim. Cosmochim. Acta.* 60: 2789–2799.

Peryea F.J. and Burrows R.L. 1999. Soil acidification caused by four commercial nitrogen fertilizer solutions and subsequent soil pH rebound. Commun. *Soil Sci. Plant Anal.* 30: 525–533.

Pigna M. and Violante A. 2003. Adsorption of sulfate and phosphate on Andisols. *Commun Soil Sci. Plant Anal.* 34: 2099–2113.

Sadzawka A., Grez R., Mora M.L., Saavedra N., et al. 2000. Métodos de análisis recomendados para los suelos chilenos. Comisión de Normalización y Acreditación, Sociedad Chilena de la Ciencia del Suelo. Available in http://www.inia.cl/cobertura /platina/pubycom/metodos2000-sep.doc

Schnitzer M. 1978. Humic substances: chemistry and reactions, In: *Soil Organic Matter*. M. Schnitzer and S.U. Khan (eds.). Elsevier Scientific, New York, NY, pp. 1–64.

Stumm W., Kummert R., and Sigg L. 1980. A ligand exchange model for the adsorption of inorganic and organic ligands at hydrous oxide interfaces. *Croat. Chem. Acta.* 53: 291–312.

Sumner M.E. 1993. Sodic Soils—New Perspectives. *Aust. J. Soil. Res.* 31(6): 683–750.

Tabatabai, M. A. 1982. Sulfur. In *Methods of Soil Analysis.* Part 2*: Chemical and Microbiological Properties* (2nd ed.). A.L. Page, R.H. Miller, and D.R. Keeney (eds.). *Agronm.* 9: 501–538.

Turner L.J. and Kramer J.R. 1991. Sulfate ion binding on goethite and hematite. *Soil Sci.* 152: 226–230.

Violante A. and Gianfreda L. 1993. Competition in adsorption between phosphate and oxalate on an aluminum hydroxide montmorillonite complex. *Soil Sci. Soc. Amer. J.* 57: 1235–1241.

Violante A., Colombo C., and Buondonno A. 1991. Competitive adsorption of phosphate and oxalate by aluminum oxides. *Soil Sci. Soc. Amer. J.* 55: 65–70.

Weaver G.T., Khanna P.K., and Beese F. 1985. Retention and transport of sulfate in a slightly acid forest soil. *Soil Sci. Soc. Amer. J.* 49: 746–750.

Wolt J.D., Hue N.V., and Fox R.L. 1992. Solution sulfate chemistry in 3 sulfur-retentive Hydrandepts. *Soil Sci. Soc. Amer. J.* 56 (1): 89–95.

14

Modeling Boron Adsorption Kinetics in Benchmark Soils of Punjab, India

Sanjay Arora* *and* **D.S. Chahal**

Abstract

As plants respond primarily to the boron (B) activity in soil solution, understanding the mechanism of B adsorption on soil materials is vital. Batch studies were conducted to investigate the adsorption behavior of B in 21 surface soils representing major soil series of Punjab. Six mathematical models, viz. zero order, first order, second order, Elovich, Power function, and Parabolic diffusion, were used to describe B adsorption. Boron adsorption pattern was characterized by an initial fast reaction followed by a slow process and was completed in 24 hours of equilibrium. The Elovich equation was best to describe the rate of B adsorption followed by Power function. Adsorption of B on all soils conformed to the Freundlich adsorption isotherm and adsorption capacity constant (K) ranged between 1.086 and 5.445 $\mu g\ g^{-1}$, which correlated significantly with clay content (r = 0.569**) and CEC (r = 0.639**) of the soils. B adsorption data also conformed well to the Langmuir adsorption isotherm in 5 of the 21 soils while in 15 soils the Langmuir adsorption isotherm was curvilinear and significant and excellent when curves were resolved in two linear parts, suggesting two types of reactions of B with soils. In one soil, representing the Ghabdan series, Langmuir fits were applicable over a limited concentration range.

1 INTRODUCTION

Boron is unique among the essential micronutrients because it is present as a nonionized molecule over the pH range suitable for plant growth. It is vital to understand the mechanism of B adsorption on soil materials as plants respond primarily to the boron activity in soil solution. Critical to the success of accurately

**Corresponding author*: Dr. Sanjay Arora. Division of Soil Science and Agricultural Chemistry, Sher-e-Kashmir University of Agriculture Main Campus—Chatha, JAMMU (J & K) Pin code: 180 009, India. E-mail: arorapau@rediffmail.com predicting the reactions and mobility of B in soils is the understanding and quantification of adsorption processes. Boron applied to soils is adsorbed to a variable extent and an equilibrium exists between B in the solid and liquid phases (Hingston, 1964). The element can be toxic for plants at relatively low concentrations in soils and the range between levels causing deficiency and toxicity is narrow (Berger, 1949). The equilibrium between B in liquid and solid phases in soil has often been described by either the Langmuir adsorption isotherm (Griffin and Burau, 1974; Elrashidi and O'Connor 1982; Evans, 1987; Borkakti, 1988; Krishnasamy et al., 1997; Dekamedhi et al., 1998) or the Freundlich adsorption isotherm (Singh, 1971; Bhatnagar et al., 1979; Elrashidi and O'Connor, 1982; Evans, 1987; Borkakti, 1988; Krishnasamy et al., 1997). A variety of soil properties have been identified as affecting the behavior of B in soils. Soil pH (Bingham et al., 1971; Keren and Gast, 1983; Gu and Lowe, 1990; Goldberg et al., 1993; Keren and Sparks, 1994), organic matter (Yermiyahu et al., 1988; Marzadori et al., 1991; Yermiyahu et al., 1995; Krishnasamy et al., 1997), clay content or soil texture (Nicholaichuk et al., 1988; Goldberg and Glaubig, 1986), sesquoxides (Sims and Bingham, 1968; Rhoades et al., 1970), and calcium carbonate content (Saha and Singh, 1998) have been reported to influence the solubility and adsorption of boron in soils. Elrashidi and O'Connor (1982) reported that boron sorption reactions were complete in 12 h whereas adsorption equilibrium was established in less than 2 h after boron was added, albeit shaken for 24 h (Keren and Gast, 1983). Mezuman and Keren (1981) used 24 h whereas, Krishnasamy (1996) together with his colleagues (1997) reported that adsorption of boron was almost complete after 24 h in all 11 agricultural soils of Tamil Nadu (southern India). B adsorption in these soils was best described by parabolic diffusion and Elovichian kinetics. Though the adsorption of boron conformed to both the Langmuir and Freundlich adsorption isotherms, all the data fitted well with the Freundlich adsorption isotherm over the concentration range of 0 to 100 mg B kg^{-1} soil as boric acid.

Adsorption of boron is one of the most important factors determining the release and fixation of applied B and thus deciding the efficiency of B fertilization. Knowledge on the kinetics of adsorption reaction may be essential for the sound prediction of nutrient availability to plants. A number of studies have been done on the kinetics of K and P reactions in soils (Sparks, 1985), but information on the adsorption kinetics of B in Indian soils is very limited and none too clear. Hence, studies were undertaken to assess the kinetics of B adsorption and the time required for B equilibration and obtain information on the behavior of B in alluvium derived benchmark soils of Punjab, India. A number of models have been used to describe the kinetics of the boron adsorption process.

2 MATERIALS AND METHODS

2.1 Soils

Twenty-one surface soil samples (0–15 cm) representing major soil series of Punjab were collected, air-dried, and passed through a 2-mm sieve. The homogenized soil samples were then analyzed for selected physical and chemical properties (see Table 14.1) by the methods described by Jackson (1973).

2.2 Rate of Reaction

The 21 selected Punjab surface (0–15 cm) soils were studied in batches employing a background electrolyte (20 ml; 0.01 M $CaCl_2$) containing 40 mg ml^{-1} as boric acid (H_3BO_3) added to 10 g soil in 50 ml polypropylene centrifuge tubes in triplicate. The contents were allowed to equilibrate with intermittent shaking at 25 ± 1°C for different intervals ranging from 1 to 72 h. At the end of the reaction time, the suspension was centrifuged and filtered. The B concentration in the equilibrium solution was determined colorimetrically using Azomethine-H (Wolf, 1974). The amount of B adsorbed was calculated from the difference between concentration of the initial solution and the equilibration solution. The amount of B adsorbed at various time intervals were fitted to different kinetic models.

2.3 Modeling of Rate Kinetics

The kinetic models used to describe B adsorption from twenty-one soils during 72 h were tested for goodness of fit by least-square regression analysis. The models used were:

(i) Zero Order: $qt = q_o + K_o t$, where K_o is the zero order rate constant [mg B kg^{-1} s^{-1}]
(ii) First Order: $\ln qt = \ln q_o + K_1 t$ where K_1 is the first order rate constant [s^{-1}].
(iii) Second Order: $1/qt = 1/q_o - K_2 t$, where K_2 is the second order rate constant [(mg B kg^{-1})s^{-1}].
(iv) Parabolic diffusion: $qt = \alpha + Kd\, t^{1/2}$, where Kd is the diffusion rate constant [(mg B kg^{-1})$s^{-1/2}$].
(v) Elovich: $qt = q_o + (1/\beta)\ln(\alpha\beta) + (1/\beta)\ln t$, where α is the initial B adsorption rate [(mg B kg^{-1})h^{-1}] and β is B desorption constant [(mg B kg^{-1})h^{-1}].
(vi) Power function : $\ln y = \ln a + b \ln t$, where y is the quantity of B adsorbed/desorbed at time t and a and b are constants (Dalal, 1974).

In all the equations, q_o and qt are the amounts of B adsorbed (mg B kg^{-1}) at time zero and t respectively. To determine the equation that best described the adsorption of B in soils, a standard error of estimate was calculated for each. A relatively high value (approaching 0.9 or above) of the coefficient of determination (R^2) and low standard error of estimate (SE less than 1) were used as criteria for the best fit (Chien and Clayton, 1980). The standard error was calculated as follows

$$SE = [\Sigma (q - q')^2 / (N - 2)]^{0.5}$$

where q and q′ are the measured and calculated amounts of B in soil respectively, at time t, and N is the number of measurements (Steel and Torrie, 1960).

2.4 Adsorption Studies

Solubility of boron in soil is mainly governed by adsorption process to soil surfaces. Adsorbed boron acts as a buffer pool between solid and solution phase. Adsorption experiments were conducted with twenty-one surface (0–15 cm) soils. To determine adsorbed boron, a 20-g sample in a 50-ml polyethylene centrifuge tube was equilibrated with 20 ml solution of 0.01 M $CaCl_2$ containing varying amounts of B (1, 2, 5, 10, 20, 40, 60, 80 and 100 μg ml^{-1}) using boric acid (H_3BO_3) for 24 h at 25 ± 1°C in an incubator with intermittent shaking. After equilibration, the suspension was centrifuged and the supernatant solution filtered through Whatman No. 42 filter paper following the procedure outlined by Elrashidi and O'Connor (1982). Boron concentration in the filtrate was determined by the Azomethine-H method (Wolf, 1974) using spectronic-20 spectrophotometer. Boron adsorbed by the soil was calculated as the difference between the initial B and equilibrium B concentrations.

Boron adsorption data were first calculated according to Langmuir adsorption isotherm. The linear form of the isotherm used was :

Langmuir equation: $$C/(x/m) = 1/b\,(C) + 1/k\,b$$

where C is the equilibrium concentration (μg B ml^{-1}), (x/m) the adsorbed B (mg B g^{-1} soil), b the adsorption maxima (μg B g^{-1} soil), and k = a constant related to bonding energy (ml μg^{-1}). Data were also described by the Freundlich adsorption isotherm. The linear form of the isotherm used was:

Freundlich equation: $$\text{Log}\,(x/m) = 1/n \log C + \log K$$

where 1/n is the slope of the regression line and K and n are empirical constants.

Simple correlation coefficients were made to determine whether any relationship existed between various soil properties and B adsorption parameters.

3 RESULTS AND DISCUSSION

3.1 Soil Properties

The study included soil representing major soil series of Punjab, varying widely in texture (sandy loam to clay loam), pH from 6.10 to 10.24, E.C. from 0.13 to 2.82 dS m^{-1}, organic carbon content from 0.16 to 1.10 %, $CaCO_3$ content from 0.00 to 3.40 % and cation exchange capacity from 4.80 to 12.91 c mol(p+) kg^{-1}. Thus the soils possessed widely varying physical and chemical properties which provided sufficient scope for the study of B adsorption. The details of the soil properties for each soil studied are presented in Table 14.1.

3.2 Soil Mineralogy

All the soils contained mixed mineralogy. Dominant minerals in the sand and silt fractions of these soils were quartz, micas and feldspars. Quartz was the

Table 14.1: Physical and chemical characteristics of the soils

Soils	pH	EC (dS m^{-1})	OC	$CaCO_3$	CEC (cmol kg^{-1})
			%		
Garhi baghi	8.49	0.17	0.65	3.40	7.50
Jodhpur ramana	8.80	0.21	0.16	1.20	5.63
Mansa	8.40	0.26	0.53	1.53	8.93
Kaheru	8.62	0.22	0.50	1.98	8.54
Isri	8.52	0.63	0.49	0.30	7.02
Langrian	8.79	0.28	0.55	2.28	8.50
Balewal	8.29	0.21	0.57	0.15	6.32
Sunairhari	8.59	0.38	0.63	0.70	10.63
Rajpura	8.48	0.35	0.72	0.00	12.39
Patiala	8.38	0.34	0.75	0.45	12.91
Tulewal	8.96	0.45	0.46	0.00	7.83
Nabha	8.41	0.23	0.57	0.23	8.06
Ballowal saunkhari	8.34	0.13	0.48	0.63	6.80
Dhar	8.10	0.23	1.04	0.90	9.01
Gurdaspur	7.84	0.18	0.61	0.00	9.54
Naushera	8.14	0.48	0.49	0.95	7.06
Jagjitpur	9.10	0.41	0.51	1.93	9.93
Khiranwali	9.76	1.30	0.45	1.73	8.72
Chamror	6.10	0.40	1.10	0.00	9.65
Samana	8.40	0.13	0.46	0.00	4.80
Ghabdan	9.50	2.82	0.18	2.60	10.83

most abundant mineral and its content decreased with decrease in particle size. Feldspar included both plagioclase and alkali species. Mica species comprised both muscovite and biotite. Heavy sand fractions of these soils predominantly showed various ferromagnesian silicates such as amphiboles, chlorite, and biotite. Illite was the most abundant clay mineral in these soils. Other minerals present in the clay fraction were kaolinite, smectite, and chlorite, which were also present in the fine silt fractions of these soils (Sharma et al., 1997).

3.3 Kinetics of Boron Adsorption

In all the soils boron adsorption was characterized by an initial fast reaction followed by a slow process. Similar reaction rates of B adsorption have been reported by Krishnasamy (1996) and Krishnasamy et al. (1997). Rapid rate of reaction can be attributed to chemical reaction and slow reaction to diffusion of B into micropores of inorganic and organic compounds (Krishnasamy, 1996). Initial adsorption by soils was fastest from Chamror, followed by Rajpura, Patiala, Sunairhari, Mansa, Khiranwali, Langrian, Jagjitpur, Ghabdan, Dhar, Nabha, Isri, Garhi baghi, Gurdaspur, Samana, Balewal, Kaheru, Jodhpur ramana, Naushera, Ballowal saunkhari and Tulewal in descending order of rate of reaction.

Adsorption of boron was almost complete after 24 h in all the soils, though complete equilibrium was not attained until 32 h has elapsed. Mezuman and Keren (1981) used 24 h whereas, Elrashidi and O'Connor (1982) adopted 12 h of equilibrium time for their experiment on boron adsorption by soils.

Concentration of boron in the equilibrium solution ranged from 15.4 to 34.0 μg ml^{-1} at completion of the adsorption reaction (Table 14.2). The equilibrium concentration of boron for Garhi baghi, Jodhpur ramana, Mansa, Kaheru, Isri, Langrian, Balewal, Sunairhari, Rajpura, Patiala, Tulewal, Nabha, Ballowal saunkhari, Dhar, Gurdaspur, Naushera, Jagjitpur, Khiranwali, Chamror and Samana soil series at 24 h was 26.5, 34.4, 23.3, 29.8, 28.0, 23.7, 28.8, 22.2, 15.5, 20.4, 30.0, 25.9, 31.0, 24.7, 30.6, 29.8, 25.5, 21.5, 19.3 and 33.2 μg ml^{-1}, respectively. The finer the texture, the lower was the concentration of boron in the equilibrium solution. Soil texture and CEC of the soils influenced the boron concentration in the equilibrium solution. The boron equilibrium concentration correlated negatively with clay content (r = – 0.476**) and positively with sand (r = 0.632**), which supports the foregoing statements. This is also consistent with the previously published literature (Elrashidi and O'Connor, 1982), which showed that fine textured soils usually had higher boron adsorption capacities than coarse textured soils. Maximum boron was adsorbed by the Rajpura soil series

Table 14.2: Concentration of B (μg ml^{-1}) in solution with respect to reaction time*

Soils	Time (h)									
	1.0	2.0	4.0	8.0	12.0	16.0	24.0	32.0	48.0	72.0
Garhi baghi	34.3	32.0	30.3	28.6	27.5	27.0	26.5	26.6	26.5	26.5
Jodhpur ramana	37.5	36.1	35.4	35.1	34.9	34.5	34.4	34.2	34.2	34.1
Mansa	29.9	27.7	26.2	25.8	24.8	23.7	23.3	23.2	23.1	23.2
Kaheru	36.7	34.1	33.2	32.1	31.0	30.0	29.8	29.7	29.5	29.5
Isri	33.9	32.4	30.9	29.4	28.9	28.6	28.0	28.0	27.9	27.8
Langrian	30.8	28.3	27.9	26.7	24.6	24.0	23.7	23.5	23.5	23.5
Balewal	36.4	34.2	32.5	30.8	29.2	28.8	28.7	28.5	28.5	28.4
Sunairhari	28.8	27.0	25.0	24.6	23.2	22.4	22.2	22.0	22.0	21.7
Rajpura	27.2	22.0	20.2	18.5	16.5	16.0	15.5	15.4	15.5	15.4
Patiala	27.8	25.9	24.4	23.8	22.0	21.2	20.4	20.4	20.2	20.2
Tulewal	37.8	36.0	34.8	33.6	32.7	30.8	30.0	29.4	29.4	29.2
Nabha	33.6	32.3	30.8	28.4	27.2	26.5	25.9	25.7	25.8	25.5
Ballowal saunkhari	37.6	36.0	34.3	33.5	32.0	31.6	31.0	31.0	31.0	30.8
Dhar	32.2	30.9	28.6	27.9	26.6	25.6	24.7	24.5	24.6	24.4
Gurdaspur	34.4	33.6	32.6	31.8	31.5	31.0	30.6	30.5	30.4	30.4
Naushera	37.6	35.7	34.4	32.5	31.0	30.2	29.8	29.6	29.5	29.4
Jagjitpur	31.2	29.9	28.7	26.5	25.9	25.4	25.5	25.2	25.3	25.2
Khiranwali	30.3	26.5	25.0	24.0	22.8	22.0	21.5	21.2	21.0	20.8
Chamror	25.6	24.2	22.4	20.8	20.0	19.6	19.0	19.2	19.0	19.0
Samana	35.8	35.0	34.6	34.2	33.9	33.5	33.2	33.2	33.0	33.0
Ghabdan	31.8	29.6	28.2	26.6	25.2	24.8	24.5	24.5	24.3	24.3

* Means of three replications.

followed by Chamror and Patiala, while minimum boron was adsorbed by the Jodhpur ramana soil series.

Among the six kinetic models delineated above for describing B adsorption in soils representing major soil series of Punjab, that model which had the highest coefficient of determination (R^2) and the lowest standard error of estimates (SE) was considered the best fit.

The first order kinetic equation could not describe the adsorption of boron by soils as R^2 values were very low (0.475 to 0.862) and S.E. values large (1.402 to 3.658). Zero order and second order equations also did not describe B adsorption sufficiently in all the soils studied, as evidenced by large SE values of zero order model (1.576 to 3.884) and small R^2 (0.585 to 0.846) and markedly large S.E. (1.334 to 2.870) values for the second order kinetic model.

The Elovich equation was best of the various kinetic equations studied to describe the rate of B adsorption, as evidenced by the overall highest values of coefficient of determination (R^2 = 0.868 to 0.965) and lowest values of standard error (SE = 0.314 to 0.963) of estimate (Table 14.3; Fig. 14.1) over the entire time range. A similar conformity was reported earlier by Krishnasamy et al. (1997) who studied B adsorption kinetics in the soils of Tamil Nadu. In the present investigation the R^2 value was maximal (0.965) in Tulewal soil series followed by 0.956 in Gurdaspur soil series.

Table 14.3: Parameters of Elovich kinetic model for boron adsorption

Soils	Elovichian parameters		
	β	R^2	SE
Garhi baghi	0.5348	0.902	0.906
Jodhpur ramana	1.3478	0.909	0.346
Mansa	0.6268	0.927	0.657
Kaheru	0.5971	0.920	0.718
Isri	0.6804	0.927	0.604
Langrian	0.5485	0.920	0.789
Balewal	0.5216	0.898	0.951
Sunairhari	0.6048	0.917	0.732
Rajpura	0.4409	0.868	0.963
Patiala	0.5265	0.948	0.658
Tulewal	0.4575	0.965	0.609
Nabha	0.4838	0.927	0.852
Ballowal saunkhari	0.6126	0.919	0.713
Dhar	0.5019	0.954	0.645
Gurdaspur	0.9968	0.956	0.314
Naushera	0.4850	0.940	0.768
Jagjitpur	0.6627	0.892	0.775
Khiranwali	0.4760	0.920	0.911
Chamror	0.6202	0.901	0.788
Samana	0.6168	0.921	0.701
Ghabdan	0.5508	0.913	0.826

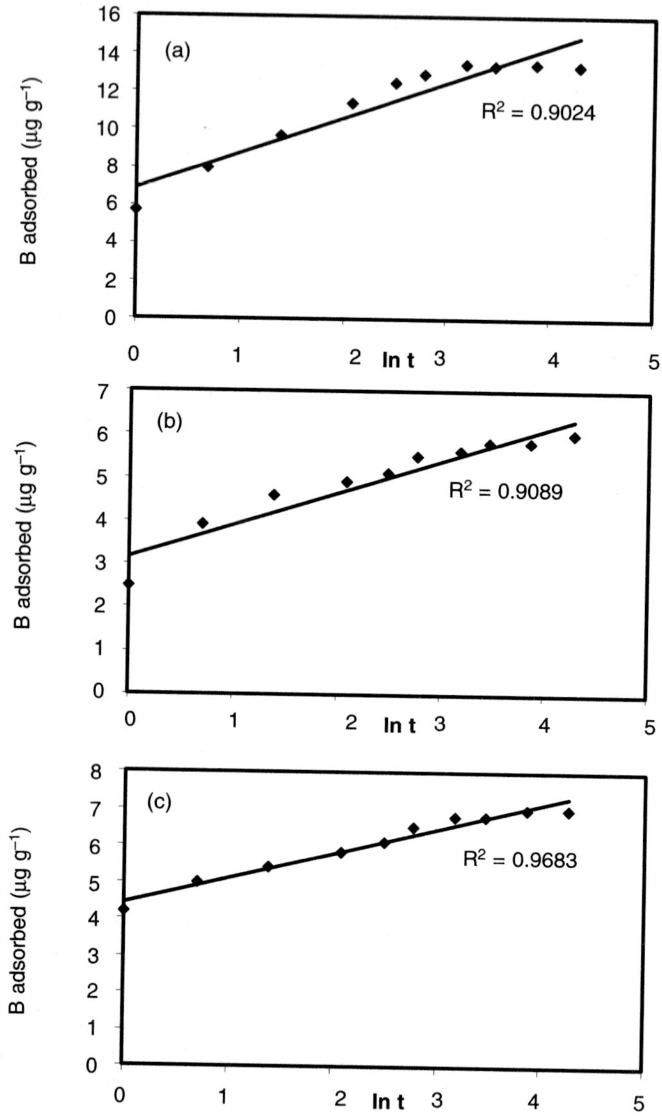

Fig. 14.1: Analysis of kinetic data in Elovichian model for B adsorption by (a) Garhi baghi, (b) Jodhpur ramana, and (c) Samana soils.

The power function was nearly as good as the Elovich equation in describing boron adsorption kinetics in all the soils (Table 14.4). The R^2 values were quite high and SE values low. The parabolic diffusion law did not satisfactorily describe the B adsorption for the soils, as is apparent from the relatively lower R^2 values (R^2 = 0.671 to 0.853; Table 14.4).

Table 14.4: Parameters of parabolic diffusion and power function models for boron adsorption kinetics

Soils	Parabolic diffusion		Power function	
	R^2	SE	R^2	SE
Garhi baghi	0.686	0.023	0.856	0.116
Jodhpur ramana	0.716	0.007	0.829	0.117
Mansa	0.749	0.020	0.902	0.056
Kaheru	0.733	0.018	0.846	0.154
Isri	0.726	0.016	0.891	0.082
Langrian	0.749	0.023	0.899	0.068
Balewal	0.682	0.002	0.840	0.165
Sunairhari	0.723	0.023	0.895	0.056
Rajpura	0.677	0.051	0.816	0.095
Patiala	0.792	0.024	0.934	0.045
Tulewal	0.853	0.016	0.925	0.152
Nabha	0.740	0.024	0.907	0.094
Ballowal saunkhari	0.716	0.017	0.855	0.179
Dhar	0.798	0.021	0.933	0.067
Gurdaspur	0.778	0.009	0.935	0.050
Naushera	0.759	0.020	0.877	0.184
Jagjitpur	0.671	0.021	0.876	0.070
Khiranwali	0.754	0.027	0.864	0.836
Chamror	0.685	0.027	0.886	0.047
Samana	0.724	0.014	0.927	0.112
Ghabdan	0.708	0.024	0.881	0.081

3.4 Adsorption of Boron

Surface soils representing 21 soil series of Punjab were used for the study of B adsorption. The results of boron adsorption clearly showed that all the soils have an affinity for boron adsorption. The amount of boron adsorbed by each soil was plotted against the equilibrium boron concentration to obtain the adsorption isotherm, indicating the effect of B concentration on boron adsorption. The adsorption isotherms indicated that although the adsorption of B increased with increasing concentration in the equilibrium solution, the percentage of adsorbed B decreased. This may be attributed to an increase in the ratio of adsorbate to adsorbent. Bloesch et al. (1987) also reported that adsorption of boron increased with increasing concentration of boron in solution. The amount of boron adsorbed differs for different soils. The shape of adsorption isotherms was classified according to the classification given by Giles et al. (1960). The adsorption isotherms were of L-shape in all the soils studied. The isotherms for Garhi baghi, Jodhpur ramana, and Samana soils are presented in Figure 14.2. The initial curvature of L-shape isotherms indicates that more sites in the substrate were occupied and hence the solute molecule faced great difficulty in finding a vacant site. This leads to the suggestion that either the adsorbed

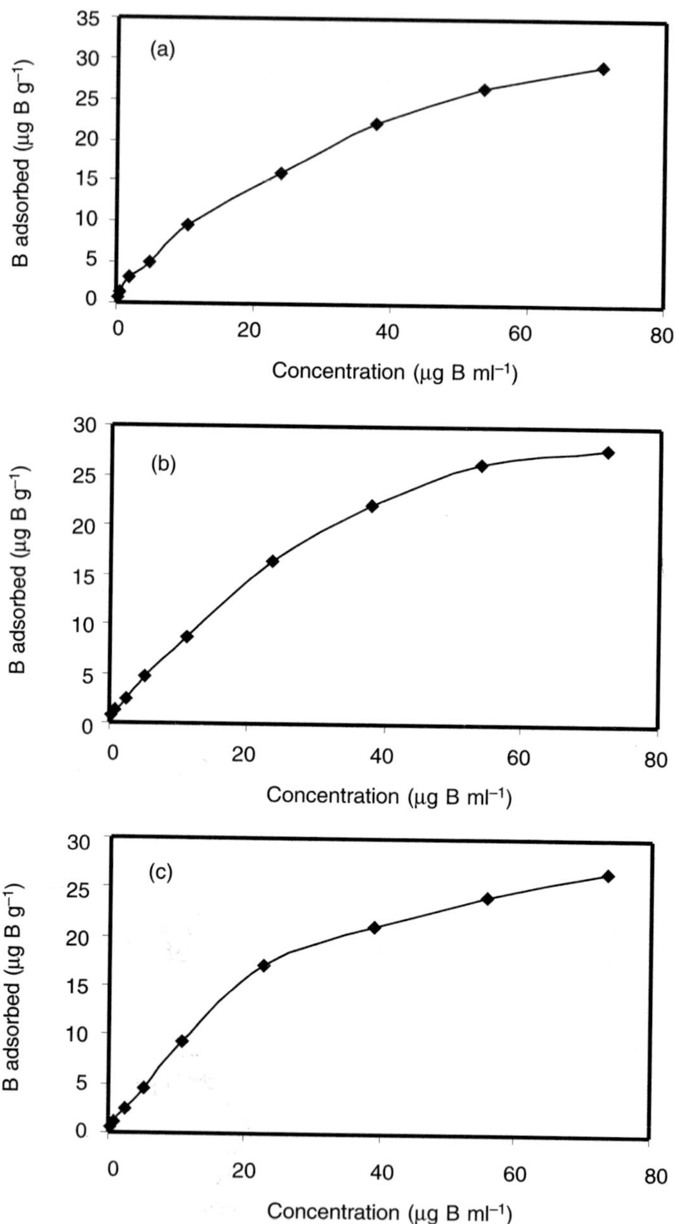

Fig. 14.2: Boron adsorption by (a) Garhi baghi, (b) Jodhpur ramana, and (c) Samana soils.

molecule was not vertically oriented or that there is no strong competition from the solvent. L-shape adsorption isotherms for boron adsorption were also obtained by Krishnasamy et al. (1997) in the soils of Tamil Nadu.

Table 14.5: Amount of boron adsorbed in different soils treated with different concentration of boron at 25 ± 1°C

Soils	B added (µg ml^{-1})								
	1	2	5	10	20	40	60	80	100
Garhi baghi	0.74	1.39	3.13	5.13	9.52	15.90	22.30	26.50	29.20
Jodhpur ramana	0.67	1.24	2.54	4.80	8.60	16.40	22.00	26.20	27.80
Mansa	0.74	1.40	3.32	5.60	10.05	15.70	19.80	23.55	26.54
Kaheru	0.56	1.10	2.48	4.35	7.50	13.10	16.50	19.20	21.85
Isri	0.59	1.14	2.65	4.60	7.12	12.66	18.12	22.08	25.13
Langrian	0.70	1.36	3.12	5.50	10.70	18.30	23.12	27.50	32.00
Balewal	0.72	1.31	3.16	5.78	9.89	16.00	19.35	22.30	24.00
Sunairhari	0.62	1.22	3.04	5.92	11.64	21.75	29.50	32.80	35.30
Rajpura	0.72	1.42	3.25	6.02	9.92	18.16	24.58	30.56	36.12
Patiala	0.74	1.44	3.46	5.89	10.98	19.20	25.22	30.24	35.46
Tulewal	0.62	1.21	2.50	4.08	7.50	13.16	17.00	19.60	20.20
Nabha	0.67	1.28	3.02	5.98	11.12	17.70	21.98	25.70	28.81
Ballowal saunkhari	0.68	1.25	2.72	4.36	8.25	15.82	19.20	22.00	23.80
Dhar	0.92	1.83	4.54	8.90	17.30	27.40	33.50	38.00	42.00
Gurdaspur	0.68	1.32	3.02	5.95	11.36	21.00	29.10	34.00	36.22
Naushera	0.52	0.98	2.24	3.56	6.82	12.24	16.30	19.40	20.50
Jagjitpur	0.73	1.42	3.19	5.46	10.15	17.80	24.70	27.23	29.15
Khiranwali	0.77	1.45	3.36	6.00	11.76	17.11	21.70	25.16	27.70
Chamror	0.58	1.15	2.87	5.56	10.90	20.84	28.20	35.50	40.23
Samana	0.66	1.20	2.42	4.60	9.25	17.14	21.00	24.15	26.54
Ghabdan	0.60	1.22	2.98	5.62	10.54	18.15	23.88	32.39	36.24

The data on boron adsorption by different soils are presented in Table 14.5. They reveal that the adsorption capacity of different soils for boron differed and the behavior of adsorption of boron by different soils was not uniform througout the concentration range.

Adsorption of boron was maximum in the Rajpura soil series, which may be attributed to the high amount of clay content and cation exchange capacity (CEC) of the soil. Jodhpur ramana soil series showed the least adsorption capacity for boron, which is probably due to its sandy loam texture and low CEC. From these results it could be concluded that adsorption of boron was mainly governed by the clay content and cation exchange capacity of the soil. Wild and Mazaheri (1979) also found that the loamy clay soil adsorbed more than four times as much boron as the sandy loam soil.

The data of boron adsorption (Table 14.5) when analyzed statistically show that adsorption of boron by these soils was correlated positively and significantly with the clay content ($r = 0.672^{**}$), CEC ($r = 0.735^{**}$), and organic carbon content ($r = 0.545^{**}$). The clay content and CEC of the soil also correlated significantly with boron adsorbed by ten soils of New Mexico (Elrashidi and O'Connor, 1982). Elrashidi and O'Connor (1982) also showed positive relation between organic matter content and B adsorbed, but were not able to establish a precise correlation. However, in the present study a nonsignificant effect of soil pH and $CaCO_3$ content on B adsorption was observed due to the narrow range or less variability in the values of these soil parameters.

3.5 Freundlich Adsorption Isotherm

The boron adsorption data were also fitted to the Freundlich equation. The linear form of the equation is

$$\log x/m = 1/n \log C + \log K$$

where C is the equilibrium boron concentration (μg B ml^{-1}), x/m the amount of B adsorbed per unit weight of soil (μg g^{-1}), K and 1/n are constants which depend upon the nature of the adsorbate and the adsorbent. The values of the constants K and 1/n can be determined by plotting log x/m versus log C. The plot should give a straight line if the data conform to the Freundlich equation. K can be obtained from the intercept at unit concentration and 1/n is the slope of the plot. K and 1/n provide the estimates of adsorbent capacity and intensity of adsorption. The values of 1/n also indicate the degree of nonlinearity between solution concentration and adsorption. Although Freundlich equation is empirical, it could be used to explain the adsorption of compounds on soils and is best suited for a heterogeneous system like the soil-nutrient water system (Krishnasamy et al., 1997).

A plot of boron adsorbed against equilibrium boron concentration on a log-log scale gave a linear relationship in all the soils. The concentration ranged from 1 to 100 μg B ml^{-1} indicating that boron adsorption data conform to the Freundlich equation (Fig. 14.3). Many workers have reported that B adsorption data could be successfully described by Freundlich model over the entire concentration range (Elrashidi and O'Connor, 1982; Saha and Singh, 1997; Dekamedhi et al., 1998).

The Freundlich model assumes multilayer adsorptions and heterogeneity of the sites compared to the monolayer adsorption and homogenity of sites in the Langmuir equation. At higher concentration of applied B, multilayer adsorption and/or precipitation reaction appears to have occurred, thereby resulting in a better fit of data on the latter than on the former model. These results accord with those reported by Nicholaichuk et al. (1988), Krishnasamy et al. (1997), Saha and Singh (1997), and Dekamedhi et al. (1998).

Freundlich K values ranged widely among the soils. The maximum value was 5.445 μg g^{-1} for the Dhar soil series followed by 2.241 for the Khiranwali soil series (Table 4.6). Dekamedhi et al. (1998) reported a range in K values from 0.86 to 1.29 and 1.23 to 2.19 in Haplaquept and Fluvaquept groups of Assam respectively. Freundlich K values for the adsorption of boron for eleven soils of Tamil Nadu range from 0.18 to 1.96 mg kg^{-1} (Krishnasamy et al., 1997).

Perusal of the data indicates that by and large the difference in values of K was associated with the magnitude of variation in clay content of the soil. The highly significant relation of the K values with clay content ($r = 0.569^{**}$)

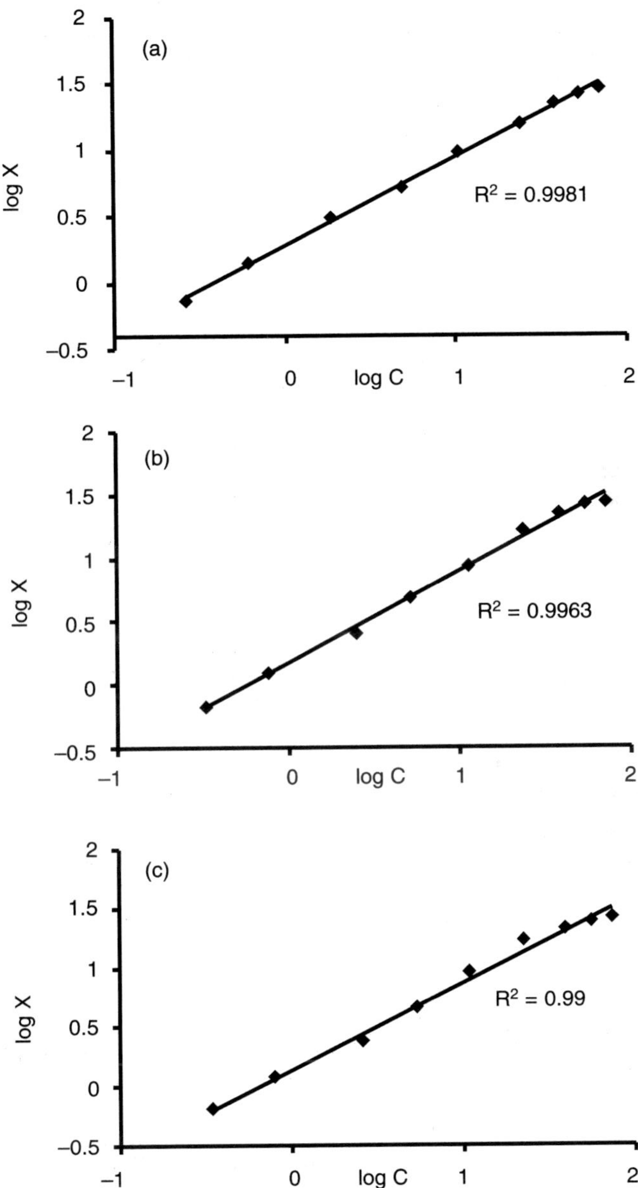

Fig. 14.3: Freundlich isotherms for boron adsorption by (a) Garhi baghi, (b) Jodhpur ramana, and (c) Samana soils. log X = adsorbed boron; log C = equilibrium concentration.

substantiates these observations. Maximum adsorption capacity appeared to increase with the fineness of the soil texture (Biggar and Fireman, 1960).

Table 14.6: Freundlich coefficients of boron adsorption by different soils

Soils	1/n	K (μg g^{-1})	R^2
Grahi baghi	0.661	1.910	0.998
Jodhpur ramana	0.720	1.475	0.996
Mansa	0.624	2.062	0.991
Kaheru	0.701	1.175	0.994
Isri	0.705	1.253	0.996
Langrian	0.710	1.835	0.992
Balewal	0.634	1.893	0.985
Sunairhari	0.814	1.644	0.983
Rajpura	0.704	2.013	0.997
Patiala	0.692	2.182	0.995
Tulewal	0.663	1.317	0.994
Nabha	0.701	1.786	0.982
Ballowal saunkhari	0.671	1.513	0.994
Dhar	0.579	5.445	0.916
Gurdaspur	0.779	1.829	0.992
Naushera	0.734	1.086	0.996
Jagjitpur	0.675	1.995	0.994
Khiranwali	0.628	2.241	0.986
Chamror	0.868	1.396	0.994
Samana	0.727	1.385	0.990
Ghabdan	0.561	1.862	0.992

The values of Freundlich constant 1/n are less than unity in all the soils, indicating an L-shape isotherm. The maximum value of 1/n (0.868) was obtained in Chamror soil series and the minimum (0.579) in the Dhar soil series (Table 14.6). The Freundlich 1/n values ranged from 0.703 to 1.230 in Haplaquepts and 0.611 to 0.703 in Fluvaquepts of Assam (Dekamedhi et al., 1998). Freundlich 'n' values, designated as the bonding energy related term (Huang, 1980), were found to be lowest for Calciorthent soil but equal for Haplustert, Haplustalfs, and Ustochrept soils (Saha and Singh, 1997).

3.6 Langmuir Adsorption Isotherms

The data of boron adsorption were next calculated according to the Langmuir adsorption equation. The linear form of the equation used was:

$$C/(x/m) = 1/b\ (C) + 1/k\ b$$

where C is the equilibrium concentration (μg B ml^{-1}), x/m is the amount of boron adsorbed per unit weight of soil (μg B g^{-1} soil), b the adsorption maxima (μg B g^{-1} soil), and k a constant related to binding energy (ml g^{-1}). A plot of C/x/m versus C will give a straight line with the slope of 1/b and an intercept of 1/kb. The data were fitted to the Langmuir equation and the Langmuir constants for boron adsorption by soils were calculated from the slopes and intercepts of the curves.

A straight line (linear curve) was obtained in Balewal, Nabha, Sunairhari, Dhar and Chamror soil series (Fig. 14.4). This showed that the adsorption data for these soils conform to the Langmuir adsorption over the entire range of the

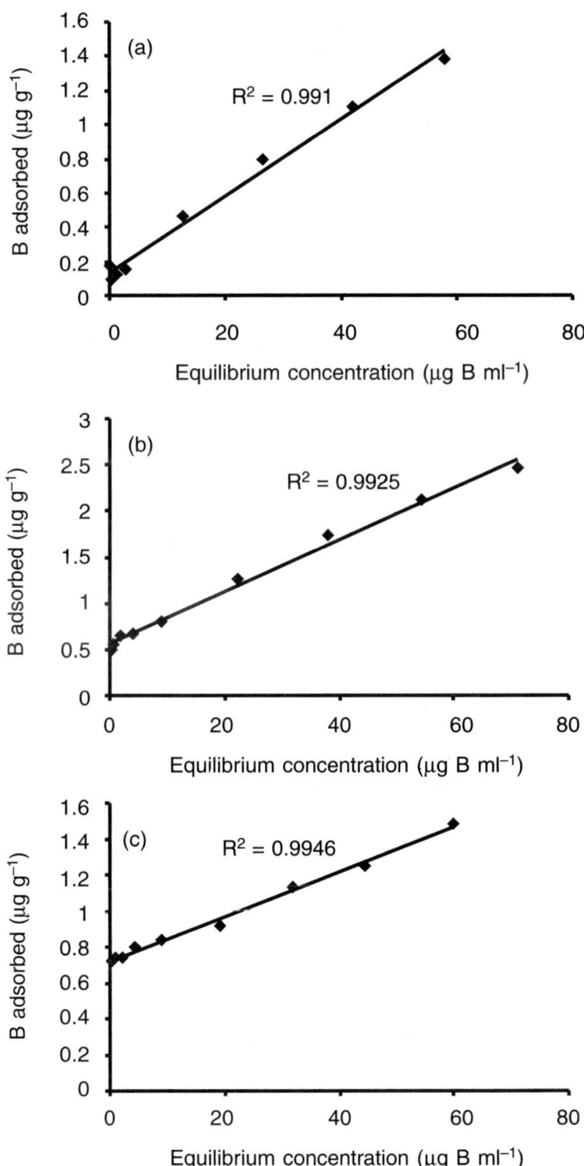

Fig. 14.4: Langmuir isotherms for boron adsorption by (a) Garhi baghi, (b) Jodhpur ramana, and (c) Samana soils.

equilibrium boron concentration. The adsorption maxima (b) value was found to be maximum for Chamror (80.0 µg B g^{-1} soil) followed by Sunairhari (54.945), Nabha (35.714), Dhar (28.089), and Balewal (20.089 µg B g^{-1} soil) (Table 14.7). The bonding energy constant (k) was 0.065 ml g^{-1} for both Balewal and Dhar soils and 0.049, 0.031 and 0.017 ml g^{-1} for Nabha, Sunairhari, and Chamror soil series. Boron adsorption data of two great soil groups of Assam also conformed to the Langmuir isotherm (Dekamedhi et al., 1998) with adsorption maxima and bonding energy coefficient ranged from 12.04 to 14.49 g kg^{-1} and 0.046 to 0.098 L mg^{-1} in Haplaquepts and 18.18 to 29.41 g kg^{-1} and 0.058 to 0.089 L mg^{-1} in Fluvaquepts, respectively. In the Ghabdan soil series, a straight line relationship for boron adsorption was observed up to the limited concentration of boron (Fig. 14.5). Thus the Langmuir equation proved applicable only up to the concentration of 40 µg B ml^{-1} in the equilibrium solution. The b and k values were 40.984 µg B g^{-1} soil and 0.037 ml g^{-1} soil, respectively (Table 14.7). A similar relationship for boron adsorption in Ghabdan soil was also reported by Sharma (1984).

In the rest of the soils representing Garhi baghi, Jodhpur ramana, Mansa, Kaheru, Isri, Langrian, Rajpura, Patiala, Tulewal, Ballowal saunkhari, Gurdaspur, Naushera, Jagitpur, Khiranwali, and Samana soil series, the fit of data in the Langmuir adsorption isotherm was curvilinear and significant and excellent when the curves were resolved into two linear parts. The type of isotherms obtained for Garhi baghi, Jodhpur ramana and Samana soils are

Table 14.7: Langmuir coefficients of boron adsorption by different soil series

Soil series	b1 (µg B g^{-1})	k1 (ml $µg^{-1}$)	R^2	b2 (µg B g^{-1})	k2 (ml $µg^{-1}$)	R^2
Grahi baghi	7.968	0.363	0.993	46.083	0.0244	0.991
Jodhpur ramana	4.545	0.509	0.997	46.296	0.0225	0.985
Mansa	9.862	0.298	0.995	34.602	0.0386	0.985
Kaheru	10.020	0.134	0.995	32.051	0.0256	0.993
Isri	11.261	0.132	0.998	54.945	0.0114	0.990
Langrian	11.614	0.199	0.994	46.083	0.0302	0.989
Balewal	20.089 (b)	0.065 (k)	0.991			
Sunairhari	54.945 (b)	0.031 (k)	0.982			
Rajpura	15.873	0.160	0.990	70.422	0.0158	0.989
Patiala	11.836	0.272	0.993	59.910	0.0284	0.988
Tulewal	5.294	0.360	0.992	28.985	0.0312	0.985
Nabha	35.714 (b)	0.049 (k)	0.992			
Ballowal saunkhari	6.693	0.321	0.991	36.364	0.0264	0.983
Dhar	28.089 (b)	0.065 (k)	0.991			
Gurdaspur	9.042	0.253	0.999	57.803	0.0291	0.988
Naushera	6.863	0.169	0.995	36.101	0.0178	0.986
Jagjitpur	9.302	0.306	0.994	42.373	0.0335	0.991
Khiranwali	9.141	0.366	0.991	35.087	0.0470	0.988
Chamror	80.00 (b)	0.017 (k)	0.995			
Samana	4.168	0.531	0.994	41.322	0.0257	0.983
Ghabdan	40.984 (b)	0.037 (k)	0.995			

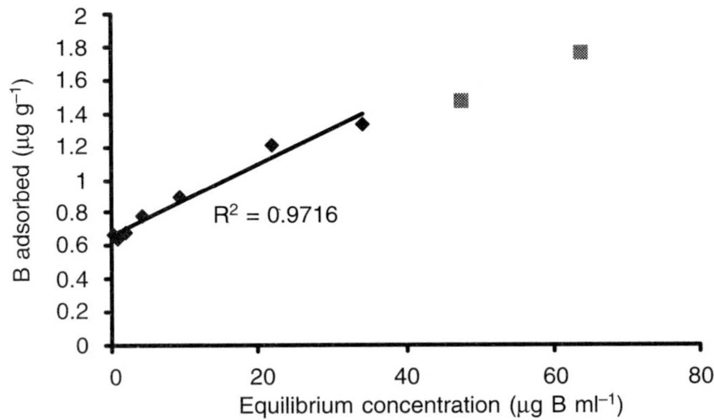

Fig. 14.5: Langmuir isotherm for boron adsorption by Ghabdan soil. Gray squares indicate the shift from the straight line isotherm.

presented in Figure 14.6. Curvilinear types of relationship have also been reported by Borkakti (1988) and Sakal et al. (1994). The two portions of the curve were considered separately because the slopes and intercepts of the lower part of the isotherm differ significantly from those of the upper part. The values of adsorption maxima and bonding energy constant were designated b1 and k1 for the lower part of the isotherm and b2 and k2 for the upper part. The Langmuir constants from the isotherms for each soil are given in Table 14.7. The maximum value of b1, i.e., 15.873 μg B g^{-1} soil was observed in Rajpura soil followed by 11.836 Patiala, 11.614 Langrian, 11.261 Isri, 10.020 Kaheru, 9.862 Mansa, 9.302 Jagjitpur, 9.141 Khiranwali, 9.041 Gurdaspur, 7.968 Garhi baghi, 6.863 Naushera, 6.693 Ballowal saunkhari, 5.294 Tulewal, 4.545 Jodhpur ramana, and 4.168 μg B g^{-1} soil in Samana soil series. The bonding energy constant (k) was maximum (0.509 ml g^{-1}) in Jodhpur ramana and minimum (0.132 ml g^{-1}) in Isri soil series. Singh (1964) also found a higher k value of 71.43 ml mg^{-1} in light textured soils vis-à-vis 55.05 ml mg^{-1} in heavy textured soils.

The data for part 1 of the curves fitted very well, indicating that the Langmuir isotherm explain best the adsorption phenomenon at low concentrations of the adsorbent which of course was different for different soils.

The marked variation in the slopes and intercepts of the two parts of the isotherms suggests two types of reaction of boron with the soil. The lower part (part 1) of the isotherm results largely from monolayer adsorption of B and the upper part (part 2) from multilayer adsorption and/or precipitation reactions as well as heterogeneity of the sites at higher concentration of boron as against the monolayer reaction and homogeneity of the soil assumed in the Langmuir equation. Syers et al. (1973) considered that the two types of adsorption reaction of P in soils was the consequence of two site populations differing widely in affinity for P, each of which can be defined by the Langmuir relationship. Similar Langmuir adsorption isotherms for boron were obtained by Borkakti (1988) in

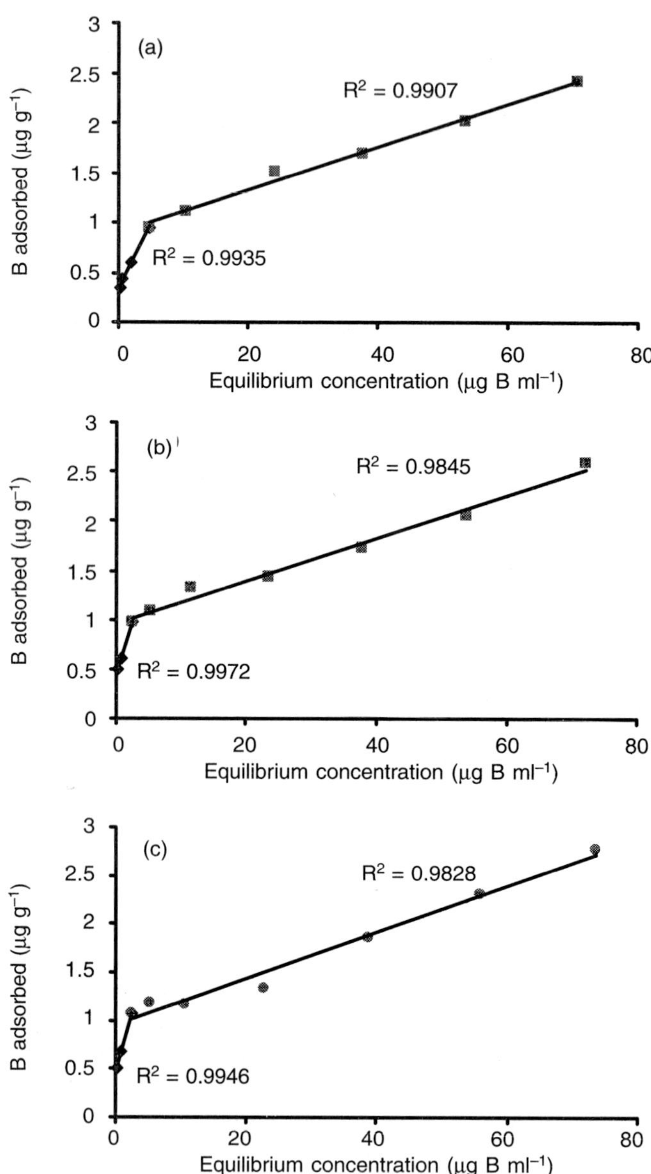

Fig. 14.6: Langmuir isotherms for boron adsorption by (a) Garhi baghi, (b) Jodhpur ramana, and (c) Samana soils. Gray squares indicate part 2 of the isotherm.

the soils of Assam. Also Shuman (1975) and Raikhy and Takkar (1981) reported the lower part as mono and the upper part as multilayer adsorption for zinc and copper respectively.

Many investigators have shown that B adsorption by soil was described by the Langmuir equation (Biggar and Fireman, 1960; Okazaki and Chao, 1968; Singh, 1971; Sulamain and Kay, 1972; Evans, 1987). Boron adsorption by seven of the ten soils from New Mexico conformed to the Langmuir sorption isotherm over limited concentration ranges. At high equilibrium solution B concentrations, sorption data deviated from the Langmuir model (Elarashidi and O'Connor, 1982). Hatcher and Bower (1958) and Griffin and Burau (1974) interpreted similar data in terms of multisite or multi-mechanism adsorption of boron.

Hingston (1964) concluded that deviations from Langmuir model at higher concentration resulted due to different reactions occurring at higher than at low boron concentrations. However, he did not give the nature of the reactions.

The difference in slopes and intercepts of the two parts of adsorption isotherms gave significantly different values of Langmuir constants. Regardless of the type of soil, the adsorption maxima (b) were higher for part 2 (upper part) while bonding energy constants (k) were higher for part 1 (lower part) of the isotherms. This suggests strong retention of boron applied in lower amounts. The strength of retention decreased with increasing concentrations of boron in the equilibrium solution as a result of precipitation reactions. Similar observations were recorded by Elrashidi and O'Connor (1982). The nature of the Langmuir isotherms and the magnitude of the constants calculated therefrom varied with soil texture and organic carbon content (Table 14.8). The values of Langmuir coefficients differed markedly for fine and coarse textured soils. The adsorption maxima b1 were much higher for fine-textured than coarse-textured soils. Singh (1964) also found that heavy soils have a high b value of 31.64 mg kg^{-1} verses 10.0 mg kg^{-1} for light-textured soils. Saha and Singh (1998) also reported that the higher B capacity of Bhopal soil over Indore soil might be due to the high clay content of the former. This is mainly due to high clay, organic matter contents, and CEC of soil, which exhibit significant correlation with b1 (r = 0.862**, 0.655**, and 0.774** respectively) (Table 14.8). Clay content, CEC

Table 14.8: Coefficient of correlation of Langmuir and Freundlich constants for boron adsorption with soil characteristics

Soil characteristics	Langmuir constants				Freundlich constants	
	b1	b2	k1	k2	n	K
Clay	0.862**	0.421	–0.374	0.198	0.254	0.569**
Sand	–0.742**	–0.613**	0.531	0.100	0.154	–0.441
CEC	0.774**	0.519	–0.508	0.059	–0.011	0.639**
pH	–0.008	–0.403	0.100	0.508	0.552**	0.218
EC	0.057	–0.156	0.090	0.432	0.474	0.260
OC	0.655**	0.465	–0.513	–0.088	–0.048	0.489
$CaCO_3$	–0.166	–0.355	0.254	0.348	0.265	0.200

**Significant at 5 % level

and organic matter content of the soils correlated positively and significantly with b1 when boron was adsorbed on Assam soils (Borkakti, 1988). Boron adsorption maxima correlated significantly with organic carbon (r = 0.96*) in soil groups of Assam (Dekamedhi et al., 1998). Evans (1987) also reported that the adsorption maxima correlated significantly with organic carbon (r = 0.872*) for boron adsorption on Ontario soils. The very poor relationship of b2 with these soil properties again indicates the multilayer adsorption and/or precipitation reaction of boron at site 2 of the adsorption isotherm.

The relative bonding energy (k) of boron adsorption by soils also varied among the soils. The k1 was higher than the respective k2 for almost all the soils, showing thereby a firmer retention of boron at part 1, i.e., at lower concentration than at higher concentration (part 2). The very low values of k2 are possibly attributable to multilayer adsorption and/or precipitation of B at higher boron concentrations while high values of k1 have resulted from the adsorption of B ions at low concentrations in the equilibrium solution. The results are in line with those obtained by Borkakti (1988).

Further, these results showed that in sandy soils low in organic matter, added B fertilizer may be expected to be readily available to plants due to the inability of the soil to adsorb B in large amounts. In fine-textured soils containg a high amount of clay content or organic carbon content, large additions of B could be made without it becoming toxic to plants because of the high adsorption capacity and bonding energy of the soils. These studies thus suggest that low and high rates of boron fertilization have to be adjusted to coarse- and fine-textured soils to ensure the level adequate for optimum crop production depending on the magnitude of these two—clay and organic matter content.

4 CONCLUSIONS

Soils not only vary in their capacity to retain B, but also in the energy with which they absorb it. These two factors affect the ability of a soil to release B and to maintain its supply in the soil solution. The study indicates that the equilibrium period of 24 h was sufficient to study the B adsorption characteristics of soils. The kinetics of B adsorption by these soils was best described by the Elovichian model. Adsorption of B conformed to Freundlich adsorption isotherms. Clay and CEC of the soils correlated significantly with K values of the Freundlich equation. The Langmuir adsorption isotherm showed two sites for B adsorption in most of the soils as it fits well at lower concentrations of added boron.

References

Berger K.C. 1949. Boron in soils and crops. *Adv. Agron* 1: 321–351.

Bhatnagar R.S., Attri S.C., Mathar G.S., and Chaudhury R.S. 1979. Boron adsorption equilibrium in soils. *Ann. Arid Zone* 18: 86–95.

Biggar J.W. and Fireman M. 1960. Boron adsorption and release by soils. *Soil Sci. Soc. Amer. Proc.* 24: 115–120.

Bingham F.T., Page A.L., Coleman N.T., and Flach K. 1971. Boron adsorption characteristics of selected amorphous soil from Mexico and Hawaii. *Soil Sci. Soc. Amer. Proc.* 35: 546–550.

Bloesch P.M., Bell L.C., and Hughes J.D. 1987. Adsorption and desorption of boron by goethite. *Austr. J. Soil Res.* 25: 377–390.

Borkakti K. 1988. Adsorption, movement and forms of occurrence of boron in major soil groups of Assam. PhD diss., P.A.U., Ludhiana.

Chien S.H. and Clayton W.R. 1980. Application of Elovich equation to the kinetics of phosphate release and sorption in soils. *Soil Sci. Soc. Amer. J.* 44: 265–268.

Dalal R.C. 1974. Desorption of soil phosphate by anion exchange resin. *Commun. Soil Sci. Plant Anal.* 5: 531–539.

Dekamedhi B., Das M.N., and Medhi, D.N. 1998. Adsorption behaviour of boron in two great soil groups of Assam. *J. Indian Soc. Soil Sci.* 46: 691–692.

Elrashidi M.A. and O'Connor G.A. 1982. Boron sorption and desorption in soils. *Soil Sci. Soc. Amer. J.* 46: 27–31.

Evans L.H. 1987. Retention of boron by agricultural soils from Ontario. *Can. J. Soil Sci.* 67: 33–42.

Giles C.H., Macewan T.H., Nakhwa S.N., and Smith D. 1960. Studies in adsorption. Part XI. A system of classification of solution adsorption isotherms and its use in diagnosis of adsorption mechanism and measurement of specific surface areas of solids. *J. Chem. Soc.*: 3973–3992.

Goldberg S. and Glaubig R.A. 1986. Boron adsorption on California soils. *Soil Sci. Soc. Amer. J.* 50: 1173–1175.

Goldberg S., Forster H.S., and Heick E.L. 1993. Temperature effects on boron adsorption by reference minerals and soils. *Soil Sci.* 156: 316–321.

Griffin R.A. and Burau R.G. 1974. Kinetic and equilibrium studies of boron desorption from soils. *Soil Sci. Soc. Amer. Proc.* 38: 892–897.

Gu B. and Lowe L.E. 1990. Studies on the adsorption of B on humic acids. *Can. J. Soil Sci.* 70: 305–311.

Hatcher J.T. and Bower C.A. 1958. Equilibria and dynamics of boron adsorption by soils. *Soil Sci.* 85: 319–323.

Hingston F.J. 1964. Reaction between boron and clays. *Austr. J. Soil Res.* 2: 83–95.

Huang P.M. 1980. Adsorption Process in soils. In: *The Handbook of Environmental Chemistry.* O. Hutzinger (ed). Springer-Verlag, Amsterdam, The Netherlands, vol. 2, Pt. A, p. 47–59.

Jackson M.L. 1973. *Soil Chemical Analysis*. Prentice Hall of India Pvt. Ltd., New Delhi.

Keren R. and Gast R.G. 1983. pH-dependent boron adsorption by montmorillonitic hydroxy aluminum complexes. *Soil Sci. Soc. Amer. J.* 47: 1116–1121.

Keren R. and Sparks D.L. 1994. Effect of pH and ionic strength on boron adsorption by pyrophyllite. *Soil Sci. Soc. Amer. J.* 58: 1095–1100.

Krishnasamy R. 1996. Kinetics of boron adsorption in soils. *J. Indian Soc. Soil Sci.* 44: 783–785.

Krishnasamy R., Jaisankar J., and Suresh M. 1997. Characterization of boron adsorption in soils of Tamil Nadu. In: *Boron in Soils and Plants.* R.W. Bell and B. Rerkasem (eds.) Kluwer Acad. Publ., Dordrecht, Netherlands pp. 255–259.

Marzadori C., Antisari L.V., Ciavatta C., and Sequi P. 1991. Soil organic matter influence on adsorption and desorption of boron. *Soil Sci. Soc. Amer. J.* 55: 1582–1585.

Mezuman U. and Keren R. 1981. Boron adsorption by soils using a phenomenological adsorption equation. *Soil Sci. Soc. Amer. J.* 45: 722–726.

Nicholaichuk W., Leyson A.J., Jame Y.W., and Campbell C.A. 1988. Boron and salinity survey of irrigation projects and the boron adsorption characteristics of some Saskatchewan soils. *Can. J. Soil Sci.* 68: 77–90.

Okazaki E. and Chao T.T. 1968. Boron adsorption and desorption by some Hawaiian soils. *Soil Sci.* 105: 255–259.

Raikhy N.P. and Takkar P.N. 1981. Copper adsorption by soils and its relations with plant growth. *Z. Pflanzenernachu Bodenk* 144: 597–612.

Rhoades J.D., Ingvalson R.D., and Hatcher J.T. 1970. Adsorption of boron by ferromagnesium minerals and magnesium hydroxide. *Soil Sci. Soc. Amer. Proc.* 34: 938–944.

Saha J.K. and Singh M.V. 1997. Boron adsorption-desorption characteristics of some major soil groups of India. *J. Indian Soc. Soil Sci.* 45: 271–274.

Saha J.K. and Singh M.V. 1998. Effect of calcium carbonate on B adsorption-desorption characteristics of swell-shrink soils. *J. Indian Soc. Soil Sci.* 46: 304–306.

Sakal R., Singh S.P., Singh A.P., and Bhogal N.S. 1994. Adsorption of boron by some calcareous soils. *Ann. Agric. Res.* 15: 14–21.

Sharma B.D., Sidhu P.S., Kumar R., and Sawhney J.S. 1997. Characterization, classification and landscape relationships of Inceptisols in N-W India. *J. Indian Soc. Soil Sci.* 45 (1): 167–174.

Sharma H.C. 1984. Boron distribution, adsorption and release in the salt-affected soils of Punjab. PhD. thesis, P.A.U., Ludhiana.

Shuman L.M. 1975. The effect of soil properties on zinc adsorption by soils. *Soil Sci. Soc. Amer. Proc.* 39: 454–458.

Sims J.R. and Bingham F.T. 1968. Retention of boron by layer silicates, sesquioxides and soil minerals, II. Sesquioxide. *Soil Sci. Soc. Amer. Proc.* 32: 364–369.

Singh M. 1971. Equilibrium adsorption of boron in soils and clays. *Geoderma* 5: 207–217.

Singh S.S. 1964. Boron adsorption equilibrium in soils. *Soil Sci.* 95: 383–387.

Sparks D.L. 1985. Kinetics of ionic reactions in clay minerals and soils. *Adv. Agron.* 38: 231–266.

Steel R.D.G. and Torrie J.H. 1960. *Principles and Procedures of Statistics.* Mc-Graw Hill Co., New York. 481 pp.

Sulaiman W. and Kay B.D. 1972. Measurement of the diffusion coefficient of boron in soil using a single cell technique. *Soil Sci. Soc. Amer. Proc.* 36: 746–752.

Syers J.K., Browman M.G., Smillie G.W. and Corey R.B. 1973. Phosphate sorption by soils evaluated by the Langmuir adsorption equation. *Soil Sci. Soc. Amer. Proc.* 37: 358–363.

Wild A. and Mazaheri A. 1979. Prediction of the leaching rate of boric acid under field conditions. *Geoderma* 22: 127–136.

Wolf B. 1974. Improvement in the azomethine—A method for the determination of boron. *Commun. Soil Sci. Plant Anal.* 5: 39–44.

Yermiyahu U., Keren R., and Chen Y. 1988. Boron sorption on composted organic matter. *Soil Sci. Soc. Amer. J.* 52: 1309–1313.

Yermiyahu U., Keren R., and Chen Y. 1995. Boron sorption by soil in the presence of composted organic matter. *Soil Sci. Soc. Amer. J.* 59: 405–409.

Part IV
Metals and Metalloids

15

Adsorption/Desorption Processes of Arsenate in Soil Environments

A. Violante*, M. Pigna, *and* S. Del Gaudio

Abstract

Studies were carried out on the adsorption/desorption of arsenate on/from soils, phyllosilicates, and various crystalline or short-range ordered metal oxides as affected by i) pH (4.0–8.0); ii) the presence and concentration of inorganic (phosphate, sulfate) and low molecular weight organic acids (LMWOAs; e.g. malic, oxalic, citric and succinic acids), iii) surface coverage of arsenate on selected sorbents, and iv) residence time.

We found that Mn and Fe oxides and phyllosilicates particularly rich in Fe (nontronite) were more effective in adsorbing arsenate than phosphate after 24–48 hours of reaction; the opposite was true for gibbsite, boehmite, and allophane. The adsorbed arsenate/adsorbed phosphate molar ratios (*rf*) increased by decreasing pH and increased with the residence time of oxyanions. Usually, the kinetics of arsenate adsorption onto some soil components and soils was slower than that of phosphate, even on sorbents for which arsenate showed a greater affinity than phosphate. The order of anion addition and nature of the sorbents also strongly influenced the adsorption of arsenate and phosphate. Phosphate partly desorbed arsenate previously fixed on goethite. Desorption of arsenate was effective mainly in the first days of reaction.

Sulfate showed a very poor influence in preventing arsenate adsorption on or in removing arsenate from soil minerals and soils, particularly at pH $\geq$ 6.0, and even at an initial sulfate/arsenate molar ratio of 4.0. At an initial sulfate/arsenate molar ratio < 10, sulfate did not prevent but simply retarded arsenate adsorption on ferrihydrite. Silicic acid seems to prevent arsenate adsorption on soil components more than sulfate.

**Corresponding author:* Dr. A. Violante, Dipart. Scienze del Suolo, della Pianta e dell' Ambiente, Università di Napoli Federico II, 80055 Portici (Napoli), Italy, E-mail: violante@unina.it

Organic ligands inhibited arsenate adsorption mainly at ligand/arsenate molar ratio > 1, when added before arsenate and more in acidic than in neutral or alkaline environments. The efficiency of LMWOAs in preventing arsenate sorption on the Andisol at pH 4.0 was as follows: citric > malic > oxalic >> succinic acid. Oxalate replaced greater amounts of phosphate than arsenate from goethite.

1 INTRODUCTION

Arsenic has both metallic and nonmetallic properties, is a toxic element for humans, animals and plants, and can be present in all natural environments (Smith et al., 1998; Berg et al., 2001; Frankenberger, 2002). Its high toxicity and increased appearance in the biosphere has triggered public and political concern.

The presence of arsenate in the environment is attributable to both the parent materials and anthropogenic waste. Arsenic may lead to contamination of both agricultural soils and surface waters and sediments in areas where mining and smelting are present (Smith et al., 1998; Francesconi and Kuehnelt, 2002).

In many countries arsenic-bearing agrochemicals (pesticides, herbicides) have been widely used in agricultural practice. Indiscriminate use of arsenical pesticides until the mid-1900s led to extensive contamination of soil worldwide. Developing countries are the most severely affected by the arsenic crisis but the situation is especially alarming in south and southeast Asia (Berg et al., 2001).

Sorption of arsenic by soil constituents controls the mobility and bioavailability of arsenic in soil-water-plant systems and potentially attenuates toxic soil solution As concentration reducing contamination to groundwaters (Barrow, 1974; Livesey and Huang, 1981; Nriagu, 1994; Sun and Doner, 1996; Sadiq, 1997; Raven et al., 1998; Goldberg, 2002).

Elevated arsenic concentrations are present in agricultural drainage waters from soils of arid regions. Arsenic concentrations in excess of 10 $\mu g\ L^{-1}$ in drinking waters are considered be hazardous to the welfare of human and domestic animals (Goldberg, 2002). Arsenic is also phytotoxic: an average toxicity threshold of 40 mg kg^{-1} has been established for crop plants (Sheppard, 1992; Matera and Le Hécho, 2001).

In nature arsenic is found in –3, 0, +3 and +5 oxidation states and is most commonly present as As[III] (arsenite) and As[V] (arsenate). Both arsenic forms are often present in either reduced or oxidized environments due to their relatively slow redox transformation (Sadiq, 1997; Smith et al., 1998). Arsenite is more mobile and toxic than arsenate. In aerated soils arsenite is readily oxidized to form arsenate. Metals and metalloids, which exist in anionic form in solution, can be retained in soils primarily by chemisorption at reactive sites of variable charge mineral surfaces and edges of phyllosilicates (Goldberg et al., 1996; Cornell and Schwertmann, 1996; Arai and Sparks, 2002). These types of anion adsorption are referred to as specific adsorption, which differs distinctly from nonspecific adsorption. Specific adsorption is inherently less reversible than nonspecific adsorption (Huang and Germida, 2002). It has been established that the chemical behavior of arsenate is similar to that of phosphate (Hingston et al., 1971; Manning and Goldberg, 1996; Smith et al., 1998). These two anions are specifically adsorbed on soil minerals, mainly on the variable charge minerals (Al-, Fe- and Mn-oxides, allophanes, imogolite), forming inner-sphere complexes (Sun and Doner, 1996; Grossl et al., 1997; Smith et al., 1998; Violante and Pigna, 2002). These oxyanions may form three different surface complexes on inorganic soil components: a monodentate complex, a bidentate-binuclear complex, and a bidentate-mononuclear complex in different proportions, depending on surface coverages (Lumsdon et al., 1984; Hsia et al., 1994; Sun and Doner, 1996; Fendorf et al., 1997; Smith et al., 1998; O'Reilly et al., 2001; Liu et al., 2001). Studies on the factors which affect arsenate removal from soil constituents are scant. Phosphate has been reported to displace adsorbed arsenate from soils. Some authors demonstrated that large additions of phosphate to As-polluted soils may displace great amounts of arsenic (Woolson et al., 1973; Davenport and Peryea, 1991; Peryea, 1991; Smith et al., 1998). On the contrary, other authors (Melamed et al., 1995; O' Reilly et al., 2001; Violante and Pigna, 2002) have shown that arsenate may be partly removed from soil colloids by phosphate, but even large amounts of phosphate are not able to desorb all the applied arsenate (Darland and Inskeep, 1997).

Recently, O' Reilly et al. (2001) showed that a significant amount of arsenate remained bound to goethite after 5 months of desorption even though the phosphate desorptive solution was three times stronger than the initial arsenate sorptive solution. Furthermore, Pierce and Moore (1982) found that once arsenate was sorbed to a soil sorbent, it was not affected by the postaddition of phosphate and sulfate. It has also been demonstrated that usually molybdate and sulfate compete poorly with arsenate (Roy et al., 1986; Myneni et al., 1997; O' Reilly et al., 2001).

Bioavailability of arsenate in the environment may also be affected by naturally occurring organic molecules, which may compete with arsenate for sorption sites (Grafe et al., 2001; Liu et al., 2001). However, there is relatively little information regarding the role of organic anions on the adsorption behavior of arsenate (Xu et al., 1988; Grafe et al., 2001; Liu et al., 2001). Humic or fulvic acids as well as low molecular-weight organic acids (LMWOAs) are specifically sorbed on variable charge minerals and soils and may greatly enhance the solubility and mobility of metals in soil (Parfitt, 1980; Oades, 1989; Cornell and Schwertmann, 1996). The sources of LMWOAs in soils include root exudates, decay of plant, animal, and microbial tissues, and microbial metabolites (Huang and Violante, 1986; Kafkafi et al., 1988; Marschner, 1995; Violante and Gianfreda, 2000; Huang et al., 2002). The competitive sorption of some chelating anions, usually present in soil environments (e.g. in the rhizosphere), is particularly important because the mobility and bioavailability of arsenate should be strongly influenced by the presence of these ligands.

In the last years, we have carried out investigations on factors which might influence the adsorption/desorption of arsenate on/from soil components. The goals here are to present our more recent findings on factors which influence the mobility of arsenate in soil environments and to provide information on sorption of arsenate onto various Fe-, Al- and Mn-oxides, phyllosilicates and soils as affected by pH, and the presence, nature, and concentration of inorganic and organic ligands, and on the desorption of arsenate, previously sorbed on variable charge minerals, by inorganic and organic anions.

2 MATERIALS AND METHODS

2.1 Synthetic Metal Oxides

Goethite 1 and gibbsite 1 were obtained by precipitating 0.1 mol L^{-1} $Al(NO_3)_3$ or 0.1 mol L^{-1} $Fe(NO_3)_3$ by 0.5 mol L^{-1} NaOH up to pH 7.0 or 12.0 respectively. The suspensions were aged for 7 days at room temperature and a further 20 days at 65°C, then washed free of salts by dialysis (molecular weight cut-off of 15,000) in deionized water. Goethite2 was obtained by slowly adding 200 mL of 2.5 mol L^{-1} NaOH at 50 g $Fe(NO_3)_3 \cdot 9\, H_2O$ solubilized in 825 mL of deionized water. The final suspension was aged 6 days at 60°C, washed free of salts by dialysis, freeze dried, and lightly ground to pass through a 0.16-mm sieve (Liu et al., 2001). Gibbsite2 was obtained from Aldrich Chemical Company (Milwaukee, WI, USA). Both goethites consisted of acicular crystals under transmission electron microscopy (TEM) observations. Gibbsite1 was mainly in the form of hexagonal plates but also showed the presence of noncrystalline materials under TEM observations, whereas gibbsite2 of the commercial preparation likewise showed the presence of noncrystalline material plus a low amount of bayerite.

For goethite1 and goethite2, the surface area determined by the retention method of ethyleneglycol monoethylether (EGME; Eltanawy and Arnold, 1973) were 80 and 90 $m^2\, g^{-1}$, respectively, and a point of zero salt effect (PZSE) values determined by the method of Sakurai et al. (1988) were 7.8–8.0. Gibbsite1 and gibbsite2 had surface areas (EGME) of 120 and 135 $m^2\, g^{-1}$, respectively, and a PZSE of 8.5–9.0.

Ferrihydrite and mixed Fe–Al gels were prepared according to the method described in a previous work (Colombo and Violante, 1996). Stock solutions of 0.1 mol L^{-1} $Al(NO_3)_3$ and 0.1 mol L^{-1} $Fe(NO_3)_3$ were mixed in different proportions to synthesize samples having initial Fe/Al molar ratios (R) of 1, 2, 4, 10, and ∞ (no Al present). The solutions (henceforth referred to as R1, R2, R4, R10, and R∞) were potentiometrically titrated to pH 5.5 by adding 0.5 mol L^{-1} NaOH at a feed rate of 0.5 mL min^{-1}. The final volume of all samples was adjusted to 1 L. After 7 days of aging at room temperature the suspensions were dialyzed with deionized water and freeze dried.

A synthetic allophane (AL-SI) was synthesized by mixing 80 mL of 1.1 mol L^{-1} $AlCl_3$ and 104 mL of 1.3 mol L^{-1} potassium silicate, which were neutralized up to pH 5.0 by 0.1 mol L^{-1} NaOH. The product had an EGME surface area of 717 $m^2\, g^{-1}$, a Si/Al molar ratio of 0.21, and an isoelectric point (IEP) of 5.5 (Mora and Canales, 1995). The IEP was determined by microelectrophoresis in a zeta meter (ZM-77) apparatus.

Birnessite was synthesized by adding hydrochloric acid to a hot solution of $KMnO_4$ (McKenzie, 1981). According to McKenzie (1981), the surface area of birnessite determined by BET-N_2 absorption was 40 $m^2\ g^{-1}$. The pyrolusite used in our study was obtained from Aldrich Chemical Company (Milwaukee, WI, USA).

2.2 Phyllosilicates

Three clay minerals (Wyoming montmorillonite (SWy-1), Georgia kaolinite (KGa-1), and Washington ferruginous smectite), obtained from the Clay Minerals Society Source Clay Minerals Repository, were lightly ground to pass through a 0.315-mm sieve and used without further preparation. According to Manning and Goldberg (1996), untreated kaolinite and montmorillonite used in this work had a surface area (BET N_2 adsorption) of 9.1 $m^2\ g^{-1}$ and 18.6 $m^2\ g^{-1}$.

2.3 Soil Samples

The < 2-mm fraction of an Andisol-16 (a 2C horizon of a Typic Hapludand) from Roccamonfina Volcano (Italy), and the < 2-μm fraction of a red Mediterranean soil (Rhodoxeralf-1) from Apulia (Italy) were used in this work. General descriptions of these soils are given by Vacca et al. (2003; Andisol) and Colombo et al. (1996; Rhodoxeralf-1). Selected chemical and physicochemical properties of the soil samples and their mineralogical composition are reported in Table 15.1.

Table 15.1: Selected physical and chemical properties for the soils studied

Soils	pH	Organic C	Fe_d	Fe_o	Al_d	Al_o	Clay
		$g\ kg^{-1}$					
Andisol †	5.5	2	36.1	10.9	21.6	105.1	121
Rhodoxeralf ‡	7.9	n.d.	24.0	1.8	n.d	n.d	1,000

Abbreviations: Fe_d, Al_d stand for Fe and Al extracted by Na-dithionite-citrate; Fe_o, Al_o stand for Fe and Al extracted by NH_4-oxalate

†The samples contained allophanic minerals and other short-range materials. The percentage of allophane was 42% for Andisol (Vacca et al., 2003).

‡Clay fraction of Rhodoxeralf sample was treated by hydrogen peroxide.

*n.d. not determined.

2.4 Adsorption of Arsenate and Phosphate as a Function of pH

Metal oxides, clay minerals, or soil samples were added at 50 to 300 mg to predetermined quantities of 0.1 mol L^{-1} solutions containing arsenate, phosphate, or a mixture of both oxyanions in reaction flasks (*As* + *P* systems), at pH previously adjusted values (pH 4.0 and 7.0).

The amounts of phosphate and arsenate were pipetted into the flasks to yield nearly the maximum adsorption of each oxyanion per kilogram of sample, as previously determined by adsorption isotherms (data not shown). The pH of each suspension was usually kept constant for 24 h by adding 0.1 or 0.01 mol L^{-1} HCl or NaOH. The final suspensions (20 mL) in 0.05 mol L^{-1} KCl were centrifuged at 10,000 g for 20 min, then filtered through a 0.22-μm filter.

Competition in adsorption between arsenate and phosphate was also determined at different pH values (3.0–8.5), changing the order of addition and precisely by adding arsenate 24 h before phosphate (*As before P* systems), or introducing arsenate 24 h after phosphate (*P before As* systems)

Adsorption kinetics of arsenate (and phosphate) added alone or in the presence of phosphate (arsenate) on gibbsite 2 and ferrihydrite to yield near maximum adsorption of the ligands was carried out at pH 5.0. Samples were taken periodically at 5, 15, 30 min and 1, 4, 24, 48 h and passed through a 0.22-μm filter. The filtrates were analyzed for arsenate and phosphate as described below.

2.5 Adsorption of Arsenate in Presence of Increasing Concentration of Organic and Inorganic Ligands on Minerals and Soils

Competitive adsorption between arsenate and phosphate at pH 4.0 was determined by adding to 100 mg Andisol suitable amounts of arsenate (470 mmol kg^{-1}; maximum surface coverage) in the presence of increments in phosphate concentration to achieve initial phosphate/arsenate molar ratios ranging from 0 to 2.

The final suspensions were centrifuged at 10,000 g min^{-1} and filtered through a 0.22-μm filter. Arsenate was determined in the supernatant as described below.

Competition in adsorption between arsenate and organic ligands (oxalate, malate, or citrate) was also determined by adding to 100 mg Andisol suitable amounts of arsenate (416 mmol kg^{-1}; 90% of surface coverage) in the presence of increasing concentrations of oxalic or malic or citric acid to achieve initial organic ligand/arsenate molar ratios ranging from 0 to 10.

Adsorption kinetics of arsenate on ferrihydrite in the presence of sulfate and on gibbsite2 in the presence of malate was carried out as previously described. Arsenate uptake from the bulk solution was monitored continuously at 0.02–100 h.

The final suspensions were centrifuged at 10,000 g for 20 min and filtered through a 0.22-μm filter. Arsenate was determined in the supernatant as described below.

2.6 Competitive Adsorption of Arsenate and Malate on Different Sorbents at pH 4.0 in Presence of Increasing Concentrations of Malate

Experiments were carried out on competitive adsorption between arsenate and malate on different sorbents. To 100 mg goethite 1, pyrolusite, gibbsite 1, and

Andisol were added 150, 20, 150, and 420 mmol kg^{-1} arsenate respectively (90–100% surface coverage for each sample) in the presence of increments in malate concentrations to achieve initial malate/arsenate molar ratios ranging from 0 to 10.

The final suspensions were centrifuged at 10,000 g for 20 min and filtered through a 0.22-μm filter. Arsenate was determined in the supernatant as described below.

2.7 Desorption of Arsenate (by Phosphate) and Phosphate (by Arsenate) from Goethite

Forty-five mL of 0.1 mol L^{-1} KCl solution were added to 50 mg goethite in reaction flasks. Predetermined quantities of 0.02 mol L^{-1} solutions of arsenate (or phosphate) whose pH values had previously been adjusted to 3.0–8.5, were pipetted into the flasks to give 240 mmol kg^{-1} arsenate. The suspensions were shaken at 20°C and periodically adjusted to the initial pH value. After 24 h, the suspensions were centrifuged at 10,000 g for 30 min and washed twice with 0.1 mol L^{-1} KCl solutions whose pHs were similar of those of the suspensions. The solids were resuspended into 45 mL of 0.1 mol L^{-1} KCl at the same initial pH values, and 240 mmol phosphate (or arsenate) kg^{-1} added in the reaction flasks. The suspensions were shaken for 24 h and periodically adjusted to the initial pH. Phosphate (or arsenate) desorption by arsenate (or phosphate) at pH 4.0 and different times of reactions (0.125–17 days) was also studied. The final suspensions (47 mL) were centrifuged and phosphate and arsenate analyzed in the supernatants.

2.8 Desorption of Arsenate from Goethite by Oxalate

Forty-five mL of 0.1 mol L^{-1} KCl solution were added to 50 mg goethite in reaction flasks. Predetermined quantities of 0.02 mol L^{-1} solution containing a mixture of arsenate and phosphate whose pH was previously adjusted (3.0–8.5), were pipetted into the flasks to give 240 mmol of each ligand per kilogram sample (nearly the maximum adsorption). The suspensions were shaken for 24 h, then centrifuged, washed, resuspended at the same initial pH, and oxalate (OX) added at an OX/As + P molar ratio of 1. The suspensions were again shaken for 24 h, after which they were centrifuged and phosphate and arsenate analyzed in the supernatants.

2.9 Arsenate and Phosphate Determination

Arsenate and phosphate present in the final solutions were determined by ion chromatography, using a Dionex DX-300 Ion Chromatograph (Dionex Co., Sunnyvale, CA), an IonPac AS11 column (4.0 mm), an eluent of 0.05 mol L^{-1} NaOH at a flow rate of 2 ml min^{-1}, and a CD20 Conductivity Detector combined with autosuppression. Average phosphate and arsenate retention times were

3.0 and 4.2 min respectively. The standard concentrations were 0.2–2 mmol L^{-1} for arsenate and phosphate. The amounts of ligands adsorbed were determined by the difference between the initial and final concentrations. The data are the mean of two or three determinations. Coefficients of variation ranged from 1.5 to 4%.

3 RESULTS AND DISCUSSION

3.1 Adsorption of Arsenate, Phosphate, Sulfate, and Organic Ligands on Soil Components

Arsenate and phosphate adsorption behavior by soil components and soils is usually similar (Hingston et al., 1971; Manning and Goldberg, 1996; Roy et al., 1986; Smith et al., 1998; Liu et al., 2001). Most of the sorbents used in this work did not differ significantly in terms of their capacity to fix these ligands after 24 h of reaction in a wide range of pH (3.0–9.0; Fig. 15.1; Table 15.2). However, some minerals and soils (e.g. kaolinite, gibbsite, and Andisol) adsorbed slightly more phosphate than arsenate after 24 h reaction (Table 15.2); with a longer reaction time (2–10 days) the amounts of arsenate and phosphate adsorbed by these sorbents were practically the same (Fig. 15.2). Usually, the kinetics of arsenate adsorption onto some metal oxides and soils was slower than that of phosphate (Fig. 15.2), even on the sorbents on which arsenate showed a greater affinity than phosphate (e.g. ferrihydrite; discussed below).

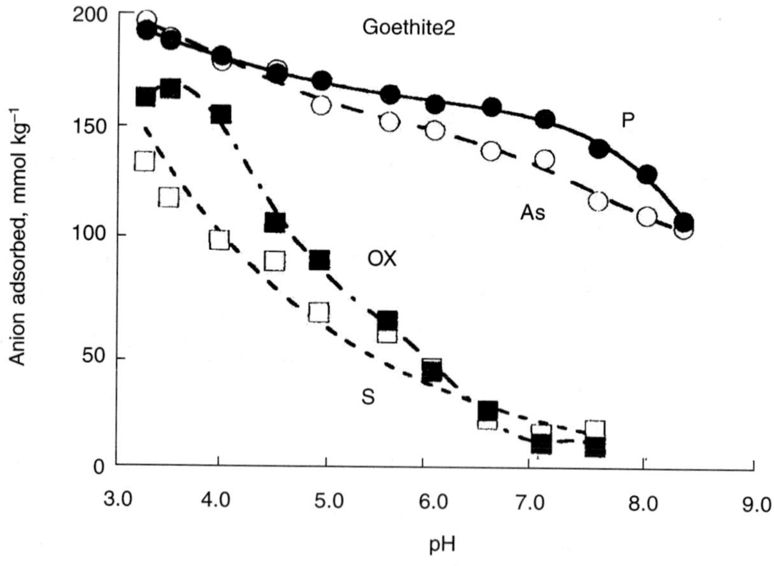

Fig. 15.1: Arsenate (As), phosphate (P), oxalate (OX), and sulfate (S) adsorption (240 mmol added kg^{-1}) on goethite2 at different pH values (3.0–8.5).

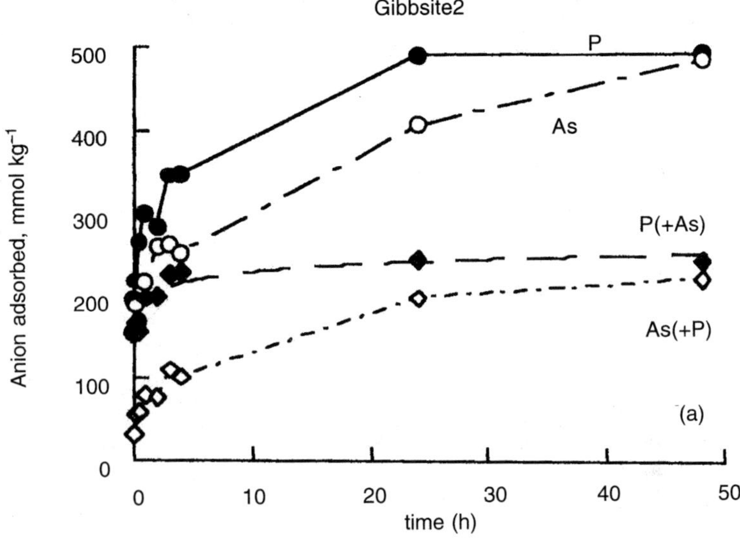

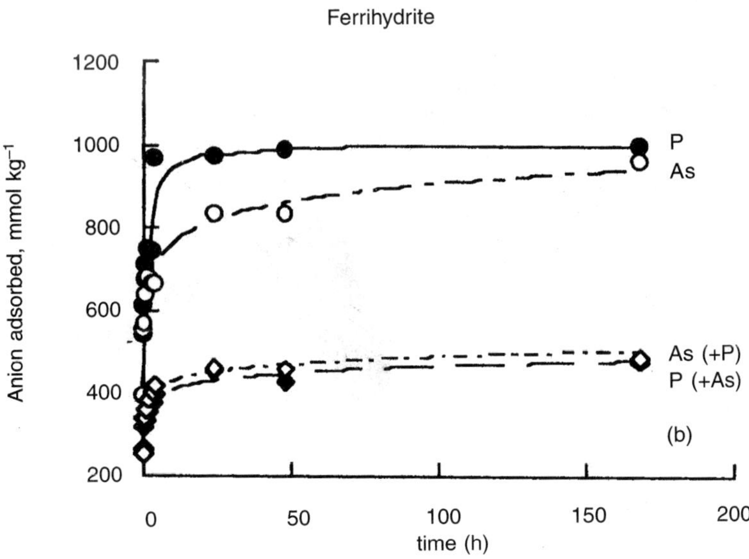

Fig. 15.2: Kinetics of arsenate (As) and phosphate (P) adsorption on gibbsite2 (a) and ferrihydrite (b). The oxyanions were added alone (As, P) or as a mixture [As (+P), P (+As)]; the initial As/P molar ratio was 1.

Table 15.2: Adsorption (mmol kg^{-1}) of arsenate (As) and phosphate (P) after 24 h reaction on metal oxides, phyllosilicates, and soils at pH 5.0, when the ligands were added alone or as a mixture (initial P/As molar ratio of 1.0) (modified from Violante and Pigna, 2002)

Samples	Anion added alone		Anion added as a mixture		
	As	P	As	P	rf*
		pH 5.0			
Birnessite	28.5	25.7	23.8	13.1	1.81
Pyrolusite	26.0	21.0	12.7	9.8	1.30
Smectite (ferr.)	21.4	19.9	17.4	14.9	1.17
Goethite 1	175	158	87	78	1.11
Montmorillonite	10.9	11.6	7.8	9.2	0.84
Rhodoxeralf	27.7	29.4	14.9	22.0	0.68
Andisol	644	806	341	738	0.46
Gibbsite1	205	280	82	198	0.41
Kaolinite	10.5	15.2	6.7	14.8	0.45
Allophane	498	917	126	880	0.14

**rf* stands for adsorbed arsenate/adsorbed phosphate molar ratio.

Arsenate and phosphate were adsorbed in larger amounts than sulfate and LMWOAs (e.g. oxalic, [Fig. 15.1], malic, tartaric, succinic acid; He et al., 1999; De Cristofaro et al., 2000) on Fe- and Al-oxides, and other variable charge minerals, clearly because arsenate and phosphate bind more strongly to iron or aluminum (Parfitt, 1980; Tabatabai, 1982), the different dissociation constants of the ligands (Parfitt and Smart, 1978), and differences in the coordinate forms of each ligand on the surfaces of the sorbents (Sparks, 1999) from LMWOAs and sulfate. Figure 15.1 shows the influence of pH on the adsorption of arsenate, phosphate, sulfate, and oxalate onto goethite2 (300 mmol kg^{-1} of each ligand added). At pH 3.0 the amounts of arsenate and phosphate adsorbed were 197 and 193 mmol kg^{-1}, while at pH 8.0 the amounts of these ligands were reduced about 40%. Adsorption of both sulfate and oxalate decreased strongly with increasing pH. Sulfate adsorption decreased from 134 to 18.5 mmol kg^{-1} as the pH increased from 3.0 to 7.0. The negligible adsorption of sulfate on soil sorbents at pH $\geq$ 6.0 has been widely documented (Chao et al., 1964; Pasricha and Fox, 1993). For oxalate adsorption on goethite, a maximum occurred near pH 3.5 (167 mmol kg^{-1}). The adsorption of oxalate and other organic ligands on clay minerals and oxides usually decreased in a wide range of pH with increments in pH (Cornell and Schindler, 1980; Violante and Gianfreda, 1993; Liu et al., 1999; He et al., 1999; De Cristofaro et al., 2000). Goldberg and Glaubig (1988) also investigated the adsorption of arsenate on a calcareous, montmorillonitic soil and calcite. These authors found that arsenate sorption on calcite increased from 6.0 to 10.0, peaked between pH 10.0 and 12.0 and decreased above pH 12.0 whereas sorption on the calcareous soil reached maximum near pH 10.5, but after removal of carbonates the maximum markedly reduced. These findings indicate that carbonates play an important role in arsenate sorption only above pH 9.0.

3.2 Competition in Adsorption between Arsenate and Phosphate

Competition in adsorption between arsenate and phosphate on selected sorbents was studied at different pH values. Table 15.2 shows the amount of arsenate and phosphate adsorbed at pH 5.0 when the oxyanions were added as a mixture at an initial phosphate / arsenate molar ratio of 1 and near their maximum surface coverage (Violante and Pigna, 2002). After 24 h reaction, arsenate and phosphate strongly competed for the surface sites of the sorbents, but a large variation in competitiveness between these ligands for the sorbents used was observed.

The adsorbed arsenate / adsorbed phosphate molar ratio (*rf*) indicated the selectivity of each sample to preferentially adsorb one of the two ligands. The *rf* values for kaolinite, gibbsite, allophane, and Andisol were usually < 0.5, indicating a much greater affinity of phosphate than arsenate for the surfaces of these minerals. The opposite was true for goethite, ferrihydrite, pyrolusite, birnessite, and ferruginous smectite; in fact, the *rf* values for these samples were usually greater than 1.

Violante and Pigna (2002) demonstrated that the *rf* values for selected clay minerals increased according to the following sequence: allophane < gibbsite ≈ noncrystalline Al precipitation product ≤ kaolinite < boehmite < illite ≅ montmorillonite < nontronite < ferruginous smectite ≅ goethite ≅ ferrihydrite < pyrolusite < anatase ≅ birnessite.

Clearly Fe- and Mn-oxides, and phyllosilicates particularly rich in Fe (ferruginous smectites) were more effective in adsorbing arsenate than phosphate. In contrast, minerals richer in aluminum (gibbsite, kaolinite) were much more effective in retaining phosphate than arsenate (Table 15.2).

Fordham and Norrish (1979, 1983) reported that clay minerals were relatively unimportant compared with Fe-oxides and, to a lesser extent, titanium oxides in the adsorption of arsenate in acidic soils.

The greater affinity of arsenate for sorbents containing Fe than for sorbents richer in Al was also demonstrated by using mixed Fe-Al oxides of different composition (Violante and Pigna, 2002). Synthetic Fe-Al oxides were formed at pH 5.5 by coprecipitating Fe and Al at Fe/Al molar ratios of 1, 2, 4, 10 and ∞ (referred to as R1, R2, R4, R10, and R_∞ in Table 15.3) and characterized by similar surface area (200–285 m^2 g^{-1}, determined by EGME) and mineralogy (ferrihydrites with different percentages of Al isomorphic substitution; Fig. 15.3) but different chemical composition (Table 15.3).

Table 15.3 shows the amounts of phosphate and arsenate adsorbed at pH 5.0 on the mixed Fe-Al oxides after different reaction times (from 5 to 96 h) when the anions were added as a mixture. In spite of the fact the Fe-Al oxides had similar surface area and mineralogy (Fig. 15.3), not only the total amount of phosphate and arsenate sorbed on the oxides differed from sample to sample (usually decreased by decreasing Al in the oxides), but a variation in

Table 15.3: Amounts of phosphate (P) and arsenate (As) sorbed on mixed Al-Fe gels when 1,000 mmol P and 1,000 mmol As kg^{-1} were added as a mixture to the Fe-Al oxides at pH 5.0

	After 5 h			After 24 h			After 96 h		
	As	P	**rf**	As	P	**rf**	As	P	**rf**
Samples†				pH 5.0 mmol kg^{-1}					
R1	386	770	**0.50**	426	830	**0.51**	554	991	**0.56**
R2	420	664	**0.63**	503	753	**0.67**	698	984	**0.71**
R4	469	586	**0.80**	610	769	**0.79**	706	804	**0.87**
R10	481	516	**0.93**	537	546	**0.98**	540	639	**0.84**
R∞	405	401	**1.01**	508	427	**1.19**	598	492	**1.21**

†R1, R2, R4, R10 and R∞ indicate the samples formed at initial Fe/Al molar ratio of 0, 1, 2, 4, 10 and ∞.

competitiveness between these anions was found. In fact, *rf* values increased with increasing amounts of Fe present in the mixed Fe-Al gels (Table 15.3).

Arsenate and phosphate adsorption also increased with time. Residence time influenced the *rf* values, which generally increased with time, clearly because the adsorption rate of phosphate is initially faster than that of arsenate (Fig. 15.2; Table 15.3). Table 15.4 and Figure 15.4 show the effect of residence time on *rf* values when arsenate and phosphate were added as a mixture on other adsorbents. On gibbsite2, the *rf* values increased from 0.20 after 0.03h to 0.81 after 24 h and 1 after 720 h (Fig. 15.4), whereas on the Andisol the *rf* values increased from 0.25 after 0.02 h to only 0.51 after 24 h (Table 15.4).

The competitiveness between arsenate and phosphate is also affected by pH. Figure 15.5 shows the *rf* values determined on different minerals and soils at pH 4.0, 5.0, and 7.0. The *rf* values usually decreased by increasing the pH of the systems. Similar findings were obtained by adding these ligands on goethite 2 at pH values ranging from 3.0 to 8.0. In *As + P* systems (arsenate and phosphate added as a mixture) the *rf* value decreased from 1.30 at pH 3.0 to 0.94 at pH 8.0 (data not shown).

These results indicate that in soil environments phosphate inhibits arsenate adsorption more in neutral and alkaline systems than in acidic systems.

3.3 Order of Anion Addition on Adsorption of Arsenate and Phosphate

The order of anion addition strongly influenced the adsorption of arsenate and phosphate on the surfaces of clay minerals and soils (Table 15.5; Fig. 15.6). Table 15.5 shows the amounts of arsenate and phosphate adsorbed on goethite2 at different pH values when arsenate was added as a mixture with phosphate (*As + P* systems), before phosphate (*As before P* systems), and after phosphate (*P before As* systems). In these experiments equimolar amounts (240 mmol kg^{-1} of each) of the ligands were added to the Fe-oxide. The quantities of each adsorbed ligand strongly decreased over all the range of pH studied compared with those

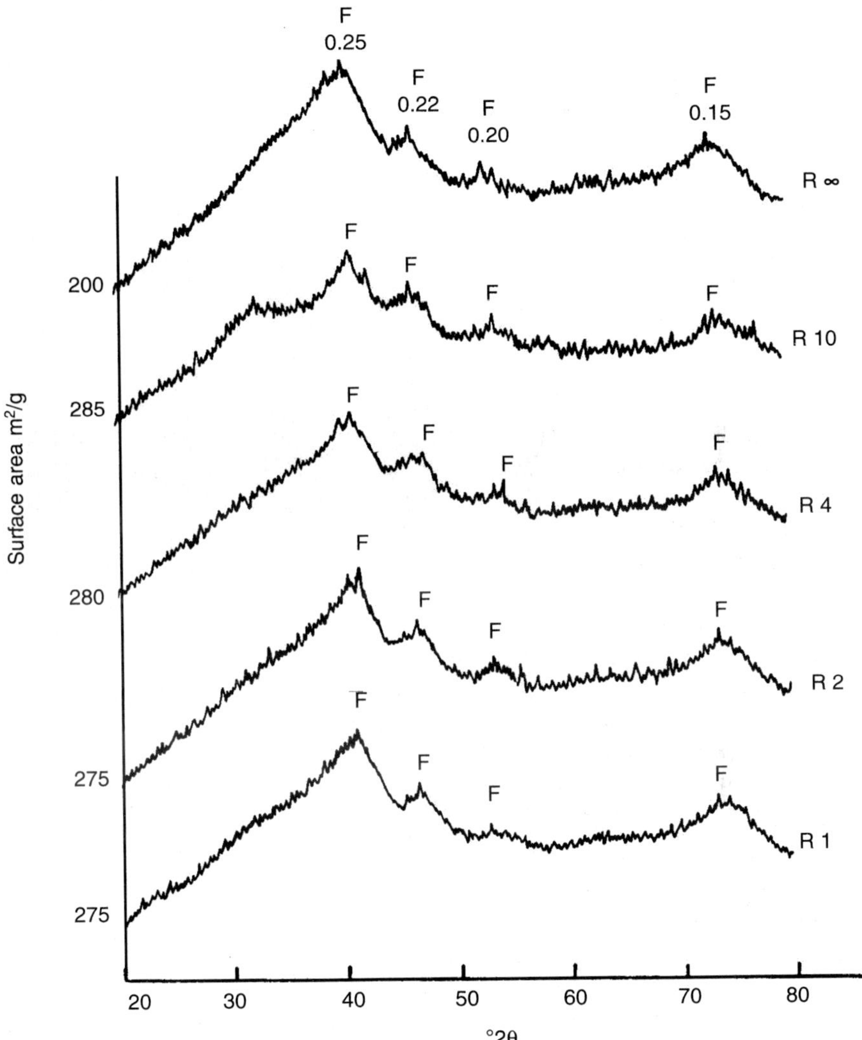

Fig. 15.3: X-ray diffraction patterns of Fe-Al oxides aged 7 days at 20°C. R1, R2, R4, R10, and R_{∞} indicate samples formed at an initial Fe/Al molar ratio of 1, 2, 4, 10 and ∞ and at pH 5.0. F = ferrihydrite. Samples R1 and R2 had the same specific surface area.

of only arsenate or phosphate systems, but the sum of arsenate + phosphate was only slightly greater than that of arsenate or phosphate added alone. Hingston et al. (1971) claimed that goethite contains adsorption sites common to both phosphate and arsenate, as well as sites that adsorb either one anion or the other, whereas Manning and Goldberg (1996) demonstrated that these anions

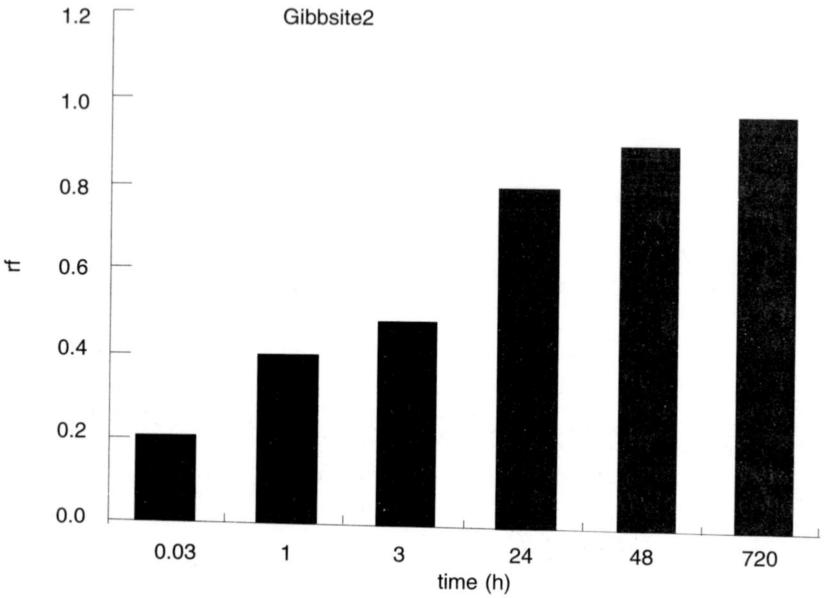

Fig. 15.4: Effect of residence time (0.03–720 h) on the arsenate adsorbed/phosphate adsorbed molar ratio (*rf*) determined on gibbsite2, when arsenate and phosphate were added as a mixture (initial arsenate/phosphate molar ratio of 1).

compete for a similar set of surface sites, though there was evidence that some sites were uniquely available for adsorption of either arsenate or phosphate.

By adding arsenate 24 h before phosphate (*As before P* systems) the amounts of arsenate adsorbed at pH 3.0–6.0 were reduced by 13–20% with respect to the quantities of arsenate adsorbed when added alone. At pH > 6.0 the amounts of arsenate adsorbed were similar to those obtained when arsenate alone was added to the oxide. In contrast, phosphate adsorption in *As before P* systems was reduced in the pH range studied but less in alkaline systems.

When phosphate was added to goethite2 24 h before arsenate (*P before As* systems), the amounts of phosphate adsorbed were nearly constant across the range of pH studied, and were greater than those of arsenate, but were substantially lower than the quantities of phosphate adsorbed when phosphate alone was added to goethite 2.

The *rfs* in *As + P*, *As before P*, and *P before As* systems are reported in Table 15.5. It is particularly interesting that *rf* values were lower than 1 only in *P before As* systems and at pH > 6.0 in *As + P* systems. Furthermore, in *P before As* systems, *rf* values indicate clearly that phosphate was not as efficient in

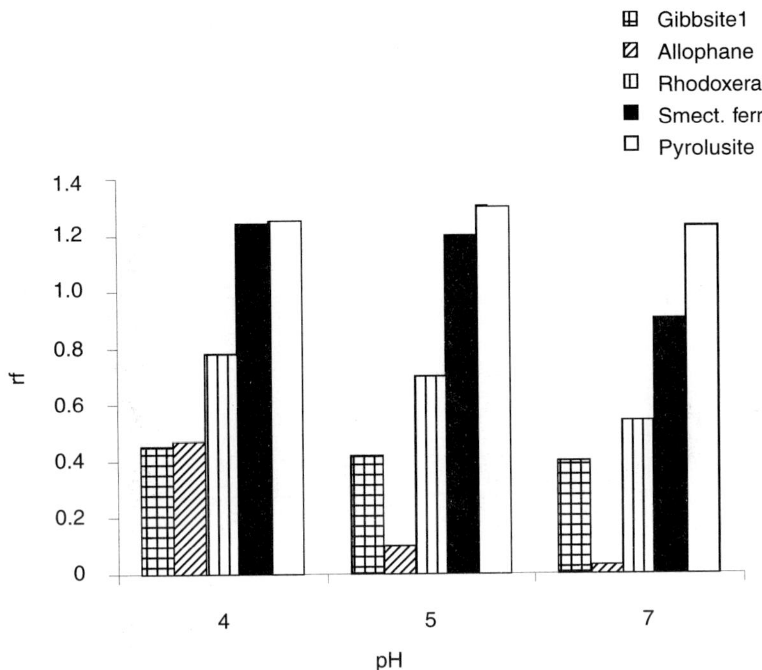

Fig. 15.5: Arsenate adsorbed/phosphate adsorbed molar ratio (*rf*) determined on gibbsite1, allophane, Rhodoxeralf soil, ferruginous smectite, and pyrolusite at pH 4.0, 5.0, and 7.0 when arsenate and phosphate were added as a mixture (initial arsenate/phosphate molar ratio of 1) near their maximum surface coverage and after 24 h of reaction.

preventing arsenate adsorption at pH < 7.0 as arsenate in preventing phosphate adsorption in *P before As* systems, as discussed above.

Not only the order of anion addition, but also the nature of the sorbents strongly influenced the competitive adsorption of arsenate and phosphate. Figure 15.6 shows the amounts of arsenate and phosphate adsorbed at pH 5.0 on gibbsite (Fig. 15.6A) and ferrihydrite (Fig. 15.6B) as affected by the order of anion addition. In these experiments equimolar amounts of the ligands were added (1,000 mmol kg^{-1} of each oxyanion on ferrihydrite and 500 mmol kg^{-1} on gibbsite).

The *rf* values in all the systems were much greater on ferrihydrite than on gibbsite, showing the much greater affinity of arsenate for the Fe-oxide than for the Al-hydroxide. In fact, *rf* values ranged from 2.18 (*As before P* systems) to 0.70 (*P before As* systems) using ferrihydrite as sorbent and from 0.34 (*As before P* systems) to 0.21 (*P before As* systems) using gibbsite.

On ferrihydrite the efficiency of phosphate in preventing arsenate adsorption (efficiency of P in preventing As adsorption, % = [1– [As adsorbed

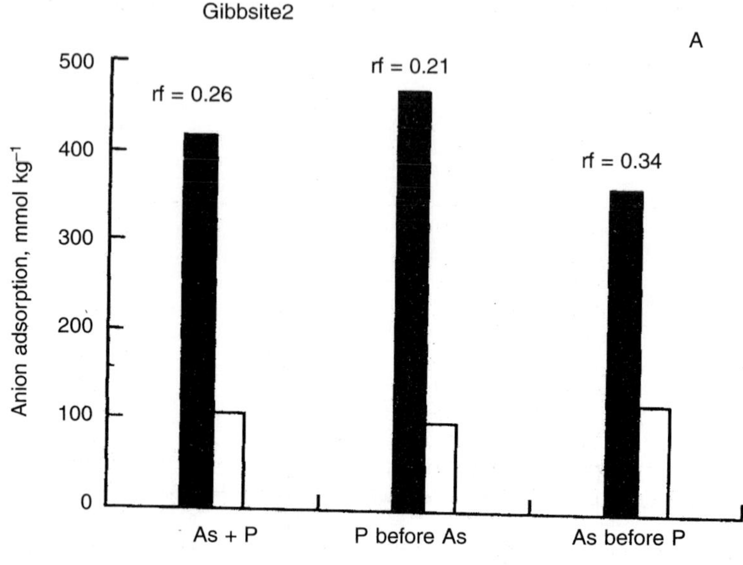

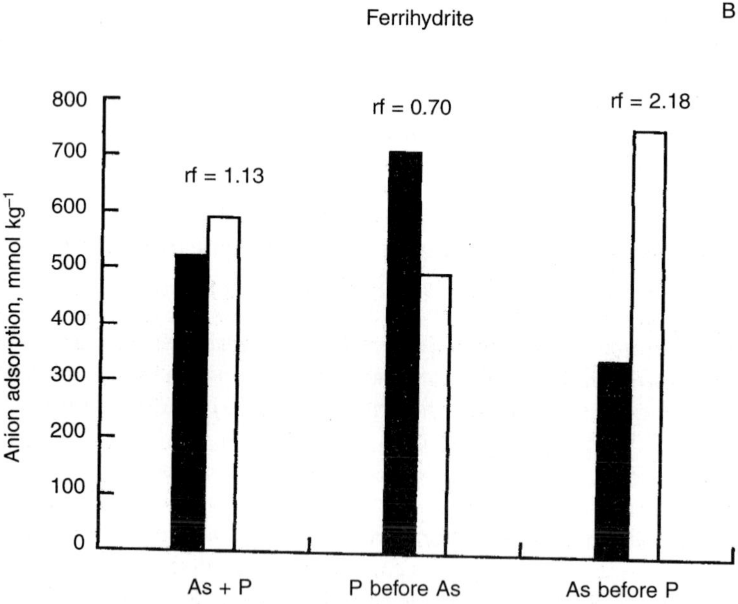

Fig. 15.6: Effect of the order of oxyanion addition on the adsorption of arsenate (As; □) and phosphate (P; ■) and the arsenate adsorbed/phosphate adsorbed molar ratio (*rf*) on gibbsite2 (A) and ferrihydrite (B) at pH 5.0. The oxyanions were added as a mixture in equimolar amounts at about 100% of their maximum surface coverage.

Table 15.4: Effect of contact time on the sorption of arsenate (As) and phosphate (P) on the Andisol at pH 6.0. The anions were added alone (1,000 mmol kg^{-1}) or as a mixture (initial As/P molar ratio of 1)

Contact time (h)	Anion added alone		Anion added as a mixture		**rf***
	As	P	As	P	
	mmol kg^{-1}				
0.02	104.4	226.4	38.4	154.8	**0.25**
0.08	219.2	308.8	58.4	200.4	**0.29**
0.17	242.0	368.9	76.0	212.0	**0.36**
0.25	285.6	429.1	82.0	247.9	**0.33**
0.5	317.6	464.0	96.1	277.2	**0.35**
1	344.8	502.2	120.1	302.0	**0.40**
3	394.6	574.1	141.4	337.1	**0.42**
24	557.2	751.8	255.2	498.0	**0.51**
48	585.0	803.0	n.d	n.d	**n.d**

* rf stands for adsorbed arsenate/adsorbed phosphate molar ratio.

Table 15.5: Arsenate (As) and phosphate (P) adsorption on goethite2 at different pH values (3.0–8.0), when 240 mmol of As or P were added per kg of goethite or when mixtures of 240 mmol of each ligand per kg were added as a mixture (*As + P* system), or when P was added before As (*P before As* system), or when As was added before P (*As before P* system) (from Liu et al., 2001)

pH	*As or P alone*		*As + P*			*P before As*			*As before P*		
	As	P	As	P	**rf**	As	P	**rf**	As	P	**rf**
						mmol kg^{-1}					
3.0	197.3	192.7	119.0	91.4	**1.30**	98.6	125.2	**0.76**	160.4	47.1	**3.40**
3.5	189.9	188.6	119.8	88.8	**1.35**	88.5	127.7	**0.69**	162.5	45.0	**3.61**
4.0	179.9	182.1	111.0	89.6	**1.24**	81.7	128.5	**0.64**	156.9	46.8	**3.35**
4.5	176.5	174.5	100.9	87.3	**1.16**	76.1	128.6	**0.59**	147.2	46.7	**3.15**
5.0	160.9	172.0	100.7	89.4	**1.13**	73.1	127.1	**0.57**	147.0	46.3	**3.17**
5.5	154.0	166.5	96.3	89.9	**1.07**	70.3	132.3	**0.53**	140.8	46.0	**3.06**
6.0	150.7	162.3	90.8	89.9	**1.00**	57.9	130.5	**0.44**	140.7	43.7	**3.21**
6.5	141.9	161.8	86.4	89.9	**0.96**	52.9	131.5	**0.44**	139.1	41.8	**3.32**
7.0	138.7	156.8	78.7	89.1	**0.88**	52.0	128.2	**0.40**	130.6	40.6	**3.22**
7.5	120.4	144.0	75.9	79.3	**0.96**	46.1	126.0	**0.37**	123.6	38.6	**3.20**
8.0	114.0	133.0	69.4	74.0	**0.94**	36.0	121.8	**0.29**	111.6	35.1	**3.18**

in the presence of P/ As adsorbed alone]) changed tremendously by modifying the sequence of oxyanion addition and was of 8% in *As before P* systems, 29% in *As + P* systems, and 40% in *P before As* systems. Contrarily, on gibbsite the efficiency of phosphate was much greater in all the systems than on ferrihydrite, but remained practically similar, ranging from 76% in *P before As* systems to 70% in *As before P* systems. Analogous results were found by Pierce and Moore (1982), who demonstrated that once arsenate was fixed on a poorly crystalline iron oxide, arsenate was not affected by the postaddition of phosphate and sulfate, but was influenced by the prior addition of phosphate and sulfate to the system.

3.4 Adsorption of Arsenate in the Presence of Increasing Concentrations of Phosphate

Adsorption of arsenate decreased in the presence of increasing amounts of phosphate. Figure 15.7 shows the amounts of arsenate adsorbed on the Andisol in the presence of increasing concentrations of phosphate (phosphate/arsenate molar ratio ranging from 0 to 2) at pH 4.0. The amount of arsenate added to the soil was 60% of its maximum surface coverage. Increasing concentrations of phosphate decreased adsorption of arsenate but even at an initial phosphate/arsenate molar ratio of 2, about 30% of arsenate initially added was adsorbed on the soil, indicating that large amounts of phosphate did not completely prevent arsenate fixation. Phosphate was more effective than organic ligands (citrate, malate, and oxalate) in suppressing arsenate adsorption (as discussed below). Violante and Pigna (2002) found that increasing concentrations of phosphate inhibited arsenate adsorption more on gibbsite than on goethite. In fact, at an initial phosphate/arsenate molar ratio of 1.5, 72% arsenate was adsorbed on goethite and only 54% on gibbsite, strengthening the observation that arsenate is adsorbed more strongly on Fe- than Al oxides. These authors also demonstrated that the effectiveness of increasing concentrations of

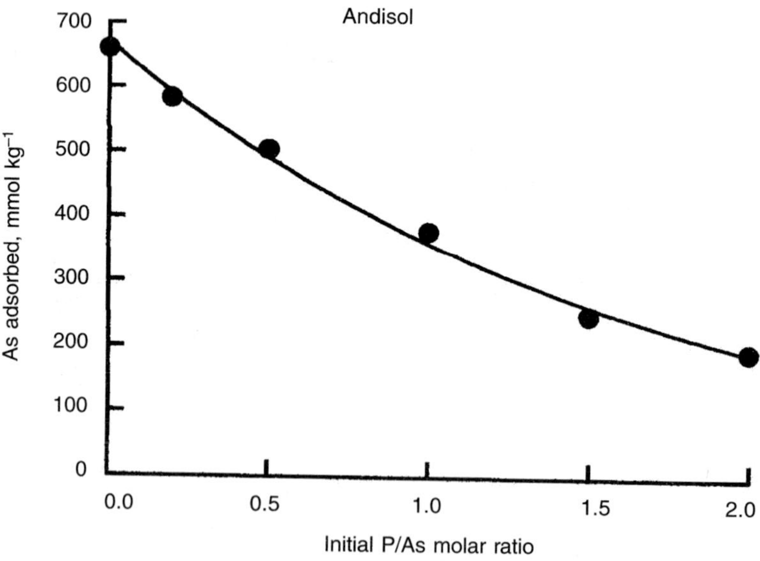

Fig. 15.7: Adsorption of arsenate (As) on the Andisol in the presence of increasing concentrations of phosphate (P) at pH 4.0. Arsenate was added at about 60% of its maximum surface coverage.

phosphate in preventing the adsorption of arsenate was greater in neutral and alkaline environments and by increasing the surface coverage of arsenate on the sorbents.

3.5 Desorption of Arsenate by Phosphate

Few studies to date have investigated in detail the desorption of arsenate from minerals and soils. Heavy addition of phosphate to arsenic polluted soils has been reported to displace approximatively 77% of the total arsenic in the soil, with arsenic being redistributed to lower depths in the soil profile (Woolson et al. 1973). Peryea (1991) found that desorption of arsenate was dependent on the soil type. Removal of arsenate from a volcanic soil was relatively poor due to the presence of allophanic minerals which have high anion-fixing capacities (Smith et al., 1998). Furthermore, Darland and Inskeep (1997) found that large amounts of phosphate did not desorb all of the applied arsenate, regardless of whether the arsenate was applied concurrently or prior to phosphate addition.

Liu et al. (2001) recently carried out experiments on the desorption of arsenate (and phosphate) by phosphate (arsenate) from a goethite. Figure 15.8 shows the desorption of arsenate and phosphate by equimolar amounts of phosphate and arsenate respectively as affected by pH. A total of 240 mmol of

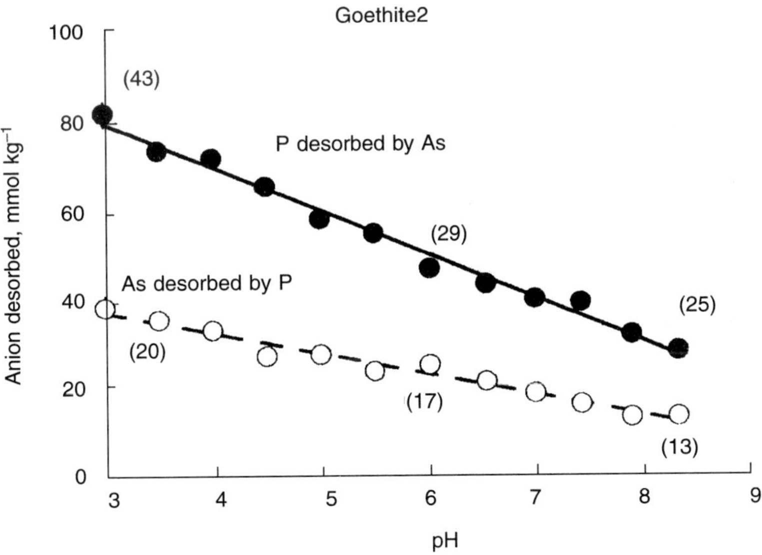

Fig. 15.8: Desorption of arsenate (phosphate) by phosphate (arsenate) from goethite2 at pH 3.0–8.5 (modified from Liu et al., 2001). Numbers in parentheses indicate the percentage of phosphate or arsenate adsorbed.

arsenate or phosphate were first added to 1 g goethite2 at different pH values (3.0–8.5). After 24 h the suspensions were centrifuged, washed, and 240 mmol of phosphate or arsenate respectively added to the solids, which had previously been resuspended at the same initial pH values. In the pH range studied, phosphate was desorbed by arsenate more than arsenate by phosphate, mainly in acidic systems. The percentage of phosphate desorbed at a given pH with respect to phosphate adsorbed on goethite actually decreased from 43% at pH 3.0 to 29% at pH 6.0 and 25% at pH 8.0. Contrarily, the amounts of arsenate desorbed by phosphate were relatively low and not strongly influenced by pH, ranging from 20% at pH 3.0 to 17% at pH 6.0 and 13% at pH 8.0.

The desorption of arsenate by phosphate from goethite was influenced by time (Liu et al., 2001; O'Reilly et al., 2001; Arai and Spark, 2002). Figure 15.9 shows that at pH 4.0 and after 3 h, only 10.5% arsenate was desorbed by phosphate versus 21.3% phosphate by arsenate. After 2 and 6–17 days, phosphate was able to desorb 27 and 32% arsenate, whereas arsenate replaced 50 and 55% phosphate. These data indicate that desorption of arsenate (or phosphate) by phosphate (or arsenate) was effective in the first few days of reaction.

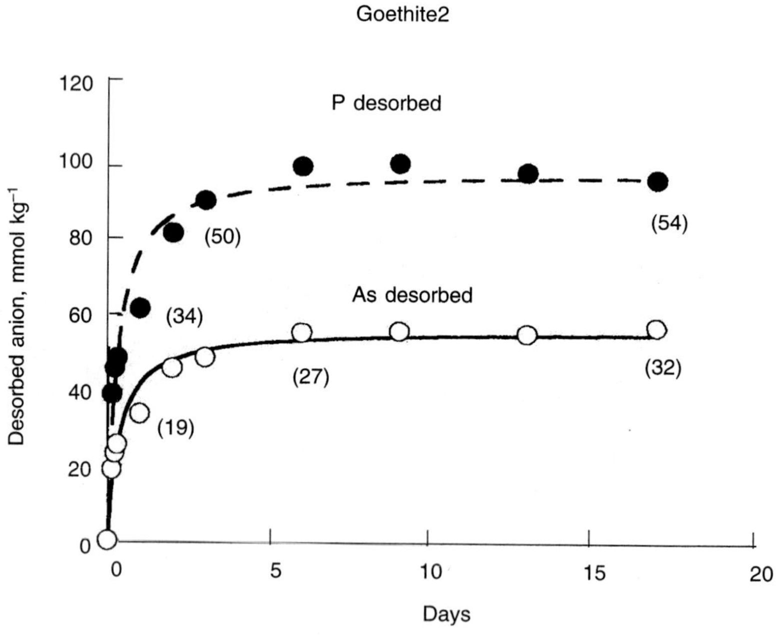

Fig. 15.9: Desorption of arsenate (by phosphate) and phosphate (by arsenate) from goethite2 as affected by different times of reactions (0.125–17 days) (modified from Liu et al., 2001). Numbers in parentheses indicate the percentage of phosphate or arsenate adsorbed.

Desorption kinetic studies carried out by O'Reilly et al. (2001) also showed that when aging was increased, there was no significant change in the amount of arsenate desorbed from goethite by phosphate. They found that desorption was initially quite rapid with > 35% of the total arsenate desorbed within 24 h at pH 6.0. After the initial rapid desorption, only a small amount of additional desorption occurred at longer times. A significant amount of arsenate remained bound to goethite after 5 months of desorption even though the phosphate added was three times greater than the initial arsenate sorptive solution. Similar results were obtained by Arai and Sparks (2002) using an Al-oxide as sorbent. According to these authors, surface transformations such as a rearrangement of surface complexes and / or a conversion of surface complexes into aluminum arsenate-like precipitates might be important chemical factors responsible for the decrease in As(V) reversibility with aging.

Residence time also has an influence on the desorption of arsenate. A low amount of arsenate added alone at pH 4.0 to Andisol (250 mmol kg^{-1}; 30% surface coverage of arsenate) was completely adsorbed within 1–2 h (Table 15.6). Experiments were carried out on the influence of large amounts of phosphate (P / As molar ratio = 3) on the desorption of arsenate. Phosphate was added from 1 h to 60 days after arsenate addition and the suspensions then kept for a further 24 h under agitation. The percentage of arsenate that remained adsorbed increased from 64% when phosphate was added after 1 h to 78% when phosphate was added after 48 h; it remained substantially constant (79–83%) when phosphate was added 3–60 days after arsenate (Table 15.6). A similar behavior was recorded when pyrolusite was used as the sorbent (50% surface coverage of arsenate (data not shown).

An explanation of these findings appears to stumble. Desorption kinetic studies seem to indicate that during the first 24–48 h, the molecular structure of arsenate sorbed on mineral surfaces reorganizes in part (Grossl et al., 1997), rendering desorption of arsenate ligands relatively more difficult. However, O'Reilly et al. (2001) employing EXAFS saw no clear difference in the bonding mechanism of arsenate onto goethite (mainly a bidentate binuclear bond forming

Table 15.6. Effect of residence time on adsorption of arsenate (As) on Andisol (250 mmol kg^{-1}) in the absence or presence of phosphate (P) at pH 4. 0. Phosphate was added 1–1,440 h after arsenate and the suspensions kept to react for a further 24 h

Contact Time (h)	As added alone	After P addition (P / As molar ratio 3)
	As sorbed mmol kg^{-1}	
1	245 (98)	161 (64.4)
5	250 (100)	189 (75.6)
24	250 (100)	193 (77.2)
240	250 (100)	205 (82.0)
1,440	250 (100)	203 (81.2)

* Data in parentheses indicate the percentage of arsenate which remains adsorbed.

between the arsenate and goethite surface) among samples reacted for different times over a period of 1 month.

Considerable evidence seems to demonstrate that the effectiveness of phosphate in desorbing arsenate is due to many concomitant factors, e.g. surface coverage of sorbed arsenate, pH, nature of sorbent, and amount of phosphate added (P/As molar ratio).

3.6 Adsorption/Desorption of Arsenate in the Presence of Sulfate and Silicate

Sulfate seems to have little influence in preventing arsenate adsorption on, or in removing it from soil minerals and soils, particularly at pH ≥ 6.0, even at an initial sulfate/arsenate molar ratio >> 1.0. We found that at an initial sulfate/arsenate molar ratio of 4–10, sulfate did not prevent but merely retarded arsenate adsorption on ferrihydrite. Figure 15.10 shows the kinetics of arsenate (500 mmol As kg^{-1}; 80% surface coverage) at pH 4.0 in the absence or presence of sulfate (sulfate/arsenate molar ratio = 4), when sulfate was added 24 h before arsenate. Whereas in the absence of sulfate arsenate was completely adsorbed within 5 h, in the presence of sulfate 70% of the arsenate added initially was adsorbed after 5 h, and only after 4 days was it completely adsorbed on the Fe-oxide.

The effect of sulfate in retarding arsenate adsorption is pH dependent. Xu et al. (1988) found that sulfate partially inhibited arsenate adsorption on alumina

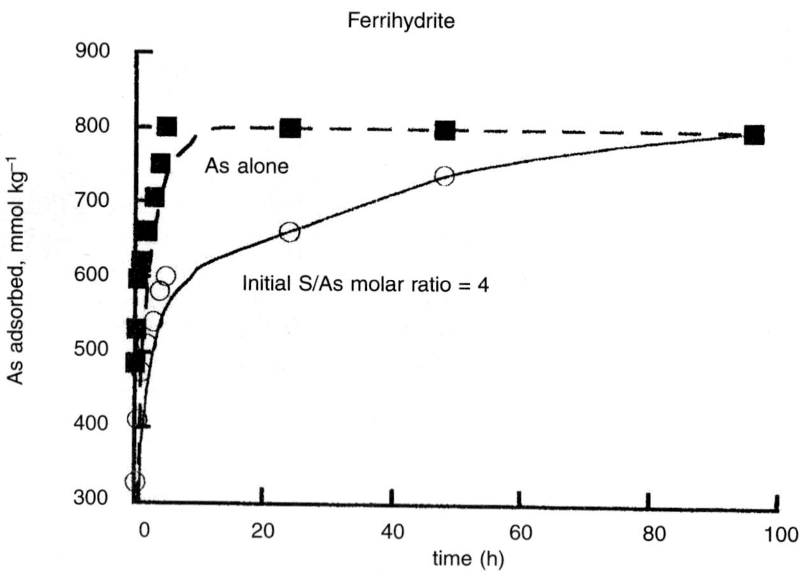

Fig. 15.10: Kinetics of arsenate (As) adsorption on ferrihydrite at pH 4.0 in the absence or presence of sulfate (S). Sulfate was added 24 h before As at initial sulfate/arsenate molar ratio of 4.0.

only in acidic systems (pH < 5.5), whereas at pH > 6.0 its influence was negligible. These authors observed that at pH < 7.0, sulfate decreased the adsorption of arsenate, but increasing the sulfate concentration from 20 to 80 mg L^{-1} had little influence on arsenate adsorption. Some experiments carried out in our laboratory confirmed this trend. Clifford and Ghurye (2002) showed equilibrium isotherms for adsorption of arsenate onto activated alumina in the absence or presence of very high concentrations of chloride and sulfate (15 meq L^{-1}) at pH 6.0. The isotherms indicated that at an equilibrium arsenate concentration of 1 mg L^{-1}, arsenic loading was reduced by 16% in the presence of chloride compared with 50% in the presence of sulfate.

The relatively stronger influence of sulfate to prevent arsenate (and phosphate; Pigna and Violante, 2003) sorption on variable charge minerals and soils at pH < 5.0 may be due to the formation of some inner-sphere complexes of sulfate ions on the surfaces of minerals. According to some authors (Peak et al., 2001), sulfate forms on variable charge minerals only outer-sphere surface complexes at pH 6.0 or above, and forms a mixture of outer-sphere and inner-sphere complexes below pH 6.0. Usually, the lower the pH, the greater the percentages of inner-sphere complexes formed by sulfate on the surfaces of metal oxides.

Some evidence seems to demonstrate that silicic acid inhibits the adsorption of arsenate more than sulfate (Violante; unpubl. data, 2003). Silicic acid (pK_1 = 9.77) is a ligand ubiquitous in natural systems and is strongly sorbed on variable charge minerals. However, information on the influence of silicic acid on arsenate adsorption is scant (Swendlund and Webster, 1999; Meng et al., 2000; Waltham and Eick, 2002). Clifford and Ghurye (2002) reported that silicic acid strongly reduced the sorption of arsenate on alumina at pH 7.5 and, furthermore, its influence decreased as pH decreased due to decreasing ionization of the silicic acid. According to Waltham and Eick (2002), only the rate and not the total quantity of arsenate was reduced in the presence of silicic acid. Furthermore, the rate of arsenate adsorption decreased as pH and silicic acid concentration increased.

3.7 Competitive Adsorption of Arsenate and Organic Ligands

Complexing organic acids are widely distributed in the environment and are low and variable in concentration. A wide variety and considerable amounts of organic compounds are released by plants, especially in the rhizosphere. Root exudates comprise both high and low molar-weight substances, such as mucilages, polysaccharides, proteins, carbohydrates, phenolics, phytosiderophores, amino acids and so on. Rhizosphere carbon flow has been estimated to account for a major fraction, up to 40%, of plant primary production. The population number and dynamics of microorganisms in the rhizosphere are significantly greater than in bulk soil (Huang and Germida, 2002). Microorganisms also release many biomolecules so that the amounts of biomolecules in the rhizosphere is particularly high. Acetic, butyric, oxalic, malic, tartaric, citric, propionic and succinic acids are the most abundant aliphatic acids present in the rhizosphere (Marschner, 1995; Jones, 1998; Huang and Germida, 2002). Biomolecules produced by plants and microorganisms may strongly influence the mobility of both nutrients and pollutants at soil-root interface (Marschner, 1995; Huang et al. 2002; Violante et al., 2002a,b). Xu et al. (1988) observed a reduction in arsenate adsorption on alumina in the presence of fulvic acid under varying fulvic acid to adsorbent concentration ratios.

We found that LMWOAs (malic, oxalic, or citric acid) inhibited arsenate adsorption particularly at organic ligand/arsenate molar ratios > 1 (Fig. 15.11). The efficiency of LMWOAs in preventing arsenate sorption on Andisol at pH 4.0 was as follows: citric acid > malic acid > oxalic acid >> succinic acid (not shown). Organic ligand/As molar ratio of 4 oxalic, malic, and citric acid inhibited arsenate adsorption respectively of 15, 40, and 55% (Fig. 15.11).

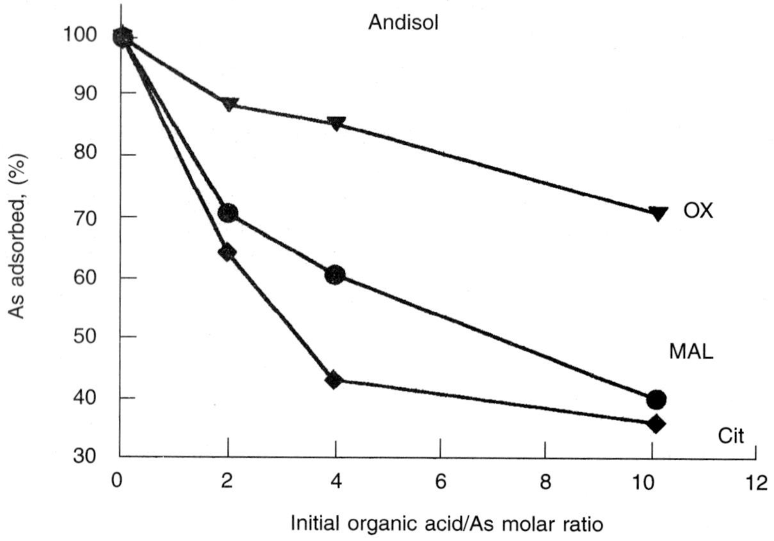

Fig. 15.11: Amounts (%) of arsenate (As) adsorbed on the Andisol at pH 4.0 in the presence of increasing concentrations of oxalate (OX) malate (Mal), and citrate (Cit) (416 mmol As per kg soil sample were added initially).

This behavior may be due to the different affinity of the LMWOAs studied for Al. In fact, the citric-, malic- and oxalic acid-Al complexes (at stoichiometry of the complex Al-ligand ratio 1:1) have stability constants of 7.9, 6.1, and 6.0 respectively (Jones, 1998). However, at a LMWOA/arsenate molar ratio > 1 (Fig. 15.11) partial dissolution of Al from the surfaces of the allophanic materials of the Andisol may also be responsible for the great decrease of arsenate sorption.

Organic ligands reduced arsenate adsorption on Andisol more when added before than when added as a mixture with arsenate or after arsenate (data not shown).

The efficiency of organic anions in preventing arsenate adsorption was influenced by the nature and surface properties of variable charge minerals and soils. Figure 15.12 shows the adsorption of arsenate on goethite, pyrolusite, gibbsite, and Andisol at pH 4.0 and in the presence of increasing concentrations of malic acid (malic acid / As molar ratio ranging from 0 to 10). Malate did not inhibit or very poorly prevented arsenate adsorption on goethite and pyrolusite, even at an initial malic acid / arsenate molar ratio of 10. Vice versa, the organic ligand strongly inhibited arsenate adsorption on gibbsite and Andisol, indicating that because arsenate shows a greater affinity for sorbents containing Fe or Mn, it was not easily desorbed even in the presence of large concentrations of malate. In these experiments dissolution of Al from noncrystalline materials present in gibbsite1 and from allophane in the Andisol influenced the reduction of arsenate fixation. Grafe et al. (2001) found that arsenate adsorption on goethite was reduced by humic and fulvic acid, but not by citric acid.

We also found that LMWOAs (e.g. malic acid) prevented adsorption of arsenate more on gibbsite than on ferrihydrite, whereas the opposite was true for phosphate (Fig. 15.13). Strongly chelating LMWOAs also showed a strong influence on the kinetics of adsorption of arsenate onto minerals and soils. Figure 15.14 shows the kinetics of adsorption of arsenate on gibbsite at pH 5.0 in the absence or presence of malate (initial malate / arsenate molar ratio of 1).

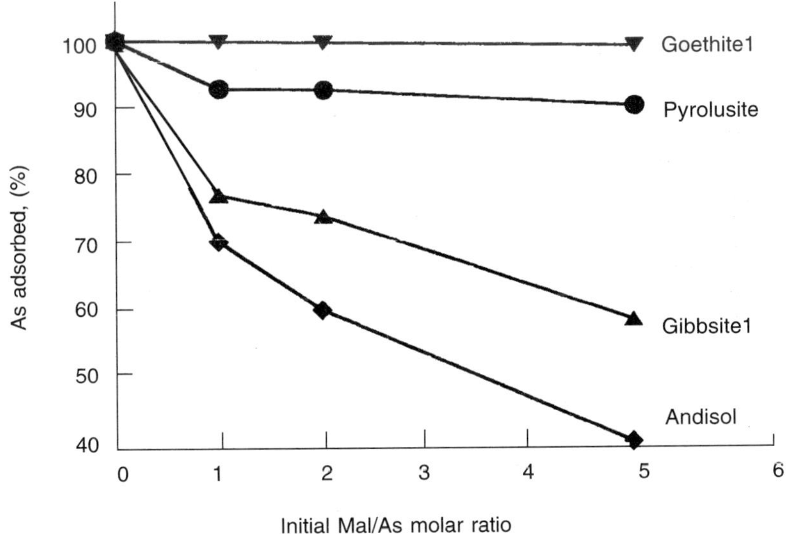

Fig. 15.12. Amounts (%) of arsenate (As) adsorbed on goethite 1, pyrolusite, gibbsite1, and Andisol at pH 4.0 in the presence of increasing concentrations of malate (150, 17, 150, and 416 mmol As per kg goethite1, pyrolusite, gibbsite1, and Andisol, respectively, were added initially).

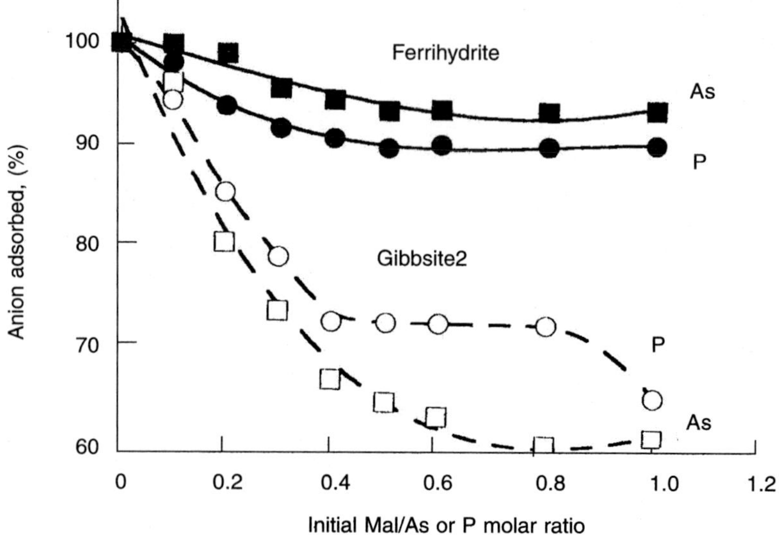

Fig. 15.13: Arsenate (As) and phosphate (P) adsorption onto ferrihydrite and gibbsite2 at pH 5.0 and in the presence of increasing concentrations of malate (Mal). Inorganic ligands were added at their maximum surface coverage.

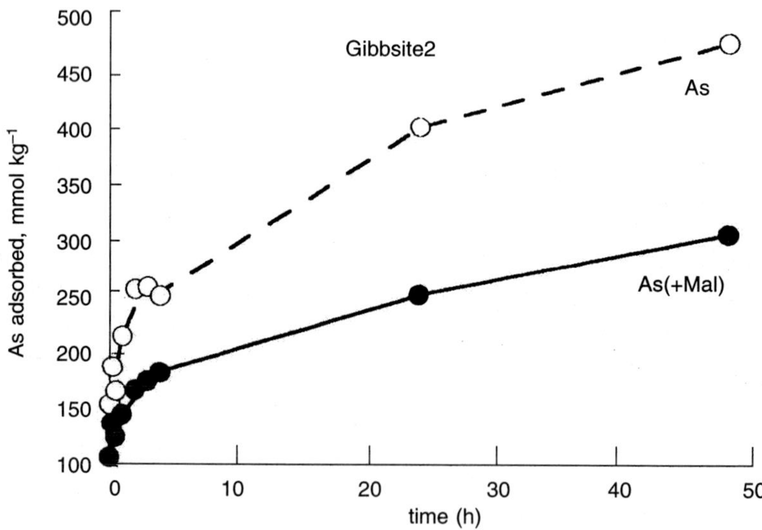

Fig. 15.14: Kinetics of arsenate (As) adsorption on gibbsite2 at pH 5.0 in the absence or presence of malic acid (Mal). Arsenate was added at its maximum surface coverage; initial Mal/As molar ratio of 1.

3.8 Desorption of Arsenate and Phosphate by Oxalate

The influence of LMWOAs on the desorption of arsenate from clay minerals and soils has received scant attention. Liu et al. (2001) recently studied the effect of oxalic acid (OX) on the desorption of arsenate and phosphate from goethite at different pH values (Fig. 15.15). Arsenate and phosphate were previously added as a mixture to the iron oxide; after 24 h, the samples were centrifuged, washed, resuspended at the same initial pH, and oxalate added (OX/As + P molar ratio of 1).

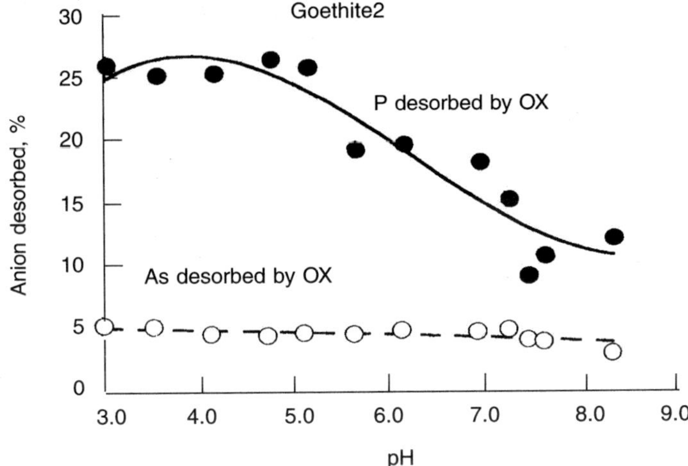

Fig. 15.15: Desorption (%) of arsenate (As) and phosphate (P) from goethite 2 at pH 3.0–8.5 in the presence of oxalic acid (initial OX/As+P molar ratio of 1). Inorganic anions were near their maximum surface coverage.

Negligible quantities of arsenate (5.3–3.9%) were desorbed by oxalate in the entire range of pH studied whereas relatively greater percentages of the phosphate (26–10%) were replaced by the organic ligand, mainly in acid systems, although slightly higher amounts of arsenate than of phosphate were adsorbed at pH < 6.0 in the *As* + *P* systems (Table 15.5). The greater efficiency of oxalate in replacing phosphate from the goethite surfaces at pH < 6.0 than in neutral or alkaline environments is not surprising. Previous studies (Violante et al., 1991; Violante and Gianfreda 1993, 2000) have demonstrated that organic ligands prevent phosphate adsorption only in acidic environments.

The effect of LMWOAs on the removal of heavy metals from soil components deserves more attention for understanding the dynamics of metals and metalloids, including arsenic, in the rhizosphere.

4 CONCLUSIONS

Many factors influence the adsorption-desorption processes of arsenate in soil environments (pH, nature and concentration of inorganic and organic anions, chemical and mineralogical properties of the sorbents, residence time, and surface coverage). Our research demonstrated that phosphate (mainly) and strongly chelating LMWOAs inhibit arsenate adsorption on clay minerals and soil samples, whereas sulfate and silicate do not prevent arsenate retention. Phyllosilicates particularly rich in iron, and Fe- and Mn-oxides were more effective in adsorbing arsenate than phosphate, whereas minerals richer in aluminum were more effective in adsorbing phosphate than arsenate. However, competition between arsenate and phosphate was affected by pH, residence time, and order of oxyanion addition. Phosphate affected arsenate adsorption more in near neutral and alkaline than in acidic environments, whereas LMWOAs have an influence only at low pH values. However, even large amounts of phosphate or LMWOAs do not completely prevent arsenate adsorption.

Desorption of arsenate from clay minerals and soils by phosphate is due to many concomitant factors such as pH, nature of the sorbent, surface coverage of adsorbed arsenate, and residence time. To be specific, the percentages of arsenate that remained adsorbed on a variable charge soil (Andisol) increased from 64% when a large amount of phosphate was added after 1 h to 83% when phosphate was added after 60 days. Inorganic (phosphate, silicate, and sulfate) and organic (malate, citrate, oxalate) ligands have a strong influence on the kinetics of adsorption of arsenate.

To predict the fate, mobility, and potential toxicity of arsenic in soil environments, studies on the factors which influence the adsorption-desorption processes appear of great importance. Research on the influence of biomolecules, usually released in the rhizosphere as root exudates, on the adsorption/desorption of arsenate on/from soil colloids are particularly promising for understanding the factors which affect the mobility and toxicity of arsenate at the soil-root interface.

Aknowledgments

This study was supported by the Italian Research Program of National Interest (PRIN), year 2002, DISSPA Number 0041.

References

Arai Y. and Sparks D.L. 2002. Residence time effects on arsenate surface speciation at the aluminum oxide-water interface. *Soil Sci.* 167: 303–314.

Barrow N.J. 1974. On the displacement of adsorbed anions from soil: 2. Displacement of phosphate by arsenate. *Soil Sci.* 117: 28–33.

Berg M., Tran H.C., Nguyen T.C., Pham H.V., Schertenleib R., and Giger W. 2001. Arsenic contamination of groundwater and drinking water in Vietnam: A human health threat. *Environ. Sci. Tech.* 35: 2621–2626.

Chao T.T., Haward M.E., and Fang S.C. 1964. Iron or aluminum coatings in relation to sulfate adsorption characteristics of soils. *Soil. Sci. Soc. Amer. Proc.* 28: 632–635.

Clifford D.A. and Ghurye G.L. 2002. Metal-oxide adsorption, ion exchange, and coagulation-microfiltration for arsenic removal from water. In: *Environmental Chemistry of Arsenic.* W.T. Frankenberger, Jr. (ed.). Marcel Dekker, New York, NY, pp. 217–245.

Colombo C. and Violante A. 1996. Effect of time and temperature on the chemical composition and crystallization of mixed iron and aluminum species. *Clays Clay Miner.* 45: 113–120.

Colombo C., Terribile F., Monti G.M., and Violante A. 1996. Red Mediterranean soils: impact of the land use. *S. It. E. Atti.* 17: 229–233.

Cornell R.M. and Schindler P.W. 1980. Infrared study of the adsorption of hydroxy-carboxylic acids on α-FeOOH and amorphous Fe(Ill)-hydroxide. *Colloid Polymer. Sci.* 258: 1171–1175.

Cornell R.M. and Schwertmann U. 1996. *The Iron Oxides. Structure, Properties, Reactions and Uses.* VCH Publ., New York, NY, 573 pp.

Darland J.E. and Inskeep W.P. 1997. Effects of pH and phosphate competition on the transport of arsenate. *J. Envir. Qual.* 26: 1133–1139.

Davenport J.R. and Peryea F.J. 1991. Phosphate fertilizers influence leaching of lead and arsenic in a soil contaminated with lead arsenate. *Water, Air, Soil Pollut.* 57–58: 101–110.

De Cristofaro A., He J.Z., Zhou D.H., and Violante A. 2000. Adsorption of phosphate and tartrate on hydroxy-aluminum-oxalate precipitates. *Soil Sci. Soc. Amer. J.* 64: 1347–1355.

Eltanawy I.M., and Arnold P.W. 1973. Reappraisal of ethylene glycol monoethyl ether (EGME) method for surface area estimation of clays. *J. Soil Sci.* 24: 232–238.

Fendorf S., Eich M.J., Grossl P., and Sparks D.L. 1997. Arsenate and chromate retention mechanisms on goethite. 1. Surface structure. *Environ. Sci. Tech.* 31: 315–320.

Fordham A.W. and Norrish K. 1979. Arsenate-73 uptake by component of several acidic soils and its implications for phosphate retention. *Austr. J. Soil Res.* 17: 307–316.

Fordham A.W. and Norrish K. 1983. The nature of soil particles particularly those reacting with arsenate in a series of chemically treated samples. *Austr. J. Soil Res.* 21: 455–477.

Francesconi K.A. and Kuehnelt D. 2002. Arsenic compounds in the environment. In: *Environmental Chemistry of Arsenic.* W.T. Frankenberger, Jr. (ed.). Marcel Dekker, New York, NY, pp. 51–94.

Frankenberger W.T. Jr. (ed.). 2002. *Environmental Chemistry of Arsenic.* Marcel Dekker, New York, NY, 391 pp.

Goldberg S. 2002. Competitive adsorption of arsenate and arsenite on oxides and clay minerals. *Soil Sci. Soc. Amer. J.* 66: 413–421.

Goldberg S. and Glaubig R.A. 1988. Anion sorption on a calcareous, montmorillonitic soil—Arsenic. *Soil Sci. Soc. Amer. J.* 52: 1297–1300.

Goldberg S., Davis J.A., and Hem J.D. 1996. The surface chemistry of aluminum oxides and hydroxides. In: *The Environmental Chemistry of Aluminum.* G. Sposito (ed.). Lewis Publishers, Boca Raton, FL, p. 271 (2nd ed.).

Grafe M., Eick M.J., and Grossl P.R. 2001. Adsorption of arsenate (V) and arsenite (III) on goethite in the presence and absence of dissolved organic carbon. *Soil Sci. Soc. Amer. J.* 65: 1680–1687.

Grossl P.R., Eich M.J., Sparks D.L., Goldberg S., and Ainsworth C.C. 1997. Arsenate and chromate retention mechanisms on goethite, 2. Kinetic evaluation using a pressure-jump relaxation technique. *Environ. Sci. Tech.* 31: 321–326.

He J.Z., De Cristofaro A., and Violante A. 1999. Comparison of adsorption of phosphate, tartrate, and oxalate on hydroxy aluminum montmorillonite complexes. *Clays Clay Miner.* 47: 226–233.

Hingston F.J., Posner A.M., and Quirk J.P. 1971. Competitive adsorption of negatively charged ligands on oxide surfaces. *Discuss. Faraday Soc.* 52: 334–342.

Hsia T.H., Lo S.L., Lin C.F., and Lee D.Y. 1994. Characterization of arsenate adsorption on hydrous iron oxide using chemical and physical methods. *Colloids Surf. A: Physicochem. Eng. Aspects* 85: 1–7.

Huang P.M. and Violante A. 1986. Influence of organic acids on crystallization and surface properties of precipitation products of aluminum. In: *Interactions of Soil Minerals with Natural Organics and Microbes.* P.M. Huang and M. Schnitzer (eds.). SSSA Spec. Publ. 17, Madison, WI, pp. 159–221.

Huang P.M. and Germida J.J. 2002. Chemical and biochemical processes in the rhizosphere: metal pollutants. In: *Interaction between Soil Particles and Microorganisms: Impact on the Terrestrial Ecosystem.* P.M. Huang, J.M. Bollag, and N. Senesi (eds.). John Wiley & Sons, New York, NY, pp. 381–438.

Huang P.M., Bollag J.-M., and Senesi N. (eds.). 2002. *Interaction between Soil Particles and Microorganisms: Impact on the Terrestrial Ecosystem.* John Wiley & Sons, New York, NY, 566 pp.

Jones D.L. 1998. Organic acids in the rhizosphere: a critical review. *Plant Soil* 205: 25–44.

Kafkafi V., Bar-Yosef B., Rosemberg R., and Sposito G. 1988. Phosporus adsorption by kaolinite and montmorillonite. II. Organic anion competition. *Soil. Sci. Soc. Amer. J.* 52: 1585–1589.

Liu F., He Z., Colombo C., and Violante A. 1999. Competitive adsorption of sulfate and oxalate on goethite in the absence or presence of phosphate. *Soil Sci.* 164: 180–189.

Liu F., De Cristofaro A., and Violante A. 2001. Effect of pH phosphate and oxalate on the adsorption/desorption of arsenate on/from goethite. *Soil Sci.* 166: 197–208.

Livesey N.T. and Huang P.M. 1981. Adsorption of arsenate by soils and its relation to selected chemical properties and anions. *Soil Sci.* 131: 88–94 .

Lumsdon D.G., Fraser A.R., Russell J.D., and Livesey N.T. 1984. New infrared band assignments for the arsenate ion adsorbed on synthetic goethite (α-FeOOH). *J. Soil Sci.* 35: 381–386.

Manning B.A. and Goldberg S. 1996. Modeling competitive adsorption of arsenate with phosphate and molybdate on oxide minerals. *Soil Sci. Soc. Amer. J.* 60: 121–131.

Marschner H. 1995. *Mineral Nutrition of Higher Plants.* Acad. Press Ltd., London, 889 pp.

Matera V., and Le Hécho L. 2001. Arsenic behavior in contaminated soils: mobility and speciation. In: *Heavy Metals Release in Soil.* H.M. Selim and D.L. Spark (eds.). Lewis Publ., Boca Raton, FL, pp. 207–235.

McKenzie R.M. 1981. The surface charge on manganese dioxides. *Austr. J. Soil Res.* 19: 41–50.

Melamed R., Jurinak J.J., and Dudley L.M. 1995. Effect of adsorbed phosphate on transport of arsenate through an oxisol. *Soil Sci. Soc. Amer. J.* 59: 1289–1294.

Meng X., Bang S., and Korfiatis G. 2000. Effects of silicate, sulfate, and carbonate on arsenic removal by ferric chloride. *Water Res.* 34: 1255–1261.

Mora M.L. and Canales J. 1995. Interactions of humic substances with allophanic compounds. *Commun. Soil Sci. Plant Anal.* 26: 2805–2817.

Myneni S.C., Traina S.J., Logan T.J., and Waychunas G.A. 1997. Oxyanion behavior in alkaline environments: sorption and desorption of arsenate in ettringite. *Environ. Sci. Tech.* 31: 1761–1768.

Nriagu J.O. 1994. *Arsenic in the Environment,* Part I. Adv. Environ. Sci. Technol. 26. John Wiley & Sons, New York, NY.

Oades J.M. 1989. An introduction to organic matter in mineral soils. In: *Minerals in Soil Environments.* J.B. Dixon and S.B. Weed (eds.). Soil Sci. Soc. Amer., Madison, WI (USA), pp. 89–159.

O'Reilly S.E., Strawn D.G., and Sparks D.L. 2001. Residence time effects on arsenate adsorption/desorption mechanisms on goethite. *Soil Sci. Soc. Amer. J.* 65: 67–77.

Parfitt R.L. 1980. Chemical properties of variable charge soils. In: *Soils with Variable Charge.* B.K.G. Theng (ed.). New Zealand Soc. Soil Sci., Lower Hutt, NZ, pp. 167–194.

Parfitt R.L. and Smart, St. C. 1978. The mechanism of sulfate adsorption on iron oxides. *Soil Sci. Soc. Amer. J.* 42: 48–50.

Pasricha N.S. and Fox R.I. 1993. Plant nutrient sulfur in the tropics and subtropics. *Adv. Agron.* 50: 209–269.

Peak D., Elzinga E.J., and Sparks D.L. 2001. Understanding sulfate adsorption mechanisms of iron(III) oxides and hydroxides: Results from ATR-FTIR Spectroscopy. In: *Heavy Metals Release in Soil.* H.M. Selim and D.L. Sparks (eds.). Lewis Publ., Boca Raton, FL, pp. 167–190.

Peryea F.J. 1991. Phosphate induced release of arsenic from soils contaminated with lead arsenate. *Soil Sci. Soc. Amer. J.* 55: 1301–1306.

Pierce M.L. and Moore C.B. 1982. Adsorption of arsenite and arsenate on amorphous iron hydroxide. *Water Res.* 16:1247–1253.

Pigna M. and Violante A. 2003. Adsorption of sulfate and phosphate on Andisol. *Commun. Soil Sci. Plant Analysis* 34: 2099–2113.

Raven K., Jain A., and Loeppert R.H. 1998. Arsenite and arsenate adsorption on ferrihydrite: kinetics, equilibrium, and adsorption envelopes. *Environ. Sci. Tech.* 32: 344–349.

Roy W.R., Hassett J.J., and Griffin R.A. 1986. Competitive coefficient for the adsorption of arsenate, molybdate, and phosphate mixtures by soils. *Soil Sci. Soc. Amer. J.* 50: 1176–1182.

Sadiq M. 1997. Arsenic chemistry in soils: an overview of thermodynamic predictions and field observations. *Water, Air, Soil Pollut.* 93: 117–136.

Sakurai K., Ohdate Y., and Kyuma K. 1988. Comparison of salt titration and potentiometric titration methods for the determination of zero point of charge. *Soil Sci. Plant Nutr.* 34: 171–182.

Sheppard S.C. 1992. Summary of phytotoxic levels of soil arsenic. *Water, Air, Soil Pollut.* 64, 539–550.

Smith E., Naidu R., and Alston A.M. 1998. Arsenic in the soil environment: a review. *Adv. Agron.* 64: 149–195.

Sparks D.L. 1999. Kinetic and mechanisms of chemical reactions at the soil mineral / water interface. In: *Soil Physical Chemistry*. D.L. Sparks (ed.). LLC, Boca Raton, FL, pp. 135–191.

Sun X. and Doner H.E. 1996. An investigation of arsenate and arsenite bonding structures on goethite by FTIR. *Soil Sci.* 161: 865–872.

Swedlund P.J. and Webster J.G. 1999. Adsorption and polymerization of silicic acid on ferrihydrite and its effect on arsenic adsorption. *Water Res.* 33: 3413–3422.

Tabatabai M.A. 1982. Sulfur. In: *Methods of Soil Analysis*, Part 2. A.L. Page and R.H. Miller (eds.). Agron. Monograph no. 9. ASA, Madison, WI, pp. 501–583.

Vacca A., Adamo P., Pigna M., and Violante P. 2003. Properties and classification of selected soils from the Roccamonfina volcano, Central-Southern Italy. *Soil Sci. Soc. Amer. J.* 67: 198–207.

Violante A. and Gianfreda L. 1993. Competition in adsorption between phosphate and oxalate on an aluminum hydroxide montmorillonite complex. *Soil Sci. Soc. Amer. J.* 57: 1235–1241.

Violante A. and Gianfreda L. 2000. Role of bio-molecules in the formation and reactivity toward nutrients and organics of variable charge minerals and organomineral complexes in soil enviroment. In: *Soil Biochemistry*, J.M. Bollag and G. Stotsky (eds.). Marcell Dekker, New York, NY, vol. 10, pp. 207–270.

Violante A. and Pigna M. 2002. Competitive sorption of arsenate and phosphate on different clay minerals and soils. *Soil Sci. Soc. Amer. J.* 66: 1788–1796.

Violante A., Colombo C., and Buondonno A. 1991. Competitive adsorption of phosphate and oxalate by aluminum oxides. *Soil Sci. Soc. Amer. J.* 55: 65–70.

Violante A., Huang P.M., Bollag J-M., and Gianfreda L. 2002a. *Soil Mineral-Organic Matter-Microorganism Interaction and Ecosystem Health. Dynamics, Mobility and Pollutants and Nutrients.* Developments in Soil Science 28A. Elsevier, Amsterdam, Netherlands, 459 pp.

Violante A., Huang P.M., Bollag J-M., and Gianfreda L. 2002b. *Soil Mineral-Organic Matter-Microorganism Interaction and Ecosystem Health. Ecological Significance of the Interactions among Clay Minerals, Organic Matter and Soil Biota.* Developments in Soil Science 28B. Elsevier, Amsterdam, Netherlands, 434 pp.

Xu H., Allard B., and Grimvall A. 1988. Influence of pH and organic substance on the adsorption of As(V) on geologic materials. *Water, Air, Soil Pollut.* 40: 293–305.

Waltham A.C. and Eick M.J. 2002. Kinetics of arsenic adsorption on goethite in the presence of sorbed silicic acid. *Soil Sci. Soc. Amer. J.* 66: 818–825.

Woolson E.A., Axely J.H., and Kearney P.C. 1973. The chemistry and phytotoxicity of arsenic in soils: Effect of time and phosphorus. *Soil Sci. Amer. Proc.* 37: 254–259.

16

Sequential Fractionation of Cu, Zn, and Cd in Soils in Absence and Presence of Rhizobia

Q. Huang*, W. Chen, *and* **X. Guo**

Abstract

Red soil (Ultisol) and Cinnamon soil (Alfisol) were collected from Chenzhou of Hunan Province and Gongyi of Henan Province, respectively. Soils were treated with $Cu(NO_3)_2$, $Zn(NO_3)_2$, or $Cd(NO_3)_2$ for 2 weeks. Bacteria of the *Rhizobium fredii* strain HN01 were inoculated into the two soils polluted with one of the three heavy metals. A sequential extraction method was employed to investigate the forms of copper, zinc, and cadmium (Cu, Zn, and Cd) in the examined soils in the absence and presence of rhizobia. The results showed that the total amount of solid-bound Zn decreased by 10% after the inoculation. The decrease for the amount of Zn associated with carbonate, Mn oxides, and organic matter fractions measured between 9 and 26%. No significant change was observed for the total amount of Zn combined with the solid phase of Red soil in the presence of rhizobia. However, the amount of specifically adsorbed and Mn oxide-bound Zn decreased, while the amount of exchangeable Zn increased.

Inoculation of rhizobia depressed the release of Cu into the soil solution and increased the total amount of Cu associated with the solid phase in Cinnamon soil. Increase in the amount of exchangeable Cu and Cu in fractions of carbonate, Mn oxides, and organic matter ranged from 20 to 54%. There was no significant change in the level of Cd in solution in either soils after rhizobia inoculation.

The amount of Cd in the exchangeable and organic matter fractions increased 22% and 11%, while that in the fractions of carbonate and Mn

Corresponding author: Dr. Qiaoyun Huang, College of Resources and Environment, Huazhong Agricultural University, Wuhan 430070, P.R. China. E-mail: qyhuang@mail.hzau.edu.cn oxides decreased by 14% and 29%, respectively. The different influence of rhizobia on the distribution of the three heavy metals in two soils was mainly ascribed to the growth status and pH changes exerted by the metabolites of rhizobia. These data are fundamental for further research on the chemical behavior and biogeochemical cycle of heavy metals in the environment.

1 INTRODUCTION

Heavy metals are important contaminants in soil environments and human activities have caused substantial increases in their release into soils (Selim and Sparks, 2001). Based on primary accumulation mechanisms in sediments, heavy metals can be classified into several categories: water-soluble, exchangeable, carbonate-bound, iron or Mn oxide-bound, organic matter-bound, and residual (Alef and Nannipieri, 1995; Krishnamurti et al., 1997; Ma and Rao, 1997). These forms are in kinetic equilibrium and determine the mobility and bioavailability of these elements. Quite a number of investigators have employed a sequential extraction method to differentiate the chemical forms of heavy metals in soil and sediments (Tessier et al., 1979; Ramos et al., 1994; Alef and Nannipieri, 1995; Krishnamurti et al., 1997; Ma and Rao, 1997). Many studies have been carried out on the speciation of heavy metals in different soils, the influence of organic materials on the forms of heavy metals, and the relationship between speciation of heavy metals and phytoavailability (Xian, 1987, 1989; Krishnamurti et al., 1997; Wang et al., 1999, 2000).

Soil microorganisms such as bacteria, which are the most active organic fractions in soil, are surface-charged and can secrete various organic compounds such as low-molecular-weight organic acids, carbohydrates, and enzymes. The huge numbers of bacteria as well as their large surface-area-to-volume ratio make them efficient at sorbing toxic heavy metals in soil environments (Beveridge and Schultze-Lam, 1995). The influence of bacteria on the mobility of heavy metals in soils has been recognized by soil, microbiological, and environmental scientists (Chanmugathas and Bollag, 1988; Berthelin et al., 1995; Guo et al., 2002).

Much attention has been paid recently by soil and environmental scientists to the bioremediation of heavy metal-polluted soils (Adriano et al., 1999; Gadd, 2000). In circumstances of equal quantity, bacteria could adsorb greater amounts of heavy metals than could inorganic soil components such as montmorillonite, kaolinite, or vermiculite (Kurek et al., 1982; Ledin et al., 1996). Huang et al. (2000) reported that the adsorption of Cu and Cd by soil colloids and minerals was significantly enhanced in the presence of rhizobia. However, little information is available regarding how microorganisms affect the speciation and distribution of heavy metals in soils. Studies in this respect have received scant attention.

Rhizobia are ubiquitous soil bacteria in the rhizosphere. They are symbiotic nitrogen fixation bacteria. Copper and zinc are the essential micronutrients for plants and can reach toxic levels in some polluted soils. Cadmium is one of the most toxic heavy metals in soil and associated environments. The present work investigate the influence of bacterial inoculation on the distribution of Cu, Zn, and Cd in two typical soils of China. These studies are fundamental for the understanding of the chemical behavior of heavy metals in soil environments, which is the basis for the decontamination and bioremediation of soils polluted by heavy metals.

2 MATERIALS AND METHODS

2.1 Preparation of Cu-, Zn- and Cd-polluted Soils

Red soil (Ultisol) and cinnamon soil (Alfisol) were sampled from Chenzhou, Hunan and Gongyi, Henan provinces, respectively. The pertinent properties of the two soils are listed in Table 16.1. Air-dried soil samples were ground to pass through a 1 mm sieve. In a 300-ml centrifuge tube, 100 mg of soil were mixed with 200 ml of 5 mM $Cu(NO_3)_2$, 5 mM $Zn(NO_3)_2$, or 0.4 mM $Cd(NO_3)_2$ solution. The suspension was shaken (150 rpm) at 25°C for 24 h and centrifuged at 5,000 rpm for 10 min. The soil was then washed with deionized water three times and the supernatant and washings removed. The residue was air dried and ground to pass through a 0.25-mm sieve screen.

Table 16.1: Selected properties of soils examined

Soil type	Location	Depth (cm)	pH (H_2O)	Clay (g kg^{-1})	OM (g kg^{-1})	CEC (cmol kg^{-1})	Clay mineral association*
Red	Chenzhou, Hunan	0–20	4.0	48	26.7	10.5	Kao., ill., 1.4 nm int.
Cinnamon	Gongyi, Henan	0–20	7.0	19.4	12.6	13.1	Ill., mont., chl.

*Kao.—kaolinite, ill.—illite, 1.4 nm int.—1.4 nm intergrade mineral, mont.—montmorillonite, chl.—chlorite.

2.2 Selection of Rhizobial Strain

Rhizobia strains were provided by the Laboratory of Agricultural Microbiology, Huazhong Agricultural University. Selection of the test strains was described in an earlier work by Huang and colleagues (2000).

2.3 Inoculation of Rhizobia into Soil

Two grams of Cu-, Zn- or Cd-treated soil were mixed with 100 ml YMA medium in a 250 ml Erlenmeyer flask and steam sterilized twice. The soil was then inoculated with rhizobium strain HN01 and agitated at 200 rpm and 28°C for 2 weeks. After incubation, the number of viable bacterial cells was recorded by plate count and the pH of the soil solution determined. The soil suspension was centrifuged and the concentration of Cu, Zn, or Cd in the supernatant was analyzed by atomic absorption spectroscopy. The heavy metals in the solid fraction of the residue were analyzed by the sequential extraction method. All experiments were designed with or without rhizobium treatment in triplicate.

2.4 Sequential Extraction of Cu, Zn, and Cd in Soils

Procedures developed for the sequential extraction of Cu, Zn, and Cd in soils by Alef and Nannipieri (1995) were applied as outlined in Table 16.2. The concentration of each element in the extraction supernatant was determined by atomic absorption spectroscopy.

Table 16.2: Successive steps for extraction of heavy metals in soils (from Alef and Nannipieri, 1995)

Step	Species	Reagent	Shaking time and condition
i.	Exchangeable	1 M NH_4NO_3, 50 ml	24 h, 25°C
ii.	Carbonate-bound	1 M NH_4OAc, 50 ml, pH 6.0	24 h, 25°C
		1 M NH_4NO_3, 25 ml	10 min, 25°C
iii.	Mn oxide-bound	0.1 M NH_2OH-HCl, 50 ml	30 min, 25°C
		1 M NH_4OAc (25 ml, pH 6.0)	10 min, 25°C
iv.	Organic matter-bound	25 mM NH_4-EDTA (50 ml, pH 4.6)	90 min, 25°C
		1 M NH_4OAc (25 ml, pH 4.6)	10 min, 25°C
v.	Noncrystalline Fe oxide-bound	0.2 M NH_4-oxalate (50 ml, pH 3.25)	60 min, 25°C, darkness
		0.2 M NH_4-oxalate (25 ml, pH 3.25)	10 min, 25°C, darkness
vi.	Crystalline Fe oxide-bound	0.1 M ascorbic acid +	
		0.2 M NH_4-oxalate (50 ml, pH 3.25)	30 min, 96°C
		and 0.2 M NH_4-oxalate (25 ml, pH 3.25)	10 min, 25°C, darkness
vii.	Residual	Digested with HNO_3-$HClO_4$	

3 RESULTS

3.1 Effect of Rhizobial Inoculation on Soil pH

After inoculation of rhizobium and incubation for 2 weeks, the pH values for heavy metal-treated soils decreased variously (Table 16.3). Vis-à-vis the uninoculated soil (control), inoculation caused decreases of 0.5, 0.5, and 0.4 pH units respectively for Cu-, Zn-, and Cd-treated Red soil. Inoculation in Cinnamon soil lowered the pH by 0.5, 1.2, and 0.7 units for Cu-, Zn-, and Cd-treated soil. The decrease in soil pH after inoculation was due to the release of various metabolic products of the bacteria, such as H^+ ions and organic acids. Inoculation of rhizobia had a more significant influence on the pH change in Cinnamon soil than in red soil. For Cinnamon soil, Cu treatment caused the least decrease in pH among the three treatments. The greatest pH decrease was observed for zinc-treated Cinnamon soil. The decrease in soil pH after inoculation accorded

Table 16.3: pH values of soils in absence and presence of rhizobia

Soil	Cu treatment		Zn treatment		Cd treatment	
	-rhizobia	+rhizobia	-rhzobia	+rhizobia	-rhizobia	+rhizobia
Red	5.51	5.00	5.67	5.25	5.76	5.38
Cinnamon	6.99	6.56	6.89	5.72	6.98	6.25

with the growth status of the rhizobia in the two soils examined. The number of active bacteria in Cinnamon soil was two orders of magnitude greater than that in Red soil. Zinc-treated Cinnamon soil had a rhizobial population one order of magnitude greater than that in Cu-treated Cinnamon soil.

3.2 Distribution of Zink (Zn) in Soils

The content of zinc in each fraction of the control Red soil followed the sequence: organic matter-bound (100 mg kg^{-1}) > crystalline Fe oxide-bound (83.7 mg kg^{-1}) > residual (75.9 mg kg^{-1}) > Mn oxide-bound (45.5 mg kg^{-1}), specifically adsorbed (45.7 mg kg^{-1}) > exchangeable (20.3 mg kg^{-1}) (Fig. 16.1). The content of water-soluble Zn was negligible. Inoculation of rhizobia increased exchangeable Zn by 9.7 mg kg^{-1} and the amount of Zn bound with the other fractions decreased 3 to 11 mg kg^{-1}. No marked changes were observed for the concentration of water-soluble Zn and the total amount of Zn in the solid fractions of Red soil. As for Cinnamon soil (noninoculated), the amount of Zn

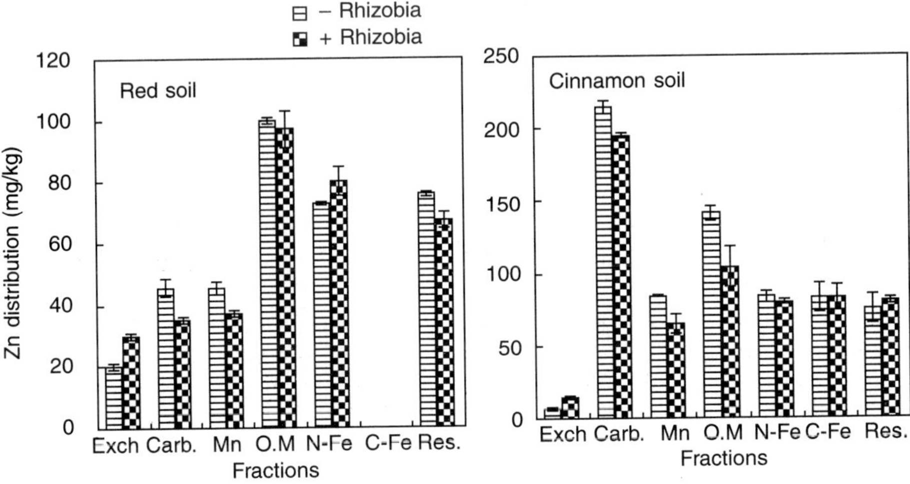

Fig. 16.1: Distribution (mg kg^{-1}) of Zn in various chemical fractions, Red vs. Cinnamon soil (Exch—Exchangeable Zn; Carb—Carbonate-bound or specifically adsorbed Zn; Mn—Manganese oxide-bound Zn; OM—Organic matter-bound Zn; N-Fe—Noncrystalline iron oxide-bound Zn, C-Fe—Crystalline iron oxide-bound Zn; Res—Zinc in residual fraction).

was in the order of carbonate fraction (215 mg kg^{-1}) > organic matter-bound (141 mg kg^{-1}) > Mn oxide-bound (84.5 mg kg^{-1}), noncrystalline Fe oxide bound (83.7 mg kg^{-1}), crystalline Fe oxide-bound (82.8 mg kg^{-1}) > residual (75.4 mg kg^{-1}) > exchangeable (7 mg kg^{-1}) > water-soluble (0.2 mg kg^{-1}). Zinc in the carbonate fraction accounted for the highest percentage of total Zn in the solid phase. Inoculation of rhizobia increased water-soluble Zn by 1 mg kg^{-1} and exchangeable Zn by 7.4 mg kg^{-1}. The amount of Zn bound by carbonate, Mn oxide, organic matter and noncrystalline Fe oxide decreased 4 to 36 mg kg^{-1} after inoculation. No remarkable changes were observed in the amount of Zn in crystalline and residual fractions. A 10% reduction in the total amount of Zn in the solid fractions of Cinnamon soil was observed after inoculation treatment compared to control.

3.3 Distribution of Copper (Cu) in Soils

Figure 16.2 shows that in the noninoculated Red soil, the order for the amount of Cu in each was as fraction follows: organic matter-bound (74.8 mg kg^{-1}) > residual (55.4 mg kg^{-1}) > crystalline Fe oxide-bound (50.7 mg kg^{-1}) > specifically adsorbed (35.3 mg kg^{-1}) > noncrystalline Fe oxide-bound (32 mg kg^{-1}) > Mn oxide-bound (10.3 mg kg^{-1}) > exchangeable (8.7 mg kg^{-1}). No Cu was detected in the solution of noninoculated Red soil. Except for a slight increase for the amount of exchangeable Cu, no significant differences were found in the amount of other forms of Cu in inoculated soil.

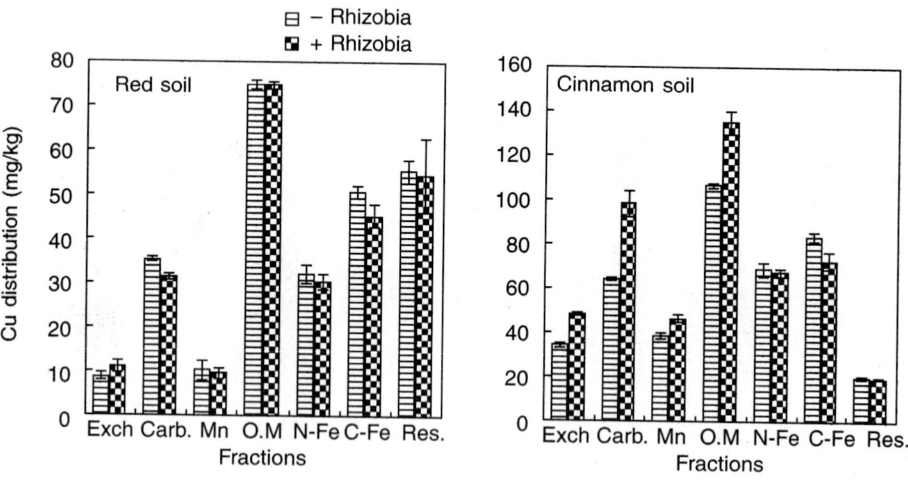

Fig. 16.2: Distribution (mg kg^{-1}) of Cu in various chemical fractions, Red vs. Cinnamon soil (Exch—Exchangeable Cu; Carb—Carbonate-bound or specifically adsorbed Cu; Mn—Manganese oxide-bound Cu; OM—Organic matter-bound Cu; N-Fe—Noncrystalline iron oxide-bound Cu; C-Fe—Crystalline iron oxide-bound Cu; Res—Copper in residual fraction).

In the noninoculated Cinnamon soil, the amount of Cu in the solid fractions was: organic matter-bound (107 mg kg^{-1}) > crystalline Fe oxide-bound (83.2 mg kg^{-1}) > noncrystalline Fe oxide-bound (68.8 mg kg^{-1}) > carbonate-bound (64 mg kg^{-1}) > Mn oxide-bound (38.6 mg kg^{-1}) > exchangeable (34.2 mg kg^{-1}) > residual (20 mg kg^{-1}). The content of water-soluble Cu was 5.2 mg kg^{-1}. Inoculation of rhizobia increased the total amount of Cu in solid fractions by 73 mg kg^{-1} (18% increment) and decreased water soluble-Cu by 3.9 mg kg^{-1}. Distribution of Cu in the solid fraction was remarkably altered, and the order was organic matter-bound > carbonate-bound > crystalline Fe oxide-bound > noncrystalline Fe oxide-bound > exchangeable > Mn oxide-bound > residual.

3.4 Distribution of Cadmium (Cd) in Soils

Figure 16.3 shows that in the noninoculated Red soil, the amount of specifically adsorbed Cd was 11 mg kg^{-1}, while Cd in the other fractions ranged from 2.3 to 3.7 mg kg^{-1}. Water-soluble Cd was not detected. Compared to control, inoculation of rhizobia decreased by 1 to 1.5 mg kg^{-1} specifically adsorbed Cd and Mn oxide-bound Cd and increased by 0.4 to 0.5 mg kg^{-1} exchangeable and organic matter-bound Cd.

The amount of carbonate-bound Cd in noninoculated Cinnamon soil was 28.7 mg kg^{-1}, and Cd in the other fractions was in the range of 1 to 4 mg kg^{-1}. Inoculation of rhizobia had no significant effect on the distribution of Cd in the solid fraction of Cinnamon soil.

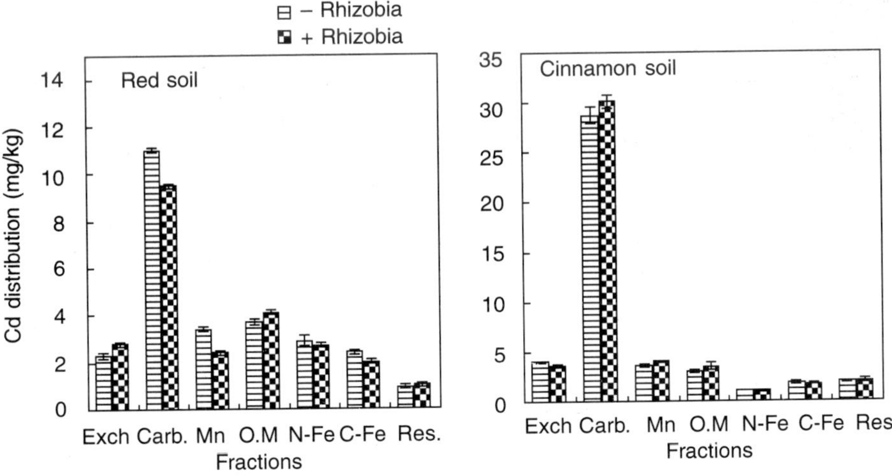

Fig. 16.3: Distribution (mg kg^{-1}) of Cd in various chemical fractions, Red vs. Cinnamon soil (Exch—Exchangeable Cd; Carb—Carbonate-bound or specifically adsorbed Cd; Mn—Manganese oxide-bound Cd; OM—Organic matter-bound Cd; N-Fe—Noncrystalline iron oxide-bound Cd, C-Fe—Crystalline iron oxide-bound Cd; Res—Cadmium in residual fraction).

4 DISCUSSION

The results obtained in this study clearly showed that inoculation of rhizobia in Red soil and Cinnamon soil changed, to some extent, the distribution of Cu, Zn, or Cd in the solid fractions as well as the release of these elements into the soil solution. The effects exerted by rhizobia may be attained by (1) lowering of soil pH due to the secretion of protons, amino acids, and organic acids; (2) complexation of heavy metals, clay minerals, and metal oxides with organic ligands from bacterial exudation (Gadd, 2000). For both Red soil and Cinnamon soil, the pH of inoculated soil was 0.5 units lower than that of noninoculated soil. Xian (1989) reported that the amount of exchangeable Zn increased while that of carbonate-, and Fe and Mn oxide-bound Zn decreased for a paddy soil when the pH decreased from 7.0 to 4.5. The amount of organic matter-bound and residual Zn remained unchanged. Rhizobia can secrete a variety of metabolic products such as organic acids and amino acids (You et al., 1987). Inoculation of rhizobia into Cinnamon soil lowered the pH by 1.2 units and resulted in a release of 10% Zn from the solid fraction (see Fig. 16.1). The range of decrease in amount of Zn bound by carbonate, Mn oxide, and organic matter was 9 to 26%. However, inoculation induced only a 0.5 pH unit decrease in the Cu-treated Cinnamon soil and the transformation of Cu in the solid fractions was relatively less than that in the Zn-Cinnamon soil.

Bacteria may also bind heavy metals via adsorption and precipitation processes on the cell walls (Tobin et al., 1984, Flemming et al., 1990; Guo et al., 2002). Attractive forces and the affinity between the metal and the adsorptive site due to Van der Waals forces can be relatively weak and nonspecific, resulting in easily exchangeable cations, or they can be strong and specific, resulting in less readily remobilized cations. Precipitation is usually formed as the hydroxy or carbonate species on carboxyl or phosphoryl groups of bacterial surfaces (Flemming et al., 1990). Interactions between bacteria and soil colloidal components have a stronger ability to adsorb heavy metals (Beveridge and Schultze-Lam, 1995). The mobility of heavy metals immobilized by micro-organisms is differ from that bound by the inorganic and organic soil colloidal components (Flemming et al., 1990). Bacteria may influence the mobility of heavy metals in soil positively or negatively. On the one hand, bacteria may enhance the mobility of heavy metals through dissolution and desorption by the secretion of protons and various ligands. Consequently more heavy metals are transformed to the more mobile fractions. On the other hand, bacteria could adsorb, immobilize, and complex heavy metals via surface interaction and secretion of chelating agents, which results in their transformation to more stable forms and reduces the mobility of heavy metals. For the three heavy metals studied, Cu has the strongest complexing ability and often forms stable complexes with organic components, whereas Zn tends to be adsorbed by organic molecules (Ramos et al., 1994). The more pronounced stabilizing behavior of Cu in the soil after inoculation suggested that chelation and binding overwhelmed desorption and dissolution between heavy metals and the soil-rhizobia composite. For example, inoculation of rhizobia into Cinnamon soil increased by 18% the total amount of Cu in the solid fractions. The amount of exchangeable, carbonate-bound, Mn oxide-bound, and organic matter-bound Cu increased by 20 to 54%. The lesser influence of rhizobia on the fractionation of Cd in cinnamon soil may be ascribed to the stalemate between mobilization and immobilization of heavy metals in the presence of rhizobia. The lower content of Cd could be another possibility for this phenomenon.

Inoculation of rhizobia had a lesser effect on the distribution of Zn and Cu in Red soil than in Cinnamon soil. This was attributed to the different status of bacterial growth in the two soils. Red soil had a lower pH and the metabolism of rhizobia was relatively weak. It is assumed that in the Red soil the sites for binding heavy metals on the bacterial cells as well as the amount of bacterial exudation in the soil were less than that in Cinnamon soil. Therefore, transformation of Cu and Zn in the Red soil was not significantly affected by inoculation of rhizobia. Appropriate acidity and adequate nutrition for microorganisms can promote the microbial dissolution of heavy metals (You et al., 1987). It is obvious that the mobility of heavy metals was governed by the activity of bacteria in soils, i.e., vigorous metabolism of microorganisms may result in significant transformation of heavy metals in soils.

Our results indicate that the inoculation of rhizobia mainly influences the species of heavy metals on exchangeable sites and bound on carbonate (specifically adsorbed), Mn oxide, and organic fractions. The content of Fe oxide-bound heavy metals was not significantly altered. It is generally accepted that the availability of exchangeable and specifically adsorbed heavy metals as well as those bound by Mn oxide and organic matter in soil and sediments is relatively higher. Our results suggest that the influence of bacteria on heavy metals was mainly on the mobile forms, which are highly bioavailable. For soils contaminated by heavy metals, attention should be paid to the effect of bacteria on the distribution and mobility of these toxic metals. In bioremediation practices, the bacterial biomass and the metabolic activity of bacteria should be properly adjusted in these soils to increase or decrease the mobility of heavy metals. Furthermore, the influence of bacteria varies to some extent with soil type and heavy metals, so specific approaches should be considered in terms of a particular soil or metal. Studies on the environmental factors controlling bacterial transformation of heavy metals in soils merit close attention.

Acknowledgment

The authors are grateful to the National Natural Science Foundation of China for financially supporting this work (Project No. 20077010).

References

Adriano D.C., Bollag, J.-M., Frankenberger, Jr. W.T. et al. 1999. *Bioremediation of Contaminated Soils*. Agron. Monograph 37, Soil Sci. Soc. Amer., Madison, WI, 772 pp.

Alef K. and Nannipieri P. 1995. *Methods in Applied Soil Microbiology and Biochemistry*. Acad. Press, London, UK, pp. 100–104.

Berthelin J., Munier-Lamy C., and Leyval C. 1995. Effect of microorganisms on mobility of heavy metals in soils. In: *Environmental Impact of Soil Component, Mineral-Organic and Microorganisms Interaction on Environmental Quality*. P.M. Huang, J. Berthelin, J.-M. Bollag, et al. (eds.). CRC Press/Lewis Publ., Boca Raton, FL, pp. 3–17.

Beveridge T.J. and Schultze-Lam S. 1995. Detection of Anionic sites on bacterial walls, their ability to bind toxic heavy metals and form sedimentable flocs and their contribution to mineralization. In: *Metal Speciation and Contamination of Soil*. H.E. Allen, C.P. Huang, G.W. Bailey, et al. (eds.). CRC Press/Lewis Publ., Boca Raton, FL, pp. 183–205.

Chanmugathas P. and Bollag J-M. 1988. Microbial role in immobilization and subsequent mobilization of cadmium in soil. *Arch. Environ. Contam. Toxicol*. 17:229–235.

Flemming C.A., Ferris F.G., Beveridge T.J. et al. 1990. Remobilization of toxic heavy metals adsorbed to bacterial wall-clay composites. *Appl. Environ. Microbiol*. 56: 3191–3203.

Gadd G.M. 2000. Bioremedial potential of microbial mechanisms of metal mobilization and immobilization. *Curr. Opinions Biotechnol*. 11: 271–279.

Guo X., Huang Q., and Chen W. 2002. Effect of microbial activities on the mobility of heavy metals in soil environments. *Chin. J. Appl. Environ. Biol*. 8:105–110.

Huang Q., Wu J., Chen W., and Li X. 2000. Adsorption of Cd by soil colloids and minerals in presence of rhizobia. *Pedosphere* 10:299–307.

Krishnamurti G.S.R., Huang P.M., Kozak L.M. et al. 1997. Distribution of cadmium in selected soil profiles of Saskatchewan, Canada: speciation and availability. *Can. J. Soil Sci*. 77: 613–619.

Kurek E., Czaban J., and Bollag J.-M. 1982. Sorption of cadmium by microorganisms in competition with other soil constituents. *Appl. Environ. Microbiol*. 43: 1011–1015.

Ledin M., Krantz-Rulcker C., and Allard B. 1996. Zn, Cd and Hg accumulation by microorganisms, organic and inorganic soil components in mult-compartment systems. *Soil. Biol. Biochem*. 28: 791–799.

Ma L.Q. and Rao G.N., 1997. Chemical fractionation of Cd, Cu, Ni and Zn in contaminated soils. *J. Environ. Qual*. 26: 259–264.

Ramos L., Hernandez L.M., and Gonzalez M.J. 1994. Sequential fractionation of Cu, Pb, Cd and Zn in soils from or near Donana National Park. *J. Environ. Qual*. 23: 50–57.

Selim H.M. and Sparks D.L. 2001. *Heavy Metals Release in Soils*. CRC Press/Lewis Publ., Boca Raton, FL, 328 pp.

Tessier A., Campbell P.G.C., and Bisson M. 1979. Sequential extraction procedure for the speciation of particulate trace metals. *Anal. Chem*. 51: 844–851.

Tobin J., Cooper D., and Neufeld R. 1984. Uptake of heavy metal ions by *Rhizopus arrhizus* biomass. *Appl. Environ. Microbiol*. 47: 821–824.

Wang G., Chen J., Gao S., et al. 1999. Influence of application of rice straw and Chinese milk vetch on the species and ecological effect of added copper in two soils. *Acta Ecol. Sinica* 19: 551–556.

Wang G., Li Y., Yang P., et al. 2000. Study on Cd speciation and availability in soil solution as influenced by organic materials. *Chin. J. Environ. Sci*. 20: 621–626.

Xian X. 1987. Chemical partitioning of Cd, Zn, Pb and Cu in soils near smelters. *J. Environ. Sci. Health* A, 6: 527–541.

Xian X. 1989. Effect of chemical forms of Cd, Zn, Pb and Cu in polluted soils on their uptake by cabbage plants. *Plant Soil* 113: 257–264.

You C., Jiang Y., and Song H. 1987. *Biological Nitrogen Fixation*. Science Press, Beijing.

17

Oxalic Acid Perturbation of Imogolite Formation and the Impact on Cadmium Adsorption

M. Tani*, C. Liu, *and* **P.M. Huang**

Abstract

The influence of oxalic acid, predominant in soil solutions of Andisols and Spodosols, on formation of imogolite and noncrystalline aluminosilicates (initial Si concentration of 1.6 mmol L^{-1}, Si/Al molar ratio of 0.5, and OH/Al molar ratio of 2.0), and their morphological features as observed by atomic force microscopy (AFM) were investigated. Subsequent effects of oxalate perturbation on the surface properties and Cd adsorption of these clay materials were also studied.

Oxalic acid perturbed the formation of imogolite remarkably at higher concentrations (oxalate/Al molar ratios of 0.08 and 0.10). However, effective impedance of the interaction of hydroxy-Al ions with orthosilicic acid was much lower compared with citric acid and other strong complexing organic acids, as indicated by the results of amounts of H^+ released into solutions during the reaction, x-ray diffractograms, and IR absorption spectra. The morphological features of synthetic imogolite and x-ray noncrystalline aluminosilicates formed as affected by oxalate were distinctly observed through contact-mode AFM. Optimum ultrasonification and sufficient air drying of imogolite on the sample holder at room temperature before AFM scanning are critical for precluding breakdown of imogolite particles and etching of a sample surface. At the oxalate/Al molar ratios of 0 and 0.02, the imogolite appeared in the AFM three-dimensional deflection images as both straight and curved threads. With increase in the oxalate/Al molar ratio, the thready structures, which are characteristic

---

**Corresponding author*: Dr. Masayuki Tani, Dept. Agro-Environmental Science, Obihiro University of Agriculture and Veterinary Medicine, Inada, Obihiro, Hokkaido 080-8555, Japan.
E-mail: masatani@obihiro.ac.jp

of imogolite, decreased and spheroidal and ill-defined amorphous particles increased in the AFM images. At the oxalate/Al molar ratio of 0.10, no thready structure was observed. The AFM study also showed that oxalic acid induced a change in size of the thready structures of imogolite and spheroidal particles of noncrystalline aluminosilicates. The length and width of thready structures decreased while the diameter of spheroidal particles increased with increment in the oxalate/Al molar ratio from 0 to 0.10.

Kinetic studies of Cd adsorption on imogolite and noncrystalline aluminosilicates were conducted at pH 5.0, using 10^{-5} mol L^{-1} $Cd(NO_3)_2$ solution. Adsorption of Cd by these particles was a very rapid reaction and actually completed within 10 min. After 240 min, 9.4% of the added Cd was removed from the solution of the imogolite suspension system. Less Cd was adsorbed by the perturbed imogolites than by imogolite. The percent adsorption of Cd decreased with increase in oxalate/Al molar ratio from 0.02 to 0.10. At the highest oxalate/Al molar ratio of 0.10, no more than 3.2% of the added Cd was adsorbed on the aluminosilicate. From a pedological viewpoint, the incorporation of oxalate in the noncrystalline aluminosilicates would hamper the retention of Cd in the surface and illuvial horizons of Andisols and Spodosols.

The present results thus suggest that the influence of oxalic acid on the shape and size of particles, the subsequent surface properties, and Cd adsorption of imogolite and poorly ordered aluminosilicates merits further attention in understanding the dynamics of Cd in the environment.

1 INTRODUCTION

Imogolite is a naturally occurring hydrous aluminosilicate polymer with a unique tubular structure (Cradwick et al., 1972; Farmer et al., 1983; Wada, 1989) and has been extensively isolated from the B horizons of Andisols (Parfitt and Henmi, 1980; Wada, 1989) and Spodosols (Brydon and Shimoda, 1972; Ross and Kodama, 1979; Farmer et al., 1980; Anderson et al., 1982; Gustafsson et al., 1995). Allophane and imogolite are negligible in the surface horizons of these soils, which are rich in organic matter (Tokashiki and Wada, 1975; Tait et al., 1978). Experimental evidence has shown that soil organic components, such as humic acid and low-molecular-weight organic acids, might retard formation of allophane and imogolite (Inoue and Huang, 1984, 1986, 1987; Lou and Huang, 1989).

In studies of synthetic allophane and imogolite formed in a solution containing hydroxy-Al ions and orthosilicic acid, the perturbation effect of relatively strong complexing organic acids, especially citric acid, on the formation of noncrystalline aluminosilicate has been substantiated (Inoue and Huang, 1984, 1985, 1986; Lou and Huang, 1989). Recent studies on low-molecular-weight organic acids in soils and soil solutions, determined by HPLC, showed that oxalic acid and formic acid were predominant in the soil solutions of Spodosols (Fox and Comerford, 1990; Tani and Higashi, 1999; Tani et al., 2001) and Andisols (Tani et al., 1995, 1996). Oxalic acid is commonly present in root exudates, metabolites of microbes, and leachates from surface litter layers in the forest floor (Stevenson, 1967; Fox, 1995). In forest ecosystems, oxalic acid is by far the most abundant organic acid produced by fungi and mycorrhizal fungi (Sollins et al., 1981; Malajczuk and Cromack, 1982).

Atomic force microscopy (AFM) is a new and powerful technique for observing the surface morphology of mineral colloids and has a distinctive advantage compared with electron microscopy in that the samples do not have to be coated and subjected to ultrahigh vacuum condition (Nagy and Blum, 1994; Bertsch and Hunter, 1998). Scanning probe microscopy such as scanning tunneling microscopy and AFM, represents an ever-increasing class of microscopic techniques that provide three-dimensional images of solid surfaces at high resolution (Rugar and Hansma, 1990; Nagy and Blum, 1994; Bertsch and Hunter, 1998). AFM imaging of uncovered samples at room temperature in ambient air is appropriate for environmental studies (Maurice, 1998). However, the surface feature of imogolite has not yet been studied by AFM.

Imogolite exerts significant effects on ion exchange properties, surface acidity, water adsorption, and physicochemical properties of soils (Farmer et al., 1983; Wada, 1989). Since the walls of a tube unit consist of curved gibbsite-like sheets with SiOH groups on the inside and AlOH groups on the outside, many nutrients and pollutants such as heavy metal cations, phosphate, and arsenate can form chemical bonds with the outer-surface of imogolite (Cradwick et al., 1972; Parfitt et al., 1974). Its structure and surface property, therefore, deserve special attention. Of all the nonessential heavy metals, Cd has perhaps attracted the most attention in soil science and plant nutrition because of its potential toxicity to humans and relative mobility in the soil-plant system (McLaughlin and Singh, 1999). The major hazard to human health from Cd is its chronic accumulation in the kidneys where it can cause dysfunction (Alloway, 1995). Retention of Cd by soil mineral colloids should have a substantial impact on Cd dynamics and subsequent food chain contamination since food is the main route by which Cd enters the human body. However, little attention has been paid to the adsorption of Cd by imogolite.

We examined the morphological features of synthesized imogolite under ambient conditions by AFM. We also investigated the influence of oxalic acid on formation of imogolite and the morphological features of synthesized imogolite and noncrystalline aluminosilicates affected by oxalic acid during their formation, using contact-mode AFM and related methods. Furthermore, we examined the Cd adsorption by imogolite and noncrystalline aluminosilicates formed under the influence of different concentrations of oxalic acid.

2 MATERIALS AND METHODS

2.1 Synthesis of Precipitation Products

X-ray noncrystalline aluminosilicate materials and imogolite were prepared from solutions containing monomeric orthosilicic acid, $AlCl_3$, and various amounts of oxalic acid using the method of Inoue and Huang (1984). Monomeric orthosilicic acid was obtained by passing Na-metasilicate solution through an H^+-saturated Dowex 50W-X8 cation exchange resin column (Alexander, 1953). To avoid polymerization of Si, the initial concentration of silicic acid was maintained below 2 mmol L^{-1} (Iler, 1979). The silicic acid was mixed with $AlCl_3$ solution and oxalic acid to give Si/Al molar ratio of 0.5 and oxalate/Al molar ratios 0 to 0.10. The solution was then titrated with 0.1 mol L^{-1} NaOH with continuous stirring at the rate of 0.5 mL min^{-1} to an OH/Al molar ratio of 2.0. The final concentration of Si was 1.6 mmol L^{-1} and the pH values of the solutions were 4.35 ± 0.01. Solutions were heated at 98°C for 110 h. After cooling to room temperature, the pH of the suspensions was determined. Precipitates were collected by ultrafiltration using a membrane filter of 0.1-μm pore size. The precipitates were dialyzed (molecular weight cutoff 3,500) against distilled deionized water until Cl^- free, freeze dried, then ground gently to pass through a 0.5-mm sieve.

2.2 Examination of Precipitation Products

The contents of Si and Al in the precipitates were determined according to the methods of Weaver et al. (1968) and Hsu (1963) respectively following dissolution of the precipitate with 0.5 mol L^{-1} NaOH and destruction of organic C with H_2SO_4 and HNO_3 (Higashi and Ikeda, 1974). Organic carbon content of the precipitates was determined by using a CN-corder (MT-500, Yanaco, Tokyo). X-ray diffraction analysis was carried out on a Rigaku D/Max-RBX x-ray diffractometer (Rigaku Co., Tokyo) with FeKα radiation, using a parallel-oriented sample dried on a glass slide. Infrared (IR) spectra were recorded on a Perkin-Elmer 983 infrared spectrometer (Perkin-Elmer Co., Buckinghamshire, England) with KBr disks incorporating 1 mg of freeze-dried material. PZSE of the precipitates was determined by potentiometry (Lou and Huang, 1989). The specific surface area was determined by the BET method (Autosorb-1, Quantachrom Corp., NY) using N_2-gas adsorption.

2.3 Atomic Force Microscopy of Precipitation Products

To obtain distinct and reliable images of imogolite by contact-mode AFM, sample preparation and mounting procedures are critical. Optimum ultrasonification and sufficient air drying of imogolite on the sample holder at room temperature before AFM scanning are imperative to preclude breakdown of imogolite particles and etching of a sample surface.

Precipitates were dispersed by using ultrasonification at 150 W for 5, 15, 30, 60, and 120 s in an ice bath. Two drops of the extremely diluted x-ray noncrystalline aluminosilicate materials and imogolite suspension (20 mg L^{-1}) were deposited on a watch glass and dried for 1 day at room temperature (23.5 ± 0.5°C). A replicate sample was dried for 1 month under the same conditions. The watch glass was fastened to a magnetized stainless steel disk with double-sided tape. Lastly, AFM three-dimensional images of the precipitates on the uncovered watch glass were taken at room temperature using a NanoScope III atomic force microscope (Digital Instrument Inc., Santa Barbara, CA). The scanner was type 1881E and the scanner size 15 μm. The image was recorded for each of 1 to 100 scans. The image obtained from the 1st scan was labeled the image obtained after the tip moved from one edge of the frame to the other edge. The silicon nitride cantilever with spring constant of 0.12 N m^{-1} was used in contact-mode. The scanning rate was 2.2 Hz. The images were neither filtered nor modified. Deflection images of 10 × 10 μm or 15 × 15 μm scanning size were displayed to describe the morphological features of the imogolite and perturbed noncrystalline aluminosilicates affected by oxalic acid in this study. The intensity of gray coloration in deflection images reflected the difference between neighboring height values. The methodology described by Liu and Huang (1999) was followed in the AFM.

2.4 Kinetic Studies of Cd Adsorption

A 50-mg sample of imogolite and noncrystalline aluminosilicates was pre-equilibrated with 50 mL of 1 mmol L^{-1} KNO_3 solution at pH 5.0 overnight. The pH was readjusted to 5.0 with diluted HNO_3 or KOH. Suspensions were mixed with 50 mL solution containing 0.02 mmol L^{-1} $Cd(NO_3)_2$ and 1 mmol L^{-1} KNO_3 adjusted to pH 5.0, after which the final concentration of Cd before adsorption was 0.01 mmol L^{-1}. Suspensions were agitated at 25°C for 4 h. An aliquot of 10 mL suspension was taken out at 10, 20, 40, 60, 120, and 240 min after initiation of agitation and filtered through a 0.1-μm membrane filter. The concentration of Cd-remaining in the filtrate was determined by flame atomic absorption spectroscopy. The experiment was conducted in duplicate.

3 RESULTS AND DISCUSSION

3.1 Reaction and Characteristics of Precipitates

In the absence of oxalic acid, the final pH of the solution was 2.82 and the amount of proton released during its formation 1.46 mmol L^{-1}. Contrarily, in the presence of oxalic acid, the final pH of the solutions ranged from 2.96 to 3.38 and increased with increasing oxalate/Al molar ratio from 0.02 to 0.10. The perturbation effect of the oxalate ligand on the crystallization of imogolite was evidenced by the decreasing amount of proton released with the increasing oxalate/Al molar ratios. When the amounts of proton released in both the presence and absence of oxalate are compared with those of citrate given by Lou and Huang (1989), the impedance effect of oxalic acid on the formation of imogolite is not as strong as citric acid (Fig. 17.1). This could be due to the difference in the stability constants ($\mathrm{Log}K_{Al}$) of Al-oxalate and Al-citrate complexes, which are 6.10 and 7.98 respectively (Fox, 1995).

The precipitate formed from the solution without oxalic acid was predominantly imogolite as indicated by x-ray diffraction peaks at 2.20, 0.89,

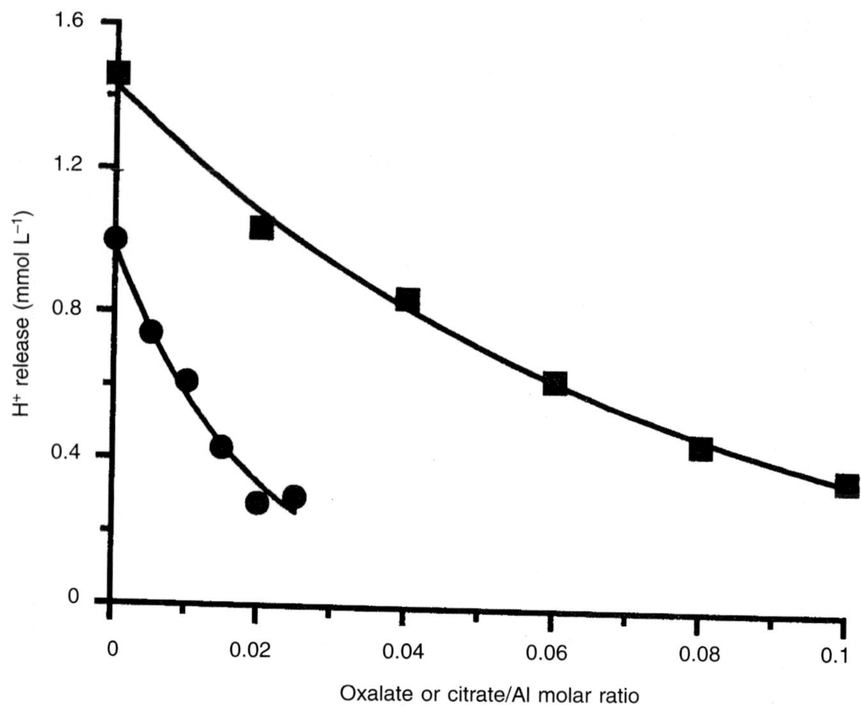

Fig. 17.1: Proton release in the oxalate/citrate systems: oxalate/Al molar ratio 0 to 0.10 and citrate/Al molar ratio 0 to 0.025. Data of citrate taken from Lou and Huang (1989). ■, oxalate; ●, citrate.

and 0.62 nm (Fig. 17.2a) (Wada et al., 1979; Farmer et al., 1983) and IR absorption bands at 987, 935, 695, 564, 488, 423, and 341 cm^{-1} (Fig. 17.3a) (Russell et al., 1969; Wada, 1989). The presence of small amounts of pseudoboehmite in the precipitate is indicated by the d-value at 0.47 nm (Fig. 17.2a) and IR absorption band at 1,069 cm^{-1} and a shoulder around 1,170 cm^{-1} (Fig. 17.3a). The contents of Si and Al in the precipitate were 221 and 546 g kg^{-1} as SiO_2 and Al_2O_3 respectively. The SiO_2/Al_2O_3 molar ratio was 0.69 and lower than the theoretical ratio of 1 for imogolite (Table 17.1) This may be due to the presence of bayerite and pseudoboehmite in the precipitate.

In the presence of low concentration of oxalic acid (oxalate/Al molar ratio of 0.02), the x-ray diffraction pattern showed sharp and high peaks, indicating the abundance of imogolite in the precipitate (Fig. 17.2b). At the oxalate/Al molar ratio of 0.04, peaks at 2.20, 0.89, and 0.62 nm could still be observed; however, the amount of imogolite must have decreased, as indicated by the abrupt decrease in peak intensities and broadening of these peaks (Fig. 17.2c). In the presence of high concentration of oxalic acid, the broad peak at 0.89 nm gradually disappeared and the peak at 0.62 nm shifted to 0.61 nm, indicating

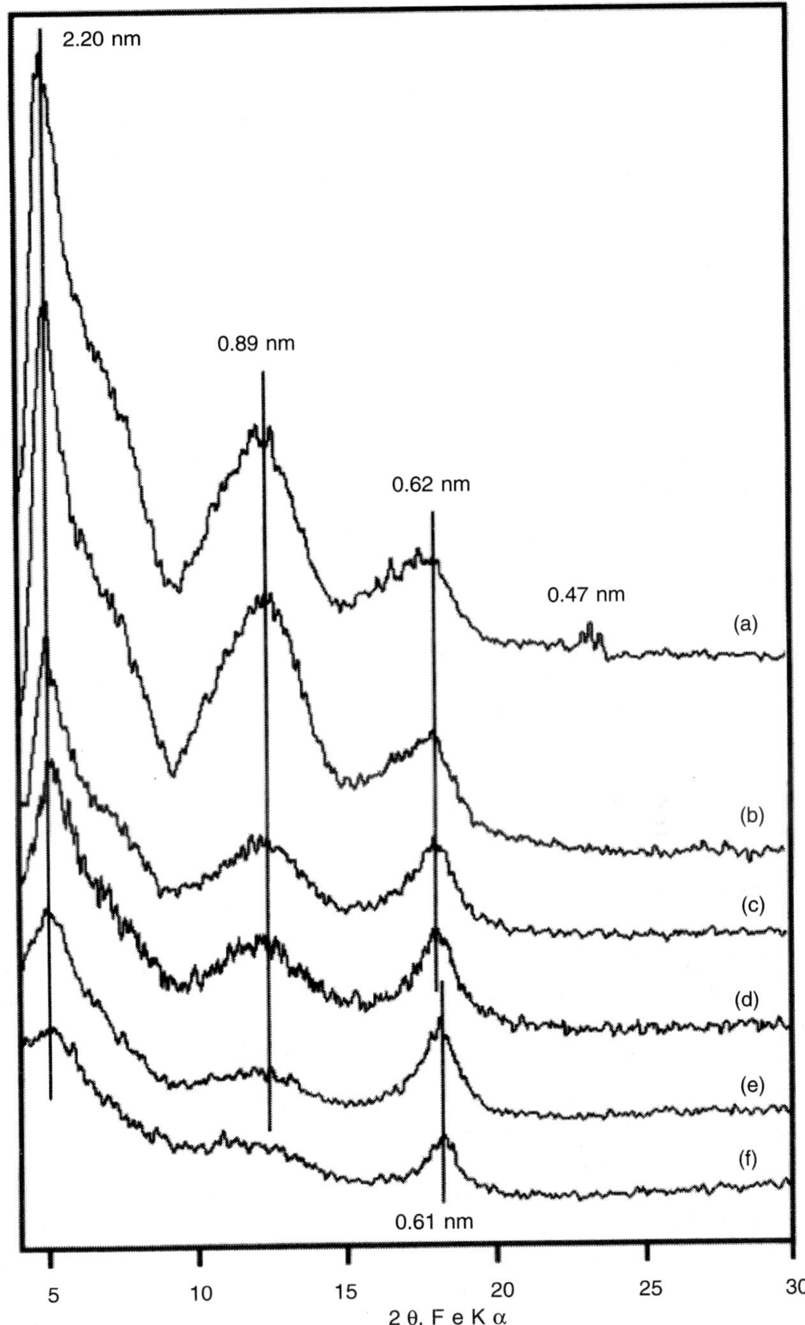

Fig. 17.2: X-ray diffractograms of the synthetic imogolite and precipitates formed both in the absence and presence of oxalic acids. Initial oxalate/Al molar ratio: (a) 0, (b) 0.02, (c) 0.04, (d) 0.06, (e) 0.08, and (f) 0.10.

Table 17.1: Composition of synthetic imogolite and precipitates formed in both the absence and presence of oxalic acid

Oxalate/Al Molar ratio	SiO_2	Al_2O_3	Oxalate	Si/Al molar ratio
		(g (kg^{-1})†		
0.00	221	546	0	0.69
0.02	175	579	33.9	0.51
0.04	155	627	81.0	0.42

†On the basis of oven-dried weight.

a decrease in imogolite and increase in pseudoboehmite (Hsu, 1989) (Fig. 17.2.d–f). At oxalate/Al molar ratio of 0.10, the precipitate was a mixture of poorly crystalline pseudoboehmite and noncrystalline aluminosilicates, as revealed by broad XRD peaks at 2.20 and 0.61 nm (Fig. 17.2f).

In the presence of oxlic acid, the IR spectra show absorption bands at 1,719 and 1,700 cm^{-1}, due to C=O stretching vibrations of COOH; at 1,490 and 1,422 cm^{-1}; due to COO$^-$ symmetric and asymmetric stretching vibrations; and at 1,288 and 1,254 cm^{-1} due to C–O stretching vibrations of COOH (Fig. 17.3b–f) (Schnitzer, 1978; White and Roth, 1986). Intensity of the absorption bands at these wave numbers increased with increase in oxalate/Al molar ratio from 0.02 to 0.10 (Fig. 17.3b–f), which was attributable to the increased incorporation of oxalic acid in the precipitates as indicated by the increasing oxalate contents in the precipitates (Table 17.1). Absorption bands at 423 and 695 cm^{-1} disappeared at oxalate/Al molar ratios of 0.04 and 0.08 respectively (Fig. 17.3d, e), and absorption doubtlets at 987 and 935 cm^{-1}, mainly due to Si-O and/or Si(Al)-O stretching vibrations of synthesized imogolite (Russell et al., 1969; Wada, 1989), disappeared at oxalate/Al molar ratio of 0.10 and showed an increasing band at 957 cm^{-1} (Fig. 17.3f), indicating the perturbing effect of oxalic acid on formation of imogolite and increase in noncrystalline aluminosilicates in precipitates. The IR spectra also indicate the presence of a small amount of pseudoboehmite at 1,069 cm^{-1} as an impurity in the precipitates and its increase with increment in the oxalate/Al molar ratio (Fig. 17.3a–f).

The present study suggests that oxalic acid significantly inhibits the interaction of hydroxy-Al ions with orthosilicic acid and the subsequent coprecipitation as indicated by the decrease in SiO_2 contents and Si/Al molar ratios with the increase in oxalate/Al molar ratio (Table 17.1), leading to formation of structurally disordered and noncrystalline materials incorporated with oxalic acid, as shown in the x-ray diffractograms (Fig. 17.2) and the IR absorption spectra (Fig. 17.3). However, the perturbation effect of oxalic acid on formation of imogolite was not as strong as citric acid, representatively used in previous studies (Inoue and Huang, 1984, 1985; Lou and Huang, 1989), due to the difference in their complexing abilities with Al, as indicated by the stability constants of Al-organic complexes (Fox, 1995; Vance et al., 1996).

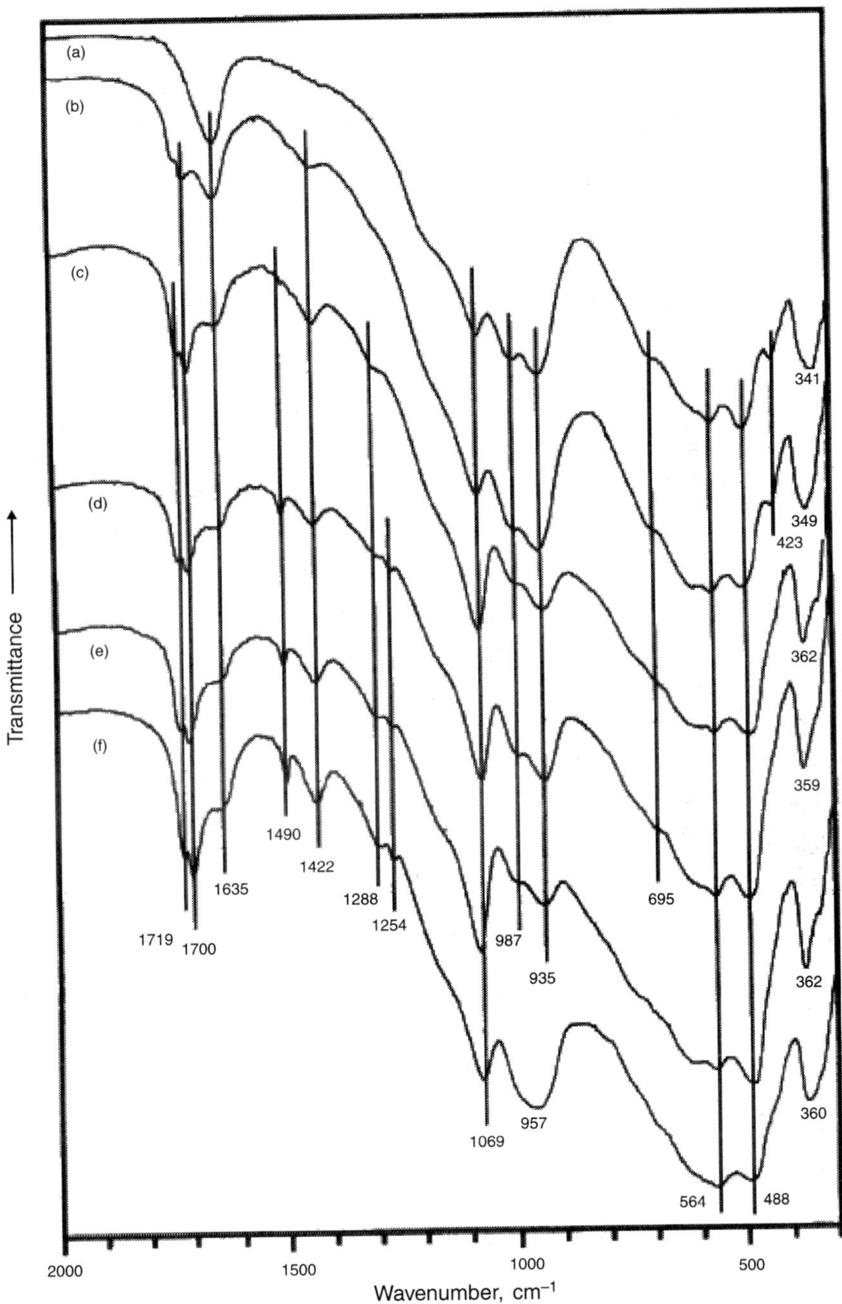

Fig. 17.3: Infrared (IR) spectra of the synthetic imogolite and precipitates formed both in the absence and presence of oxalic acid. Initial oxalate/Al molar ratio: (a) 0, (b) 0.02, (c) 0.04, (d) 0.06, (e) 0.08, and (f) 0.10.

3.2 AFM Images of Precipitates

The AFM three-dimensional deflection images in Fig. 17.4 show the effect of time of ultrasonification on the shape and structure of imogolite. Without ultrasonification, no distinctive structure could be seen in the AFM image (Fig. 17.4a). This is evidently due to aggregation of imogolite particles, as found in the AFM studies on illite, smectite, and goethite (Blum, 1994; Fischer et al., 1996). After ultrasonification for 5 s, imogolite appeared in the AFM image as curved threads (Fig. 17.4b). Many electron microscopic studies have already shown the thready tubular structures of imogolite (Yoshinaga and Amomine, 1962; Russell et al., 1969; Inoue and Huang, 1984). After ultrasonification for 15 s to 2 min, the thready structures, which are characteristic of imogolite, were still observed (Fig. 17.4c–f). With increase in ultrasonification time, the diameter of thready structures of imogolite decreased negligibly, contrary to their length which decreased significantly. The imogolite threads after ultrasonification for 5 s were three times as long as at 2 min. This had to have resulted from the thready structures being cut by the physical force of ultrasonic dispersion. This study shows that ultrasonification of the suspension is necessary to disperse imogolite particles, but ultrasonification time is critical to preclude alteration of imogolite structure.

After deposition of the imogolite suspension on a watch glass, the solution was allowed to evaporate for 1 day at 23.5°C. At the initial scanning, distinctive thereby structure of imogolite was observed in the AFM image (Fig. 17.5a). However, after repeating the scanning 5 and 10 times, the surface features were perturbed and the thready morphology of smaller particles disappeared (Fig. 17.5b,c). After scanning 50 and 100 times, no thready structures were observed; the AFM images showed spheroidal and poorly defined, irregularly shaped particles (Fig. 17.5d,e). When the deflection image after 100 scans with scanning size 15 × 15 μm was displayed to describe the morphological features of the imogolite, the build-up of eroded materials along the edges of the scanning spot was observed (Fig. 17.5f). Maurice (1998) reported that frictional forces, adhesive forces, and chemical interactions between tip and sample could result in etching of a sample surface along microtopographic features such as a step edge during the scanning process. Delawski and Parkinson (1992) also showed that the edges of pits were eroded parallel to scan direction, as found in this study, and the etching process was related to relative humidity.

However, after the imogolite suspension was deposited on a watch glass and air dried for 1 month at 23.5°C, scanning times (1 to 100) had no effect on the morphological features of imogolite (Fig. 17.6a–e). Even when scanning was repeated 100 times, the typical thready structure of imogolite was still observed in the AFM deflection image (Fig. 17.6e). Although a small accumulation of eroded material was observed along the edges of the scanning site (Fig. 17.6f) where the scanning size of 15 × 15 μm was displayed, the effect of etching processes was almost negligible. These results suggest that it takes considerable time to evaporate moisture from the suspension and leave the

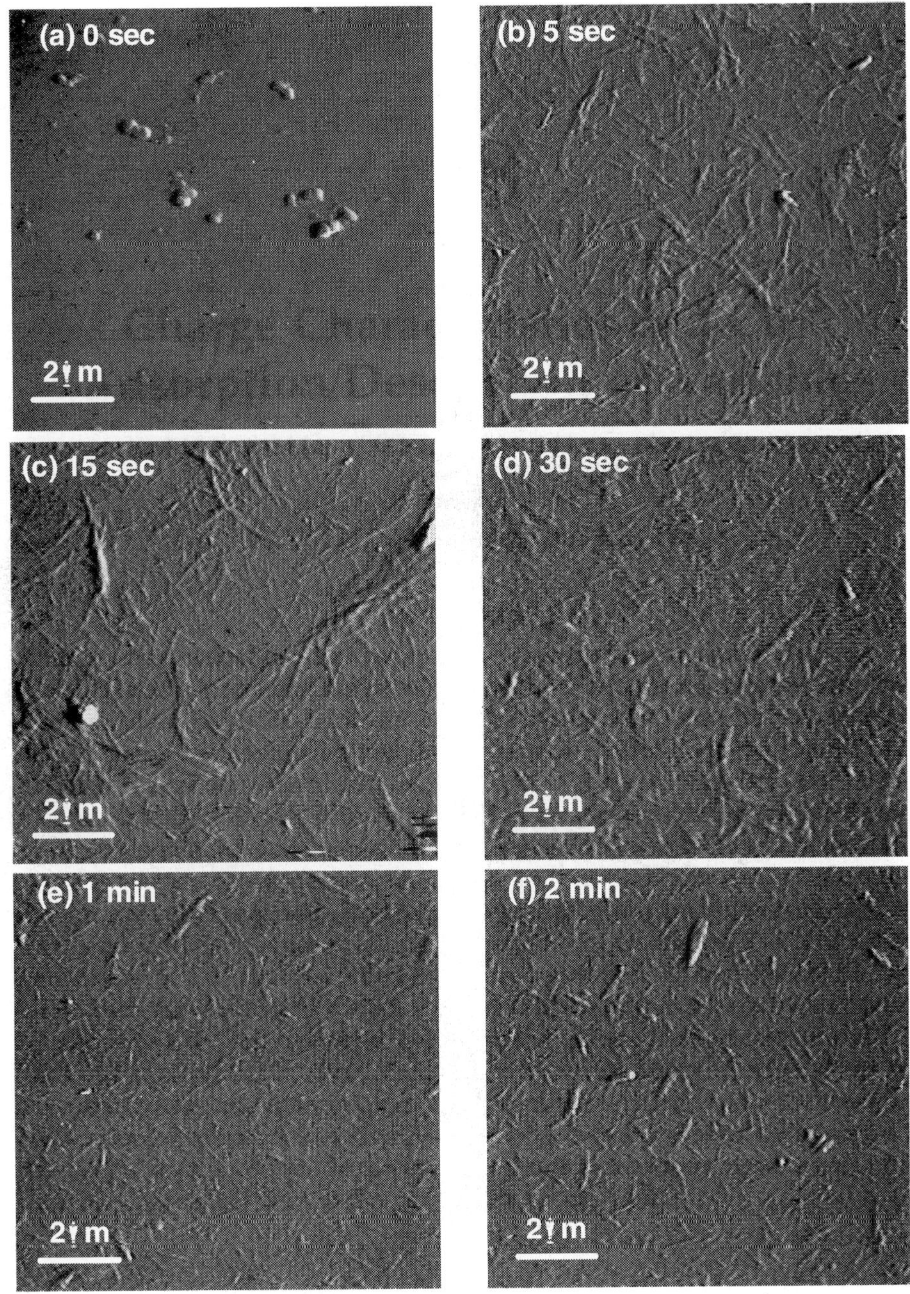

Fig. 17.4: AFM three-dimensional deflection images of the synthetic imogolite after different times of ultrasonification (a) 0 s, (b) 5 s, (c) 15 s, (d) 30 s, (e) 1 min, and (f) 2 min.

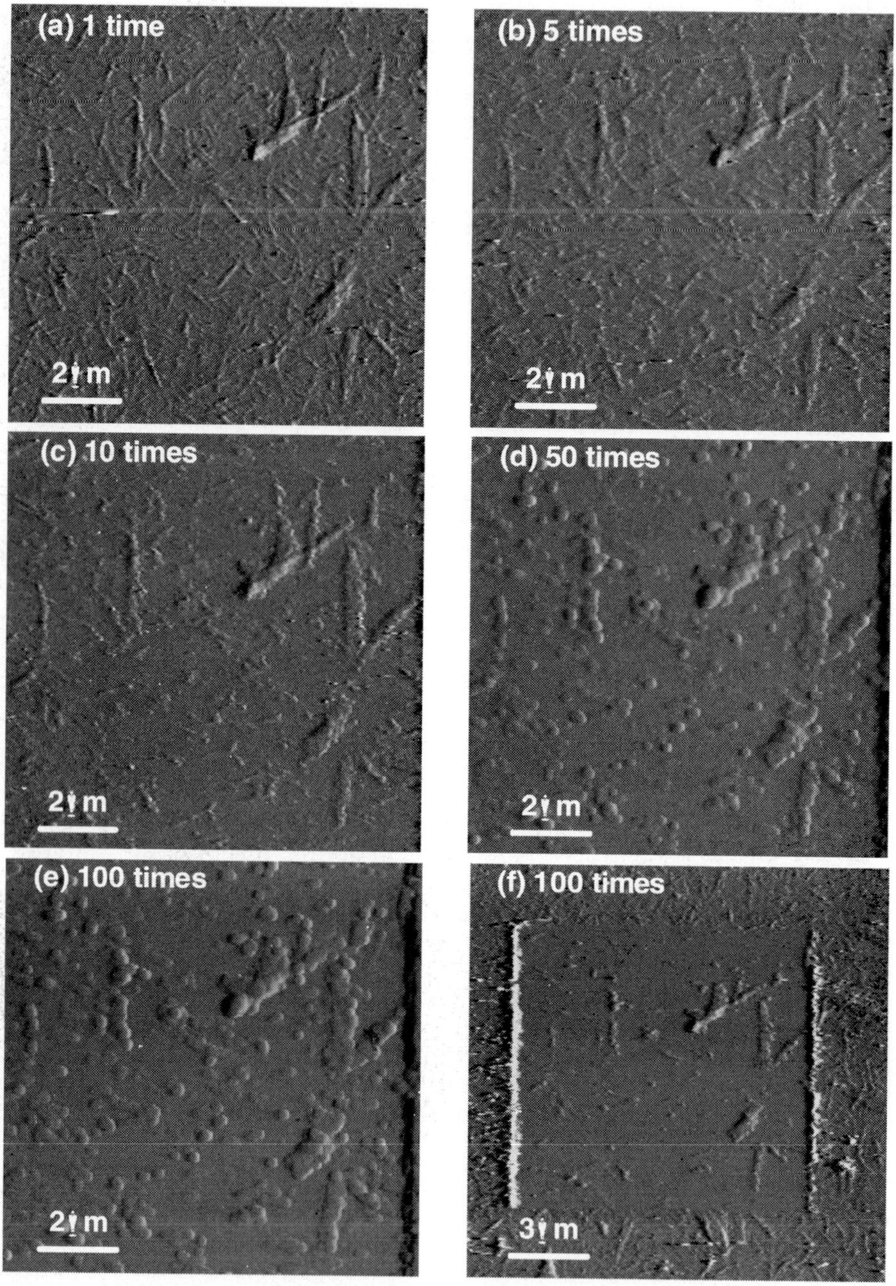

Fig. 17.5: AFM three-dimensional deflection images of the synthetic imogolite dried 1 day on the holder after different times of scanning: (a) 1 time, (b) 5 times, (c) 10 times, (d) 50 times, (e) 100 times (full scale of 10 μm), and (f) 100 times (full scale of 15 μm).

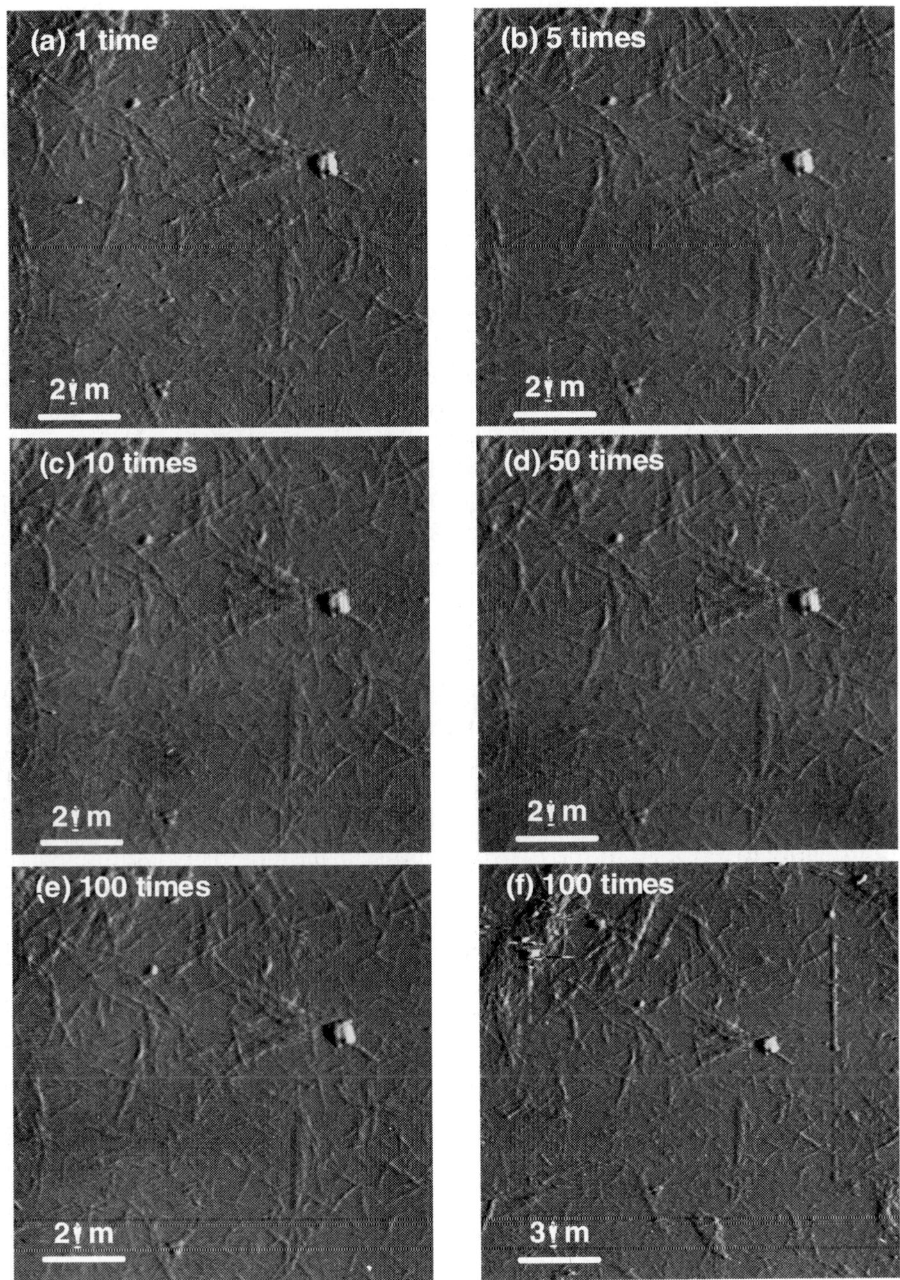

Fig. 17.6: AFM three-dimensional deflection images of the synthetic imogolite dried 1 month on the holder after different times of scanning: (a) 1 time, (b) 5 times, (c) 10 times, (d) 50 times, (e) 100 times (full scale of 10 μm), and (f) 100 times (full scale of 15 μm).

particles firmly attached to the surface in the case of imogolite, which has a high affinity for water. The relative humidity of the ambient air was observed to affect the etching rate of the substrate (Delawski and Parkinson, 1992). Thus, in applying contact-mode AFM to investigate the fragile imogolite structure, the moisture conditions of the sample and the ambient air are critical in precluding erosion of the sample surface.

In the absence of oxalic acid, the AFM three-dimensional deflection image showed the unique thready structure of imogolite (Fig. 17.7a). In the presence of a low concentration of oxalic acid (oxalate/Al molar ratio of 0.02), the thready structures, which are characteristic of imogolite, were still observed (Fig. 17.7b). At oxalate/Al molar ratios of 0.04 and 0.06, although the AFM images still showed the presence of fine and short imogolite particles, spheroidal and ill-defined particles were also observed (Fig. 17.7c,d). The distinctive thready structures were scarce at the oxalate/Al molar ratio of 0.08 and spheroidal particles predominant (Fig. 17.7e). At the oxalate/Al molar ratio of 0.10, no thready structures were observed (Fig. 17.7f). This is consistent with the XRD and FTIR results (Fig. 17.2f and Fig. 17.3f).

The average values of the diameter and length of the filamentous imogolite structures at the different oxalate/Al molar ratios were obtained from 15 different imogolite particles and statistical analysis was performed by the Fisher-LSD test (Table 17.2). With increments in the oxalate/Al molar ratio from 0 to 0.06, both the diameter and length decreased significantly. Imogolite threads at the oxalate/Al molar ratio of 0.00 were three times as long as those of 0.06. Contrarily, the average diameter of spheroidal particles of perturbed imogolite increased significantly with the increased in oxalate/Al molar ratio from 0.06 to 0.10 (Table 17.3). Lou and Huang (1989) suggested that the occupation of the coordination sites of Al in the hydroxy-aluminosilicate ions by citrate ligands perturbed the regular linkage of the hydroxy-aluminosilicate fragments into imogolite tubes, leading to the formation of structurally disordered material.

In the present study the disordered materials were spheroidal to irregularly shaped particles whose sizes increased with increase in effectiveness of complexing organic ligands to retard formation of imogolite, which accords with an earlier study that reported the morphology of the precipitation products affected by citrate was characterized by irregularly shaped particles and their high electron density aggregates (Inoue and Huang, 1984).

3.3 Dynamics of Cd Adsorption on Precipitates

The time function of Cd adsorption on imogolite and noncrystalline aluminosilicates which formed under the influence of oxalic acid, at the initial Cd concentration of 1×10^{-5} mol L^{-1}, showed that most Cd adsorption reached equilibrium within 10 min (Fig. 17.8). Inskeep and Baham (1983) found that the adsorption reaction of Cd by Na-montmorillonite monitored with an ion-selective electrode method was so rapid that constant millivolt readings were established within 2 min of initiation of the reaction and remained constant for

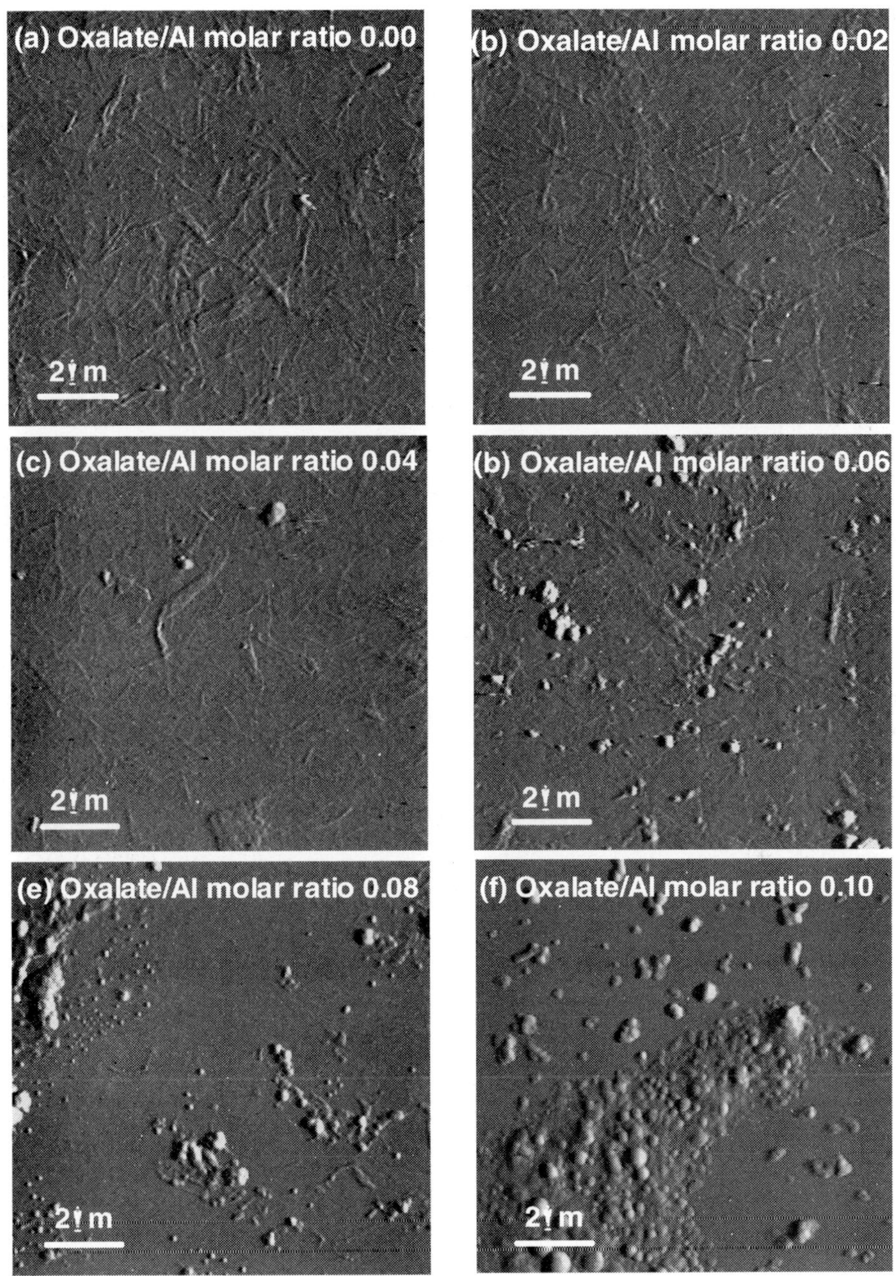

Fig. 17.7: AFM three-dimensional deflection images of synthetic imogolite and precipitates formed in the absence and presence of oxalic acid. Initial oxalate/Al molar ratio: (a) 0, (b) 0.02, (c) 0.04, (d) 0.06, (e) 0.08, and (f) 0.10. Precipitates were dispersed using ultrasonification at 150 W for 5 s in an ice bath. Suspensions of the precipitates were deposited on a watch glass and dried for 1 month at room temperature (23.5 ± 0.5°C).

Table 17.2: Diameter and length of thready structures of synthetic imogolite and precipitates as affected by oxalic acid

Oxalate/Al molar ratio	Diameter (nm)		Length (μm)	
	Avg. S.D.	Range	Avg. S.D.	Range
0.00	78.3 ± 15.5 d†	(43.8–97.9)	1.56 ± 0.40 d	(0.78–2.28)
0.02	65.7 ± 12.9 c	(43.8–83.0)	1.26 ± 0.4 c	(0.76–2.00)
0.04	53.4 ± 13.2 b	(27.7–70.6)	0.83 ± 0.2 b	(0.60–1.58)
0.06	44.2 ± 11.1 a	(20.0–61.9)	0.57 ± 0.1 a	(0.37–0.85)
0.08	N.A.		N.A.	
0.10	N.A.		N.A.	

N.A., Not applicable.
†Value within columns followed by the same letter did not differ significantly at the 0.05 level using the Fisher-LSD test.

Table 17.3: Diameter of spheroidal structures of synthetic imogolite and precipitates as affected by oxalic acid

Oxalate/Al molar ratio	Diameter (nm)	
	Avg. S.D.	Range
0.00	N.A.	N.A.
0.02	N.A.	N.A.
0.04	N.A.	N.A.
0.06	77.5 ± 16.2 a†	(58.7–100.0)
0.08	82.6 ± 33 a	(43.7–141.1)
0.10	133.0 ± 28.7 b	(87.5–180.4)

N.A., Not applicable.
†Values within columns followed by the same letter did not differ significantly at the 0.05 level using the Fisher-LSD test.

24 h. The specific surface area of the precipitates, measured by the Brunauer-Emmett-Teller (BET) method, was approximately 150 $m^2 g^{-1}$ and no significant difference was observed between the imogolite and perturbed imogolites (Table 17.4). PZSE values of the precipitates were around 6.7–6.8 with no difference among the precipitates (Table 17.4).

We applied first-order, second-order, and Elovich equations to determine the reaction order (Sparks, 1989). First-order plotting seemed to best fit the imogolite formed in the absence of oxalic acid, for which the rate constants of the imogolite and those of perturbed imogolite as affected by oxalic acid at the oxalate/Al molar ratios of 0.02 and 0.04, were calculated as 7.08×10^{-5} min^{-1} ($R^2 = 0.96$), 5.47×10^{-5} min^{-1} ($R^2 = 0.75$), and 1.69×10^{-5} min^{-1} ($R^2 = 0.73$), respectively. As for the perturbed imogolite samples, the Elovich equations fitted

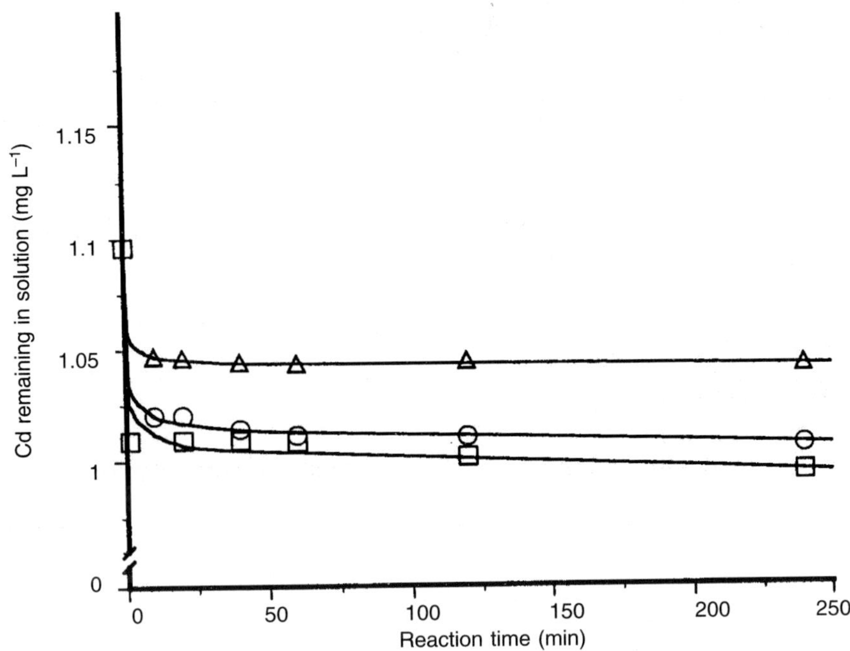

Fig. 17.8: Time function of Cd adsorption on the imogolite and noncrystalline aluminosilicates as affected by oxalic acid at the oxalate/Al molar ratios of 0.02 and 0.04. Oxalate/Al molar ratios: □, 0.00; ○, 0.02; Δ, 0.04.

for best, which the rate constants ß of the perturbed imogolite as affected by oxalic acid at the oxalate/Al molar ratios of 0.02 and 0.04 were calculated as 12.2 (R^2 = 0.92) and 38.2 (R^2 = 0.92) respectively. Since the Cd adsorption by these samples mostly finished within 10 min, the accuracy of the regression analyses was somewhat questionable. However, these equation analyses suggested that the kinetics of Cd adsorption on the imogolite and the perturbed imogolites should differ. As the Elovich equation has been used to describe the kinetics of heterogeneous sorption of various chemical species on soils and soil minerals (Sparks, 1989), it could be supposed that the Cd adsorptive sites and mechanism of the perturbed imogolites as affected by oxalic acid were more complicated than the imogolite formed in the absence of oxalic acid.

After 240 min, 3.15 to 9.35% of the Cd was adsorbed by the precipitates. The standard error values of the duplicates of the Cd adsorbed on clay minerals were generally lower than the decreasing percent adsorption of Cd by the precipitates formed at the increasing oxalate/Al molar ratio. The amount of Cd adsorbed decreased with increase in the oxalate/Al molar ratio (Table 17.5) in spite of no distinctive difference in the surface area and PZSE values. The

Table 17.4: Specific surface area measured by BET method and PZSE of the imogolite and precipitates formed at the oxalate/Al molar ratio 0.02 and 0.4

Oxalate/Al molar ratio	Specific surface area $m^2\ g^{-1}$	PZSE
0.00	146 ± 2†	6.75
0.02	153 ± 1	6.74
0.04	144 ± 0	6.82

†Standard errors of duplication.

Table 17.5: Adsorption of Cd on imogolite and perturbed imogolite as affected by oxalic acid at the oxalate/Al molar ratios 0.02, 0.04, 0.06, 0.08, and 0.10 after 4 h when the initial Cd concentration was 10^{-5} mol L^{-1}†

Oxalate/Al molar ratio	Cd remaining in solution (mg L^{-1})	Percent adsorption of Cd (%)	Cd adsorbed on clay minerals (mg kg^{-1})
0.00	0.994	9.35	205 ± 16‡
0.02	1.006	8.21	180 ± 9.9
0.04	1.042	4.97	109 ± 4.2
0.06	1.049	4.33	95 ± 1.4
0.08	1.051	4.11	90 ± 8.5
0.10	1.062	3.15	69 ± 1.4

† Initial Cd concentration of original solution was 1.096 mg L^{-1} (approximately 1×10^{-5} mol L^{-1}).
‡ Standard errors of duplication.

decrease in the adsorption rate of Cd by the perturbed imogolites observed in the present study, is attributed to the partial blocking of the reactive Al-OH sites on the outer surface of the imogolite tube by COOH groups of oxalic acid through complexation reaction. Modifications in structural properties of the imogolite by oxalic acid during its formation and subsequent changes in the adsorption of Cd by the perturbed imogolites observed in the present study, merit attention in understanding the dynamics and fate of toxic Cd in soils. Further study of Cd adsorption by poorly crystalline aluminosilicates, using lower concentrations of Cd solution and a graphite furnace atomic adsorption spectrophotometry, is essential to discussing on the more detailed dynamics of Cd in soil environments.

4 CONCLUSION

Oxalic acid, mostly present in the soil solutions of Andisols and spodosols, perturbs the formation of imogolite. However, the effectiveness of impedance of the interaction of hydroxy-Al ions with orthosilicic acid and subsequent formation of imogolite is four or five times less than that of citric acid and other strong complexing organic acids, as indicated by the results of x-ray diffractograms and IR absorption spectra. The morphological features of synthetic imogolite and x-ray noncrystalline aluminosilicates formed under the influence of oxalate can be observed using contact-mode AFM. With increase in the oxalate/Al molar ratio, the thready structures characteristic of imogolite decreased and spheroidal and irregular amorphous particles increased in the AFM three-dimensional images. The present AFM study also showed that oxalic acid can induce a change in size of filamentous imogolite structures and spheroidal particles of noncrystalline aluminosilicates, decreasing the length and width of thready structures and increasing the diameter of spheroidal particles with increase in oxalate/Al molar ratio.

Adsorption of Cd by these particles was a very rapid reaction and actually completed within 10 min. After 240 min, 3.15 to 9.35% of the Cd was adsorbed by the precipitates. The amount of Cd adsorbed decreased with an increase in oxalate/Al molar ratio. The decrease in the adsorption rate of Cd by the perturbed imogolites, observed in the present study, is attributed to the partial blocking of the reactive Al-OH sites on the outer surface of the imogolite tube by COOH groups of oxalic acid through complexation reaction. The present results suggest that the influence of oxalic acid on the shape and size of imogolite and noncrystalline aluminosilicates, and their subsequent surface properties and Cd adsorption on these minerals merits further attention.

Acknowledgments

This study was supported by the Ministry of Education of Japan and Research Grant GP2383-Huang of the Natural Sciences and Engineering Research Council of Canada.

References

Alexander G.B. 1953. The preparation of monosilicic acid. *J. Amer. Chem. Soc.* 75: 2887–2888.

Alloway B.J. 1995. Cadmium. In: *Heavy Metals in Soils*. B.J. Alloway (ed.). Blackie Acad. & Prof., London, UK, pp. 122–151 (2nd ed.).

Anderson H.A., Berrow M.L., Farmer V.C., Hepburn A., et al., 1982. A reassessment of podzol formation processes. *J. Soil Sci.* 33: 1125–136.

Bertsch P.M. and Hunter D.B. 1998. Elucidating fundamental mechanisms in soil and environmental chemistry: The role of advanced analytical, spectroscopic, and microscopic methods. In: *Future Prospects for Soil Chemistry*. P.M. Huang et al. (eds.). SSSA, Madison, WI (USA), pp. 103–122.

Blum A.E. 1994. Determination of illite/smectite particle morphology using scanning force microscopy. In: *Scanning Probe Microscopy of Clay Minerals. Clay Mineral Society Workshop Lectures*. K.L. Nagy and A.E. Blum (eds.). Clay Min. Soc., Boulder, CO (USA), vol. 7, pp. 171–202.

Brydon J.E. and Shimoda S. 1972. Allophane and other amorphous constituents in a podzol from Nova Scotia. *Can. J. Soil Sci.* 52: 465–475.

Cradwick P.D.G., Farmer V.C., Russell J.D., et al. 1972. Imogolite, a hydrated aluminium silicate of tubular structure. *Nature Phys. Sci.* 240: 187–189.

Delawski E. and Parkinson A. 1992. Layer-by-layer etching of two-dimensional metal chalcogenides with the atomic force microscope. *J. Amer. Chem. Soc.* 114: 1661–1667.

Farmer V.C., Russell J.D., and Berrow M.L. 1980. Imogolite and proto-imogolite allophane in spodic horizons: evidence for a mobile aluminium silicate complex in podzol formation. *J. Soil Sci.* 31: 673–684.

Farmer V.C., Adams M.J., Fraser A.R., and Palmieri F. 1983. Synthetic imogolite: properties, synthesis, and possible applications. *Clay Miner.* 18: 459–472.

Fischer L., Zur Mühlen E., Brümmer G.W., and Niehus H. 1996. Atomic force microscopy (AFM) investigations of the surface topography of a multidomain porous goethite. *Eur. J. Soil Sci.* 47: 329–334.

Fox T.R. 1995. The influence of low-molecular-weight organic acids on properties and processes in forest soils. In: *Carbon Forms and Functions in Forest Soils*. W.W. McFee and J.M. Kelly (eds.). SSSA, Madison, WI (USA), pp. 43–62.

Fox T.R. and Comerford N.B. 1990. Low-molecular-weight organic acid in selected forest soils of the southeaster USA. *Soil Sci. Soc. Amer. J.* 54: 1139–1144.

Gustafsson J.P., Bhattacharya P., Bain D.C., Eraser A.R., and McHardy W.J. 1995. Podzolisation mechanisms and the synthesis of imogolite in northern Scandinavia. *Geoderma* 66: 167–184.

Higashi T. and Ikeda H. 1974. Dissolution of Allophane by acid oxalate solution. *Clay Sci.* 4: 205–211.

Hsu P.H. 1963. Effect of initial pH, phosphate, and silicate on the determination of aluminum with aluminon. *Soil Sci.* 96: 230–238.

Hsu P.H. 1989. Aluminum hydroxides and oxyhydroxides. In: *Minerals in Soil Environment*. J.B. Dixon and S.B. Weed (eds.). SSSA, Madison, WI (USA), pp. 351–378.

Iler R.K. 1979. *The Chemistry of Silica*. John Wiley & Sons, New York, NY.

Inoue K. and Huang P.M. 1984. Influence of citric acid on the natural formation of imogolite. *Nature* 308: 58–60.

Inoue K. and Huang P.M. 1985. Influence of citric acid on the formation of short-range ordered aluminosilicate. *Clays Clay Miner.* 33: 312–322.

Inoue K. and Huang P.M. 1986. Influence of selected organic ligands on the formation of allophane and imogolite. *Soil Sci. Soc. Amer. J.* 50: 1623–1633.

Inoue K. and Huang P.M. 1987. Effect of humic and fulvic acids on the formation of allophane. *Proc. Int. Clay Conf. Denver 1985*. L.G. Schultz et al. (eds.). Clay Miner. Soc., Bloomington, IN (USA) pp. 221–226.

Inskeep W.P. and Baham J. 1983. Adsorption of Cd(II) and Cu(II) by Na-montmorillonite at low surface coverage. *Soil Sci. Soc. Amer. J.* 47: 660–665.

Liu C. and Huang P.M. 1999. Atomic force microscopy and surface characteristics of iron oxides formed in citrate solutions. *Soil Sci. Soc. Amer. J.* 63: 65–72.

Lou G. and Huang P.M. 1989. Nature and charge properties of x-ray noncrystalline aluminosilicates as affected by citric acid. *Soil Sci. Soc. Amer. J.* 53: 1287–1293.

Malajczuk N. and Cromack Jr. K. 1982. Accumulation of calcium oxalate in the mantle of ectomycorrhizal roots of *Pinus radiata* and *Eucalyptus marginate*. *New Phytologist* 92: 527–531.

Maurice P.A. 1998. Scanning probe microscopy of environmental surfaces. In: *Structure and Surface Reactions of Soil Particles*. P.M. Huang et al. (eds.). John Wiley and Sons, Chichester, UK, pp. 109–153.

McLaughlin M.J. and Singh B.R. 1999. Cadmium in soils and plants. A global perspective. In: *Cadmium in Soils and Plants*. M.J. McLaughlin and B.R. Singh (eds.). Kluwer Acad. Publ., Dordrecht, Netherlands, pp. 1–9.

Nagy K.L. and Blum A.E. 1994. *Scanning Probe Microscopy of Clay Minerals. Clay Mineral Society Workshop Lectures*. Clay Miner. Soc. Boulder, CO (USA), vol. 7.

Parfitt R.L. and Henmi T. 1980. Structure of some allophanes from New Zealand. *Clays Clay Miner.* 28: 285–294.

Parfitt R.L., Thomas A.D., Atkinson R.J., and Smart R.S.C. 1974. Adsorption of phosphate on imogolite. *Clays Clay Miner.* 22: 455–456.

Ross G.J. and Kodama H. 1979. Evidence for imogolite in Canadian soils. *Clays Clay Miner.* 27: 297–300.

Rugar D. and Hansma P. 1990. Atomic force microscopy. *Physics Today* 43: 23–30.

Russell J.D., McHardy W.J., and Fraser A.R. 1969. Imogolite: A unique aluminosilicate. *Clay Miner.* 8: 87–99.

Schnitzer M. 1978. Humic substances: Chemistry and reactions. In: *Soil Organic Matter.* M. Schnitzer and S.U. Khan (eds.). Elsevier, Amsterdam, Netherlands, pp. 1–64.

Sollins P., Cromack Jr. K., and Li C.Y. 1981. Role of low-molecular-weight organic acids in the inorganic nutrition of fungi and higher plants. In: *The Fungal Community*. D.T. Wicklow and G.C. Carroll (eds.). Marcel Dekker, New York, NY, (USA) pp. 607–619.

Sparks D.L. 1989. *Kinetics of Soil Chemical Processes.* Acad. Press Inc., San Diego, CA, (USA) pp. 4–38.

Stevenson F.J. 1967. Organic acids in soils. In: *Soil Biochemistry*. A.D. McLaren and G.H. Peterson (eds.). Marcel Dekker, New York, NY, (USA) pp. 119–146.

Tait J.M., Yoshinaga N., and Mitchell B.D. 1978. The occurrence of imogolite in some Scottish soils. *Soil. Sci. Plant Nutr.* 24: 145–151.

Tani M. and Highashi T. 1999. Vertical distribution of low molecular weight aliphatic carboxylic acids in some forest soils of Japan. *Eur. J. Soil Sci.* 50: 217–226.

Tani M., Higashi T., and Nagatsuka S. 1995. Low-molecular-weight aliphatic carboxylic acids in some Andisols of Japan. In: *Environmental Impacts of Soil Component Interactions*, Vol. 1: *Natural and Anthropogenic Organics*. P.M. Huang et al. (eds.). CRC Press, Boca Raton, FL (USA), pp. 233–241.

Tani M., Higashi T., and Nagatsuka S. 1996. Dynamics of low-molecular-weight aliphatic carboxylic acids (LACAs) in forest soils: II. Seasonal changes of LACAs in an Andisol of Japan. *Soil Sci. Plant Nutr.* 42: 175–186.

Tani M., Shida K.S., Tsutsuki K., and Kondo R. 2001. Determination of water-soluble low-molecular-weight organic acids in soils by ion chromatography. *Soil Sci. Plant Nutr.* 47: 387–397.

Tokashiki Y. and Wada K. 1975. Weathering implications of the mineralogy of clay fractions of two ando soils, Kyusyu. *Geoderma* 14: 47–62.

Vance G.F., Stevenson F.J., and Sikora F.J. 1996. Environmental chemistry of aluminum-organic complexes. In: *The Environmental Chemistry of Aluminum.* G. Sposito (ed.) CRC Press, Boca Raton, FL (USA), pp. 169–220 (2nd ed.).

Wada K. 1989. Allophane and imogolite. In: *Minerals in Soil Environment.* J.B. Dixon and S.B. Weed (eds.). SSSA, Madison, WI (USA), pp. 1051–1087 (2nd ed.).

Wada S.-I., Eto A., and Wada K. 1979. Synthetic allophane and imogolite. *J. Soil Sci.* 30: 347–355.

Weaver R.M., Syers J.K., and Jackson M.L. 1968. Determination of silica in citrate-bicarbonate-dithionite extracts of soils. *Soil Sci. Soc. Amer. J.* 32: 497–501.

White J.L. and Roth C.B. 1986. Infrared spectrometry. In: *Methods of Soil Analysis*, Part 1. *Physical and Mineralogical Methods*. A Klute et al. (eds.). SSSA, Madison, WI (USA), pp. 291–330.

Yoshinaga N. and Aomine S. 1962. Imogolite in some Ando soils. *Soil Sci. Plant Nutr.* 8: 22–29.

18

Charge Characteristics and Cu^{2+} Adsorption/Desorption of Variable-Charge and Permanent-Charge Soils

X.Y. Li*, W.T. Ling, *and* **J.Z. He**

Abstract

Charge characteristics, Cu^{2+} adsorption-desorption of variable-charge soil (Latosol) and permanent-charge soil (Brown soil) and the relation between them were studied by back-titration and the adsorption equilibrium method respectively. Results showed that the amount of variable negative charge in the variable-charge soil was much less than that in the permanent-charge soil and increased with soil pH. This is attributed to surface coating of short-range ordered sesquioxides on layer silicates. The soil PZC was higher in the variable-charge soil than in the permanent-charge soil. The amount of Cu^{2+} sorbed by the permanent-charge soil was greater than that by the variable-charge soil and increased with Cu^{2+} concentration in the equilibrium solution. The amount and percentage of Cu^{2+} ion desorbed from the permanent-charge soil by KCl was larger than those from the variable-charge soil; the opposite trend was observed for those of Cu^{2+} ion desorbed by deionized water. The increase of soil PZC with adsorption of Cu^{2+} ion varied in the two soils. Most of the Cu^{2+} ion was adsorbed by specific adsorption in both permanent-charge and variable-charge soils. A larger proportion of Cu^{2+} ions was adsorbed by specific adsorption by the later than the former. Hydrolyzable charge was more pronounced in the variable-charge soil than in the permanent-charge soil.

1 INTRODUCTION

Environmental deterioration due to increase in population, urbanization, as well as development of industry, agriculture, and economy in China has received

---

Corresponding author: Prof. Xueyuan Li, College of Resources and Environment, Huazhong Agricultural University, Wuhan 430070, P.R. China. E-mail: qyhuang@mail.hzau.edu.cn increasing attention from society (Zhou, 1996; Zhang, 1998; Lin et al., 1999). Protection of soil resources and sustainable development of agriculture are important tasks confronting farmers and agricultural scientists (He et al., 1998; Zhao and Cao, 2000). Some heavy metals are beneficial elements whereas some are toxic to the growth and development of organisms. Heavy metals are not degradable and slowly mobile. Further, they readily accumulate in soil and organisms, which affects the quality of agricultural products and threatens human health through the food chain (Liu et al., 1995). Because it is very difficult to completely remove a heavy metal from soils, remediation of heavy metals is still a hot and problematical soil environmental issue (Dai, 1997; Zhang, 1999).

The concentration, activity, bioavailability or biotoxicity of heavy metals are controlled by the adsorption-desorption phenomena of heavy metal on the surface of soil colloids (Xiong and Chen, 1990). The surface charge characteristics of soil constitute one of the essential factors governing the behavior of the adsorption-desorption process of heavy metals (Ma and Chen, 1998). The difference in charge characteristics between variable-charge soils and permanent-charge soils is remarkable. The effect of these soil characteristics on the adsorption-desorption phenomena of heavy metals likewise differs, which provides information for understanding heavy metal behavior in the two types of soils (Qiu and Xue, 1990; Chubin and Steet, 1981; Hingston and Quirk, 1967). The present study investigated the charge characteristics of a permanent-charge soil (Brown soil) and a variable-charge soil (Latosol) in relation to their adsorption-desorption of Cu^{2+} ions.

2 MATERIALS AND METHODS

2.1 Location and Properties of Soils Studied

The location and basic properties of soils studied are listed in Table 18.1 (see below). The texture, pH value, CEC, organic matter, and free iron-oxides were measured by pipette method, pH meter method (soil:H_2O = 1:2.5), ammonium acetate (pH 7) extraction method, Walkley-Black method, and dithionite-citrate-bicarbonate (DCB) extraction method (Lu, 2000), respectively.

The composition of inorganic colloids in the soils studied is given in Table 18.2. Soil inorganic colloids were prepared by mixing a certain amount of soil samples (< 60 mesh sieve) with deionized water. The suspensions were agitated for 24h. Organic matter and free $CaCO_3$ were removed by treatment of soil

Table 18.1: Location and basic properties of the soils studied

Soil	Location	Latitude	Depth	Texture	pH_{H_2O}	pH_{KCl}	CEC	OM	Free Fe_2O_3
	County Province	°N	cm				Cmol kg^{-1}		g kg^{-1}
Latosol	Xuwen, Guangdong	20.5	80–130	Clay	5–15	5.01	7.15	7.0	156.4
Brown soil	Weihai, Shandong	36.7	40–70	Clay loam	6.40	5.66	14.41	7.0	24.0

samples with H_2O_2 and acetic acid-sodium acetate solution respectively. Suspensions of treated soils were dispersed by adding 0.5 mol L^{-1} or 0.01 mol L^{-1} NaOH solution to pH 6.5–7.0, together with sonification occasionally, then passed through a 300 mesh sieve. Finally, the <2 μm inorganic colloids in soil samples were separated by the sedimentation method. The free and noncrystalline Fe(Fe_2O_3) in these colloids were extracted with DCB and ammonium oxalate (pH 3.2) solutions, respectively (McKeague and Day, 1966). The contents of Fe were determined by atomic absorption spectroscopy (AAS, Shimadzu AA-607). The composition of clay minerals in soil colloids was analyzed by the x-ray diffraction (XRD) method.

Cu^{2+} treated soils were prepared as follows: 0.01 mol L^{-1} KNO_3 solutions containing 2 mmol L^{-1} $Cu(NO_3)_2$ with pH 5.0 were added to a series of beakers containing soil samples at a soil:solution ratio of 1:5 and equilibrated for 48 h at 25 ± 1°C. The soil suspensions were agitated and mixed for 1 h per day, then centrifuged and the supernatant discarded. Soils in the centrifuge tube were washed 3–4 times with alcohol, dried at 40°C and passed through a 60 mesh sieve.

2.2 Adsorption and Desorption of Cu^{2+} Ions

Twenty milliliters of 0.5 to 4.0 mmol L^{-1} $Cu(NO_3)_2$ containing 0.01 mol L^{-1} KCl solution with pH 5.0 were added to a series of centrifuge tubes containing soil samples at soil:solution ratio of 1:10. The mixtures were equilibrated for 2 days, stirred for 1 h per day, centrifuged (5,020 g for 10 min, Bakemen J_2-Mc) then washed 2 times with alcohol. Finally the content of Cu^{2+} ion in the supernatant was determined by the AAS (Shimadzu AA 607) and the amount of Cu^{2+} ions sorbed on the soils was calculated.

The residual soil in the centrifuge tubes was washed three times with alcohol and the washings were discarded. The soil in the centrifuge tubes was again washed with alcohol and then with deionized water (pH 6.4) or with alcohol and 1 mol L^{-1} KCl (pH 6.4) respectively. The content of Cu^{2+} in the washings was determined by AAS. The amount and proportion of Cu^{2+} desorbed with deionized water or 1 mol L^{-1} KCl were calculated.

2.3 Determination of PZC and Amount of Surface Negative Charge in Soils Studied

PZC and surface variable negative charge for the soils were estimated by the salt titration method (Van Raij and Peech, 1972) and back-titration method (Martin and William, 1993) respectively.

3 RESULTS AND DISCUSSION

3.1 Properties and Inorganic Colloid Composition of Soils Studied

The pH and CEC values of the Brown soil (corresponding to Udic Luvisols) were higher than those of the Latosol (corresponding to Udic Ferriallisols) whereas the content of free iron oxides of the Brown soil was about 6 times less than that of Latosol (Table 18.1). The amount of noncrystalline iron oxides in the inorganic colloids of Brown soil (4.0 g kg^{-1}) was twice that of the Latosol (2.0 g kg^{-1}), while the content of free and crystalline iron oxides in the inorganic colloids of the Latosol was twice that of the Brown soil (Table 18.2). Kaolinite (95%) was the predominant clay mineral in the Latosol with gibbsite accounting for 5%, while the contents of kaolinite (35%), hydromica (35%), and vermiculite (30%) were similar in the Brown soil.

Table 18.2: Composition of clay minerals and iron oxides in inorganic colloids of soils studied

Soil	Fe oxides(Fe_2O_3)			Clay mineral association[a]
	Noncrystalline	Crystalline	Free	
	g kg^{-1}			
Latosol	2.0	62.4	64.4	Kao (95%), Gib (5%)
Brown soil	4.0	23.0	27.0	Kao (35%), HM (35%), Ver (30%)

[a]Kao: Kaolinite; Gib: gibbsite; HM: hydrous mica; Ver: vermiculite.

3.2 Adsorption of Cu^{2+} Ions

Adsorption data on the adsorption isotherms of Cu^{2+} ions in the soils studied at pH 5.0 and 20°C conformed to the Langmuir equation, $X = B_{max} KC/(1+KC)$, where X is the amount of Cu^{2+} adsorbed at equilibrium (mg kg^{-1}), B_{max} the maximum amount of Cu^{2+} that could be adsorbed (mg kg^{-1}), K a constant related to the adsorption energy, and C the concentration of Cu^{2+} in the equilibrium solution (mg L^{-1}). The r^2 values of the linear regression of the fit of the Langmuir equation to the data were 0.9985 for the Brown soil and 0.9956 for the Latosol. For the Brown soil, the B_{max} and K values were 1,130 mg kg^{-1} and 0.169 respectively; for the Latosol, the B_{max} and K values were 904 mg kg^{-1} and 0.081, respectively. The adsorption isotherms of both soils were of L class (Fig. 18.1), indicating that adsorption becomes increasingly difficult as the sites for adsorption are continually filled. The B_{max} of the Brown soil (1,130 mg kg^{-1}) was higher than that of Latosol (904 mg kg^{-1}). With the same concentration of Cu^{2+} ions in the equilibrium solution, the amount of Cu^{2+} ion sorbed by the Brown soil was larger than that by the Latosol. These were consistent with the differences in their colloidal composition (Tables 18.1 and 18.2) and surface charge characteristics (Table 18.3).

3.3 Desorption of Cu^{2+} Ions

The Cu^{2+} ions desorbed by 1 mol L^{-1} KCl (pH 6.4) could be mainly those adsorbed by electrostatic attraction. Figure 18.2 shows that the amount and percentage of Cu^{2+} ions desorbed by 1 mol L^{-1} KCl in the soils studied increased with

Table 18.3: PZC and amount of variable negative surface charge for soils studied at different pH levels before and after Cu^{2+} ion sorption

Soil and Treatment*	Qv (cmol. kg^{-1})								PZC
	pH 3.5	pH 4	pH 5	pH 6	pH 7	pH 8	pH 9	pH 10	
1*	0.27	0.45	1.32	1.76	2.33	3.26	5.35	7.98	5.0
2*	0.06	0.16	1.01	1.34	1.95	2.67	4.01	6.21	5.3
1–2	0.21	0.29	0.31	0.42	0.38	0.59	1.34	1.77	
1–2/1 × 100	77.78	64.44	23.48	23.86	16.31	18.10	25.05	22.18	
3*	0.60	0.90	2.09	2.61	3.31	4.73	6.97	11.28	3.0
4*	0.31	0.66	1.73	2.33	2.65	2.76	5.41	7.73	3.1
3–4	0.29	0.24	0.36	0.28	0.66	1.97	1.56	3.55	
3–4/3 × 100	48.33	26.67	17.22	12.02	19.94	41.65	22.38	31.47	

1* Latosol; 2* Latosol after Cu^{2+} ion sorption; 3* Brown soil; 4* Brown soil after Cu^{2+} ion sorption

increments in Cu^{2+} concentration in the equilibrium solution and were much higher in the Brown soil (0–396 mg kg^{-1}, 0–36.38%) than in Latosol (0–66 mg kg^{-1}, 0–8.7%). These results imply that the proportion of electrostatic adsorption in Cu^{2+} ion adsorption on Brown soil was greater than on Latosol. However, when adsorbed Cu^{2+} ions in the soils studied were desorbed by deionized water with pH 6.4 the amount and percentage of Cu^{2+} ions desorbed also increased

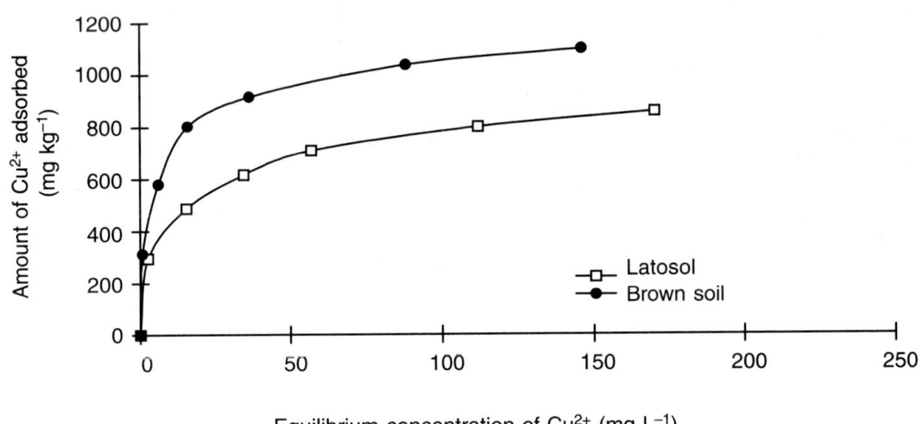

Fig. 18.1: Isotherms of Cu^{2+} ion adsorbed on the soils studied.

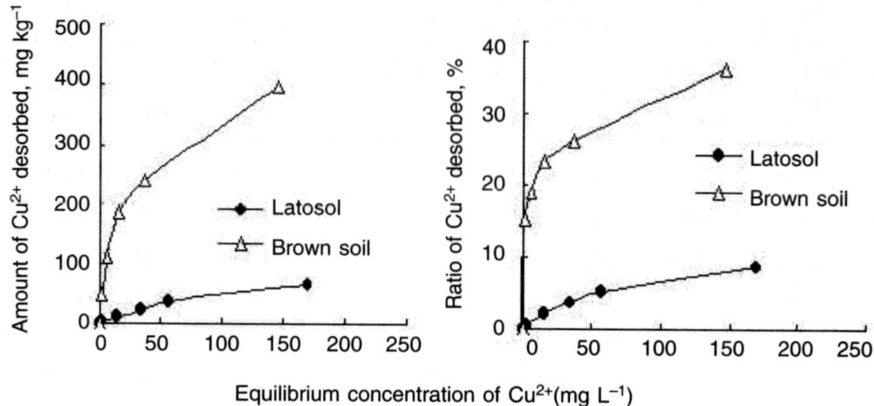

Fig. 18.2: The amount (mg kg^{-1}) and proportion (%) of Cu^{2+} ions desorbed by 1 mol L^{-1} KCl.

with increments in Cu^{2+} concentration in the equilibrium solution, and the magnitudes of their increments in the Latosol were obviously higher than those in the Brown soil (Fig. 18.3). The possible reasons could be: 1) more permanent-negative charge in the Brown soil than in the Latosol, and a considerable amount of Cu^{2+} ions sorbed on the Brown soil was caused by electrostatic adsorption, so that the Cu^{2+} sorbed would exchange with K^+ in solution when 1 mol L^{-1} KCl (pH 6.4) was used as desorptive reagent; and 2) adsorption of Cu^{2+} ions on the Latosol was largely due to specific adsorption but also due to hydrolyzable charge to a considerable extent; permanent charge played a relatively small role in Cu^{2+} adsorption in the Latosol. This was apparently due to the remarkably higher content of crystalline iron oxides in the Latosol than in the Brown soil (Tables 18.1 and 18.2). These metal oxides should have a substantial amount of hydrolyzable charge (Huang and Jackson, 1966; Huang, 1980), which would account for more Cu^{2+} desorption by deionized water for the Latosol than the Brown soil.

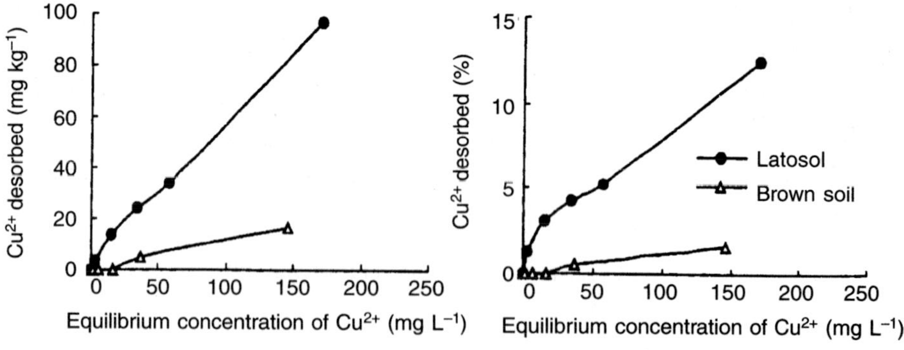

Fig. 18.3: Amount (A) and proportion (B) of Cu^{2+} ions desorbed by deinoized water.

3.4 PZC and Surface Charge of Soils Studied before and after Adsorption of Cu^{2+} Ions

Variable negative surface charge in soil suspension (QV_{ss}) and soil centrifuge solution (QV_{scs}) were determined by back-titration method, and the amount of variable negative surface charge on the soil surface (Q_v) was calculated from the equation: QV (cmol kg^{-1}) = QV_{ss} (cmol kg^{-1}) – QV_{scs} (cmol kg^{-1}).

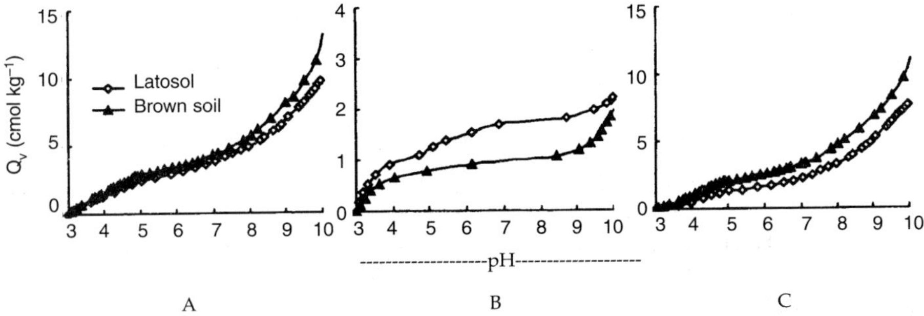

Fig. 18.4: Variable negative surface charge (QV) in soil suspension (A), soil centrifuge solution (B), and on soil surface (C) at different pH levels.

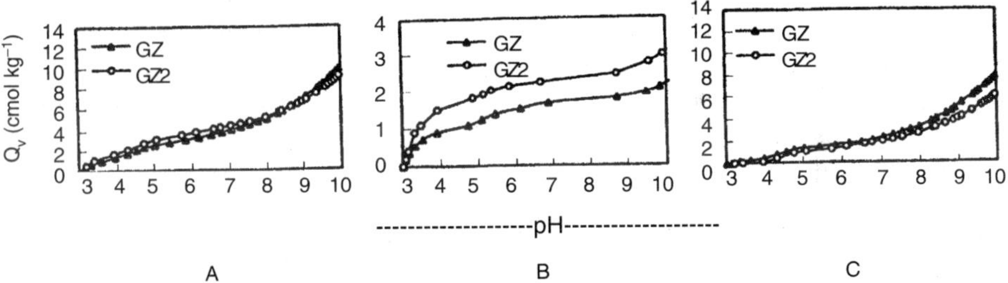

Fig. 18.5: Variable negative surface charge (QV) on Latosol before (GZ) and after (GZ2) Cu^{2+} sorption at different pH levels: A—in soil suspension, B—soil centrigue solution, C—on soil surface.

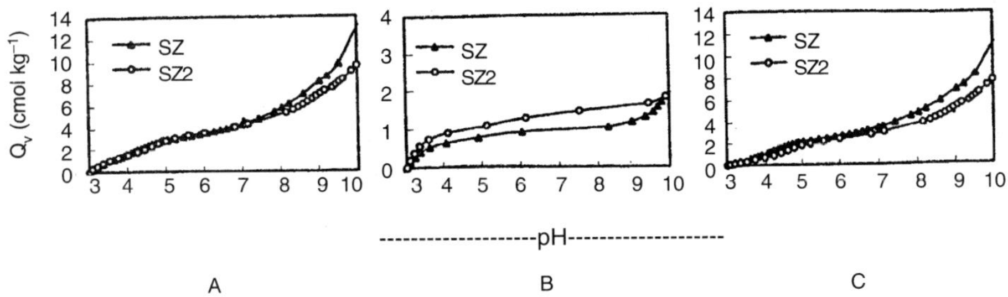

Fig. 18.6: Variable negative surface charge (QV) on brown soil before (SZ) and after (SZ2) Cu^{2+} sorption at different pH levels. A—in soil suspension, B—soil centrifuge solution, C—on soil surface.

Figures 18.4, 18.5, and 18.6 show that the amounts of Qv_{ss}, Qv_{scs} and Qv of the soils studied before and after soil adsorption of Cu^{2+} ions increased with increments in pH both soil systems. However, the ranges of the increase at different pH levels differed and varied with soil type. The increase of Qv_{ss} and Qv in the Brown soil was greater than that in the Latosol whereas the increase of Qv_{scs} in the Brown soil was less than that in the Latosol. Possible reasons for these results are: 1)the magnitude of the variable negative surface charge for different types of soils and its relation with soil pH value are determined by the composition and content of the solid phase of each soil (Wu et al., 2002), and 2) the two soils have the same organic matter content but the clay mineral composition of the Brown soil is mainly 2:1 with a small amount of iron oxides, while the clay mineral association of the Latosol is predominantly kaolinite with much greater amounts of iron oxides.

The relation between variable negative surface charge and pH of equilibrium solution for the Brown soil and the Latosol after adsorption of Cu^{2+} ions is shown in Table 18.3. Compared with the soils before adsorption of Cu^{2+} ions, the amount of Qv of the soils after adsorption of Cu^{2+} ions decreased, more so in the Brown soil (0.29–3.55 cmol kg^{-1}) than in the Latosol (0.21–1.77 coml kg^{-1}), in the range of pH 3–10 of the equilibrium solution; at the same pH, the percentage of Qv decrease in the range of pH 3–6 in the Latosol (77.8%–23.48%) was higher than that in the Brown soil (48.33%–12.02%). Further, after Cu^{2+} adsorption, the PZC of the Brown soil and the Latosol changed from 3.0 to 3.1 and from 5.0 to 5.3 respectively, which accorded with their Qv change in the range of pH < 6. The results indicate that adsorption of Cu^{2+} had a marked effect on variable negative surface charge of soil with variable charge or soil with permanent charge.

4 SUMMARY AND CONCLUSIONS

The amount of variable negative surface charge in the so called permanent-charge soil (Brown soil) was higher than that of the variable-charge soil (Latosol) and increased with increments in soil pH. This is attributed to the surface coatings of sesquioxides on layer silicates, in particular vermiculite. The PZC of the former was lower than that of the latter. These properties are closely related to the composition and content of the clay minerals and Fe and Al oxides in the soils studied. The amount of Cu^{2+} adsorbed by the permanent-charge soil was higher than that of the variable-charge soil and increased with increase in Cu^{2+} concentration in equilibrium solution. The PZC of both soils increased after adsorption of Cu^{2+}. After adsorption of Cu^{2+} the amount of variable negative surface charge in the permanent-charge soil decreased more than in the variable-charge soil. The amount and percentage of Cu^{2+} ion desorbed from the permanent-charge soil by 1 mol L^{-1} KCl was higher than that from the variable-charge soil. Both the amount and percentage of Cu^{2+} ion desorbed by deionized water were extremely low for the permanent-charge soil versus relatively higher amounts of Cu^{2+} desorption by deionized water in the variable-charge soil.

Data indicate that the permanent-charge soil and the variable-charge soil differ in proportion of electrostatic adsorption and specific adsorption of Cu^{2+} ions. Vis-à-vis the permanent-charge soil, the Cu^{2+} ions adsorbed by specific adsorption on the variable-charge soil were considerably more. However, most of the Cu^{2+} was adsorbed by specific adsorption even on the permanent-charge soil. The data also indicate that the variable-charge soil contained a higher amount of hydrolyzable charge, resulting in a greater amount and fraction of Cu^{2+} desorbed by deionized water compared with the permanent-charge soil.

Acknowledgments

This project (No. 49831005, No. 49871043) was supported by the National Natural Science Foundation of China.

References

Chubin R.G. and Steet J.J. 1981. Adsorption of cadmium on soil constituents in the presence of complexing ligands. *J. Environ. Qual.* 10: 225–228.

Dai Shu-gui. 1997. *Environment Chemistry*. High Education Publ. House, Beijing, China.

He Zhen-li, Zhou Qi-xing, and Xie Zheng-miao. 1998. *Soil Chemical Equilibrium of Pollutant and Beneficial Elements*. Chin. Environ. Sci. Publ. House, Beijing, China.

Hingston F.J. and Quirk J.P. 1967. Specific adsorption of anions. *Nature* (London) 215: 1459–1461.

Huang P.M. 1980. Adsorption processes in soils. In: *Handbook of Environmental Chemistry*. O. Hutzinger (ed.). Springer-Verlag, New York, NY, vol. 2, 47–59.

Huang P.M. and Jackson M.L. 1996. Fluoride interaction with clays in relation to third buffer range. *Nature* (London) 211: 779–780.

Lin Shao-xing, Liu Tian-ji, and Liu Yi-nong. 1999. *Introduction to Environment Protection*. High Education Publ. House. Beijing, China.

Liu Pie-ton, Xue Ji-yu, and Wang Hua-dong. 1995. Introduction to Environment Science. High Education Publ. House, Beijing, China.

Lu Yu-kun (Ed.-in-Chief). 2000. *Analytical Methods of Soil and Agrochemistry*. Agric. Sci. Tech. Publ. House China, Beijing, China (in Chinese).

Ma Yi-jie and Chen Jia-fang. 1998. Form, characteristics and function of iron oxides in red soils of China. *Soil* 30: 1–6 (in Chinese).

Martin D. and William H. 1993. Soil surface charge evaluation by back-titration. 1. Theory and method development. *Soil Sci. Soc. Amer. J.* 57: 1222–1228.

Mckeague J.A. and Day J.H. 1966. Dithionite and oxalate-extractable Fe and Al as aid in differentiating various classes of soils. *Can. J. Soil Sci.* 46: 13–22.

Qiu shao-min and Xue Jia-hua. 1990. Kinetics of Cd competitive sorption in red earth. *Environ. Chem.* 9: 1–5 (in Chinese).

Van Raij B. and Peech M. 1972. Electrochemical properties of some oxisols and alfisols of the tropics. *Soil Sci. Soc. Amer. Proc.* 36: 587–593.

Wu Jing-ming, Liu Yong-hong, Li Xue-yuan Ling, Wan-ting, and Dong Yuan-yan. 2002. Surface charge characteristics of inorganic colloids for several zonal soils in China. *Acta Pedol. Sinica* 39: 177–183.

Xiong Yi and Chen Jia-fang. 1990. *Soil Colloid*, vol. 3: *Properties of Soil Colloid*. Science Press, Beijing, China.

Zhang Gui-yin, 1998. Effect of organic acid on adsorption-desorption of cadmium by soil and minerals and its mechanism. PhD thesis, Huazhong Agric. Univ., Wuhan, China.

Zhang Nai-ming. 1999. The present situation and prospect of research on heavy metal pollution in soil–plant system. *Adv. Environ. Sci.* 7: 30–33 (in Chinese).

Zhao Qi-guo and Cao Hui. 2000. Pedosphere and its effect on global changes. Pedosphere 10: 97–106.

Zhou Dai-hua. 1996. Mechanism of Cu^{2+} adsorption on surfaces of Fe and Al oxides and effects of natural organic ligands. PhD thesis, Huazhong Agric. Univ., Wuhan, China.

19

Phosphate-Induced Cadmium Release from Soils

J.O. Onyatta *and* **P.M. Huang***

Abstract

Phosphate fertilizers are applied to correct phosphorus deficiency in acidic tropical soils of variable charge, which account for a large proportion of the world's arable land. They are commonly applied in bands or mixed with seeds. Within the vicinity of the phosphate fertilizer zone, the concentration of phosphate is high and can cause dissolution of soil minerals. Hence a study was conducted to investigate the degree to which phosphate may effect release of cadmium (Cd) from soil, rate of release, and formation of phosphate reaction products. Since phosphate fertilizers contain a wide range of Cd concentrations, perturbation of phosphate reaction product formation by Cd was also investigated. The surface soils used in this study were selected from tropical soils in Kenya varying widely in physicochemical properties. The study shows that 1M $NH_4H_2PO_4$ solution induced the release of Cd from natural soils and the soils treated with Idaho monoammonium phosphate (MAP)-fertilizer. The enhanced release of Cd by the phosphate was attributed to the combined effect of Cd introduced to the soils and the Cd released from the soils through the attack of protons and the complexation of phosphate. Phosphate-induced Cd release from natural soils and treated soils increased during the short reaction period of 0.25 to 1 h, then decreased with time, and tended to approach a plateau. Decrease in Cd concentration was apparently due to readsorption of the Cd released on the surface of the soil particles and/or formation of sparingly soluble reaction products in the solution. The amounts of Cd released by $NH_4H_2PO_4$ increased with increments in concentration of $NH_4H_2PO_4$. XRD analysis showed that NH_4-taranakite formed in the soils treated with 1M $NH_4H_2PO_4$ solution. However, taranakite did not form when monoammonium

**Corresponding author*: Dr. P.M. Huang, Dept. Soil Science, University of Saskatchewan, 51 Campus Drive, Saskatoon SK S7N 5A8, Canada.

phosphate was spiked with Cd even at 6.4 × 10^{-3} M $Cd(ClO_4)_2$, indicating that Cd perturbed taranakite formation. The study indicates that at the fertilizer granule-soil interface wherein concentration of orthophosphate is high, more Cd was mobilized compared to bulk soil. The study further shows that NH_4-taranakite can form in acidic tropical soils but the formation may be perturbed if the MAP-fertilizer applied contains significantly high Cd as an impurity.

1 INTRODUCTION

The chemical characteristics of the soil and the source of phosphate fertilizer determine soil fertilizer reactions, which influence P availability to plants (Brady, 1990; Tisdale et al., 1993). When phosphate fertilizers are applied to soils and dissolved by soil water, reactions occur among the phosphate, soil constituents, and nonphosphatic fertilizer compounds. As a result, the P in the fertilizers becomes fixed due to adsorption and precipitation. The nature of the reactions and reaction intensity of phosphate fertilizers with soil constituents should vary with distance from the fertilizer granule relative to the extent of phosphate concentration and pH. The concentration of P in saturated monoammonium phosphate (MAP) solution is 2.9 M with a pH of 3.5 (Lindsay et al., 1962). In the immediate vicinity of the fertilizer granule, the pH is quite low. When the concentrated P is released into the surrounding soil, composition of the soil solution is altered, which results in the dissolution of some minerals. Dissolution of phosphate fertilizer granules is fairly rapid in soil, even under conditions of low soil moisture (Sample et al., 1980). It has been reported that large quantities of reactive cations such as Fe^{2+}, Al^{3+}, Mn^{2+}, K^+, Ca^{2+} and Mg^{2+} can be released during the dissolution processes (Lindsay et al., 1962). Cations from exchange sites may also be displaced by the cations present in these concentrated solutions. Low and Black (1947) found that treating kaolinite with ammonium phosphate solutions (up to 1.5 M) released significant amounts of Si and Al. Similar evidence for the release of cations by solutions of monoammonium phosphate (MAP) and diammonium phosphate (DAP) has been reported by Lindsay et al. (1962).

The high concentration of phosphate and low pH may cause dissolution of soil constituents such as Al and Fe oxides, kaolinite, montmorillonite, and illite (Haseman et al., 1950; Kittrick and Jackson, 1954, 1955, 1956). According to studies by Zhou and Huang (1995), monoammonium phosphate induces release of K from soils. The combined effect of phosphate and proton on the alteration of K-bearing minerals was reported to be the major mechanism of K release from the soils tested in the $NH_4H_2PO_4$ solution. NH_4^+ has nearly the same ionic radius as the K ion and can be subjected to fixation; however, release of K by NH_4^+ is possibly due to exchange reaction. Release of K^+ by NH_4^+ is relatively minor significance since NH_4Cl releases only a fraction of the K released by MAP (Zhou and Huang, 1995).

Phosphate fertilizers added to agricultural soils are also known to form various reaction products in soils (Lindsay et al., 1962; Zhou and Huang, 1995). The phosphate concentration in a soil is a key factor in determining the nature of the reactions. Numerous phosphates are involved in these soil-fertilizer reactions and have been studied by various researchers (Bouldin et al., 1960; Lehr et al., 1959, 1967; Lindsay et al., 1962). Formation of reaction products in soils is of primary importance because plants obtain their phosphorus requirement from these reaction products (Lindsay et al., 1962).

The presence of both ammonium taranakite $[(NH_4)_3Al_3H_6(PO_4)_8{\cdot}18H_2O]$ and potassium taranakite $[K_3Al_5H_6(PO_4)_8 \cdot 18H_2O]$ in agricultural soils as P-fertilizer reaction products with soils has been reported (Lindsay et al., 1962; Taylor et al., 1963; Tamimi et al., 1964; Sakar et al., 1977; Prabhudesai and Kudrekar, 1984; Zhou and Huang, 1995). Taranakites contain N, P, and K. Formation of taranakites in soils can result in the transformation of these nutrients into forms that are slowly available. Taylor et al. (1960) investigated taranakites as a source of phosphate for plants. Formation of taranakites in soils plays an important role in nutrient dynamics, especially in the vicinity of phosphate fertilizers when applied to acidic soils.

However, little is known about the release of Cd and the formation of phosphate reaction products in phosphorus-deficient acidic soils of variable charge found in the tropics, where large amounts of phosphate fertilizers are applied to correct phosphorus deficiency. Phosphate fertilizers are applied in bands or mixed with seeds; hence the release of Cd in the region surrounding the band may differ from that of the bulk soil and consequently affect Cd phytoavailability. Therefore, the objectives of this study were to determine in selected tropical soils from Kenya: i) the degree to which phosphate may affect the release of cadmium from soils and ii) the rates of Cd release and formation of phosphate reaction products.

2 MATERIALS AND METHODS

Three surface soils (0–40 cm depth) located in Egerton (Andisol), Naivasha (Inceptisol), and Soy (Oxisol) in Kenya were used for this study. The sites from which the soils were sampled constitute major agricultural areas in Kenya that vary widely in physicochemical properties. The Egerton soils are derived from volcanic ash and pyroclastic rocks, well drained, very deep, and dark reddish-brown with humic topsoil. Naivasha soils are derived from lacustrine deposits, mainly of volcanic ash origin, imperfectly drained, moderately deep to deep, and found in the lake region (Lake Naivasha) of the Central Rift valley. Soy soils are derived from intermediate igneous rocks (syenites, trachytes, and phonolites), well drained, moderately deep to deep, dark red, friable clay over petroplinthite. All the soil samples were air dried and crushed to pass through a 2-mm sieve prior to use. Selected properties of these soils are presented in Table 19.1.

The pH of the soils in water (1:2 soil:water) was measured using a pH meter model 1820P (Fisher Scientific, Pittsburgh, PA). The cation exchange capacity (CEC) of the soils was determined after saturation with ammonium acetate solution according to the method of Jackson (1958). Mechanical analysis of the

Table 19.1: Selected properties of the soils studied (Onyatta and Huang, 1999)

Soil[†]	Parent material	pH	Mechanical analysis			Org.	Inorg.	CDB	Na-pyroph.[‡]			Ammon. Oxal[§]		Total Cd	AAAc-EDTA[¶] extractable Cd
			> 50	2–50	< 2	C	C	Fe[††]	Al	Fe	Mn	Al	Fe		
				µm											
				%					$g\ kg^{-1}$					$mg\ kg^{-1}$	
Egerton (Andisol)	Volcanic ash	6.0	20.6	26.4	53.0	26.0	1.0	35.0	1.4	0.9	0.02	1.7	6.5	0.157	0.137
Naivasha (Inceptisol)	Lacustrine	6.8	39.1	11.9	49.0	14.0	2.0	12.0	0.8	0.5	0.10	1.6	0.6	0.119	0.096
Soy (Oxisol)	Igneous rocks	5.4	34.7	11.3	54.0	20.0	2.0	17.0	1.3	0.6	0.01	1.9	0.4	0.019	0.017
$LSD_{0.05}$		0.2	0.3	0.1	0.4	0.3	0.1	0.4	0.1	0.1	0.01	0.2	0.1	0.001	0.001
$LSD_{0.01}$		0.4	0.6	0.2	0.7	0.5	0.2	0.7	0.2	0.2	0.02	0.3	0.2	0.002	0.002

[†]Surface soil sample (0–40 cm)

[††] Citrate-dithionite-hydrogen carbonate extractable Fe

[‡]Sodium pyrophosphate-extractable Al, Fe, and Mn

[§]Ammonium oxalate-extractable Al and Fe

[¶]Ammonium acetate-acetic acid-ethylene diamine tetraacetic acid extractable Cd soils was carried out based on the method outlined by him in 1979. The total and organic carbon contents of the soils were determined by heating them without prior treatment at 1,100°C and 850°C, respectively (Wang and Anderson, 1998), using the Leco CR-12 carbon analyzer (Leco Corp., St. Joseph, MI). Inorganic carbon was determined from the difference between the two values. The Fe, Mn, and Al contents of the pyrophosphate extract (McKeague, 1967), the contents of Fe extracted by the citrate-dithionite-bicarbonate (CDB) extraction method (Mehra and Jackson, 1960), and those of the Al and Fe extracted by the ammonium oxalate-oxalic acid extraction method (McKeague and Day, 1966) were determined using a Perkin-Elmer 300 atomic absorption spectrometer (Norwalk, CT). The flame types and conditions were the same as those recommended by the manufacturer. The graphite furnace parameters used for determination of Cd are presented in Table 19.2.

Table 19.2: Graphite atomic absorption parameters used for determination of Cd (Krishnamurti et al., 1994)

Steps	Temp. (°C)	Ramp time (s)	Hold time (s)	Argon flow (ml min^{-1})
Drying	200	10	20	300
Charring	300	5	20	300
Cooling	50	5	15	300
Atomization	1,600	0	10	0
Cleaning	2,700	5	5	300

For total Cd analysis the soils were ground with an agate mortar and pestle to pass through a 0.16-mm sieve and homogenized before use. The total Cd was determined after microwave digestion of the soil samples (Krishnamurti et al., 1994). The Cd availability index (CAI) of the soils (< 2 mm) was determined following the ammonium acetate-acetic acid-ethylene diamine tetraacetic acid (AAAc-EDTA) extraction procedure (Lakanen and Ervio, 1971). Cadmium in the supernatants was determined using graphite furnace atomic absorption spectrometer (GFAAS) at wavelength 228.8 nm.

2.1 Determination of Cadmium by Graphite Furnace Atomic Absorption Spectrometer

In GFAAS determination, dissolution of the sample, dissociation from the matrix, and generation of the analyte ground state occur sequentially during the following steps: drying, thermal treatment (charring), cooling, atomization and cleaning. Thus, analysis performed with GFAAS requires careful selection of step temperatures and periods to ensure that a process is carried out effectively. The furnace parameters are therefore chosen to achieve maximum accuracy and sensitivity in determination of Cd in aqueous extracts (Krishnamurti et al., 1994).

2.2 Soil Mineralogy

Soil samples were treated sequentially with 1 M NaOAc (pH 5.0) and 30% H_2O_2 prior to x-ray diffraction (XRD) analysis. Soils were not treated with dithionite-citrate-bicarbonate to remove metal oxides. The soils tested were not calcareous and the crystalline oxides present in tropical soils are known to selectively sorb heavy metals. Soil samples were fractionated to 0.2, 0.2–2, 2–5, 5–20, 20–50, and > 50 µm sizes by the sedimentation method (Jackson, 1979). Slides of 0.2, 0.2–2 and 2–5 µm size fractions were prepared for examination by Rigaku X-ray diffraction instrument (Tokyo, Japan) using monochromatic Cu-K_α radiation generated at 50 kV-150 mA. Coarse fractions (5–20, 20–50, and > 50, µm) were ground and used for powder x-ray diffraction analysis by Philips x-ray diffractometer (Model PW 1031) (Eindhoven, the Netherlands) using a Fe-K_α radiation at 35 kV-16 mA. Mineralogy of all the fractions among the soils studied showed predominance of micas and kaolinite in the 0.2 and 0.2–2 µm fractions. In the 2–5 µm fractions, micas, kaolinite, quartz and feldspars were identified and in the 5–20 µm fractions, micas, quartz, and feldspars were present. In both the 20–50 µm and > 50 µm fractions, only quartz and feldspars were identified. Oxides of Al (gibbsite), Fe (hematite) and Mn (birnessite) were detected in all the particle size fractions of the soils investigated.

The point of zero charge (PZC) of the soils was determined according to the method of Duquette and Hendershot (1993) and the specific surface area of the soil samples was determined by ethylene glycol monoethyl ether (EGME) method (Eltantawy and Arnold, 1973; Tiller and Smith, 1990) and the BET-N_2 method using the Autosorb (Quantachrome Corp., Syosset, N.Y.). These properties are presented in Table 19.3.

Table 19.3: Point of zero charge (PZC) and specific surface of the soils

Site	PZC	Specific surface ($m^2\ g^{-1}$)	
		EGME†	BET-N_2
Egerton	6.3	66.3	32.4
Naivasha	7.2	79.0	36.0
Soy	5.0	71.1	36.0
ASE‡	± 0.2	± 0.3	± 0.2

† Ethylene glycol monoethyl ether
‡ Average Standard Deviation

2.3 Incubation Studies

To each 100 g of the three surface soil samples, the Idaho MAP-fertilizer was added to give a Cd content of 62.4 mg kg^{-1} soil as calculated in Appendix I (a, b). The Cd content of Idaho MAP-fertilizer was 144 mg Cd kg^{-1}. The amount of Cd added to the soils was calculated based on the fertilizer recommendation for these soils (130 kg MAP fertilizer/ha or 67.6 kg P_2O_5/ha; fertilizer type, 11:52:0) (Annual Farmers' Recommendations, 1993), the Cd content of the fertilizer, and the amount of soil that would be in contact with fertilizer granules (Appendix 1b). The amount of MAP fertilizer added to 100 g soil used in the incubation study was calculated as described in Appendix I(c). Natural soil samples and soils treated with the Idaho MAP fertilizer were placed in plastic containers in two replicates and water added to field capacity, and the samples allowed to incubate at room temperature (25°C) for 100 days. During incubation moisture was maintained at field capacity.

2.4 Kinetics of Cadmium Release from the Soils

The concentration of orthophosphate in a saturated solution of $NH_4H_2PO_4$ fertilizer can be as high as 2.90 M (Lindsay et al., 1962). Phosphate fertilizers are often mixed with seeds or side banded; hence the concentration of orthophosphate will be high at the fertilizer granule-soil interface and especially at the soil rhizosphere. In order to simulate the effect of orthophosphate on Cd release from the soils in the vicinity of the fertilizer granules, 1.00 M $NH_4H_2PO_4$ solution was chosen for the kinetic study.

At the end of the 100-day incubation period, the treated soils were washed with 10 mL of deionized distilled water three times to remove any free salts prior to their use. In total, 30 mL of deionized distilled water was used in the leaching process. The amount of Cd removed by leaching process was not determined since incubation had been a pretreatment. One gram of each soil (natural or Idaho MAP-fertilizer treated soil), in duplicate, was placed in a polypropylene centrifuge tube and 10.0 mL of 1.00 M $NH_4H_2PO_4$ solution added. The centrifuge tube containing a soil suspension was placed in a water bath shaker at 25°C, with an agitation speed of 60 cycles min^{-1}. At the end of each reaction period (0.25, 0.5, 0.75, 1.0, 2.0, 7.0, 15.0, and 24.0 h), the suspensions were filtered through a 0.45-μm pore-size membrane under vacuum. Separate duplicate samples were used during each equilibration time. The pH of the suspensions was monitored at the end of each reaction period.

Cadmium in the filtrates was determined using a Perkin Elmer 2280 graphite furnace atomic absorption spectrometer (Norwalk, CT) at 228.8 nm using a pyrolytically coated tube. The Cd determined in the filtrates (Cd_r) was the Cd released from the soil during reaction time, t. The Cd remaining in the sample (Cd_t) at different reaction times was the difference between the total Cd present in the soil samples and the Cd released in the solution.

Under field conditions the concentration of phosphate should decrease gradually with increasing distance from the phosphate fertilizer zone. In order to investigate the influence of phosphate concentration on Cd release, 0.10 M $NH_4H_2PO_4$ and 0.01 M $NH_4H_2PO_4$ solutions were also used in the study of Cd release from the soils according to the procedure described above.

2.5 Examination of Phosphate Reaction Products in Soils and Perturbation by Cadmium

One gram of each of the natural soil samples (Egerton, Naivasha and Soy) was placed in a 50-mL polypropylene centrifuge tube and treated with 10 mL of $NH_4H_2PO_4$ solution. Soil suspensions were allowed to incubate for a period of 100 days at room temperature (25°C). Separate samples of the soils were also treated with only water using the same procedure to serve as control. The effect of Cd on formation of phosphate reaction products was also investigated by reacting the natural soils with 1 M $NH_4H_2PO_4$ solution in the presence of 6.4×10^{-3} M $Cd(ClO_4)_2 \cdot 6H_2O$ and allowed to incubate for 100 days as described above. The concentration of 6.4×10^{-3} M Cd was chosen after trials using a lower Cd concentration in the range of 10^{-4} M. The trials had indicated that at 10^{-4} M Cd concentration, formation of phosphate reaction product was not perturbed; hence a concentration of 10^{-3} M was used. The choice of $Cd(ClO_4)_2$ concentration at this magnitude was based on the calculation for the incubation studies which depended on the rate of MAP fertilizer application for the three soils studied and the Cd content of the Idaho MAP fertilizer (Appendix Ia and Id). Cadmium concentration in the fertilizer granule-soil interface calculated for Soy soil at a field moisture capacity of 20% was 2.8×10^{-3} M (Appendix Id). This was adjusted to Cd concentration of 6.4×10^{-3} M, namely, the concentration at which notable perturbation on phosphate reaction formation by Cd was observed during the trials. Hence using a Cd concentration level of 6.4×10^{-3} M is justified, especially since the concentration of Cd at the immediate fertilizer granule-soil interface should be high. At the end of each reaction period, the soil suspensions were filtered through a 0.45-µm pore-size membrane under vacuum.

The reacted soil samples were leached with 10 mL deionized distilled water four times to remove any free salts in the soil samples. The amount of Cd removed by leaching prior to the preparation of soil samples for XRD and IR analysis after the equilibration period was not determined. This would indeed affect calculation of the amount of Cd remaining; however, XRD and IR analyses were carried out only to identify the phosphate reaction products formed in the soils. It may be assumed that the Cd remaining after leaching and subsequently released by MAP were from the sorbed phase. The amount of Cd remaining after leaching would be essential in calculating the fraction or % Cd released but in this study, it is the actual amounts of Cd released that are relevant to food chain contamination. Each washed soil sample was air dried and ground to < 5 µm with an agate mortar and pestle and used for XRD and IR analysis. The ground soil samples were directly mounted on a glass slide by adding a few drops of acetone prior to x-ray diffraction analysis by Rigaku D/MaX-RBX x-ray diffraction instrument (Tokyo, Japan) using monochromatic Fe-K_α radiation generated at 40 kV-130 mA. For IR analysis, 1 mg soil sample (< 5 µm) was mixed with 250 mg KBr and then pressed into a pellet. The KBr pellet with the sample was examined under a Perkin-Elmer 983 infrared absorption spectrometer (Buckinghamshire, England).

3 RESULTS AND DISCUSSION

3.1 Amounts of Cadmium Released

The amounts of Cd released from the natural Egerton, Naivasha and Soy soils during the 15-min reaction period as influenced by 1 M $NH_4H_2PO_4$ solution were very low and varied with the soils (Table 19.4). The addition of Idaho MAP fertilizer to the soils influenced the amount of Cd released from them (Table 19.5). No Cadmium was detected in the deionized distilled water in which control soils were soaked.

Soil pH is often regarded as the major factor in controlling plant uptake of Cd from soils (Chaney and Hornick, 1978). The pH effect on Cd availability in soils is largely ascribed to the marked effect pH has on retention of Cd^{2+} by soil surfaces. The initial pH of 1 M $NH_4H_2PO_4$ solution was 4.04. The pH of the soil-$NH_4H_2PO_4$ suspension at the end of a 15-min reaction period was in the order:

Table 19.4: Amount of Cd released from control soils at the end of a 15-min reaction period by 1 M $NH_4H_2PO_4$ solution

Soil†	pH‡	Cd released (mg kg^{-1} soil)	% Cd released
Egerton	4.97	0.04	25.4
Naivasha	5.09	0.03	25.2
Soy	4.66	0.01	52.6
$LSD_{0.05}$	0.08	0.01	
$LSD_{0.01}$	0.15	0.02	

†After 100-day incubation period with deionized distilled water at the field capacity.

‡pH of soil-$NH_4H_2PO_4$ suspension.

Table 19.5: Amount of Cd released from treated soils at the end of a 15-min reaction period by deionized distilled water and 1 M $NH_4H_2PO_4$ solution

Soil	Treatment†	DDW‡		1 M $NH_4H_2PO_4$	
		pH	Cd_r††	pH	Cd_r
		mg kg^{-1} soil		mg kg^{-1} soil	
Egerton	Idaho MAP fertilizer	5.43	0.14	4.68	1.30
Naivasha	Idaho MAP fertilizer	5.04	0.13	4.46	1.10
Soy	Idaho MAP fertilizer	5.45	0.13	4.60	1.30
$LSD_{0.05}$		0.04	0.02	0.06	0.05
$LSD_{0.01}$		0.07	0.03	0.10	0.09

†After 100-day incubation period with Idaho MAP-fertilizer at field capacity.

‡Deionized distilled water.

††Amount of Cd released to solution.

Naivasha (5.09) > Egerton (4.97) > Soy (4.66) for the control soils. The amounts of Cd released from the soils were not in the same order as the pH of soil-$NH_4H_2PO_4$ suspensions. The Soy soil which had the lowest soil-$NH_4H_2PO_4$ suspension pH released the lowest amount of Cd (Table 19.4). Hence, pH was not the only factor influencing the amount of Cd released from soils in the system studied. In our view, the effect of pH on Cd release is important since it is one of the factors that control plant uptake of Cd. However, in terms of percent of the total soil Cd released, the Soy soil (pH of soil-$NH_4H_2PO_4$ suspension = 4.66) released the highest percent of Cd (Table 19.4). But use of the fraction of Cd released from the soils as affected by pH merely indicates how readily Cd can be released from soils, and use of absolute amounts of Cd released from the soils (mg Cd kg^{-1} soil) is more meaningful in terms of food chain contamination. The amounts of Cd released from soils treated with the Idaho MAP fertilizer by deionized distilled water and by 1 M $NH_4H_2PO_4$ solution were 0.13–0.14 and 1.10–1.30 mg kg^{-1} soil respectively (Table 19.5). Further, it was observed that the amounts of Cd released from the treated soils by deionized distilled water were one order of magnitude lower than the amounts of Cd released by 1 M $NH_4H_2PO_4$ solution (Table 19.5), indicating that phosphate substantially induced Cd release from the soils studied. However, the effect of pH cannot be discounted due to the observed difference in hydrogen ion concentration between the distilled water and the MAP treatments. Since the amount of Cd added was 62.4 mg kg^{-1}, the difference between the amount added and the amount recovered could be attributed to the fact that, within the 15-min reaction period, only a small fraction of the Cd from the sorbed phase was removed. In the process of Cd release, other factors that may account for the release of Cd include release of adsorbed Cd (II) by NH_4^+ and the chemical form of Cd (II) in the presence of 1 M phosphate. Cd forms complexes with phosphate, e.g. $[CdH_2PO_4]^+$, $[CdH_2PO_4)_2]^0$, and others, which are likely to form particularly whenever concentrations of phosphate are orders of magnitude greater than Cd, as in the case of the present study.

The amounts of Cd released during the 0.25 to 1 h reaction period from the soils increased with concentration of $NH_4H_2PO_4$ solution and also varied with soil type (Figs. 19.1 and 19.2). Irrespective of the initial Cd content in the control soils, 0.1 M $NH_4H_2PO_4$ solution appeared to release almost the same amount of Cd from control soils (Fig. 19.1). This is attributable to the low Cd contents of the soils studied and the low P concentration used. Presented in graphic form, the difference may not be significant (Fig. 19.2) even for soils treated with Idaho MAP fertilizer. However, Table 19.5 for the 15-min reaction period for the treated soils indicated a significant difference. At a concentration of 0.01 M $NH_4H_2PO_4$ no detectable amounts of Cd release were observed for the control soils. During the 0.25 to 1 h-reaction period, the lowest Cd release by 1 M $NH_4H_2PO_4$ was observed in the natural Soy soil, which is derived from igneous rocks low in Cd (Fig. 19.1). It was further observed that the amounts of Cd released by 0.10 M $NH_4H_2PO_4$ and 1 M $NH_4H_2PO_4$ solutions in natural Soy soil were almost the

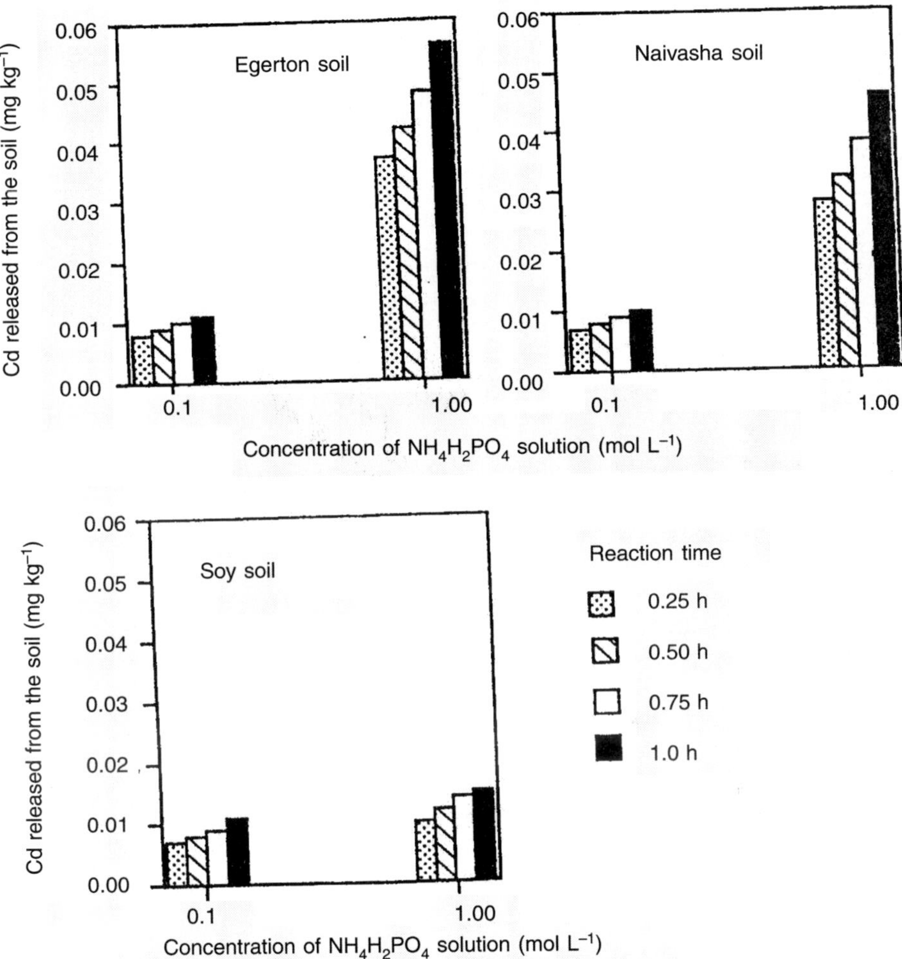

Fig. 19.1: Amount of Cd released from control soils by $NH_4H_2PO_4$ solution at various concentrations as a function of reaction period. Cd released by 0.01 M $NH_4H_2PO_4$ solution was not detectable.

same (Fig. 19.1), which is also attributable to its low Cd content. However, appreciable amounts of Cd were released when 0.01 M $NH_4H_2PO_4$ solution was reacted with the soils treated with the Idaho MAP fertilizer (Fig. 19.2). Results showed that the release of Cd by $NH_4H_2PO_4$ solution of the Idaho monoammonium phosphate-treated soils was significantly higher than in control soils (Figs. 19.1 and 19.2). Although ionic strength was not determined during the study, in this case the phosphate-induced Cd release to the MAP-treated soils and the Cd impurity in the MAP apparently contributed markedly to the difference in Cd release. The results further showed that the phosphate-induced

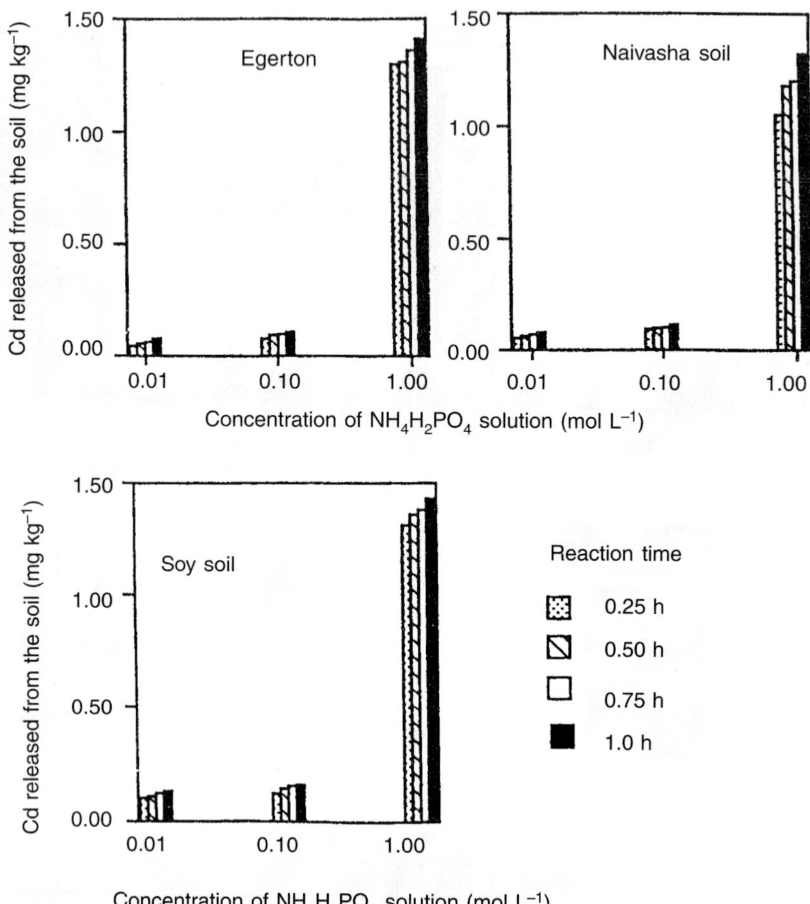

Fig. 19.2: Amount of Cd released from Idaho monoammonium phosphate (MAP)-fertilizer treated soils by $NH_4H_2PO_4$ solution at various concentrations as a function of reaction period.

Cd release increased during the short reaction period of 0.25 to 1 h and then decreased with time and tended to reach plateau in both the natural soils and soils treated with Idaho MAP fertilizer (Fig. 19.3). The decrease in Cd released could be due to readsorption of released Cd onto the surface of soil particles and/or formation of insoluble reaction products in solution.

Mortvedt and Osborn (1982) suggested that the chemical form of Cd in P-fertilizers was $Cd(H_2PO_4)_2$, $Cd(HPO_4)$, or a mixture of these salts. However, in solution Cd can precipitate as $Cd_3(PO_4)_2$ ($K_{sp} = 2.5 \times 10^{-33}$) (Dean, 1992). Formation of sparingly soluble Cd compound in solution could reduce the amount of Cd released to solution with time.

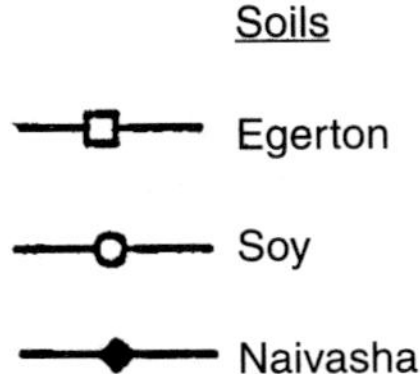

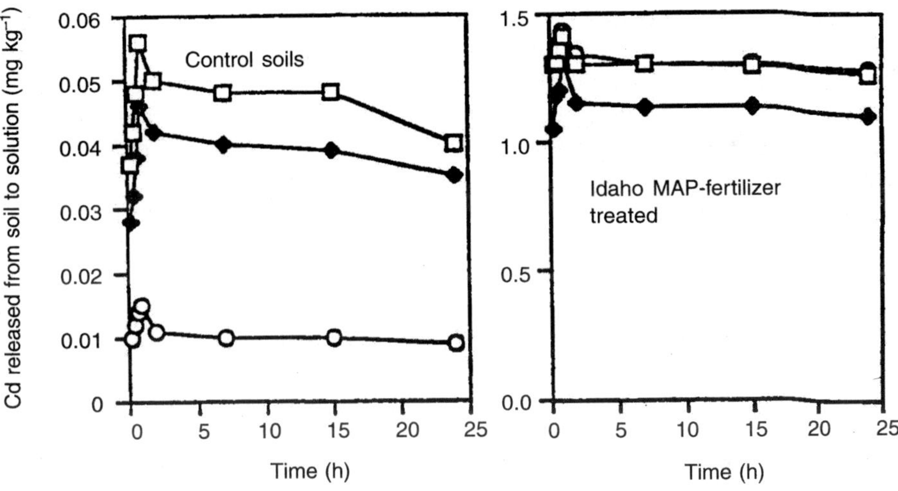

Fig. 19.3: Changes in amounts of Cd released by 1 M $NH_4H_2PO_4$ solution during the 0.25 to 24 h-reaction period from control soils and soils treated with Idaho MAP-fertilizer.

In natural soils, Cd release (Fig. 19.3) varied with Cd content (Table 19.1). For example, the Cd content of the surface horizon of the Soy soil was one order of magnitude lower than those of the Egerton and Naivasha soils. However, when the same amount of Cd was added to the soils, the Naivasha soil released the lowest amount of Cd (Fig. 19.3). The Naivasha soil compared with the other two soils retained more strongly the Cd added and thus limited its release. The manganese content of the surface horizon of the Naivasha soil was one order of magnitude higher than that of the Soy soil (Table 19.1). Heavy metals are strongly adsorbed on Mn oxides (McKenzie, 1980). Manganese oxides are very reactive components of soils; they strongly sorb cations (Huang, 1991). The added Cd was apparently adsorbed by the Mn oxides present in the Naivasha soil, which thus limited its release (Fig. 19.3). Retention of Cd by the Naivasha soil could also be attributed to the high specific area of the Naivasha soil (Table 19.3). The other soil components, e.g. organic C, oxides of Al and Fe which could bind Cd in the surface horizon of Naivasha soil, were not higher than those of the surface horizons of the Egerton and the Soy soils (Table 19.1).

3.2 Kinetics of Cadmium Release

The short reaction period of 0.25 to 1 h was used to establish the kinetics of Cd release by 1 M $NH_4H_2PO_4$ solution from the soils. The reaction period 0.25 to 1 h was chosen for kinetic study to avoid significant disintegration of soil aggregates caused by agitation. Different kinetic equations were used to fit the data on Cd remaining in the sample (Cd_t) or Cd released to solution (Cd_r) versus the reaction time (t) (Sparks, 1989). The kinetic equations examined in this study were zero-order, first-order, and parabolic diffusion (Table 19.6). The degree of fit of each equation to the data was evaluated based on the r^2 (correlation coefficient), and p value (probability). The choice of kinetic models for describing the release of Cd from the soils was based on these parameters. The closer the r^2 value to 1.0 and the lower the p value, the better the equation fitted the kinetic data. For the kinetic study, absolute amounts (instead of fractions) of the Cd remaining in the sample (Cd_t) or Cd released to solution (Cd_r) were used. The use of absolute amounts of Cd released (mg Cd kg^{-1} soil) instead of fractions (% Cd released) of Cd released in calculating the rate constants is more relevant to the amount of Cd entering the food chain. Based on the r^2 and p values, the degree of fit for the zero-order and first-order equations were generally found to fit the data better than parabolic diffusion equation, as clearly evidenced from the r^2 and p values presented in Table 19.6. Since the zero-order and first-

Table 19.6: Comparison of degree of fit of the kinetic equations to the data of Cd release from the soils by 1 M $NH_4H_2PO_4$ solution

Soil	Treatment†	Kinetic equations‡					
		Zero-order $Cd_t = a-kt$		First-order $\ln Cd_t = a_1-k_1t$		Parabolic diffusion $Cd_r = a_2+Dt^{1/2}$	
		r^2	p	r^2	p	r^2	p
Egerton	Control	0.988	6.0×10^{-3}	0.982	9.0×10^{-3}	0.773	1.2×10^{-1}
	Idaho MAP fertilizer	0.937	3.2×10^{-2}	0.935	3.2×10^{-2}	0.878	6.3×10^{-2}
Naivasha	Control	0.978	1.1×10^{-2}	0.968	1.6×10^{-2}	0.734	1.4×10^{-1}
	Idaho MAP fertilizer	0.939	3.1×10^{-2}	0.937	3.1×10^{-2}	0.939	3.1×10^{-2}
Soy	Control	0.978	1.0×10^{-2}	0.994	3.0×10^{-3}	0.903	5.0×10^{-2}
	Idaho MAP fertilizer	0.974	1.2×10^{-2}	0.974	1.2×10^{-2}	0.968	1.6×10^{-2}

†After 100-day incubation period at field capacity.

‡Cd_t is the amount of Cd remaining in the soil sample at time t; Cd_r the amount of Cd released to solution from the soil at time t; a, a_1, and a_2 are constants in respective equations; k and k_1 are rate constants; and D is the overall diffusion coefficient.

order equations fitted the data virtually equally well, the zero-order equation was selected to establish the kinetics of Cd release during the short-reaction period of 0.25 to 1 h.

The rate constants of Cd release from the soils in the 1 M $NH_4H_2PO_4$ solution and deionized distilled water were calculated from the zero-order kinetic equation. The data obtained are shown in Table 19.7. The rate of Cd release varied with the soils. The amount of Cd released based on the rate of Cd released from the soils during the 0.25 to 1 h-reaction period (Table 19.7) was lower than the amount of Cd detected in the solution at the end of 1 h-reaction period (Fig. 19.3). This indicates that the rate of Cd release in the soil system before 0.25 h was extremely fast, which accounts for the differences between the amount of Cd released after 1 h and the amount of Cd released based on the rate constant calculated from the zero-order equation. The rate of phosphate-induced Cd release from the control soils followed the order: Egerton (0.03 mg kg^{-1} h^{-1}) > Naivasha (0.02 mg kg^{-1} h^{-1}) > Soy (0.01 mg kg^{-1} h^{-1}), while for soils treated with Idaho MAP fertilizer the order was: Naivasha (0.33 mg kg^{-1} h^{-1}) > Egerton (0.15 mg kg^{-1} h^{-1}) = Soy (0.15 mg kg^{-1} h^{-1}). The rate constants are taken as a measure of the rate of Cd release from the soils. The rate of Cd release from the Idaho MAP fertilizer-treated soils, as indicated by the rate constants, was higher for the Naivasha soil than for either the Egerton or the Soy (Table 19.7). The relatively high amounts of Mn oxides and high surface area of Naivasha soil (Tables 19.1 and 19.3) apparently contributed to the strong retention of Cd, which limited its release (Fig. 19.3). However, the rate constant of Cd release in the Naivasha soil treated with Idaho MAP fertilizer as influenced by $NH_4H_2PO_4$ solution was higher than the other two soils (Table 19.7). When $NH_4H_2PO_4$ solution was

Table 19.7: Rate for phosphate-induced Cd release from soils by 1 M $NH_4H_2PO_4$ during the 0.25 to 1 h-reaction period

		Rate constants (mg kg^{-1} h^{-1})	
Soil	Treatment†	DDW‡	$NH_4H_2PO_4$
Egerton	Control	ND‡‡	0.03
	Idaho MAP fertilizer	0.02	0.15
Naivasha	Control	ND	0.02
	Idaho MAP fertilizer	0.02	0.33
Soy	Control	ND	0.01
	Idaho MAP fertilizer	0.01	0.15
$LSD_{0.05}$		0.01	0.02
$LSD_{0.01}$		0.12	0.03

†After 100-day incubation period with Idaho MAP fertilizer at the field capacity.
‡Deionized distilled water.
‡‡Not detectable.

added, dissolution of Mn oxides occurred, which resulted in release of Cd from the Naivasha soil. The ease with which dissolution of Mn oxide occurs would mean that the rate of Cd release would be higher. The strength with which Cd is bound to metal oxides depends on the metal of the adsorbent. The higher the valence : coordination ratio of the adsorbent metal, the lower the strength of adsorption of trace metals (McBride, 2000). Mn is octahedrally coordinated to six O atoms and has a valence of + 4; it thus donates a + 0.67 charge to each O atom. This creates a greater pull on the electrons between the O and Mn atoms, resulting in a large distance between Cd and the electrons in Mn oxides. This would result in a weaker bond with Cd and could, therefore contribute to the higher rate constant of Cd release in the Naivasha soil treated with Idaho MAP fertilizer, which retained Cd due to its Mn oxide content. In the natural soils, the rate constant of Cd release by 1 M $NH_4H_2PO_4$ (Table 19.7) from the Soy soil was lower than that from the Naivasha or Egerton soil. This indicates that more Cd would be released from the Naivasha and Egerton soils than from the Soy. The rate constants of Cd release from the treated soils by the deionized distilled water were much lower than those of Cd released by the monoammonium phosphate solution (Table 19.7), indicating that the application of the phosphate fertilizer substantially enhanced the rate of Cd released from the soils.

3.3 Mechanisms of Cadmium Release

Poorly crystalline Al, Fe, and Mn oxides, present in highly weathered tropical soils, are known to adsorb Cd selectively (Christensen and Huang, 1999). The mineralogy of these soils shows the presence of kaolinite and oxides of Al, Fe, and Mn. Dissolution of Al, Fe, and Mn oxides and kaolinite in these soils by phosphate would therefore enhance the release of Cd from the soils through formation of Cd-complexes such as $[CdH_2PO_4]^+$ (log $K[CdH_2PO_4]^+ = 2.91$; Martell et al., 1997). H^+ can protonate the hydroxyl groups and oxygen atoms at the broken edges or surfaces of the soil minerals to weaken the Mn–O, Fe–O, and Al–O bonds resulting in increased dissolution of Cd by phosphate. Phosphate is a strong complexing ligand. It can exchange with the OH and OH_2 groups in soil minerals to form complexes with the surface structural cations such as Al, Fe, and Mn. In the presence of H^+, phosphate ligands can thus attack these bonds more easily and replace OH_2 or OH groups to form complexes with structural cations. The mechanism of dissolution of minerals controlled by surface complexation was explained by Bloom and Nater (1991). The mechanism of enhanced Cd release from the soils by 1 M $NH_4H_2PO_4$ solution was deemed to be the combined effect of Cd introduced to soils and the Cd released from the soils through attack of the protons, the complexation effect of phosphate and the resultant solubilization effect of Al, Fe, and Mn oxides and kaolinite by the $NH_4H_2PO_4$ solution.

3.4 Reaction Products and Significance

Crystalline reaction products were found in the phosphate-treated soils by x-ray diffraction analysis at the end of the 100-day reaction period. The main d-spacings of the phosphate reaction products were: 15.88, 7.92, 7.49, 7.20, and 5.83 Å in both the Egerton and the Naivasha soils (Fig. 19.4). The d-spacings compared well with the d-spacings of the standard taranakites (NH_4-taranakite and K-taranakite) synthesized by Taylor and Gurney (1961), Frazier and Taylor, (1965), and Zhou and Huang (1995).

The standard NH_4-taranakite [$(NH_4)_3Al_5H_6(PO_4)_8 \cdot 18H_2O$] and K-taranakite [$K_3Al_5H_6(PO_4)_8 \cdot 18H_2O$] have identical main d-spacings. The peak at 6.47 d-spacing which consistently appeared in both control and treated soils indicates the presence of feldspars (Jackson, 1979). The peak at d-spacing of 3.19–3.23 Å also indicates the presence of feldspars, e.g. K-feldspars or plagioclase feldspars (Berry, 1974; Jackson, 1979). The peak at 3.74 Å, present in both control and

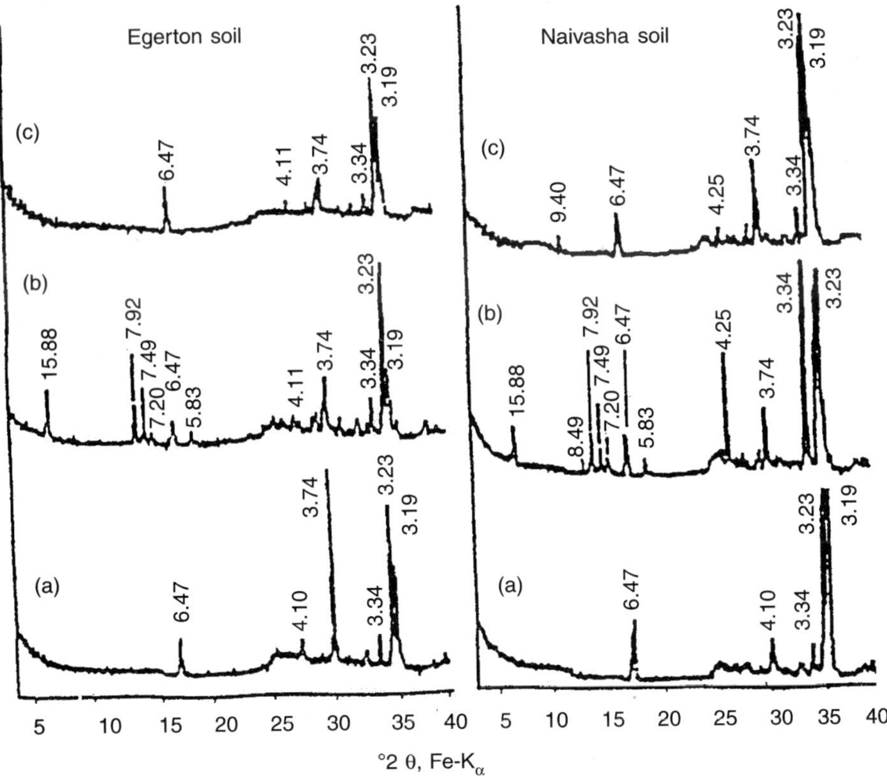

Fig. 19.4: X-ray diffractograms of soils treated with 1 M $NH_4H_2PO_4$ solution both in the presence and absence of 6.4 x 10^{-3} M $Cd(ClO_4)_2$ at 25°C for 100 d. (a) Control, (b) 1 M $NH_4H_2PO_4$ solution, and (c) 1 M $NH_4H_2PO_4$ + 6.4 × 10^{-3} M $Cd(ClO_4)_2$. Characteristic d-spacings of NH_4-taranakite: 15.88, 7.92, 7.49, 7.20, and 5.83 Å.

treated soils, could also indicate the presence of feldspars (Berry, 1974) (Fig. 19.4). It has been reported that the quantities of feldspars in soils would vary with the nature of the parent material and the stage of weathering (Somasiri et al., 1971). In moderately weathered soils, there are usually considerable quantities of K-feldspars, while in strongly weathered soils, such as those in humid tropics, K-feldspars are often present in small amounts (Prabhudesai and Kudrekar, 1984).

In the Egerton soil, the peak at 4.11 Å, which appeared even in the control soils, could be goethite. Peaks at 3.34 Å indicate the presence of quartz. In the treated Naivasha soil, the peak at 4.25 Å, which did not appear in the control soil, could be due to a phosphate reaction product. Prabhudesai and Kudrekar (1984) reported that variscite ($AlPO_4 \cdot 2H_2O$) can form in soils as a monoammonium phosphate reaction product in soils and peak at 4.22 Å.

The peak observed in the x-ray diffractogram of the Naivasha soil at 9.40 Å could arise from a phosphate reaction product. Frazier and Taylor (1965) reported the occurrence of $NH_4AlPO_4OH \cdot 2H_2O$ with a d-spacing of 9.74 Å. The peak at 8.49 Å observed in the Naivasha soil (Fig. 19.5) could be a compound of ammonium aluminum phosphate hydrate $[(NH_4)_2AlH(PO_4)_2 \cdot 4HO_2)]$ (Frazier and Taylor, 1965). After heating at 110°C, the 8.49 Å peak disappeared (Fig. 19.5). Formation of ammonium aluminum phosphate compounds in soils when ammonium phosphate fertilizers are applied was reported by Frazier and Taylor

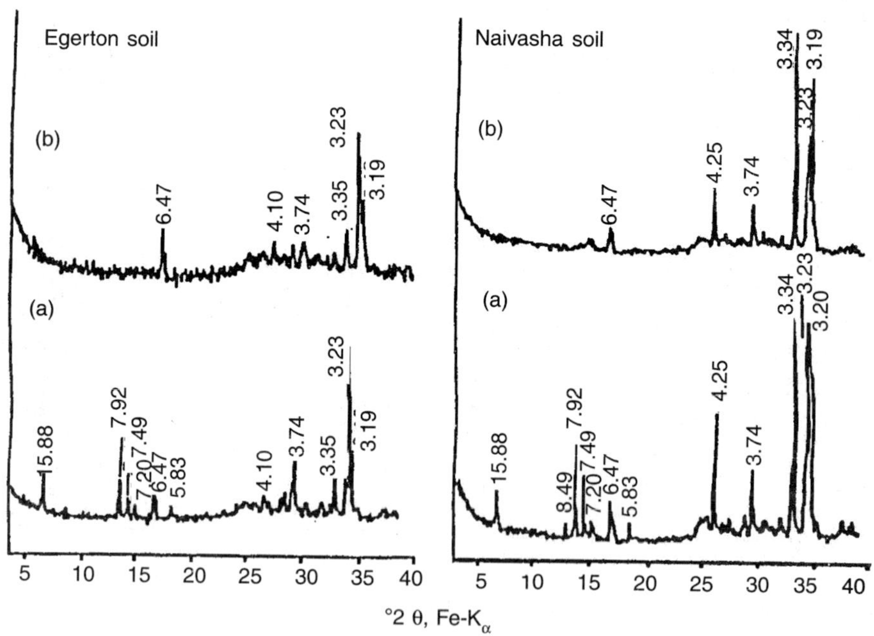

Fig. 19.5: X-ray diffractograms of NH_4-taranakite formed in soils treated with 1 M $NH_4H_2PO_4$ solution at 25°C for 100 d. (a) Before heating and (b) after heating in air at 110°C for 2 h. Characteristic d-spacings of NH_4-taranakite are: 15.88, 7.92, 7.49, 7.20, and 5.83 Å.

(1965). The 8.49 Å peak was not observed in the Egerton soil, which indicates that formation of phosphate reaction products also varies with soil type (Fig. 19.5).

When the taranakite formed was heated in air at 110°C for 2 h and then examined by XRD, the characteristic d-spacings of taranakites (15.88, 7.92, 7.49, 7.20, and 5.83 Å) disappeared, indicating that the K-taranakite transforms to new crystalline product whereas NH_4-taranakite transforms to a noncrystalline state (Fig. 19.5). The product formed in the present study was therefore identified as NH_4-taranakite. The reacting solution contained 1 M NH_4^+ and 1 M phosphate; hence if the soils released enough Al^{3+}, NH_4-taranakite would readily form. In control soils (treated only with deionized distilled water) no taranakite was detected at the end of the 100-day reaction period. Evidence of release of Cd to solution over time (§ 3.1) indicates that application of MAP fertilizers to soil should increase the Cd concentration in the soil solution whereas formation of ammonium taranakite (Figs. 19.4 and 19.5) may lead to conversion of some of the nitrogen and phosphate in the fertilizer to slowly available forms significant in plant growth. An evaluation of taranakites as sources of phosphate for plants was carried out by Taylor et al. (1960). Therefore, formation of taranakites plays an important role in nutrient dynamics of soils, especially in the vicinity of the phosphate fertilizer zone.

3.5 Perturbation of Taranakite Formation by Cadmium

Ammonium taranakite [$(NH_4)_3Al_5H_6(PO_4)_8{\cdot}18H_2O$] and potassium taranakite [$K_3Al_5H_6(PO_4)_8{\cdot}18H_2O$] have the same structure (Frazier and Taylor, 1965), although their composition differs. Lindsay et al. (1962) reported that both NH_4- and K-taranakites may be present in agricultural soils as reaction products of phosphate fertilizers. The present study shows the formation of ammonium taranakite in selected tropical soils in Kenya. When taranakites are formed, the aluminum in soil minerals becomes the source of Al in the taranakites. Phosphate fertilizers contain a wide range of Cd (0.2–345 mg kg^{-1}) as a contaminant (Alloway and Steinnes, 1999). Since large amounts of phosphate fertilizers are applied to tropical soils to correct phosphorus deficiency, the possible perturbation of Cd on taranakite formation was also investigated in this study. It was found that in the presence of 6.4×10^{-3} M $Cd(ClO_4)_2$, no taranakite formed (Fig. 19.4c), indicating that even at a concentration of 6.4×10^{-3} M $Cd(ClO_4)_2$, the formation of taranakite was perturbed.

3.6 IR Analysis

To investigate further the nature of the reaction products, infrared analysis (IR) was conducted using a Perkin-Elmer 983 infrared absorption spectrophotometer (Buckinghamshire, England), as described in § 2.5. The infrared spectra of the reaction product are shown in Figures 19.6 and 19.7 for the Egerton and Naivasha soils respectively. Ammonium ion (NH_4^+) has absorption bands that occur in

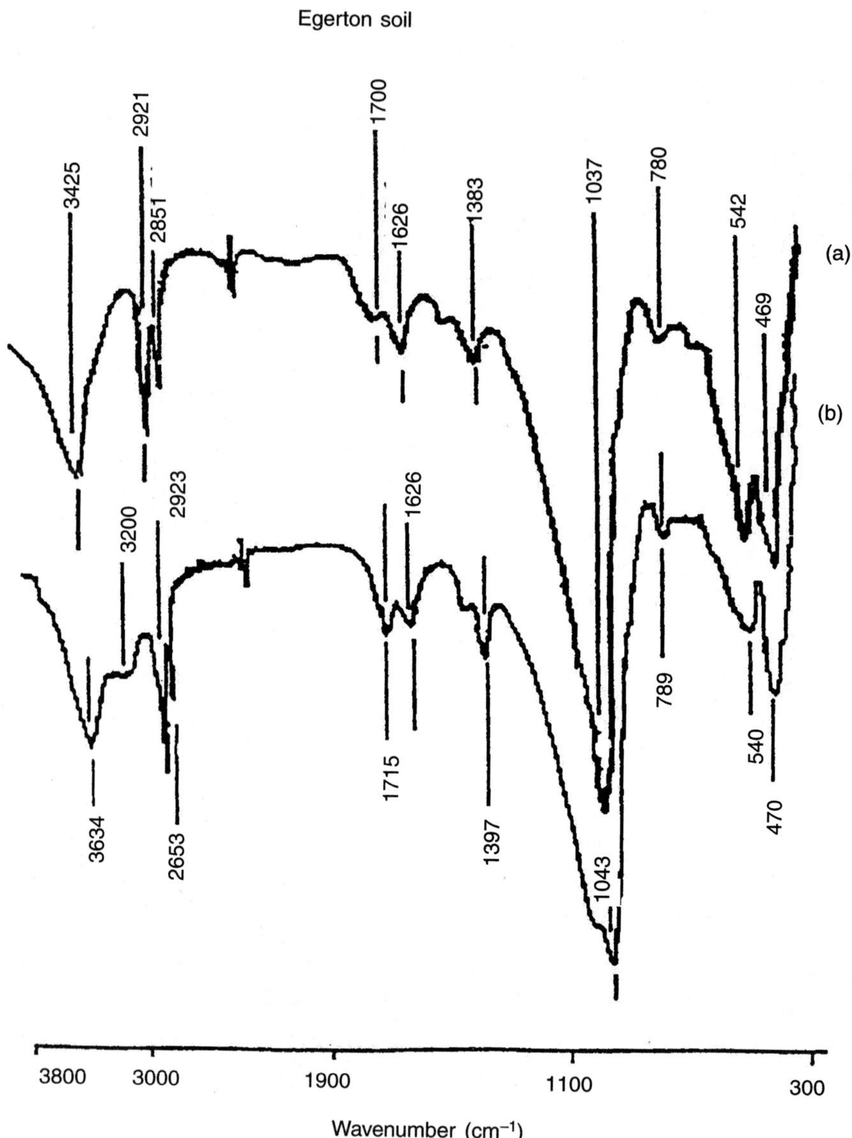

Fig. 19.6: Infrared spectra of Egerton soil before and after treatment with 1 M $NH_4H_2PO_4$ for 100 days at 25°C. (a) Control and (b) treated soil.

the region of 3,350–3,050 cm^{-1} (Gadsden, 1975) and may be obscured or confused with absorptions arising from the presence of water. Absorption bands between 3,600 and 2,900 and 1,639–1,626 cm^{-1} (Figs. 19.6 and 19.7) indicate the presence

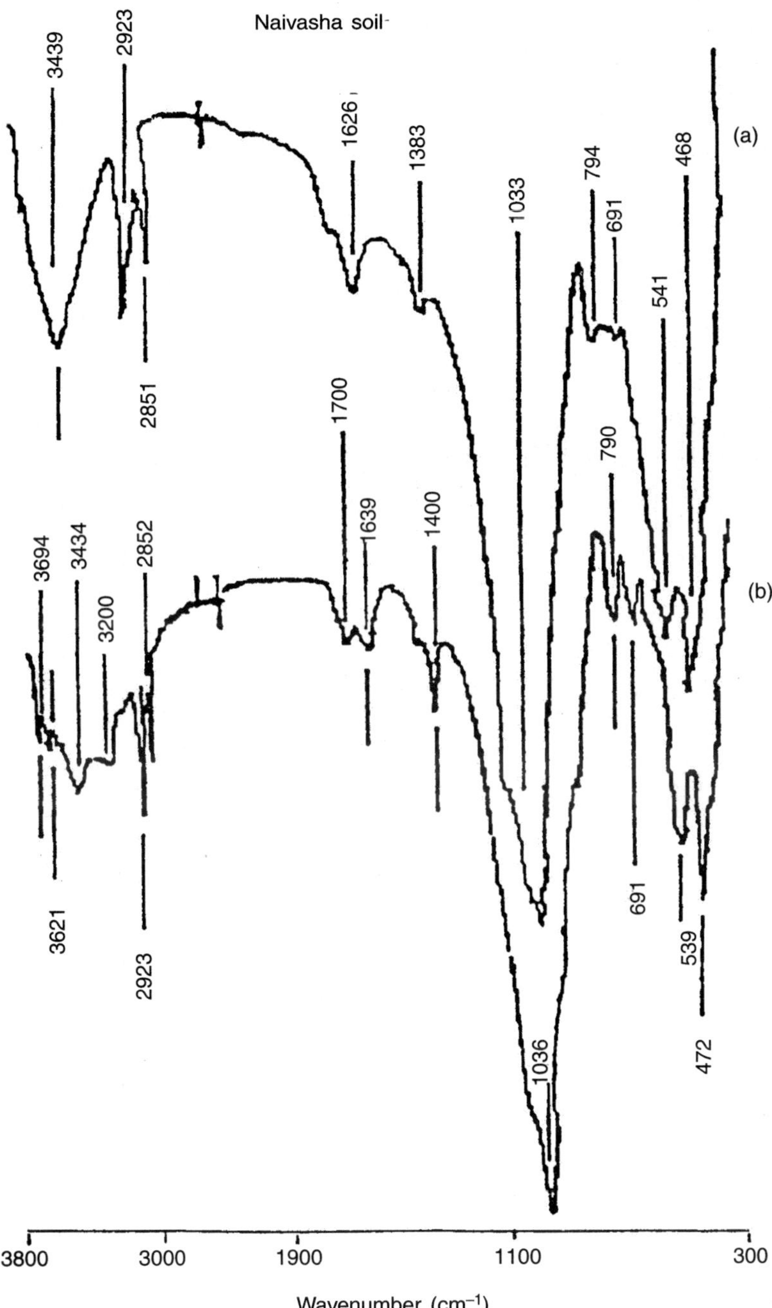

Fig. 19.7: Infrared spectra of Naivasha soil before and after treatment with 1 M $NH_4H_2PO_4$ for 100 days at 25°C. (a) Control and (b) treated soil.

of water and hydroxyl ion (Frazier and Taylor, 1965; Zhou, 1995). The characteristic absorption band at 3,200 cm^{-1} (Figs. 19.6b and 19.7b) was attributed to NH_4-taranakite (Zhou, 1995). Bands less than 1,300 cm^{-1} can be attributed to the phosphate (Frazier and Taylor, 1965; Gadsden, 1975; Zhou, 1995). Heating did not affect the bands related to phosphate in the structure of taranakite (Zhou, 1995). Infrared analysis data of the samples before and after heating in air at 110°C for 2 h are presented in Figs. 19.8 and 19.9 for the Egerton and Naivasha soils respectively. After heating at 110°C for 2 h, the absorption band at 3,200

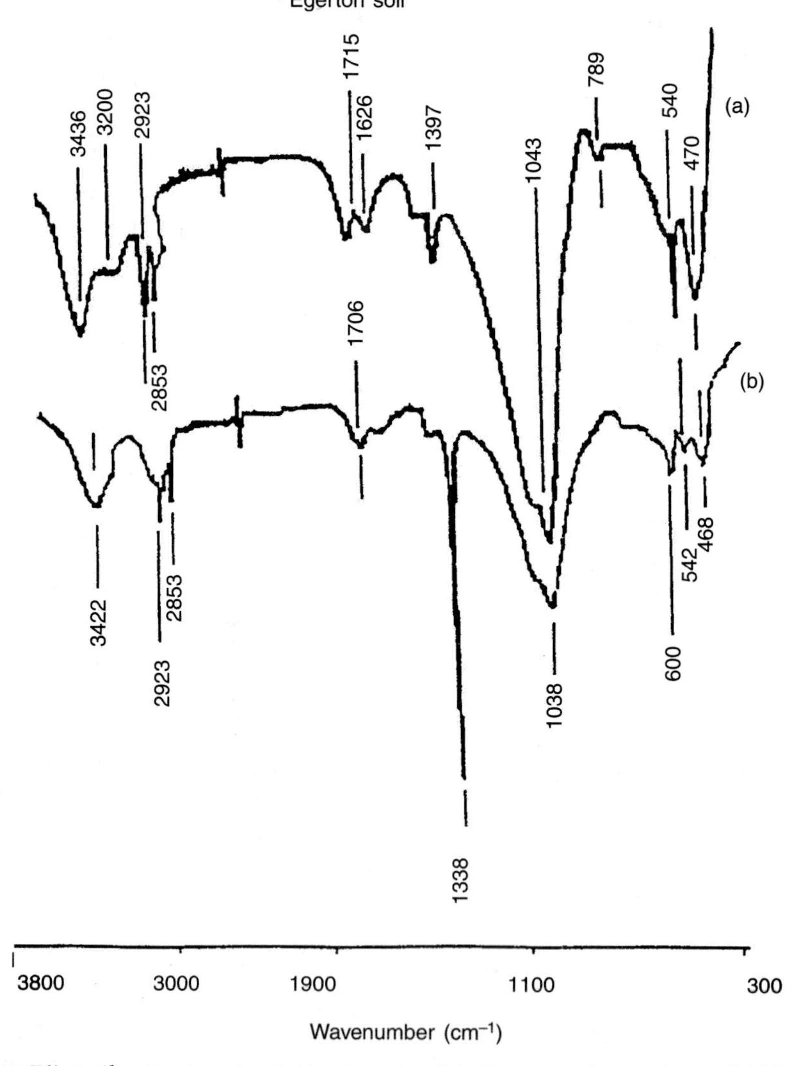

Fig. 19.8: Effect of heat treatment on infrared spectra of the Egerton soil treated with 1 M $NH_4H_2PO_4$ for 100 days at 25°C. (a) Control soil and (b) heat treatment at 110°C in air for 2 h.

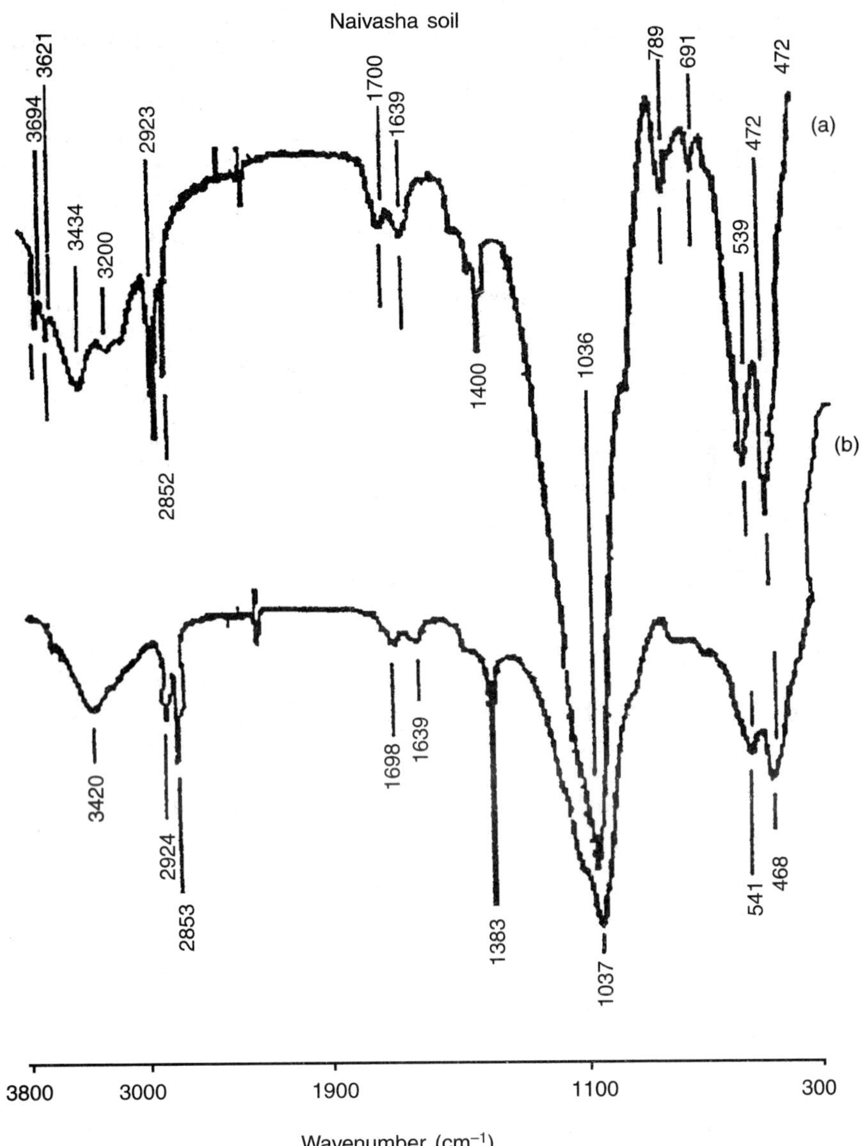

Fig. 19.9: Effect of heat treatment on infrared spectra of Naivasha soil treated with 1 M $NH_4H_2PO_4$ for 100 days at 25°C. (a) Control soil and (b) heat treatment at 110°C in air for 2 h.

cm^{-1} disappeared (Figs. 19.8 and 19.9), indicating that the ammonium taranakite structure was destroyed (Zhou, 1995). This IR evidence was in accord with the x-ray evidence that NH_4-taranakite was formed in the reaction products.

4 SUMMARY AND CONCLUSIONS

The amounts of Cd released from natural soils and Idaho MAP fertilizer-treated soils were significantly enhanced by $NH_4H_2PO_4$ solution. The rate constants for Cd released by 1 M $NH_4H_2PO_4$ solution from soils treated with Idaho MAP fertilizer were one order magnitude higher than the rate constants for Cd released by deionized distilled water from soils with the same treatments. This indicates that the rate of Cd release from the soils was greatly enhanced by application of the Idaho MAP fertilizer. The combined effect of attack by proton, Cd present in Idaho MAP fertilizer, and dissolution by phosphate would enhance the rate of Cd release from the soils.

Increase in the amounts of Cd released with increments in the concentration of $NH_4H_2PO_4$, would imply that at the fertilizer granule-soil interface in the soil rhizosphere where the concentration of orthophosphate is expected to be high, more Cd would be released compared to the bulk soil.

In the immediate vicinity of the fertilizer granule the phosphate concentration is saturated. High concentrations of phosphate can cause substantial alterations of soil constituents and subsequent release of metal ions. Phosphate fertilizers contain Cd as a contaminant and their application to soils is known to lead to Cd accumulation in soils. Application of phosphate fertilizers to soils may promote Cd release from soils, thus increasing the Cd content in soil solution. This may lead to increased Cd uptake by plants and adverse effects on human health. Using x-ray diffraction analysis, NH_4-taranakite was identified in the Egerton and Naivasha soils treated with 1 M $NH_4H_2PO_4$ solution for 100 days. However, taranakite did not form when monoammonium phosphate was spiked with Cd even at 6.4×10^{-3} M $Cd(ClO_4)_2$, indicating that Cd perturbs the formation of taranakites in soils. The nature of the soil-fertilizer reaction products depends largely on the acidity developed during dissolution of the applied fertilizer and may vary with distance from the fertilizer granule because of change of phosphate concentration. Since plants obtain most of their fertilizer phosphorus from the reaction products and not from the fertilizer per se, identification of reaction products is of primary importance in understanding the nature and behavior of phosphorus in soils.

The results presented in this study highlight the influence of the Idaho MAP fertilizer applied to agricultural soils on Cd release. Cd mobilized in the soil rhizosphere can be transported to the plant roots and then be taken up by the plant. The study also shows that NH_4-taranakite can form in the acidic tropical soils when monoammonium phosphates is applied to these soils. However, in the fertilizer-granule interface, Cd may perturb formation of taranakite, if the applied MAP fertilizer contains significantly high Cd as impurity.

Acknowledgments

This study was supported by funding from the Canadian International Development Agency (CIDA) and Research Grant GP 2383-Huang of the Natural Sciences and Engineering Research Council of Canada.

References

Alloway B.J. and Steinnes E. 1999. Anthropogenic additions of cadmium to soils. In: *Cadmium in Soils and Plants*. M.J. McLaughlin and B.R. Singh (eds.). Kluwer Acad. Publ., Dordrecht, Netherlands, pp. 97–123.

Annual Farmers' Recommendations. 1993. Kenya Agricultural Research Institute, Plant Breeding Center, Njoro, Kenya.

Berry L.G. (ed.) 1974. Selected powder diffraction data for minerals. Search Manual. Publication M-123. Joint Committee on Powder Diffraction Standards. Swarthmore, PA, USA.

Bloom P.R. and Nater E.A. 1991. Kinetics of dissolution of oxide and primary silicate minerals. In: *Rate of Soil Chemical Processes*. D.L. Sparks and D.L. Suarez (eds.). Soil Sci. Soc. Amer., Madison, WI, SSSA Special Publ. 27, pp. 151–189.

Bouldin D.R., Lehr J.R., and Sample E.C. 1960. The effect of associated salts on transformations of monocalcium phosphate monohydrate at the site of application. *Soil Sci. Soc. Amer. Proc.* 24: 464–468.

Brady N.C. 1990. *The Nature and Properties of Soils*. McMillan Publishing Co., New York, NY (10th ed.)

Chaney R.L. and S.B. Hornick. 1978. Accumulation and effects of cadmium on crops. *Proc. 1st Int. Cd. Conf., San Francisco,* Metals Bull. Ltd., London, pp. 125–140.

Christensen T.H. and Huang P.M. 1999. Solid phase cadmium and the reactions of aqueous cadmium with soil surfaces. In: *Cadmium in Soils and Plants*. M.J. McLaughlin and B.R. Singh (eds.) Kluwer Acad. Publ., Dordrecht, the Netherlands, pp. 65–96.

Dean J.A. 1992. *Lange's Handbook of Chemistry*. McGraw-Hill, Inc., New York, NY (14th ed.).

Duquette M. and Hendershot W. 1993. Soil surface charge evaluation by back-titration: 1. Theory and method development. *Soil Sci. Soc. Amer. J.* 57: 1222–1234.

Eltantawy I.M. and Arnold P.W. 1973. Reappraisal of ethylene glycol monoethyl ether method for surface area estimation of clays. *J. Soil Sci.* 24: 232.

Frazier A.W. and Taylor A.W. 1965. Characterization of taranakites and ammonium aluminum phosphates. *Soil Sci. Soc. Amer. Proc.* 29: 545–547.

Gadsden J.A. 1975. *Infrared Spectra of Minerals and Related Inorganic Compounds*. Butterworth and Co. Ltd., London, UK, 277 pp.

Haseman J.F., Brown E.H., and Whitt C.D. 1950. Some reactions of phosphate with clays and hydrous oxides of iron and aluminum. *Soil Sci.* 70: 257–271.

Huang P.M. 1991. Kinetics of redox reactions on manganese oxides and its impact on environmental quality. In: *Rates of Soil Chemical Processes*. D.L. Sparks and D.L. Suarez (eds.). SSSA Spec. Publ. no. 27, Soil Sci. Soc. Amer., Madison, WI, USA, pp. 191–230.

Jackson M.L. 1958. *Soil Chemical Analysis*. Prentice Hall, Englewood Cliffs, NJ (USA), 498 pp.

Jackson M.L. 1979. *Soil Chemical Analysis, Advanced Course*. Author published, Univ. Wisconsin, Madison, WI (USA), 895 pp.

Kittrick J.A. and Jackson M.L. 1954. Electron microscope observations of the formation of aluminum phosphate crystals with kaolinite as the source of aluminum. *Science* 120: 508–509.

Kittrick J.A. and Jackson M.L. 1955. Rate of phosphate reaction with soil minerals and electron microscope observations on the reaction mechanism. *Soil Sci. Soc. Amer. Proc.* 19: 292–295.

Kittrick J.A. and Jackson M.L. 1956. Electron microscope observations of the reactions of phosphate with minerals, leading to a unified theory of phosphate fixation in soils. *J. Soil Sci.* 7: 81–88.

Krishnamurti G.S.R., Huang P.M., Van Rees K.C.J., Kozak L.M., and Rostard H.P.W. 1994. Microwave digestion technique for the determination of total cadmium in soils. *Comm. Soil Sci. Plant Anal.* 25: 615–625.

Lakanen E. and Ervio R. 1971. A comparison of eight extractants for the determination of available micronutrients in soils. *Acta Agric. Fenn.* 123: 223–232.

Lehr J.R., Brown W.E., and Brown E.H. 1959. Chemical behavior of monocalcium phosphate monohydrate in soils. *Soil Sci. Soc. Amer. Proc.* 23: 3–7.

Lehr J.R., Brown E.H., Frazier A.W., Smith J.P. and Thrasher R.D. 1967. Crystallographic properties of fertilizer compounds. *Chem. Eng. Bull.* 6. Tennessee Valley Authority, Muscle Shoals, AL USA.

Lindsay W.L., Frazier A.W., and Stephenson H.F. 1962. Identification of reaction products from phosphate fertilizers in soils. *Soil Sci. Soc. Amer. Proc.* 26: 466–452.

Low P.F. and Black C.A. 1947. Phosphate induced decomposition of kaolinite. *Soil Sci. Soc. Amer. Proc.* 12: 180–184.

Martell A.E., Smith R.M., and Motekatis R.J. 1997. *NIST Critical Stability Constants of Metal Complexes, Database.* Texas & M Univ., College Station, TX, USA.

McBride M.B. 2000. Chemisorption and precipitation reactions. In: *Handbook of Soil Science.* M.E. Summer (ed). CRC Press, Boca Raton, Fl (USA), pp. B265–B302.

McKeague J.A. 1967. An evaluation of 0.1 M pyrophosphate and pyrophosphate-dithionite in comparison with oxalate as extractants of the accumulation products in podzols and some other soils. *Can. J. Soil Sci.* 47: 95–99.

McKeague J.A. and Day J.H. 1966. Dithionite and oxalate extractable Fe and Al as aids in differentiating various classes of soils. *Can. J. Soil Sci.* 46: 13–22.

McKenzie R.M. 1980. The adsorption of lead and other heavy metal ions on oxides of manganese and iron. *Austr. J. Soil Res.* 18: 61–73.

Mehra O.P. and Jackson M.L. 1960. Iron oxide removal from soils by dithionite-citrate system buffered with sodium bicarbonate. *Clays Clay Miner.* 7: 317–327.

Mortvedt J.J. and Osborn G. 1982. Studies on the chemical form of cadmium contaminants in phosphate fertilizers. *Soil Sci.* 134: 185–192.

Onyatta J.O. and Huang P.M. 1999. Chemical speciation and bioavailability index of cadmium for selected tropical soils in Kenya. *Geoderma.* 91: 87–101.

Prabhudesai S.S. and Kudrekar S.B. 1984. Reaction products from fertilizer phosphorus in lateritic soils of Konkan region. *J. Indian Soc. Soil Sci.* 32: 52–56.

Sample E.C., Soper R.J., and Racz G.J. 1980. Reactions of phosphate fertilizers in soils. In: *The Role of Phosphorus in Agriculture.* F.E. Khasawneh, E.C. Sample, and E.J. Kamprath (eds.). Amer. Soc. Agron., Crop Sci. Soc. Amer., Soil Sci. Soc. Amer. Madison, WI (USA), pp. 263–310.

Sarkar D., Sarkar M.C., and Ghosh S.K. 1977. Reaction products in red soils of West Bengal. *J. Indian Soc. Soil Sci.* 25: 141–149.

Somasiri S., Lee S.Y., and Huang P.M. 1971. Influence of certain pedogenic factors on potassium reserves of selected Canadian prairie soils. *Soil Sci. Soc. Amer. Proc.* 35: 500–505.

Sparks D.L. 1989. *Kinetics of Soil Chemical Processes.* Acad. Press, Inc., New York, NY.

Tamimi Y.N., Kanehiro Y., and Sherman G.D. 1964. Reactions of ammonium phosphate with gibbsite and with montmorillonitic and kaolinitic soils. *Soil Sci.* 98: 249–255.

Taylor A.W. and Gurney E.L. 1961. Solubilities of potassium and ammonium taranakites. *J. Phys. Chem.* 65: 1613–1616.

Taylor A.W., Gurney E.L., and Lindsay W. 1960. An evaluation of some iron and aluminum phosphates as sources of phosphate for plants. *Soil Sci.* 90: 25–31.

Taylor A.W., Lindsay W., Huffman E.O., and Gurney E.L. 1963. Potassium and ammonium taranakites, amorphous aluminum phosphate and variscite as sources of phosphate for plants. *Soil Sci. Soc. Amer. Proc.* 27: 148–151.

Tiller K.G. and Smith L.H. 1990. Limitations of EGME retention to estimate the surface area of soils. *Austr. J. Soil Res.* 28: 1–28.

Tisdale S.L., Nelson W.J., Beaton J.D., and Havlin J.L. 1993. *Soil Fertility and Fertilizers.* McMillan Publ. Co. New York, NY, pp. 176–271.

Wang D. and Anderson D.W. 1998. Direct determination of organic carbon content in soils by the Leco 12 carbon analyzer. *Soil Sci. Plant Anal.* 29: 15–21.

Zhou J.M. 1995. Kinetics and mechanisms of phosphate-induced potassium release from selected K-bearing minerals and soils. PhD Thesis, Univ. Saskatchewan, Saskatoon, SK, Canada.

Zhou J.M. and Huang P.M. 1995. Kinetics of monoammonium phosphate-induced potassium release from selected soils. *Can. J. Soil Sci.* 75: 197–203.

Appendix I

(a) Calculation for the amount of cadmium added to the soil in the incubation study:

Recommended rate of monoammonium phosphate (MAP) fertilizer application

= 130 kg kg^{-1}

Cd content of Idaho MAP fertilizer = 144 mg kg^{-1}

Amount of Cd introduced to soil from the fertilizer: 130 kg ha^{-1} MAP × 144 mg kg^{-1}

= 18,720 mg Cd/ha^{-1}

= 18.72 g Cd/ha^{-1}

Amount of fertilizer in 1 m^2 plot = 130 kg 10^4 = 0.013 kg

= 13 g of fertilizer

(b) Amount of soil in contact with fertilizer granules:

For a spherical fertilizer granule with diameter of 3 mm (0.3 cm), i.e., radius of 1.5 mm (0.15 cm)

Volume of a fertilizer granule = $4/3\ \pi\ r^3$

= 4/3 (3.14)$(0.15\ \text{cm})^3$

= 0.0141 cm^3

Thickness of soil around fertilizer granule

= 1 mm = 0.1 cm

Radius of fertilizer granule and soil around it

= 0.15 cm + 0.1 cm = 0.25 cm

Total volume (fertilizer + soil around it)

= 4/3 × 3.14 $(0.25\ \text{cm})^3$ =0.0654 cm^3

Vol. of soil around fertilizer granule

= 0.0654 cm^3 – 0.0141 cm^3

= 0.0513 cm^3

Bulk density of soil $= 1.49 \text{ g cm}^{-3}$

Weight of soil around 1 fertilizer granule

$$= 0.0513 \text{ cm}^3 \times 1.49 \text{ g cm}^{-3}$$

$$= 0.0764 \text{ g soil}$$

Weight of MAP fertilizer in 1 m^2 plot = 13 g (equivalent to 392 MAP fertilizer granules)

Therefore, wt. of soil in 1 m^2 plot around granules

$$= 0.0764 \text{ g} \times 392 = 30 \text{ g}$$

Amount of soil around fertilizer granules in 1 ha

$$= \frac{30\,\text{g} \times 10^4}{1000}$$

$$= 300 \text{ kg soil}$$

Hence 18.72 g Cd (from Idaho MAP fertilizer) is present in 300 kg soil

Amount of Cd in 1 kg soil $= \dfrac{18.72}{300}$

$$= 0.0624 \text{ g Cd kg}^{-1} \text{ soil}$$

Hence amount of Cd used in the incubation study

$$= 62.4 \text{ mg Cd kg}^{-1} \text{ soil}$$

$$= 6.24 \text{ mg Cd}/100 \text{ g soil}$$

(c) Calculation for amount of MAP added to 100 g soil

Wt. of soil around MAP granules in 1 m^2 plot calculated as 30 g.
Wt of MAP fertilizer added to 100 g soil calculated as:
30 g soil around 13 g MAP fertilizer granules

Hence, 100 g soil =

$$\frac{100\,\text{g} \times 13\,\text{g MAP fertilizer granules}}{30\,\text{g}}$$

$$= 43.3 \text{ g MAP fertilizer}$$

This is equivalent to 6.24 mg Cd/100 g soil as calculated in Appendix I (b).

(d) Sample calculation for solution Cd concentration in fertilizer granule-soil interface:

Field capacity of soils: Egerton: 30%, Naivasha 29%, and Soy 20%.

$Cd(ClO_4)_2{\cdot}6H_2O$ FW = 419.3

Therefore, weight of Cd perchlorate added to 100 g soil

$$= \frac{419.3 \text{ g} \times 0.0062 \text{ g}}{112.4 \text{ g}}$$

= 0.0231 g cadmium perchlorate

For example, solution Cd concentration at 20% field capacity

$$= \frac{0.0231 \text{ g} \times 1{,}000 \text{ g}}{419 \text{ g} \times 20 \text{ mL}}$$

$= 2.8 \times 10^{-3}$ M

Part V
Rhizosphere Processes

20

Organic Compounds Associated with Water Repellency in Sandy Soils

S.H. Doerr*, P. Douglas, C.P. Morley, C.T. Llewellyn, K.A. Mainwaring, C. Haskins, L. Johnsey, C.J. Ritsema, F. Stagnitti, A.J.D. Ferreira, *and* A.K. Ziogas

Abstract

Water repellency occurs in the root zone of a wide range of soils, and is thought to be caused by hydrophobic organic compounds. In this study, we separated, characterized, and compared organic compounds extracted from a range of repellent and wettable sandy soils, and also evaluated different extraction methods for effectiveness. Extraction efficiency was determined by: mass of material extracted; repellency pre- and post-extraction and after sample drying at 20°C and 105°C; relative amounts of aliphatic C-H units removed, using DRIFT analysis (Diffuse Reflectance Infrared Fourier Transform spectroscopy); and ability of extracts to cause repellency when applied to acid-washed sand (AWS). Tetrahydrofuran-soluble portions of selected extracts were analyzed by Gas Chromatography-Mass Spectrometry.

The key findings were these. (i) Sample drying at 105°C provides additional information on extraction effectiveness. (ii) Isopropanol:aqueous ammonia was the most effective extraction solvent. (iii) The mass of material extracted was poorly related to sample repellency, and extracts from wettable control soils also induced repellency in AWS. (iv) The amount of aliphatic C–H units as determined by DRIFT is not related to repellency of a sample, and hence DRIFT analysis is not well suited for indirect determination of the water repellency of a soil. (v) Both wettable and water repellent samples contained hydrophobic compounds (i.e., long chain carboxylic acids (C_{16}–C_{24}), amides (C_{14}–C_{24}), alkanes (C_{25}–C_{31})). Our conclusion: the presence of hydrophobic compounds is not the

**Corresponding author*: Dr. S.H. Doerr, Dept. Geography, University of Wales Swansea, Singleton Park, Swansea SA28 PP, UK. E-mail: s.doerr@swansea.uk only requirement for the expression of water repellency by a soil. Its relative abundance and structural arrangement are also critical.

1 INTRODUCTION

Water repellency (or hydrophobicity) is often a feature of dry or moderately moist soils under a wide range of vegetation types. Its consequences include poor plant growth, as commonly observed for example for turf grass during dry periods, reduced seed germination, accelerated surface runoff and soil erosion by wind and water, and enhanced preferential flow in the vadose zone of soils, which in turn can lead to accelerated leaching of nutrients and agrichemicals (see reviews by DeBano, 2000; or Doerr et al., 2000). Other effects of water repellency include enhanced soil aggregate stability (Hallet and Young, 1999) and an increased stabilization of soil organic carbon (Piccolo et al., 1999). It has been proposed that water repellency is caused by the accumulation of hydrophobic organic compounds which are naturally abundant in the biosphere and may originate from root exudates (Dekker and Ritsema 1996; Doerr et al., 1998), fungal or microbial byproducts (Savage et al., 1972; Jex et al., 1985), or directly from decomposing organic matter (McGhie and Posner, 1981). These compounds are generally thought to be present as a coating on mineral or aggregate surfaces (Bisdom et al., 1993; Wallis and Horne, 1992; Doerr et al., 2000), but might also be as part of the organic interstitial matter (Franco et al., 2000). Few studies to date have attempted to isolate and characterize these compounds.

The first step in isolating compounds causing repellency has generally been removal of organic material from the soil. A number of extraction techniques have been used and their relative efficiencies discussed (e.g. Roberts and Carbon, 1972; Ma'shum et al., 1988; Horne and McIntosh, 1994; Hudson et al., 1994; Roy et al., 1999; Franco et al., 2000). The types of organic compounds suggested to cause repellency include plant and cuticular waxes (McIntosh and Horne, 1994), alkanes (Savage et al. 1972; Ma'shum et al. 1988; Roy et al. 1999; Horne and McIntosh, 2000) fatty acids and their salts and esters (Schnitzer and Preston 1987; Ma'shum et al., 1988; Franco et al., 2000; Hudson et al., 1994; Roy et al., 1999; Horne and McIntosh, 2000), and also phytanes, phytols, and sterols (Franco et al., 1995).

Current knowledge in extraction and characterization of the compounds responsible for water repellency in soils exhibits the following shortcomings: (i) a focus on just a small range of samples from one or two regions; (ii) wettable soils were not always included as a control; (iii) incomplete extraction of hydrophobic compounds from water-repellent soils, i.e., soils were not always rendered wettable after extraction; (iv) the substances isolated were not always shown to be capable of causing soil water repellency; and (v) although generic chemical classes have been proposed or determined for these materials, their precise chemical structure has not yet been identified (Roy et al., 1999; Doerr et al., 2000).

The study reported here sought to address these shortcomings by: (i) investigating a wide range of water-repellent soils from different countries; (ii) including also wettable 'control' samples; (iii) undertaking detailed assessment of extraction method efficiencies; (iv) examining the ability of extracts to induce water repellency in wettable acid-washed sand (AWS); and (v) undertaking further separation and characterization of compounds contained in the extracts of some selected soil samples.

2 MATERIALS AND METHODS

2.1 Soil Sampling and Sample Preparation

Soils known to exhibit either water-repellent or wettable properties were sampled in five countries comprising a range of locations differing in climate and vegetation cover. The sample sites selected in the Netherlands (NL) and Wales (UK) have an oceanic humid-temperate climate with rainfall occurring throughout the year. The sites in Greece (GK) are also temperate, but with a summer dry season, whereas sites in Portugal (PT) and Australia (AU) exhibit a warmer Mediterranean type climate with prolonged dry periods during the summer months. Samples were taken at a range of depths within the root penetration depth of the soils, oven dried at 20 °C, and passed through a 2-mm sieve prior to further analysis. All soils are of medium sand texture, with a clay content of < 0.1%. Overall, three water repellent and one wettable sample were taken in each of the aforementioned countries (n = 20). Site locations, sampled depths, and sample characteristics are summarized in Table 20.1.

To reduce variability between subsamples, the coning-and-quartering method was used (Jackson, 1958). All material of each soil sample (~2 kg) was thus shaped into a cone with a funnel and flattened at the top. The cone was then divided into four equal quarters and two opposite segments removed. These segments were combined and shaped into a second cone. The resultant cones were then coned and quartered until each subsample was 40–50 g in size.

2.2 Water Repellency Assessments

After equilibrating the samples in a controlled atmosphere of 20°C and 45–55% relative humidity for 24 h to preclude any influence of changing atmospheric conditions on measurement results (Doerr et al., 2002), water repellency was assessed using the Water Drop Penetration Time (WDPT) method. This test involves placing 5 drops of distilled water (~80 μL) on the sample surface and recording the time for complete droplet penetration (Letey, 1969). Repellency values were recorded according to distinct repellency classes (Table 20.2) and based on the median class of the 5 drops. Additional WDPT tests were carried out after heating samples at 105°C for 24 h and then equilibrating them as described above. This procedure was included since previous studies had shown that some soil samples, either taken directly from the field or after extraction,

Table 20.1: Sample codes and origin, mean particle diameter and distribution width, and their water repellency levels (see Table 20.2 for respective WDPT classes) before extraction. Samples with the letter C denote wettable control soils

Sample code	Location (UTM coordinates)	Land use/ vegetation type	Sampled depth (cm)	Mean diameter and distribution width (mm)	WDPT and repellency rating‡ (dried at 20°C)	
PT 1	40°32′N, 8°46′W	agriculture	0–10	0.57; 0.23	180	strong
PT 2	40°19′N, 8°46′W	eucalyptus forest	0–10	0.46; 0.16	18,000	extreme
PT 3	40°20′N, 8°47′W	pine forest	0–10	0.47; 0.16	60	slight
PTC	40°20′N, 8°47′W	pine forest	0–10	0.50; 0.17	< 5	nonrepellent
NL1	51°48′N, 3°54′E	permanent pasture	0–10	0.27; 0.22	180	strong
NL2	51°48′N, 3°54′E	permanent pasture	10–20	0.23; 0.10	3,600	severe
NL3	51°48′N, 3°54′E	permanent pasture	20–30	0.22; 0.08	18,000	extreme
NLC	51°48′N, 3°54′E	permanent pasture	30–40	0.22; 0.07	< 5	nonrepellent
$UK1_a$†	51°35′N, 4°06′W	dune herbs & grasses	0–5	no data	600	strong
$UK1_b$†	51°35′N, 4°06′W	dune herbs & grasses	0–5	0.33; 0.08	900	severe
UK2	51°35′N, 4°06′W	grassland (sports turf)	0–5	0.30; 0.08	300	strong
UKC	51°35′N, 4°06′W	dune, unvegetated	0–5	0.39; 0.12	< 5	nonrepellent
AU1	36°26′S, 140°40′E	agriculture	0–10	0.25; 0.16	3,600	severe
AU2	36°26′S, 140°40′E	agriculture	0–10	0.29; 0.23	180	strong
AU3	36°26′S, 140°41′E	agriculture	0–10	0.23; 0.11	18,000	extreme
AUC	36°30′S, 140°42′E	agriculture	0–10	0.24; 0.14	< 5	nonrepellent
GK1	41°07′N, 25°07′W	permanent pasture	0–12	0.45; 0.32	600	strong
GK2	40°57′N, 20°19′W	permanent pasture	0–5	0.47; 0.24	600	strong
GK3	40°56′N, 24°59′W	dune herbs & grasses	0–19	0.35; 0.14	180	strong
GKC	40°55′N, 24°53′W	agriculture	0–26	0.70; 0.36	< 5	nonrepellent
AWS		Supplier: Riedel-de Haën	n.a.	0.27; 0.07	<5	nonrepellent

† Samples $UK1_a$ and $UK1_b$ were taken from the same location, but sampled 10/14/99 and 11/24/99 respectively.
‡ For repellency rating, see Table 20.2.

Table 20.2: WDPT class increments used in this study, log midpoint WDPT values for each class, and corresponding descriptive repellency rating. Numbers denote the upper time limits (in seconds) for individual repellency classes

WDPT classes	≤ 5	10	30	60	180	300	600	900	3,600	18,000	>18,000
log midpoint WDPT	0.40	0.88	1.30	1.65	2.08	2.38	2.65	2.88	3.35	4.03	4.56‡
Repellency rating†	wettable		slight			strong			severe		extreme

† after Bisdom et al. (1993).
‡ arbitrarily taken as 2 × lower limit.

which were wettable after drying at 20°C, developed water repellency after heating to 105°C (Ma'shum and Farmer, 1985; Dekker et al., 1998).

2.3 Initial Assessment of Extraction Procedures Using a Range of Solvents

A range of solvents employed in previous studies on water repellent soils was used (Ma'shum et al., 1988; Horne and McIntosh, 1994; Franco et al., 1995). These included, in order of increasing polarity, hexane, toluene, dichloromethane, isopropanol, and isopropanol: aqueous ammonia (0.88 specific gravity) (7:3 v:v). Unless stated otherwise, extractions were carried out using a Soxhlet apparatus for 24 h with 80 g soil and 800 mL of solvent using Whatman cellulose thimbles ca. 11.5 cm in length and 3 cm in diameter. After extraction, the solvent was filtered and the liquid concentrated by rotary evaporation under reduced pressure at 45°C. A fraction of the extract was retained as liquid for later use and the remainder taken to dryness on a hot water bath. A blank consisting of a clean, empty cellulose thimble was also extracted to provide a correction factor for the presence of residual organics in the solvents and/or Soxhlet glassware.

2.4 Extract Reapplications Following Isopropanol/Ammonia Extraction

Following the procedures of Ma'shum et al. (1988), extracted material was filtered, taken to dryness, redissolved in chloroform ($CHCl_3$), and reapplied to 5 g wettable (WDPT < 5 s) acid-washed sand (AWS; HCl-washed quartz sand supplied by Riedel-de Haën). Based on laser particle size analysis (Malvern Mastersizer), the AWS used had a mean diameter of 270 μm and distribution width of ±70 μm, which are intermediate values with respect to the soils investigated here (see Table 20.1). Since dried extracted material did not always fully redissolve, extract that had not been taken to dryness was also applied directly to AWS as an additional exploratory procedure. Note that this was in effect application from an isopropanol/water (ca. 7:3 v:v) mixture, because by this stage in the process there was little, if any, ammonia left in the extraction mixture solvent. The ratio of sand to extract was so chosen that the extract was reapplied at the same mass ratio as it was extracted.

As for the unextracted soils, all samples were equilibrated for 24 h under a controlled humidity of 45–55% at 20°C before WDPT tests were carried out.

2.5 Extraction and Separation Procedure for Further Analyses

After initial evaluation of extraction procedures, samples from the Netherlands (NL) were selected for further analysis. These samples consisted of three soils with different degrees of water repellency and one wettable control soil. Since all four samples were taken from different depths from the same soil profile, they were considered to be particularly well suited for detailed study, as other variables such as vegetation cover, climate, and to some degree sample texture could be considered constant. Thus 24 h Soxhlet extractions were carried out with 240 g soil using a 2.4 L isopropanol: aqueous ammonia (0.88 specific gravity) (7:3 v:v) solvent mixture. It was decided to prewet the samples with the solvent mixture for 15 min prior to refluxing as ammonia is lost during the extraction procedure. The extracted material dissolved in the solvent was then filtered, concentrated under reduced pressure on a rotary evaporator at 45°C, transferred to a preweighed porcelain evaporating dish and taken to dryness on a hot water bath.

The dried extract was dissolved in a 200 ml chloroform and water mixture (1:1 v:v). The phases were separated and washed with 2 × 100 mL aliquots of the other solvent. The aqueous phase contained amphipathic and polar material and the chloroform phase the lipid compounds (Horne and McIntosh, 2000). Both phases were taken to dryness on a hot water bath. The dried aqueous phase was then dissolved in methanol. The small portion of probably highly polar material, which did not dissolve in the methanol, was filtered off. The filtrate was taken to dryness on a water bath.

2.6 Characterization Techniques

Diffuse Reflectance Infrared Fourier Transform (DRIFT) spectroscopic analysis

DRIFT spectra were recorded using a Mattson Satellite FTIR Spectrometer equipped with a Spectra Tech Diffuse Reflectance Unit (0030-033). Soil samples were placed into sample cups with a 2.5 mm diameter and depth of 2 mm and flattened using a glass microscope slide. For each sample 1,024 scans were recorded from 4,000 to 400 cm^{-1} with a resolution of 4 cm^{-1}. Depending on sample-to-sample reproducibility, up to twelve independent samples were measured per soil. Dried KBr was used as background.

Gas Chromatography (GC) and Gas Chromatography-Mass Spectrometry (GC-MS) analysis

The dry, chloroform-soluble phase of extracts was dissolved in tetrahydrofuran (THF; chosen in preference to chloroform because it dissolved more material), filtered through tightly packed glass wool to remove insoluble particles, and analyzed by GC and GC-MS. Chromatograms were obtained using a Hewlett Packard 5890 series II gas chromatograph equipped with a flame ionization detector (FID) and a ZB5 5% phenyl polysiloxane coated capillary column (30 m, 0.32 mm i.d., 1.0 μm df). Samples (2 μL) were injected splitlessly (0.6 min) and helium used as the carrier gas. The oven temperature was programmed from 210°C to 280°C at 2°C min^{-1} and held constant for 2.5 min followed by a second program of 15°C min^{-1} to a final temperature of 310°C (held for 80 min). The injection port was set at 250°C. A Fisons GC8000 gas chromatograph interfaced directly with a Fisons Masslab MD800 low-resolution GC-MS instrument was used to obtain Electron Impact (EI) spectra. The Fisons GC8000 gas chromatograph also contained a ZB5 5% phenyl polysiloxane capillary column (30 m, 0.32 mm i.d., 0.25 μm df). Splitless, 1 μL injections with hydrogen carrier gas and a temperature program of 40°C isothermal for 2 min, then ramped at 10°C/min^{-1} to 300°C and held for 30 min, were used. The injection port was set at 250°C. Compounds were identified based on retention times, mass spectral interpretation and use of the NIST mass spectral search program and NIST/EPA/NIH mass spectral library v.2.0.

Microanalysis

Microanalysis of extracts was performed by Butterworths Laboratories Limited at Teddington in Middlesex using a Leeman Labs Inc. Model 44 CHN Elemental Analyzer.

Nuclear Magnetic Resonance Spectroscopy (NMR)

^{1}H and ^{13}C NMR spectra of chloroform extracts were recorded in deuterated chloroform ($CDCl_3$) at 400 MHz on a Bruker Avance 400 Spectrometer.

2.7 Experimental Protocol

To select the most effective extraction procedure, an initial exploratory evaluation of a range of solvents employed in previous studies described in § 2.3 was carried out using the strongly water repellent sample $UK1_a$ (WDPT 600 s) and the wettable sample UKC (WDPT < 5 s) as control (see Table 20.1 for details). A 100 g mass of dry soil and 500 mL of solvent were used for these experiments. After extraction, the soils were air dried at 20°C and tested for repellency, and the amounts of material extracted were determined. Isopropanol/ammonia was found to be the most effective solvent and was used in extracting all soil samples as described in § 2.3.

To examine whether the compounds extracted from the soils were capable of inducing water repellency in wettable soils, extracts were reapplied on AWS as described in § 2.4 above, and sub-samples were then dried at 20°C and 105°C respectively and tested for water repellency.

DRIFT analysis, as described in § 2.6, was carried out on all samples before and after the main extraction protocol. This allowed determination of the relative amounts of aliphatic CH present before extraction and left on the soil after extraction.

Subsequently, using also isopropanol/ammonia Soxhlet extraction, extracts of further sample material of soils NL were obtained, separated, and analyzed using GC, GC-MS, and NMR techniques as described above (§ 2.6) to determine the chemical nature of the compounds present. Microanalysis was carried out on the three separated fractions (chloroform-, methanol-, and water-soluble) to determine relative C, H, and N contents.

3 RESULTS

3.1 Initial Assessment of Extraction Procedures Using a Range of Solvents

Results of initial evaluation of solvents, both in terms of amount extracted and sample repellency level after the extraction, are given in Table 20.3. Extraction

Table 20.3: Efficiency of Soxhlet extractions with different solvents on sample UK1$_a$ (bold; WDPT class 600 s) and the wettable control sample UKC (italics; WDPT <5 s)

Solvent	Hexane†	Dichloromethane†	Toluene†	Isopropanol†	Isopropanol/ ammonia
Amount extracted	**368**	**418**	**881**	**1240**	**2024‡**
(mg kg^{-1})	*54*	*18*	*46*	*590*	*226**
WDPT class (s)	**18000**	**18000**	**18000**	**18000**	**< 5**
after extraction	*< 5*	*< 5*	*< 5*	*< 5*	*< 5*

† Extraction duration: 60–67 hours ‡ Extraction duration: 48 hours *Extraction duration: 24 hours

with the organic solvents dichloromethane, hexane, isopropanol, and toluene was considerably less effective than with isopropanol/ammonia. Isopropanol/ammonia extracted the most material and rendered the soil wettable. Surprisingly, the other four solvents rendered the soil more water repellent, despite removing some organic material. The control sample UKC retained its wettability (WDPT < 5 s) after all extractions. Isopropanol/ammonia (0.88 specific gravity) (7:3 v:v) was therefore used for all subsequent extractions.

3.2 Extraction with Isopropanol/Ammonia

Results of isopropanol/ammonia extractions with respect to water repellency levels and amounts extracted for all samples are given in Table 20.4. Of the 15 repellent samples investigated[1], all but two were rendered wettable after extraction for both drying temperatures, and all five control samples remained wettable. For samples UK1$_b$ and GK2, water repellency was reduced from severe and strong respectively, to a slight repellency after extraction and drying at 20°C. Drying at 105°C caused a slight increase in repellency for GK2 (WDPT from 30 to 60 s) and a large increase in repellency for UK1$_b$ (WDPT from 30 to 3,600 s).

Extraction of empty Soxhlet thimbles was also carried out to demonstrate that there was no significant contribution to the extract from the apparatus used. Only trace quantities of material (< 3 mg per extraction) were obtained in this way.

Amounts extracted ranged from ca. 250 to 9,800 mg kg^{-1}. The lowest value derives from a wettable dune sand with a low organic matter content (UKC), while the highest value came from the top layer of an organic-rich, grass-covered sand (NL1), which was one of the most water-repellent soils investigated here. Since hydrophobic compounds are of organic origin, this finding might suggest a relationship between sample repellency and the amount extracted. There was a very rough correlation between WDPT and amount extracted in that no soils showed a high amount extracted and low WDPT, while a few showed a low amount extracted and high WDPT. In the general region of a moderate amount

[1] Sample material UK1$_a$ had been consumed by this stage and was replaced with UK1$_b$ sampled later at the same location.

Table 20.4: WDPT classes (s) (as defined in Table 20.2) of samples before and after Soxhlet isopropanol/ammonia (IPA/NH_3) extraction, and after reapplying extracts direct (IPA/NH_3), or after drying and redissolving in $CHCl_3$, to hydrophilic acid-washed sand (AWS). For each treatment, WDPT tests were carried out on subsamples after drying at (20°C) and at (105°C). Samples with the letter C denote wettable control soils

Sample code	Pre-extraction (20°C)	Post-extraction (20°C)	Pre-extraction (105°C)	Post-extraction (105°C)	Mass extracted (g kg^{-1}) †	IPA/NH_3 extract on AWS (20°C)	IPA/NH_3 extract on AWS (105°C)	$CHCl_3$ extract on AWS (20°C)	$CHCl_3$ extract on AWS (105°C)
PT1	180	< 5	180	< 5	1.55 (0.31)	10	60	10	600
PT2	18,000	< 5	> 18,000	< 5	3.28 (0.40)	< 5	10	900	18,000
PT3	60	< 5	180	< 5	1.22 (0.55)	30	60	180	600
PTC	< 5	< 5	< 5	< 5	1.28 (0.44)	10	60	30	3,600
NL1	180	< 5	18,000	< 5	9.76 (1.50)	< 5	3,600	3,600	> 18,000
NL2	3,600	< 5	18,000	< 5	2.64 (0.33)	< 5	180	18,000	> 18,000
NL3	18,000	< 5	3,600	< 5	1.10 (0.06)	< 5	60	18,000	> 18,000
NLC	< 5	< 5	< 5	< 5	0.55 (0.23)	< 5	10	3,600	18,000
UK1$_b$	900	30	900	3,600	1.17 (0.52)	10	180	600	18,000
UK2	300	< 5	180	< 5	2.41 (0.28)	10	60	18,000	>18,000
UKC	< 5	< 5	<5	< 5	0.23 (0.03)	10	30	10	180
AU1	3,600	< 5	18,000	< 5	2.64 (0.57)	< 5	900	3,600	18,000
AU2	180	< 5	180	< 5	3.67 (1.26)	< 5	300	3,600	18,000
AU3	18,000	< 5	18,000	< 5	0.83 (0.47)	180	900	3,600	> 18,000
AUC	< 5	< 5	< 5	< 5	0.86 (0.61)	30	180	600	300
GK1	600	< 5	600	< 5	2.62 (0.79)	10	300	900	3,600
GK2	600	30	3,600	60	3.02 (1.90)	30	600	300	3,600
GK3	180	< 5	60	< 5	0.41 (0.18)	60	300	180	600
GKC	< 5	< 5	30	< 5	2.32 (0.54)	180	3,600	3,600	> 18,000

† Error estimates typically based on 5 independent measurements.

extracted and moderate WDPTs, however, the correlation between amount extracted and WDPT is very weak (Fig. 20.1a). These comments apply equally to samples of similar origin or vegetation (Table 20.4) and the complete sample pool (Figs. 20.1a, 1b, 1c, and 1d).

3.3 Extract Reapplications

WDPT results for the AWS after application of the material extracted with isopropanol/ammonia and after drying at either 20°C or 105°C are given in Table 20.4. Control experiments in which chloroform or isopropanol:aqueous ammonia alone were applied to AWS showed that these solvents did not affect the wettable nature of AWS.

Extracted material from water-repellent soils redissolved in chloroform induced water repellency in AWS in all cases, although extracted dry matter did not always fully redissolve in chloroform. Drying these AWS samples at

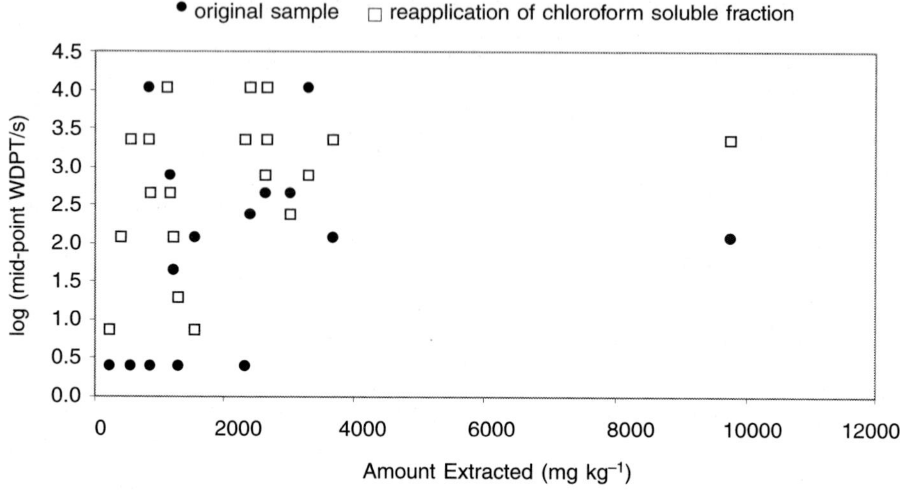

Fig. 20.1a: Comparison between log midpoint Water Drop Penetration Times (see Table 20.2) for soil preextraction and acid-washed sand after reapplication of chloroform soluble fraction dried at 20°C.

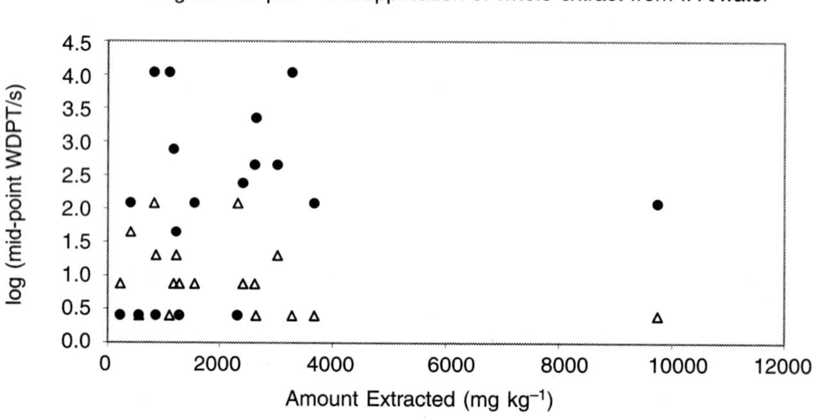

Fig. 20.1b: Comparison between log midpoint Water Drop Penetration Times (see Table 20.2) for soil preextraction and acid-washed sand after reapplication of whole extract from IPA/H_2O dried at 20°C.

105°C consistently resulted in higher WDPTs than drying at 20°C. Reapplication of the isopropanol/ammonia extract and drying at 20°C gave AWS with WDPTs lower than those for soil samples from which the extract was taken (Fig. 20.1b); reapplication using the $CHCl_3$ soluble portion of the dried isopropanol: aqueous ammonia extract followed by drying at 105°C gave WDPTs generally higher than the original samples (Fig. 20.1d). The best match between WDPTs for AWS after reapplication and the original soil was obtained using the $CHCl_3$ soluble portion with subsequent drying at 20°C (Fig. 20.1a).

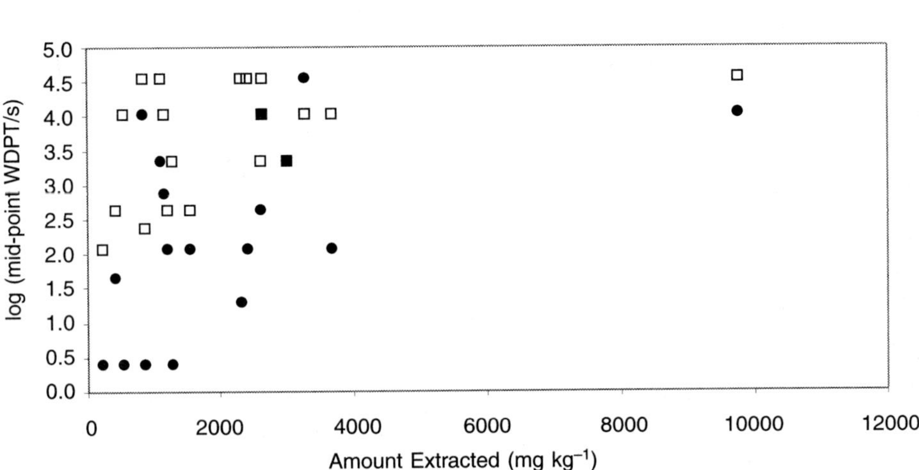

Fig. 20.1c: Comparison between log midpoint Water Drop Penetration Times (see Table 20.2) for soil preextraction and acid-washed sand after reapplication of chloroform soluble fraction dried at 105°C.

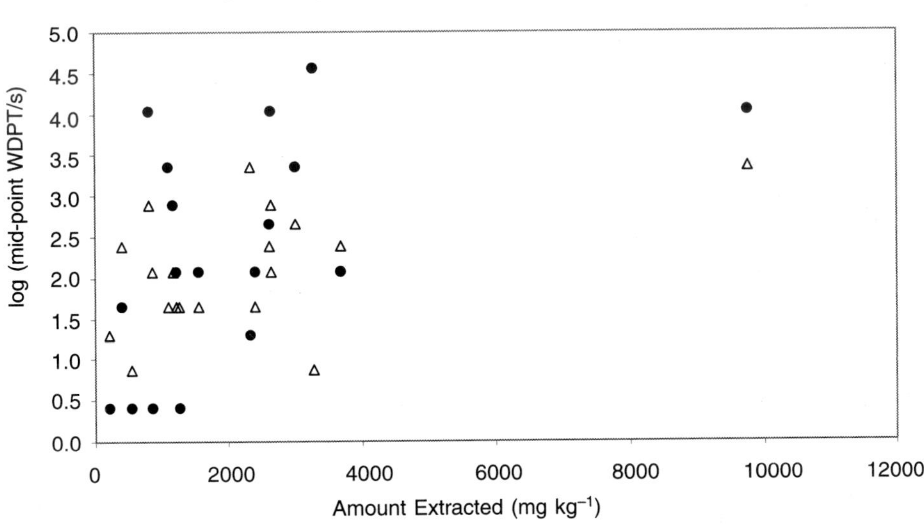

Fig. 20.1d: Comparison between log midpoint Water Drop Penetration Times (see Table 20.2) for soil preextraction and acid-washed sand after reapplication of whole extract from IPA/H_2O dried at 105°C.

Figure 20.1e gives the difference between log midpoint WDPT values for soil preextraction and AWS postreapplication following the aforesaid method for water-repellent samples (wettable controls removed from the set) as a

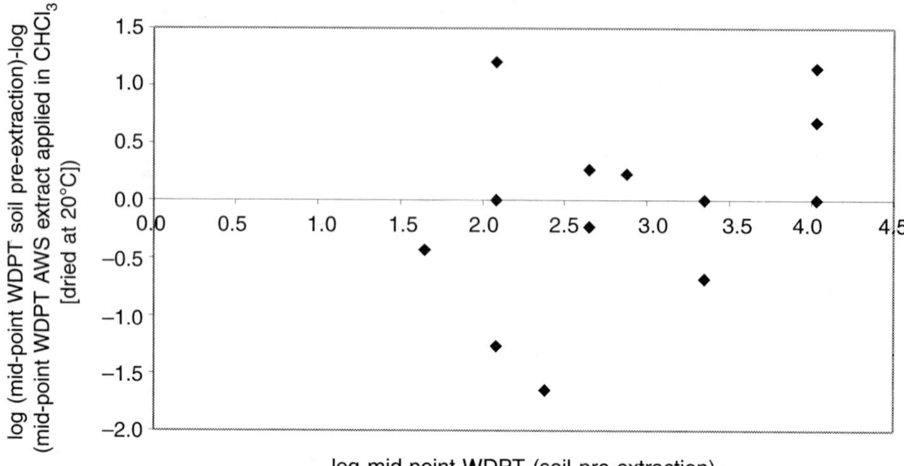

Fig. 20.1e: Difference between log midpoint Water Drop Penetration Times (see Table 20.2) for water repellent soils (control samples removed from set) preextraction, and acid-washed sand after reapplication of chloroform soluble fraction dried at 20°C vs log midpoint Water Drop Penetration Times of soil preextraction.

function of soil WDPT. It should be noted, however, that extracts from each of the control soils also induced water repellency, with levels within the range of those induced by extracts from repellent soils for both drying temperatures.

Extracted material applied directly from the extraction solvent (after loss of ammonia) without reduction to dryness and redissolution induced no, or only comparatively low levels of repellency after drying at 20°C. After drying at 105°C, however, all extracts, including those from control soils, had also induced repellency in AWS, but with WDPT classes mostly lower than those attained for extracts applied in chloroform.

3.4 DRIFT Analysis

An example of a DRIFT spectrum is given in Figure 20.2. IR absorption by minerals in the soils masks much of the spectral region which would be of help in identifying the functional groups of the organic compounds present. However, bands arising from C-H stretching in aliphatic compounds can be observed, as shown in Figure 20.3. Absorption efficiency, as measured by the area beneath the peaks in the 2,800–3,000 cm^{-1} region, is expected to be quantitatively related to the amount of aliphatic C-H present (Capriel et al., 1995). Thus, integration of the area under the spectra, as shown in Figure 20.3, provides an assessment of the relative amounts of aliphatic C-H units present in each sample studied. These values are given in Table 20.5 for all samples before and after extraction.

Comparison of the mean area values for samples before and after extraction revealed that in no case did extraction remove all the aliphatic material from

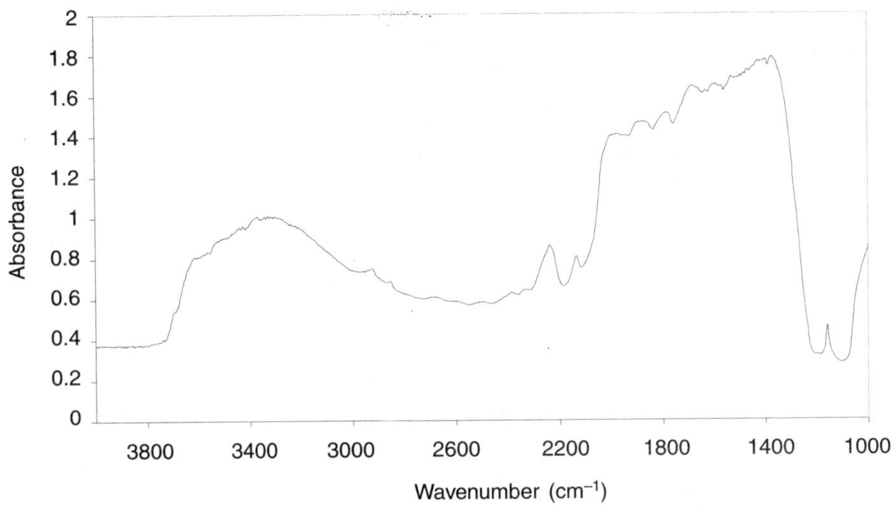

Fig. 20.2: DRIFT spectrum of the repellent soil sample UK2.

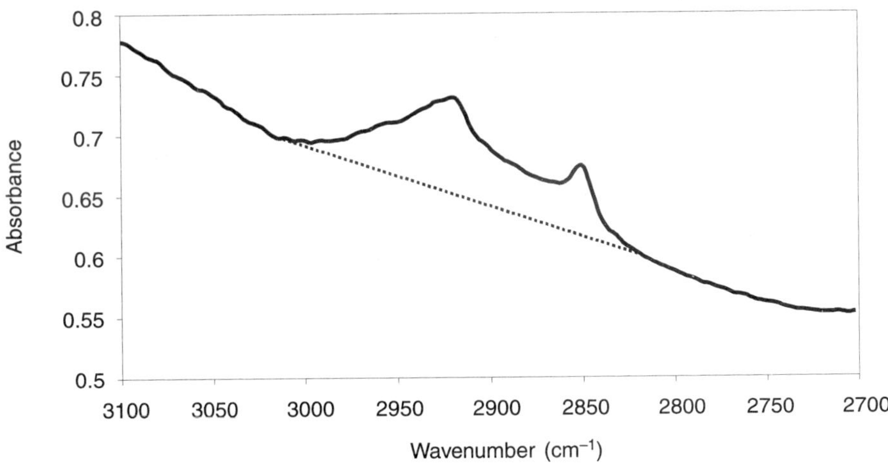

Fig. 20.3: DRIFT spectrum of sample UK2 in the spectral region of interest. Area between the absorption bands due to C-H stretches in the absorption spectrum (solid line) and background (dashed line) is a relative measure of the amount of C-H units present in the sample.

the soil. The proportion removed generally ranges from 25 to 85%, with an average removal of approximately 50%. However, for UKC, containing very little organic matter, little change in C-H absoption band intensity was recorded before and after extraction. A good correlation was found between amount extracted, as measured by change in absorption intensity, and initial amount of C-H present (Fig. 20.5). DRIFT samples soil heterogeneity on a much smaller scale than say bulk extractions, which can result in high variations in intensity

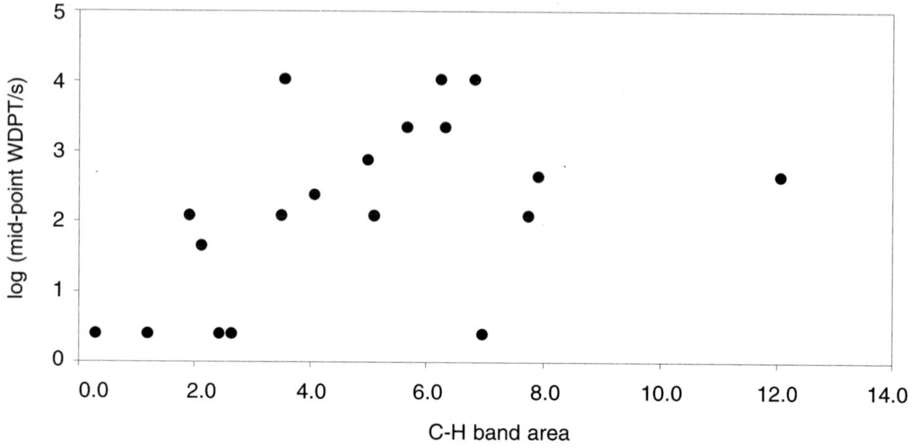

Fig. 20.4: log midpoint Water Drop Penetration Times (see Table 20.2) vs DRIFT C-H absorption band area of samples prior to extraction. Data points at log midpoint WDPT = 0.40 are from the wettable control samples.

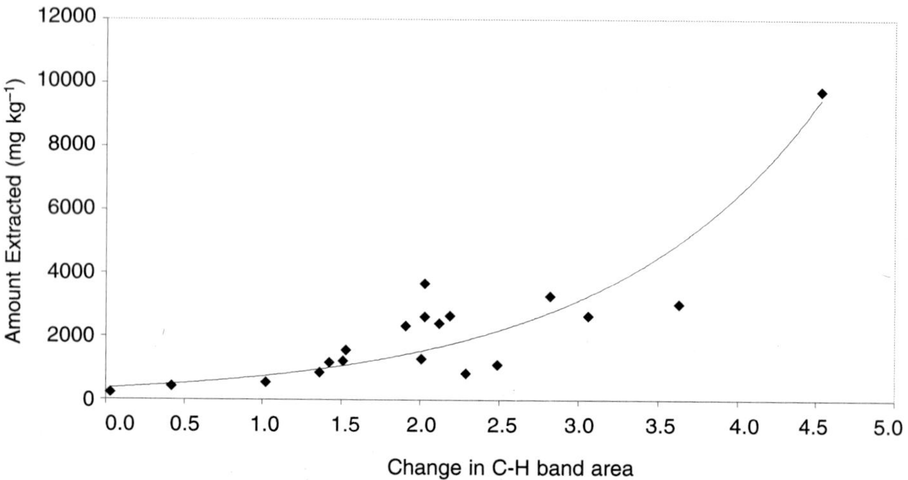

Fig. 20.5: The amount extracted vs change in DRIFT C-H absorption band area as a consequence of extraction. Trend line shown for convenience only and has no theoretical significance.

of C-H absorption bands within subsamples taken from an individual soil. High variations among subsamples (see Table 20.5) occurred despite subsampling by coning and quartering, which reflected the heterogeneity of the soils on this mm scale. Although all the soils studied were of sand texture with minimal clay content (Table 20.1), the PT, GK, and UK samples showed high sample-to-sample variations, while the AU and NL samples were more uniform. However, the differences in C-H content among the various soil samples examined here were so large that subsample variability was of little consequence.

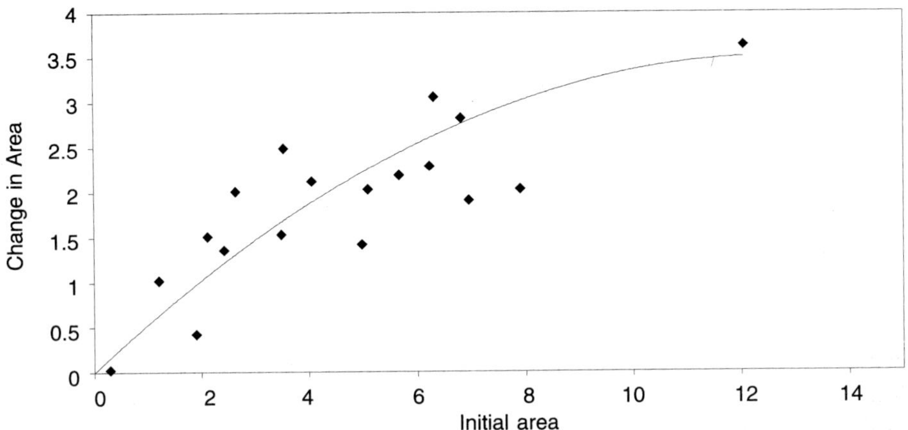

Fig. 20.6: Change in DRIFT C-H absorption band area as a consequence of extraction vs initial area for samples prior to extraction. Trend line is shown for convenience only and has no theoretical significance.

Table 20.5: Relative amounts (including error estimates) of C-H extracted as measured by DRIFT analysis and total mass of organic matter extracted. Depending upon sample-to-sample reproducibility, 6–12 independent spectra were measured using a different subsample for each sample type. Error estimates based on ± 1 standard deviation. Preextraction WDPT classes (s) of samples are also shown. WDPT classes are given in Table 20.2.

Sample code	Relative area under C-H peak				% total C-H extrated	Mass extracted (g kg^{-1})	WDPT class (s) preextraction (20°C)
	Preextraction		Postextraction				
PT1	3.5	(±1.6)	2.0	(±0.6)	44	1.55	180
PT2	6.8	(±2.1)	4.0	(±1.2)	41	3.28	18,000
PT3	2.1	(±0.4)	0.61	(±0.30)	71	1.22	60
PTC	2.7	(±1.0)	0.64	(±0.39)	76	1.28	< 5
NL1	7.7	(±1.6)	3.2	(±1.0)	59	9.76	180
NL2	6.3	(±0.8)	3.3	(±1.1)	49	2.64	3,600
NL3	3.6	(±0.4)	1.1	(±0.4)	70	1.10	18,000
NLC	1.2	(±0.3)	0.18	(±0.05)	85	0.55	< 5
UK1$_b$	5.0	(±1.2)	3.6	(±2.0)	28	1.17	900
UK2	4.1	(±0.8)	2.0	(±1.1)	52	2.41	300
UKC	0.29	(±0.08)	0.26	(±0.16)	10	0.23	< 5
AU1	5.7	(±1.1)	3.5	(±1.0)	39	2.64	3,600
AU2	5.1	(±0.6)	3.1	(±1.0)	40	3.67	180
AU3	6.2	(±0.5)	4.0	(±0.5)	37	0.83	18,000
AUC	2.4	(±0.6)	1.07	(±0.14)	56	0.86	< 5
GK1	7.9	(±3.6)	5.9	(±2.3)	26	2.62	600
GK2	12.1	(±2.7)	8.4	(±2.3)	30	3.02	600
GK3	1.9	(±0.5)	1.5	(±1.1)	22	0.41	180
GKC	6.7	(±1.7)	5.1	(±0.9)	27	2.32	< 5

Comparison of amount of aliphatic C-H units detected by DRIFT and water repellency of samples revealed no close relationship between these two variables (Fig. 20.4), although again there is some correlation in that no samples show both a high extraction amount and low C-H absorption intensity or vice-versa. There is, however, good correlation between the amount extracted and the difference in C-H absorption intensity before and after extraction (Fig. 20.6).

3.5 Separation and Characterization of Compounds Present in Soils Selected

GC-MS chromatograms of the THF soluble portions of the isopropanol/ammonia extracts obtained from sample batch NL are shown in Figure 20.7. Chloroform-soluble phases were considered vis-à-vis methanol- and water-soluble phases because it was found that, after following the procedures of Ma'shum et al. (1988) (see § 2), this fraction did indeed induce hydrophobicity. The chloroform-soluble phases were dissolved in THF rather than chloroform for GC and GC-MS analysis because of the improved solubility of the material in the latter solvent. For ease of comparison the main peaks in each chromatogram were numbered 1–37 (Fig. 20.7). Assignment of peaks (Table 20.6)

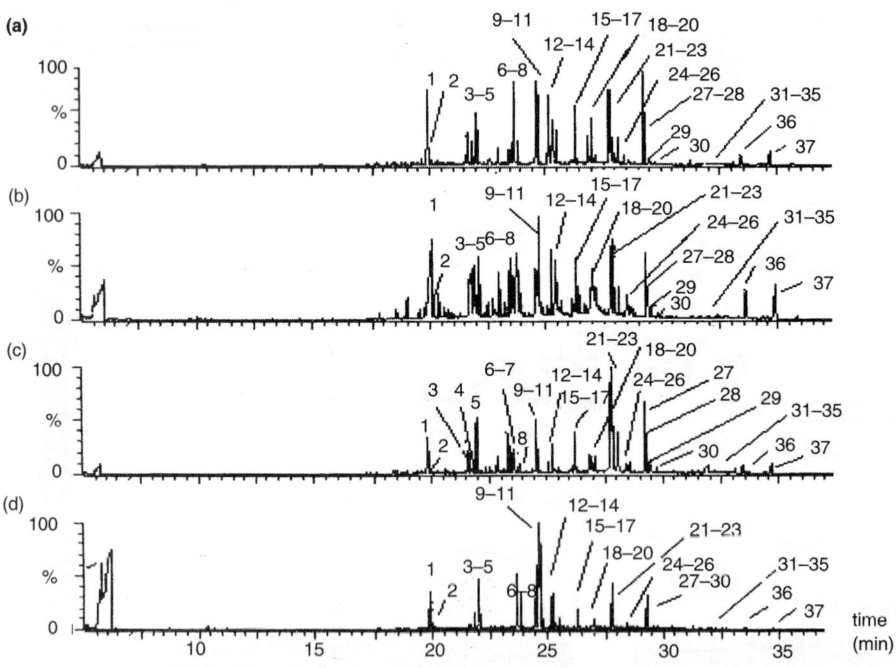

Fig. 20.7: GC-MS chromatograms of tetrahydrofuran (THF) soluble extracts from (a) NL1, (b) NL2, (c) NL3, and (d) NLC.

were based on retention times, mass spectral interpretation, NIST mass spectral search program, and NIST/EPA/NIH mass spectral library. In some cases, it was also possible to confirm the identities of the compounds by comparing the retention times and mass spectra obtained with those of purchased authentic standards. Some mass spectra of the compounds found in the soil and those of the authentic compounds are shown in Figure 20.8 to illustrate the closeness of the matches. The compounds whose identities were confirmed by the injection of purchased standards are indicated in bold font in Table 20.6.

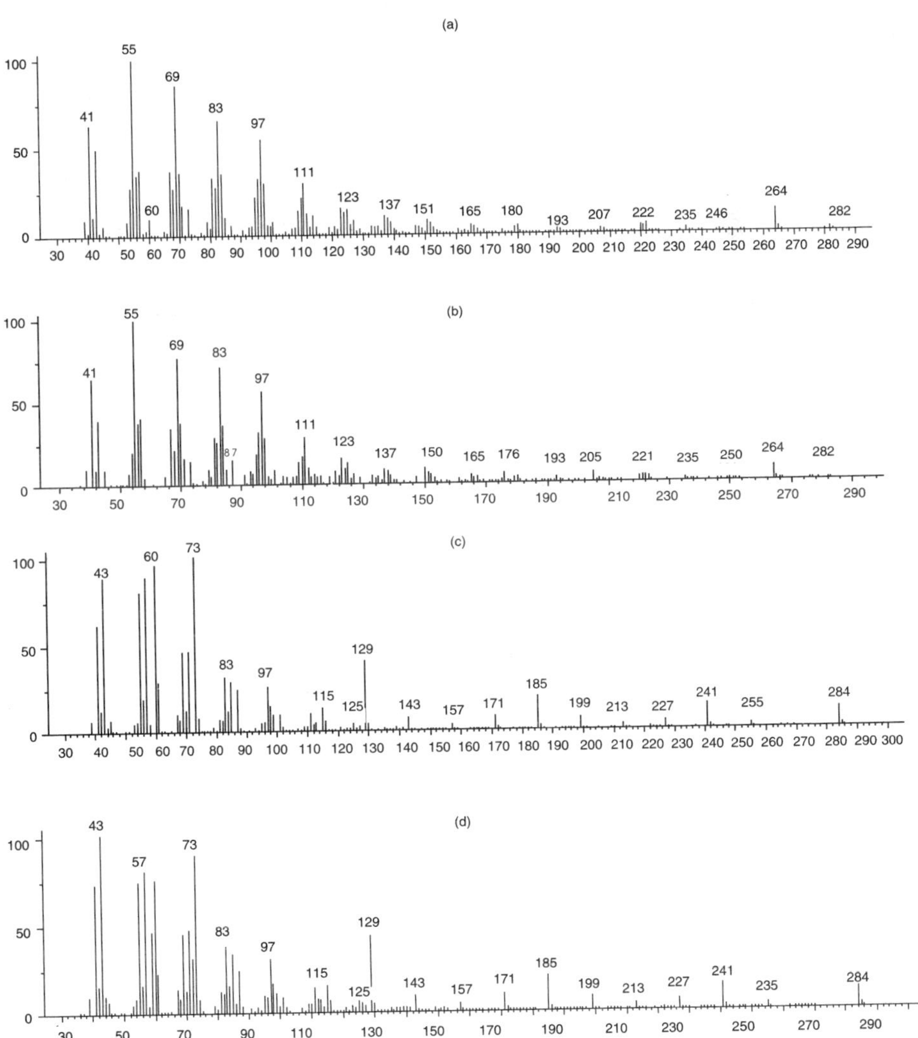

Fig. 20.8:

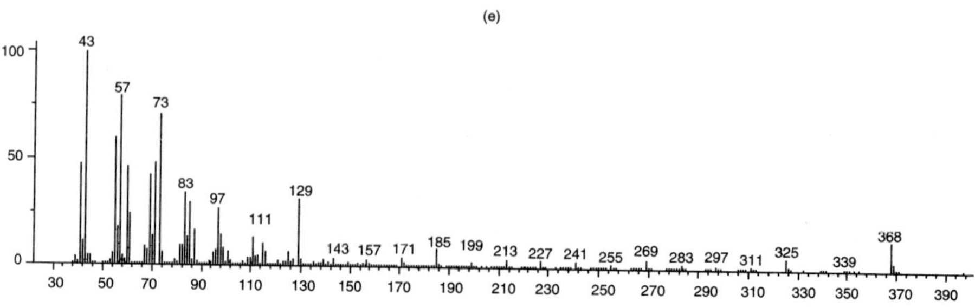

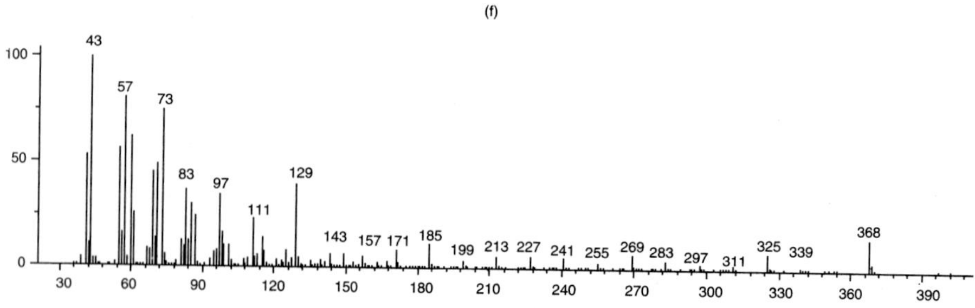

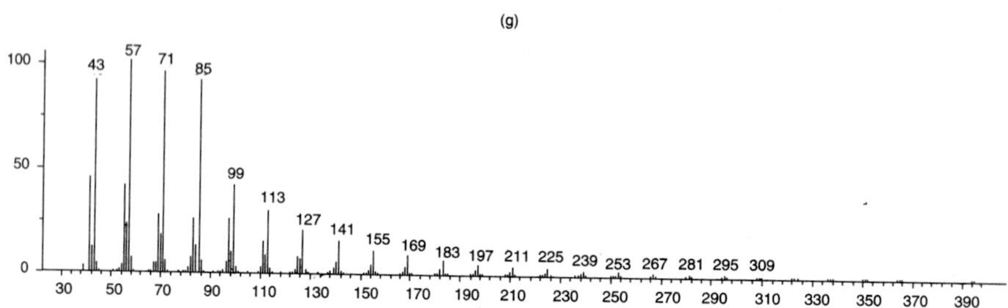

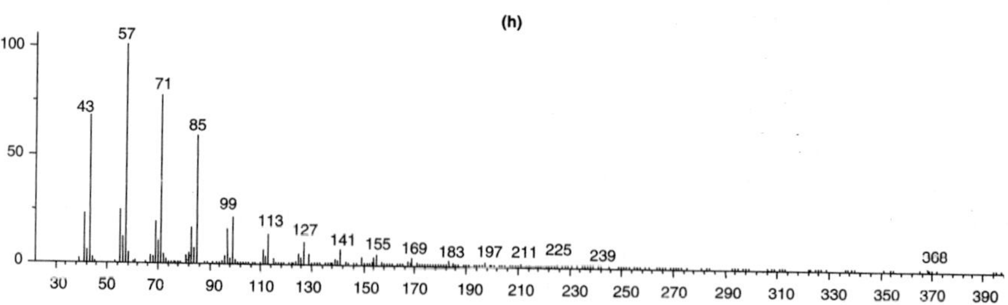

Fig. 20.8: Comparison of mass spectra of authentic compounds with those of compounds in the NL2 extract: (a) elaidic acid, $C_{18}H_{34}O_2$; (b) peak 3; (c) octadecanoic acid, $C_{18}H_{36}O_2$; (d) peak 4; (e) tetracosanoic acid, $C_{24}H_{48}O_2$; (f) peak 18; (g) octacosane, $C_{28}H_{58}$, and (h) peak 19.

	$CH_3(CH_2)_nCO_2H$		$CH_3(CH_2)_nCONH_2$		$CH_3(CH_2)_nCH_3$
(1)	n = 14	(2)	n = 12	(10)	n = 23
(4)	n = 16	(5)	n = 14	(14)	n = 24
(7b)	n = 18	(8)	n = 16	(16)	n = 25
(9b)	n = 19	(20)	n = 20	(19)	n = 26
(13)	n = 20	(25)	n = 22	(21)	n = 27
(15)	n = 21			(24)	n = 28
(18)	n = 22			(27)	n = 29

$CH_3(CH_2)_nCH{=}CH(CH_2)_nCO_2H$ (3) n = 7

$CH_3(CH_2)_nCH{=}CH(CH_2)_nCONH_2$ (7a) n = 7

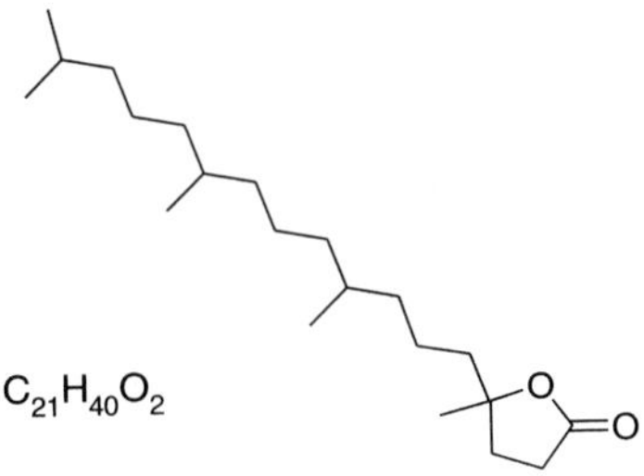

(6) $C_{21}H_{40}O_2$ 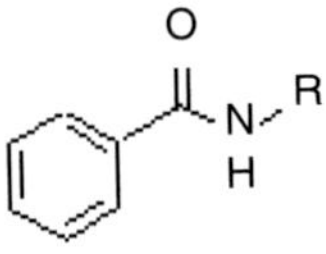 (9b) 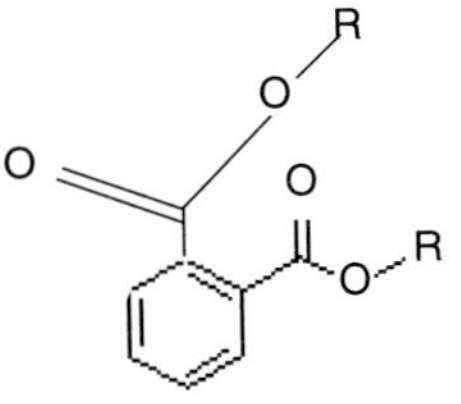 (12)

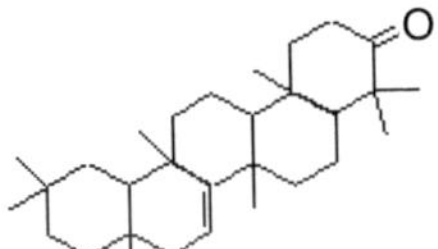

(33a) $C_{30}H_{48}O$ 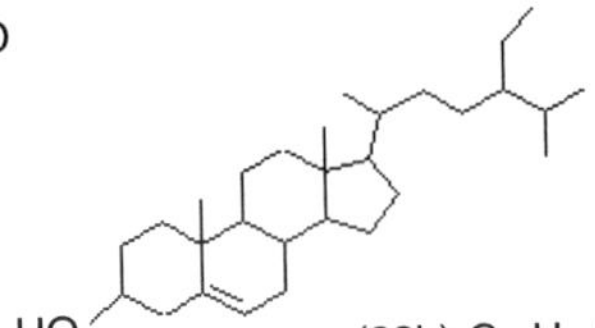 (33b) $C_{29}H_{50}O$ 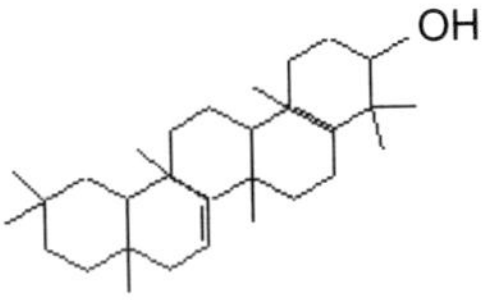 (34) $C_{30}H_{50}O$

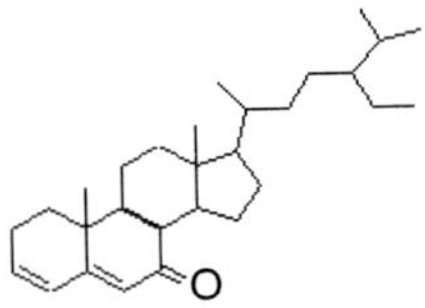

(35) $C_{29}H_{46}O$ 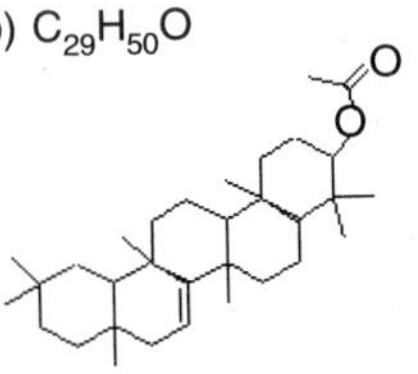 (36) $C_{32}H_{52}O_2$

Fig. 20.9: Structures of compounds identified in extracts of samples NL1, 2, 3 and NLC.

Five main types of compounds were identified by GC-MS in all four samples—long chain fatty acids (C_{16}–C_{24}), amides (C_{14}–C_{24}) of chain lengths similar to the acids, slightly higher chain length alkanes (C_{25}–C_{31}), aldehydes or ketones (C_{25}–C_{29}), and more complex annular structures. An unsaturated acid and its amide (C_{18}), and a phthalate (1,2-benzene dicarboxylate) were also present. There were indications of the presence of alkenes, alcohols, esters, and diols, but further work is needed (e.g. using chemical ionization MS) before unequivocal identification of these compounds can be made. Chemical structures of the compounds detected in the soil extracts are shown in Figure 20.9.

Table 20.6: Assignment of selected peaks present in GC-MS of the extracts (see Fig. 20.1)

Peak No.	Retn. time (min)	Identification†	Formula
1	20.0	n-hexadecanoic acid	$C_{16}H_{32}O_2$
2	20.1	tetradecanamide	$C_{14}H_{29}NO$
3	21.7	oleic acid (cis-9-octadecanoic acid) ‡	$C_{18}H_{34}O_2$
4	21.9	**octadecanoic acid**	$C_{18}H_{36}O_2$
5	22.1	hexadecanamide	$C_{16}H_{33}NO$
6	23.5	4,8,12,16-tetramethylheptadecan-4-olide	$C_{21}H_{40}O_2$
7 a	23.6	9-octadecenamide	$C_{18}H_{35}NO$
7 b	23.6	eicosanoic acid (NL2 & 3 only)	$C_{20}H_{40}O_2$
8	23.8	octadecanamide	$C_{18}H_{37}NO$
9 a	24.6	benzamide type compound	not known
9 b	24.6	heneicosanoic acid (NL1 only)	$C_{21}H_{42}O_2$
10	24.7	pentacosane	$C_{25}H_{52}$
11	24.8	currently unidentified	not known
12	25.2	phthalate type compound (1,2-benzene dicarboxylate)	not known
13	25.3	**docosanoic acid**	$C_{22}H_{44}O_2$
14	25.5	**hexacosane**	$C_{26}H_{54}$
15	26.1	tricosanoic acid	$C_{23}H_{46}O_2$
16	26.3	heptacosane	$C_{27}H_{56}$
17	26.4	aldehyde/ketone*	$C_{25}H_{50}O$
18	26.9	**tetracosanoic acid**	$C_{24}H_{48}O_2$
19	27.1	**octacosane**	$C_{28}H_{58}$
20	27.2	docosanamide	$C_{22}H_{45}NO$
21	27.8	nonacosane	$C_{29}H_{60}$
22	27.9	aldehyde/ketone*	$C_{27}H_{54}O$
23	28.2	currently unidentified	not known
24	28.5	**triacontane**	$C_{30}H_{62}$
25	28.7	tetracosanamide	$C_{24}H_{49}NO$
26	28.9	currently unidentified	not known
27	29.3	hentriacontane	$C_{31}H_{64}$
28	29.4	stigmasterol type compound	not known
29	29.5	aldehyde/ketone*	$C_{29}H_{58}O$
30	29.8	currently unidentified	not known
31	31.3	alkane type compound	not known
32	31.6	aldehyde/ketone*	not known
33 a	31.9	d-friedoolean-14-en-3-one	$C_{30}H_{48}O$
33 b	31.9	stigmasterol, 22, 23-dihydro-	$C_{29}H_{50}O$
34	32.3	taraxerol	$C_{30}H_{50}O$
35	33.2	stigmasta-3,5-dien-7-one	$C_{29}H_{46}O$
36	33.5	d-friedoolean-14-en-3-ol, acetate, (3. beta.)	$C_{32}H_{52}O_2$
37	34.8	currently unidentified	not known

†Identification of compounds in bold type confirmed by comparison of their spectra with those of authentic samples.

‡ Comparison with an authentic sample of the isomeric elaidic acid supports this assignment (see Figs. 20.2(a) and 20.2(b)).

*Although the molecular formula and the principal functional groups present have been established, the structures of these compounds could not be determined on the basis of their EI mass spectra alone.

Preliminary estimates of relative abundances can be obtained on the basis of peak heights. These allow gross differences between the chromatograms to be identified and discussed.

GC-MS analyses indicated the presence of long-chain fatty acids with 16–24 carbon atoms (C_{16}, C_{18}, C_{20}, C_{21}, C_{22}, C_{23}, C_{24}). Even-number chain acids predominate and a similar chain length distribution of amides was observed.

Alkanes were present with a slightly higher chain length distribution (C_{25-31}) than that of the acids and amides. The chain length distribution of alkanes observed is shown in Figure 20.10, which shows a predominance of compounds having an odd number of carbon atoms.

Aldehydes or ketones having 25, 27, and 29 carbon atoms were also detected. The precise structures of these compounds could not be determined on the basis of the mass spectra obtained, however. The annular compounds observed were predominantly stigmasterol derivatives.

An unsaturated acid was found in the NL1 and NL2 soil extracts. It was a 9-octadecenoic, oleic, or elaidic acid depending on the cis or trans orientation of the double bond. However, the spectrum appeared to be a better match with oleic acid (see Figs. 20.8a and 20.8b). The corresponding unsaturated amide was also identified, 9-octadecenamide, $C_{18}H_{35}NO$. The phthalate (benzene 1,2-dicarboxylate) detected (peak 12) is most likely due to contamination from the plastic bags used for initial storage of the soil samples.

Other analyses performed on the chloroform-soluble portion of the extracts reinforce the GC-MS results obtained. The NMR spectra of the extracts (data not shown) confirmed the view that they consist primarily long-chain aliphatic compounds, as the observed signals were predominantly in the expected regions for aliphatic CH_2 groups. The spectra of all the extracts are very similar with the major signals due to aliphatic H and C appearing in the 0–2.5 ppm region in the ^1H NMR spectra and the 0–45 ppm region in the ^{13}C NMR spectra. The ^1H aliphatic region is dominated by a large signal at about 1.2 ppm, which is characteristic of CH_2 in polymethylene chains. Four signals are present downfield from 1.2 ppm at 1.5, 1.9, 2.1, and 2.3 ppm and may be attributed to CH_2 groups in polymethylene chains in closer proximity to electron withdrawing groups, for example the carboxylic acid groups of long-chain fatty acids or particularly in the case of the signals at 2.1 and 2.3 ppm, perhaps a C-H group. Another common feature is a signal upfield from 1.2 ppm at 0.8 ppm, which falls into the correct region to be attributed to CH_3 groups. It is also worth noting the lack of a carboxylic acid proton resonance between 11 and 13 ppm, which may be due to the expected broadness of the signal. Weak signals were also observed at 5.2–5.3 ppm, which are in the correct region to be attributed to olefinic protons.

The ^{13}C NMR spectra of the extracts generally showed at least nine carbon environments but due to the higher noise level in the spectrum of the NL1 chloroform-soluble extract, only five environments were clearly visible. A large resonance at 77 ppm arise from the solvent. Resonances at 23.1, 27.6, 30.1, 32.3, and 43.8 ppm were in the correct region to be attributed to $C\text{-}H_2$ groups in

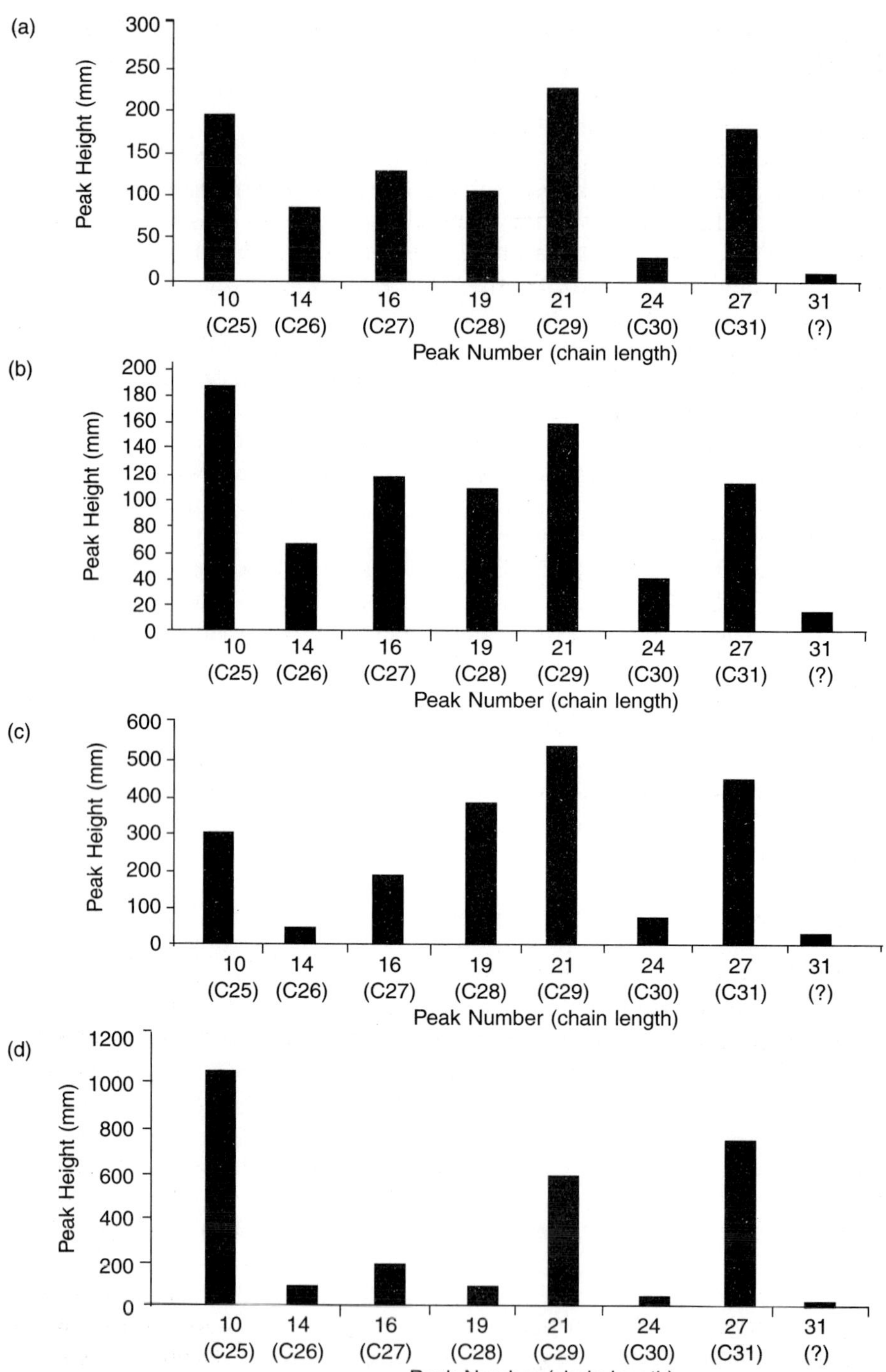

Fig. 20.10: Alkane distributions of tetrahydrofuran (THF) soluble extracts from (a) NL1, (b) NL2, (c) NL3, and (d) NLC. Peak heights normalized using octacosane (peak 19).

polymethylene chains, which is also supported by the orientation of the signals in the DEPT spectra. A signal at 14.5 ppm is in the correct region to be attributed to a $C\text{-}H_3$ group and its orientation in the DEPT spectrum supports this assignment. A downward orientation of the resonance at 28.4 ppm suggests a C-H rather than a $C\text{-}H_2$ group. Again, as in the case of the ^1H NMR spectra, the carboxylic acid resonance was absent. The spectra show that the extracts contain predominantly aliphatic components, thereby confirming the GC-MS data.

Results of the microanalysis carried out for soil NL1 are summarised in Table 20.7. Note that ~ 48% (weight) of the extract dissolved in chloroform and ~41% in water. ~76% of the water-soluble extract dissolved in methonol. The chloroform soluble fraction of the extracts contains the bulk of the carbon (68.7%) and hydrogen (9.8%), suggesting a preponderance of aliphatics in the material soluble in the chloroform and THF. The residual content of 19.7% indicates the presence of a significant fraction of oxygen-containing groups (e.g. carboxylic acid groups) while the small nitrogen content of only 1.8% suggests a small contribution of nitrogen-containing compounds, such as the amides detected by GC-MS.

Table 20.7: Microanalysis results for NL1 extract

Fraction	% Carbon expressed as C	% Hydrogen expressed as H	% Nitrogen expressed as N	% Oxygen expressed as O‡
Chloroform soluble†	68.70 (± 1.3)	9.79 (± 0.1)	1.80 (± 0.24)	19.71 (± 1.06)
Methanol soluble	27.97	5.19	13.97	52.87
Water soluble	53.17	6.51	4.15	36.17

†Values for the chloroform soluble fraction are based on the mean of three determinations.

‡ values derived by subtraction.

4 DISCUSSION

4.1 Effect of Different Drying Temperatures on Sample Repellency

Although a temperature of 105°C is outside the temperature range soils are normally exposed to under field conditions, this temperature was used in a number of previous studies for sample drying prior to water repellency assessments to either provide more consistent repellency values (Roy et al., 1999, Franco et al., 2000) or in an attempt to increase existing sample repellency to its "potential" maximum value one could expect in the field (Dekker and Ritsema, 1994). The latter effect may be associated with an enhanced alignment of hydrophobic moieties on particle surfaces caused by heating (Valat et al., 1991). More recent work (Doerr et al., 2003), has shown, however, that this treatment can also reduce repellency in some samples and these authors argue that soil sampling should ideally be carried out under conditions of low soil moisture, when soils are likely to be most water repellent.

For the samples investigated here, drying at 105°C increased repellency for five of the fifteen repellent samples, whereas for three, repellency decreased

(Table 20.4). However, only for sample NL1 did this change exceed more than one WDPT class and therefore only for this sample can the change be regarded as significant. All control samples except GKC retained their wettability after drying at 105°C. After extraction, for the two samples that had retained some level of repellency, drying at 105°C did cause an increase in repellency over several classes compared to drying at 20°C (see Table 20.2 for classes). Furthermore, a notable increase in repellency was also found for the AWS samples after extract applications (Table 20.4). Since WDPTs were consistent for the five drops applied it is concluded that for the conditions of this study, WDPT measurement of samples dried at 105°C is an unnecessary step for soils prior to extraction. In a study on repellent soils in New Zealand, Horne and McIntosh (2000) came to a similar conclusion. In contrast, measurements taken after drying at 105°C for soils after extraction, and for samples subjected to extract reapplications, provide a useful additional assessment of soil, and extract, water repellency characteristics.

4.2 Evaluation of Extraction Procedure

Of the five solvents examined in the initial evaluation, isopropanol/ammonia extracted by far the largest amount of material and was also the only solvent that rendered the soil wettable after extraction. Extraction with other less polar solvents increased sample repellency despite removing some organic material. These results accord with the studies of Ma'shum et al. (1988) and Roy et al. (1999), who evaluated a wider range of solvents in extracting several Australian and Canadian soils respectively and found that a mixture of isopropanol/ammonia was the only solvent type that fully eliminated repellency after extraction and was therefore best suited for extracting hydrophobic organic materials from soils.

The chemical characterization carried out on the extracts from the four NL soil samples shows that organic acids make up an important group of extracted compounds. The role of ammonia in extracting these compounds is probably that of a transient base. In the early stages of extraction the extraction mixture is rich in ammonia and is alkaline. This alkalinity is expected to assist in the extraction of compounds such as organic acids by bringing them out as ammonium salts. As extraction proceeds, and also in the subsequent concentration of the extract by evaporation, ammonia is lost to the atmosphere and the equilibrium between the protonated form and ammonium salts of these weak acids shift to favor the former, giving as was found, the organic acid rather than the ammonium salt in the extraction mixture.

After extraction with less polar solvents, such as hexane or dichloromethane, soils contain less nonpolar organic material and could thus expectedly be less water repellent. On the contrary, soils were more repellent after this treatment, suggesting that water repellency is not solely related to the amount of nonpolar organic material present (Roy et al., 1999). In a series of treatments with aqueous and organic solvents on repellent clayey Australian soils, McGhie and Posner

(1980) also found that treatment with nonpolar or weakly polar solvents increased sample repellency. It is feasible that such solvents cause a redistribution of the organic material left in the soil, or more specifically, a reorientation of hydrophobic moieties towards the pore spaces of the soil matrix, which leads to increased water repellency (Ma'shum and Farmer, 1985).

Results from Soxhlet extraction on the whole range of soils investigated here confirm the suitability of the chosen extraction procedure for isolating hydrophobic compounds from soils. Almost all repellent samples were rendered wettable after extraction for both drying temperatures, while the control samples retained their wettability (Table 20.3). Also, the fact that all extracts contained material that could induce repellency in AWS (Table 20.4) suggests that extraction removed at least some of the material that causes water repellency in the soils investigated, and that these substances retained their ability to bond to silicate particle surfaces and impart water repellency after subjection to the extraction procedure. Furthermore, DRIFT analysis confirmed that for all repellent samples some, but not all, of the aliphatic C-H was removed during extraction.

The two samples $UK1_b$ and GK2 did retain some level of water repellency after extraction. Initial repellency levels, amounts extracted, and effects of extract reapplication were within the range of those for all other samples investigated here (Table 20.4), but for sample GK2, the initial amount of C-H units was relatively high and a considerable proportion still remained after extraction (Table 20.4). Thus, it could be expected that repeated extraction would eliminate water repellency. However, aliphatic C-H content in $UK1_b$ is within the range of that detected in other samples (Table 20.5) despite its higher level of water repellency after extraction (Table 20.4) and it might be that the hydrophobic compounds contained in these soils either differ in structural type, or in distribution of structural types, and/or are more strongly bound to the soil mineral surfaces than in other soils.

Although extract reapplication and drying procedures provided a valuable assessment of whether extracts contained compounds capable of causing repellency, induced levels of repellency were not always consistent with the repellency levels of the original samples. Most notably, extracts from all four wettable control soils redissolved in chloroform induced repellency levels comparable to those induced by extracts from repellent soils. This demonstrates that these soils also contain hydrophobic substances capable of inducing water repellency. Also, some chloroform extracts induced higher levels of repellency than exhibited by the original sample and all chloroform extracts induce higher levels compared to extract reapplications directly from the extraction solvent mixture. This may seem surprising, since chloroform did not always redissolve all the extracted material. Differences in repellency levels between the original soil and AWS after reapplication may also be due to differences in their respective particle size and associated surface area characteristics (Table 20.1), resulting in greater or lesser coverage by organic materials. An additional factor might be that the bonding of the hydrophobic organic compounds to AWS is not identical to that in the original soil. It may well be that adsorption of the extract onto a soil sample does not take place in a single stage process and some relaxation from an initial metastable, kinetically favored arrangement must occur to reach the same "pseudoequilibrium" situation as is found in soils which have been exposed to organics over a long period of time.

4.3 Relationship of Organic Material and Sample Water Repellency

Since water repellency is caused by organic compounds, it appears reasonable to assume that the amount of organic material in a sample is related to its repellency. Some studies have indeed found a positive relationship between total organic matter content in soils and water repellency (Wallis et al., 1990; Berglund and Persson, 1996). In fact, Capriel et al. (1995) used the implied relationship of hydrophobicity to the amount of aliphatic C-H units in a study of the hydrophobicity of organic matter in arable soils. Other authors, however, found no relationship between organic matter and water repellency (Jungerius and DeJong, 1989; DeBano, 1991). For the broad range of samples investigated here, no strong relationship between water repellency level of a sample with amount of organic material extracted from a sample (Figs. 20.1a and 20.1b), or amount of aliphatic C-H as determined by DRIFT analysis (Fig. 20.4) was found, even though the extracts from both repellent and wettable soils clearly contained compounds that can cause water repellency in soils, and aliphatic C-H units must be contained in the compounds causing repellency. These results indicate that hydrophobic compounds responsible for water repellency may only represent a fraction of the extract composition. They also support the suggestions made by Horne and McIntosh (2000) and Roy and McGill (2000) that the presence of water repellency (in a dry soil) is not only a function of certain organic compounds being present in the soil, but also of their structural composition and arrangement. It is worth noting that all extract reapplications to AWS of the extract prior to its being taken to dryness (i.e., from the residual isopropanol/water mixture; see § 2) results in either hydrophilic or, at most, only slightly repellent AWS conditions after drying at 20°C. By way of contrast, reapplication of that portion of extract which is soluble in chloroform gives AWS that (ignoring wettable control samples) shows some correlation in water repellency level with the original soil (Fig. 20.5).

The observations that reapplication of similar materials from $CHCl_3$ and isopropanol/water gives sands with very different water repellencies, and that heating to 105°C increases sample repellency markedly, are compelling evidence that the simple presence or absence of specific organic compounds does not suffice to determine water repellency. Hydration effects and the intermolecular arrangement of material are likely to be important contributory factors.

4.4 Characterization of Compounds Present in the Extracts

As mentioned earlier, the four soil samples from the Netherlands were intensively invetigated. In all four samples (NL), the five main types of compounds identified by GC-MS were long-chain fatty acids (C_{16}–C_{24}), amides (C_{14}–C_{24}) of chain lengths similar to the acids, slightly higher chain length alkanes (C_{25}–C_{31}), aldehydes or ketones (C_{25}–C_{29}), and more complex ring-containing structures. With respect to long-chain fatty acids, even chain compounds predominated. This is consistent with results reported in studies of organic fractions derived from some water repellent Australian soils by Ma'Shum et al. (1988), who noted a similar distribution of mostly even-number long-chain fatty acids, but with 16 to 32 carbon atoms, and by Franco et al. (1995) and Horne and McIntosh (2000), who likewise showed the presence of long-chain fatty acids. The presence of long-chain fatty acids is not surprising, particularly in samples form an undisturbed sandy soil as investigated here, since these substances are particularly difficult to degrade (Hayes and Graham, 2000). They may originate from a range of sources including for example plant cuticles (Hayes, 1998). Fatty acids are thought to contribute to the hydrophobic behavior which humic substances display in some circumstances (Clapp et al., 1993). For amides, a similar chain length distribution was observed. For many of the long-chain acids, there appeared to be an amide of the same chain length (e.g. C_{16}, C_{18}, C_{22}, C_{24}). Alkanes were present with a slightly higher chain length distribution ($C_{25–31}$) than that of the acids and amides. Edlington et al. (1962) reported that most plant hydrocarbons are straight-chain, saturated compounds with an odd number of carbon atoms. Each chain length between 25 and 31 carbon atoms was nevertheless detected, which could suggest that not all the alkanes are of direct plant origin, but may arise from decay or microbial/fungal attack. Franco et al. (1995) and Horne and McIntosh (2000) likewise also observed the presence of alkanes. The aldehydes and ketones detected had 25, 27, and 29 carbon atoms, but their precise structures could not be determined, while the annular compounds observed were predominantly stigmasterol derivatives. As mentioned earlier, the phthalate (benzene 1,2-dicarboxylate) detected (peak 12, Fig. 20.7 and Table 20.6) was most likely due to contamination from the plastic bags used for storing the soil samples.

Since all four soil sample extracts contained alkanes in similar relative abundances (see Fig. 20.10), the observed difference in wettability between NLC (wettable) and the others (repellent) cannot be ascribed to the presence or absence of these hydrophobic compounds alone. However, the extract from the wettable sample contains only small amounts of other, larger polar compounds (see Fig. 20.11d), whereas all three of the extracts from the water repellent samples contain significant amounts of this material (Fig. 20.11a–c). This appears to be the major difference between the repellent and wettable samples. The mechanism whereby the presence of higher molecular mass polar compounds might induce hydrophobicity in soils is not known. We may nevertheless, speculate that the inherently lower solubility in water of these compounds is a key factor. When

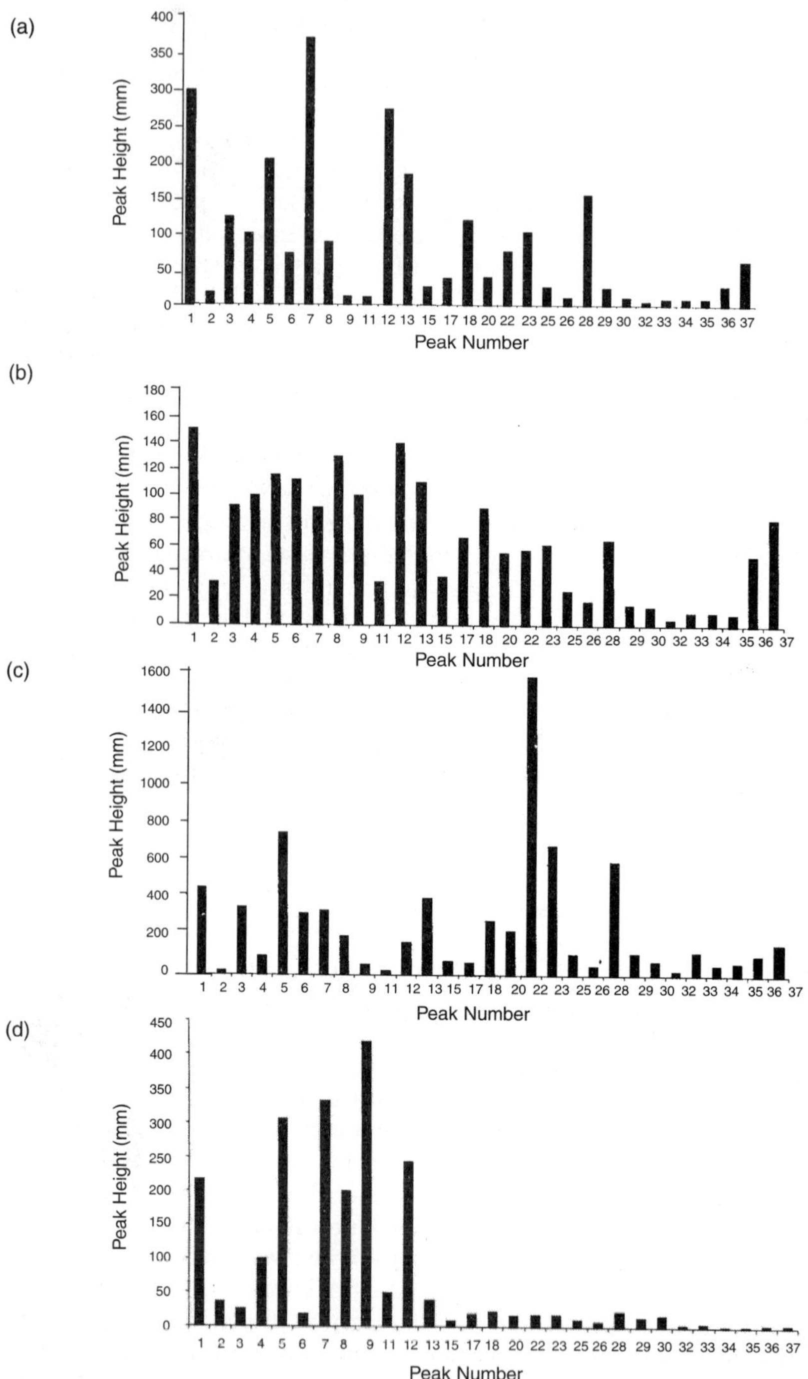

Fig. 20.11: Distribution of non-alkane compounds for tetrahydrofuran (THF) soluble extracts from (a) NL1, (b) NL2, (c) NL3, and (d) NLC. Peak heights were normalized using octadecanoic acid (peak 4).

water comes into contact with a wettable soil, the lower molecular mass polar compounds will dissolve relatively quickly and act in a detergent-like fashion to solubilise other, inherently more hydrophobic material, such as alkanes. By contrast, when water does eventually penetrate an originally repellent soil, the higher molecular mass polar compounds will dissolve relatively slowly so that repellency can be maintained longer. It is also significant that the wettable soil NLC in this study had a smaller overall organic content than the water-repellent soils. NMR spectra confirmed that the extracts contain predominantly aliphatic components. In addition, microanalysis results for the sample NL1 (Table 20.7) showed that, as expected, the chloroform-soluble fraction of the extracts contained the bulk of the carbon and hydrogen content.

5 CONCLUSIONS

In this study, procedures to extract compounds responsible for causing water repellency in sandy soils were examined using repellency testing at different drying temperatures, extract reapplications, and DRIFT analysis to evaluate the success of extractions, and further separation and analysis by GC, GC-MS, and NMR were carried out on selected extracts to characterize the compounds likely to cause water repellency. Water-repellent samples with different degrees of repellency and also wettable samples taken from five different countries were included in the extraction program, resulting in a much broader sample range than examined in previous studies. The following key findings are therefore thought to have relatively wide applicability.

1. Drying of samples at 105°C rather than room temperature prior to repellency testing and before extraction is viewed as an unnecessary procedure, provided samples are already relatively dry in the field and water repellency measurements give consistent results. Sample drying at 105°C after extraction, however, can be a valuable additional procedure in assessing effectiveness of the extraction procedure.
2. Extracting a repellent sample with a nonpolar or weakly polar solvent increased sample repellency despite removing some nonpolar hydrocarbons. This supports the notion that organic compounds, which are soluble in such weakly polar solvents, are not per se the main cause of water repellency in soils.
3. Soxhlet extraction using isopropanol/ammonia (7:3 v:v) was (i) an effective method for extracting hydrophobic compounds from the soils studied, and (ii) did not make the nonrepellent control samples water repellent. While it is not possible to demonstrate unequivocally that the extracts contained the key compounds responsible for causing water repellency in the samples investigated, the fact that they were capable of inducing considerable repellency in AWS suggests that they contained at least some key compounds responsible for repellency in the soils investigated. Thus, the procedure carried out in this study is considered suitable for extraction of water-

repellent compounds from soils for subsequent further separation and characterization.

4. The amount of organic compounds extracted from soils correlated poorly with the water repellency of a sample, indicating that hydrophobic compounds responsible for water repellency may merely represent a fraction of the extract composition and that the presence of water repellency (in a dry soil) is not only a function of certain organic compounds present in the soil, but their structural composition and arrangement.
5. Extracts from wettable control soils also contained hydrophobic substances capable of inducing water repellency. This suggests that (i) these compounds are present in too low a quantity in these soils to induce water repellency; (ii) other compounds present confer hydrophilicity even in the presence of these hydrophobic materials; or (iii) the particular molecular adsorption at the soil surface of these nominally hydrophobic materials is such as to mask their hydrophobic nature.
6. DRIFT analysis proved a useful tool in examining the amount of aliphatic C-H units present in soils. Variability of results between subsamples of these sandy soils was high despite the careful subsampling procedure. However, differences in C-H content between different samples were so large that intrasample variability was of little consequence. The intensity of absorption due to aliphatic C-H units was not related to the repellency of a sample. If we accept that the absorption intensity is related to the concentration of C-H-containing material on the soil, then we can conclude that the amount of aliphatic C-H-containing material does not, alone, determine the water repellency of a soil.

Four samples comprising three repellent and one wettable sample taken at four respective depths (0–10, 10–20, 20–30 and 30–40 cm) from a sandy soil under permanent pasture near Ouddorp in the Netherlands were subjected to further investigations. Isopropanol/ammonia extracts were separated and the compounds contained in the chloroform-soluble fraction determined. Future work on the remaining soil extracts not analyzed here, will determine whether the following key findings are likely to be of wider applicability.

7. The amount of extract obtained from the wettable control soil was significantly less than that obtained from the water-repellent soils. This may not be so surprising given the fact that this soil was taken from the greatest depth (30–40 cm). The degree of water repellency of the four samples, however, was not well reflected in the amount of extract obtained.
8. The composition of the extract from the wettable soil resembles that of the extracts from the water-repellent soils in most respects.
9. The main types of compounds identified in all four samples were long-chain fatty acids (C_{16}–C_{24}), amides (C_{14}–C_{24}) of chain lengths similar to the acids, slightly higher chain length alkanes (C_{25}–C_{31}), aldehydes or ketones (C_{25}–C_{29}), and more complex ring-containing structures. It was possible to confirm the identities of some of the compounds by comparing their reten-

tion times and mass spectra with those of purchased standards. There were also indications of alkenes, alcohols, esters and diols; further work is needed, however (e.g. involving chemical ionisation), before unequivocal identifications of the compounds can be made.

10. The fatty acid chain length distribution was predominantly composed of even-number chain acids, a phenomenon observed previously in an Australian study by Ma'shum et al. (1988). A similar distribution of chain lengths was observed for the amides. The chain length distribution of the alkanes showed a predominance of compounds with odd numbers of carbon atoms, which may reflect their plant origin.
11. Assignments made on the basis of the GC-MS chromatograms are supported by other experimental techniques such as NMR spectroscopy and microanalysis, which confirm the predominantly long-chain aliphatic character of the extracts.
12. Preliminary estimates of relative abundances on the basis of peak heights showed that the major difference between the wettable and water-repellent soils concerns high molecular mass polar compounds, such as fatty acids/amides with 23 or 24 carbon atoms and stigmasterols. These are virtually absent from the wettable control soil and this type of compound may be important in determining whether or not a soil has the potential to exhibit water repellency.

The findings listed above do not allow unequivocal identification of a specific compound or compound group as being the universal cause of water repellency in soils. This is not surprising given the fact that compounds with hydrophobic properties are ubiquitous in the environment in general, and soils in particular exhibit a high variability with respect to both physical and biochemical parameters. The results of this study do, however, provide some insight into the role of common organic compounds in the water repellency behavior expressed by the mineral particles of a wide range of soils. Achieving a fundamental understanding of the (bio)chemical origin of water repellency is critical not only in the amelioration of repellency by, for example, developing more effective and environmentally friendly wetting agents, it is also critical if we aim to balance the aforementioned detrimental effects of repellency with its beneficial effects. The latter, which may in some cases be of equal or even greater importance include an enhanced organic carbon sequestration potential (Piccolo et al., 1999), reduced evaporative water losses, and enhanced soil aggretage stability in water-repellent soils (see reviews by DeBano, 2000; Doerr et al., 2000).

Acknowledgements

The authors thank K. Oostindie, L.W. Dekker, and G. Allinson for sample shipment, I. Matthews for technical help with the GC (FID) instrument, the staff at the EPSRC National Mass Spectrometry Service Centre, Department of Chemistry, University of Wales Swansea, for advice and the use of their GC-MS instrument, M. Nettle for running the NMR spectra, S. Szajda for technical assistance, and J. Schneider for performing particle-size analysis. Financial assistance of the University of Wales Swansea in providing a postgraduate studentship (KAM) and a postgraduate bursary (CTL) is acknowledged. This study was supported by an EU grant FAIR-CT98-4027, Australian Research Council International Linkage Scheme Grant LX0211202, NERC-Advanced Fellowship NERIJISI 200200662, and Aquatrols (USA). This work does not necessarily reflect the European Commission's views and in no way anticipates its future policy in this area.

References

Berglund K. and Persson L. 1996. Water repellence of cultivated organic soils. *Acta Agric. Scand.* 46: 145–152.

Bisdom E.B.A., Dekker L.W., and Schoute J.F.Th. 1993. Water repellency of sieve fractions from sandy soils and relationships with organic material and soil structure. *Geoderma* 56: 105–118.

Capriel P., Beck A.J., Borchert H., Gronholz J., and Zachmann G. 1995. Hydrophobicity of the organic matter in arable soils. *Soil Biol. Biochem.* 27: 1453–1458.

Clapp C.E., Hayes M.H.B., and Swift R.S. 1993. Isolation, fractionation, functionalities and concepts of structures of soil organic macromolecules. In: *Organic Substances in Soil and Water: Natural Constituents and Their Influences on Contaminant Behaviour*. A.J. Beck, K.C. Jones, M.H.B. Hayes, U. Mingelgrin (eds.). Roy. Soc. Chem., Cambridge, UK, pp. 31–68.

DeBano L.F. 1991. The effect of fire on soil properties. *USDA Forest Servive Gen. Tech. Rept.* 280: 151–156.

DeBano L.F. 2000. Water repellency in soils: a historical overview. *J. Hydrol.* 231–232: 4–32.

Dekker L.W. and Ritsema C.J. 1994. How water moves in a water repellent sandy soil. 1. Potential and actual water repellency. *Water Resour. Res.* 30: 2507–2517.

Dekker L.W. and Ritsema C.J. 1996. Variation in water content and wetting patterns in Dutch water repellent peaty clay and clayey peat soils. *Catena* 28: 89–105.

Dekker L.W., Ritsema C.J., and Oostindie K. 2000. Extent and significance of water repellency in dunes along the Dutch coast. *J. Hydrol.* 231–232: 112–125.

Dekker L.W., Ritsema C.J., Oostindie K., and Boersma O.H. 1998. Effect of drying temperature on the severity of soil water repellency. *Soil Sci.* 163: 780–796.

Doerr S.H., Shakesby R.S., and Walsh R.P.D. 1998. Spatial variability of soil hydrophobicity in fire-prone eucalyptus and pine forests, Portugal. *Soil Science* 163: 313–324.

Doerr S.H., Shakesby R.A., and Walsh R.P.D. 2000. Soil water repellency: its characteristics, causes and hydro-geomorphological consequences. *Earth Sci. Rev.* 51: 33–65.

Doerr S.H., Dekker L.W., Ritsema C.J., Shakesby R.A., and Bryant R. 2002. Water repellency of soils: the influence of ambient relative humidity. *Soil Sci. Soc. Amer. J.* 66: 401–405.

Doerr, S.H., Ferreira A.J.D., Walsh R.P.D., Shakesby R.A., Leighton-Boyce G., and Coelho C.O.A. 2003. Soil water repellency as a potential parameter in rainfall-runoff modelling: Experimental evidence at point to catchment scales from Portugal. *Hydrological Processes* 17: 363–377.

Edlington G., Hamilton R.J., and Raphael R.A. 1962. Hydrocarbon constituents of the wax coating of plant leaves: a taxonomic survey. *Nature* 193: 439–442.

Franco C.M.M., Tate M.E., and Oades J.M. 1995. Studies on non-wetting sands. I. The role of intrinsic particulate organic matter in the development of water repellency in non-wetting sands. *Austr. J. Soil Res.* 33: 253–263.

Franco C.M.M., Clarke P.J., Tate M.E., and Oades J.M. 2000. Hydrophobic properties and chemical characterisation of natural water repellent materials in Australian sands. *J. Hydrol.* 231–232: 47–58.

Hallett P.D. and Young I.M., 1999. Changes to water repellence of soil aggregates caused by substrate-induced microbial activity. *Eur. J. Soil Sci.* 50: 35–40.

Hayes M.H.B. 1998. Humic substances: progress towards more realistic concepts of structures. In: *Humic Substances—Structures, Properties and Uses,* G. Davies, E.A. Ghabbour, and K.A. Khairy (eds) Roy. Soc. Chem., Cambridge, UK, pp. 1–27.

Hayes M.H.B. and Graham C.L. 2000. Procedures for the isolation and fractionation of humic substances. In: *Humic Substances—Versatile Components of Plants, Soil and Water,* E.A. Ghabbour and E. Davies (eds.). Roy. Soc. Chem., Cambridge, UK, pp. 91–110.

Horne D.J. and McIntosh J.C. 1994. Causes of repellency. II Interactions between hydrophobic compounds, other extract fractions and the soil matrix. In: *Proc. 2nd National Water Repellency Workshop,* D.J. Carter and K.M.W. Howes (eds.), Perth, Western Australia, pp. 13–17.

Horne D.J. and McIntosh J.C. 2000. Hydrophobic compounds in sands in New Zealand-extraction, characterisation and proposed mechanisms for repellency expression. *J. Hydrol.* 231–232: 35–46.

Hudson R.A., Traina S.J., and Shane W.W. 1994. Organic matter comparison of wettable and non-wettable soils from bentgrass sand greens. *Soil Sci. Soc. Amer. J.* 58: 361–367.

Jackson M.L. 1958. *Soil Chemical Analysis.* Prentice-Hall, Inc., Englewood Cliffs, NJ (USA).

Jex G.W., Bleakley B.H., Hubbel D.H., and Munro L.L. 1985. High humidity-induced increase in water-repellency in some sandy soils. *Soil Sci. Soc. Amer. J.* 49: 1177–1182.

Jungerius P.D. and De Jong, J.H. 1989. Variability of water repellence in the dunes along the Dutch coast. *Catena* 16: 491–497.

Letey J. 1969. Measurement of contact angle, water drop penetration time, and critical surface tension. *Proc. Symp. Water-Repellent Soils,* L.F. DeBano, and J. Letey (eds.). Univ. California, Riverside, CA, pp. 43–47.

Ma'shum M. and Farmer V.C. 1985. Origin and assessment of water repellency of a sandy south Australian soil. *Austr. J. Soil Res.* 23: 623–626.

Ma'shum M., Tate M.E., Jones G.P., and Oades J.M. 1988. Extraction and characterisation of water-repellent material from Australian soils. *J. Soil Sci.* 39: 99–110.

McGhie D.A. and Posner A.M. 1980. Water repellence of a heavy-textured Western Australian surface soil. *Austr. J. Soil Res.* 18: 309–323.

McGhie D.A. and Posner A.M. 1981. The effect of plant top material on the water repellence of fired sands and water-repellent soils. *Austr. J. Agric. Res.* 32: 609–620.

McIntosh J.C. and Horne D.J. 1994. Causes of repellency. I. The nature of the hydrophobic compounds found in a New Zealand development sequence of yellow-brown sands. In: *Proc. 2nd National Water Repellency Workshop,* D.J. Carter and K.M.W. Howes (eds.). Perth, Western Australia, pp. 8–12.

Piccolo A., Spaccini R., Habernauer G., and Gerzabeck M.H. 1999. Increased sequestration of organic carbon in soil by hydrophobic protection. *Naturwissenschaften* 86: 496–499.

Roberts F.J. and Carbon B.A. 1972. Water repellence in sandy soils of southwestern Australia. II. Some chemical characteristics of hydrophobic skins. *Austr. J. Soil Res.* 10: 35–42.

Roy J.L. and McGill W.B. 2000. Flexible conformation in organic matter coatings: An hypothesis about soil water repellency. *Can. J. Soil Sci.* 80: 143–152.

Roy J.L., McGill B., and Rawluk M. 1999. Petroleum residues as water-repellent substances in weathered nonwettable oil-contaminated soils. *Can. J. Soil Sci.* 79: 367–380.

Savage S.M., Osborn J., Letey J., and Heaton C. 1972. Substances contributing to fire induced water repellency in soils. *Soil Sci. Soc. Amer. Proc.* 36: 674–678.

Schnitzer M. and Preston C.M. 1987. Supercritical gas extraction of a soil with solvents of increasing polarities. *Soil Sci. Soc. Amer. J.* 51: 639–646.

Valat B., Jouany C., and Rivière L.M. 1991. Characterization of the wetting properties of air-dried peats and composts. *Soil Sci.* 152: 100–107.

Wallis M.G. and Horne D.J., 1992. Soil water repellency. In: *Advances in Soil Science.* B.A. Stewart (ed.). Springer, New York, NY, 20: 91–146.

Wallis M.G., Horne D.J., and McAuliffe K.W. 1990. A study of water repellency and its amelioration in a yellow-brown sand. 1. Severity of water repellency and the effects of wetting and abrasion. *New Zealand J. Agric. Res.* 33: 139–144.

21

Influence of Acetate, Phosphate, and Citrate on the Sorption of Acid Phosphatase on Variable-charge Minerals and Soil Clays: Implications in Rhizosphere Chemistry

Q. Huang*, M. Pigna, M.A. Rao, *and* A. Violante

Abstract

The influence of acetate, phosphate, and citrate, common inorganic and organic anions present in soils, on the sorption of acid phosphatase by kaolinite, goethite, and clay fractions of an Alfisol, Ultisol, and Oxisol of central-south China was investigated. The Alfisol and Ultisol clays had the major clay mineral composition of 1.4 nm mineral, illite, and kaolinite; the Oxisol clay mainly contained kaolinite and oxides. The samples were Na-saturated and kept in suspension or Ca-saturated and air dried in order to study the influence of aggregation on enzyme sorption, both in the absence and presence of foreign ligands.

Sorption isotherms of acid phosphatase on the soil clays and minerals examined fitted the Langmuir model. The amounts of enzyme sorbed on the Ca-saturated and air-dried samples, expressed as g per kg, were in the order Alfisol > Oxisol ≈ Ultisol > kaolinite ≈ goethite, but the sequence was different when the amounts of enzyme sorbed per m^2 were considered. Furthermore, the samples kept in suspension sorbed greater amounts of enzyme molecules than air-dried clays. In particular, goethite kept in suspension sorbed far greater amounts of enzyme than all the other samples.

Phosphate and citrate competed with the enzyme for sorption sites of soil clays and minerals, inhibiting enzyme sorption while acetate showed no

**Corresponding author:* Dr. Qiaoyun Huang, College of Resources and Environment, Huazhang Agricultural University, Wuhan 430070, P.R. China. E-mail: qyhuang@mail.hzau.edu.in influence on phosphatase sorption. The efficiency of phosphate and citrate in preventing phosphatase sorption was particularly high on goethite and soil clays containing greater amounts of oxides (e.g. Oxisol clay). This behavior was attributed to the ability of these ligands to form strong inner-sphere complexes on the surfaces of variable-charge minerals (in particular metal oxides). Citrate was more effective than phosphate in inhibiting sorption of the enzyme.

Surprisingly, high concentrations of phosphate or citrate ions (0.1 M or more) enhanced the sorption of phosphatase, mainly if introduced into the systems before the enzyme. This could likely be explained by the formation of precipitates (Fe- or Al-phosphate) or dissolution reactions in the presence of citrate, which may create new sorption sites on the surfaces of minerals as well as the formation of soluble OH-Al or OH-Fe species, which may facilitate precipitation of enzyme molecules.

The results obtained in this study demonstrate the important role that ligands present in buffer solutions may have in the sorption/desorption of enzymes on/from soil colloids. They also indicate that the nature and concentration of foreign ligands added to the systems to study the sorption of biopolymers on soil colloids and probably the determination of enzymatic activity of extracellular enzymes in soils must be critically considered.

1 INTRODUCTION

It has been demonstrated that various kinds of enzymes are released by plant roots and microorganisms (Curl and Truelove, 1986; Burns, 1986; Marschner, 1995; Violante and Gianfreda, 2000; Violante et al., 2002a). Root exudates, in fact, comprise both high and low molecular weight substances released by plant roots. The most important high molecular weight compounds are mucilage, polysaccharides and ectoenzymes, whereas the main constituents of the low molecular weight compounds are carbohydrates, organic acids, nucleic acids, amino acids, peptides and phenols (Curl and Truelove, 1986; Marschner, 1995; Huang and Germida, 2002).

The most common ectoenzymes present in the epidermal cells of both plants and roots are polyphenol oxidase and phosphatases (Marschner, 1995). The latter are of great importance in the metabolism of organic phospho-compounds in soil (Rao et al., 1996). They allow the organic phospho-fractions to produce inorganic phosphate, the only form available for plant roots and soil microorganisms. It is well known that these enzymes are released into the rhizosoil where they may be sorbed on soil components and organomineral surfaces (Theng, 1979; Burns, 1986; Boyd and Mortland, 1985, 1990; Haubling and Marschner, 1989; Dinkelaker and Marschner, 1992).

Sorption of enzyme molecules on mineral and organomineral surfaces is a common phenomenon in the soil environment. The role of clay minerals such as montmorillonite, illite and kaolinite in the sorption of enzymes has been well documented (McLaren et al., 1958; Theng, 1979; Burns, 1986; Boyd and Mortland, 1990; Gianfreda et al., 1991, 1992; Nannipieri et al., 1996; Violante and Gianfreda, 2000). The capacity of proteins to be sorbed is influenced by the pH, the surface area of clay minerals (kaolinite, illite, montmorillonite), isoelectric point of proteins, cation exchange capacity (CEC) of minerals, nature of the cation saturating the clays, and temperature (McLaren et al., 1958; Harter and Stotzky, 1973; Theng, 1979; Violante et al., 1995; De Cristofaro et al., 1999; Violante and Gianfreda, 2000).

Enzyme molecules can be sorbed on mineral surfaces through electrostatic interactions (Harter and Stotzky, 1973; Quiquampoix, 1987, 2000; Staunton and Quiquampoix, 1994; Violante et al., 1995), nonelectrostatic forces such as hydrophobic ones (Boyd and Mortland, 1985; Fusi et al., 1989; Staunton and Quiquampoix, 1994), ligand exchange (Sepelyak et al., 1984; Naidja et al., 1997), hydrogen bonding (Boyd and Mortland, 1990; Nannipieri et al., 1996), and van der Waals forces (Kobayashi and Aomine, 1967; Nannipieri et al., 1996). Nutrients (phosphate, sulfate) and low molecular mass organic ligands (LMMOLs, e.g. oxalic, malic, formic, benzoic, acetic, tartaric, citric acid, etc.) are sorbed on soil components and clay minerals (particular on variable-charge minerals and soils), forming complexes on their surfaces. Organic and inorganic ligands with a high affinity for Al and Fe (such as phosphate, selenite, arsenate, citrate, oxalate, and malate) are strongly adsorbed as inner-sphere complexes on clay minerals, and adsorption occurs through a ligand exchange mechanism; contrarily, monodentate ligands (such as acetate, formiate, and benzoate) are weakly adsorbed. Adsorption of anions usually increases with decreasing pH (Bowden et al., 1980; Barrow, 1985; Sparks, 1995; Violante et al., 2002b).

Sorption studies of proteins on soils and clay minerals and the determination of the activity of soil enzymes are usually carried out using buffer solutions that contain organic and / or inorganic anions, which may be differently held on soil clays. As a consequence, the sorption, desorption, and residual activity of soil enzymes may be profoundly influenced in the presence of foreign ligands. However, until now, scant information has been available on the influence of inorganic and organic ligands on enzyme sorption by soil colloidal components (Naidja et al., 1995; Violante and Gianfreda, 2000).

The aim of this work was to study the effect of increasing concentrations of phosphate, citrate, and acetate ligands on the sorption of acid phosphatase by kaolinite, goethite, and three soil clays (Oxisol, Ultisol, and Alfisol) from central-south China. Kaolinite and goethite were the main clay minerals in the soil clays selected. The samples were Na-saturated and kept in suspension or Ca-saturated and air dried in order to study the influence of aggregation on enzyme sorption, both in the absence and presence of foreign ligands.

2 MATERIALS AND METHODS

2.1 Soil Clays and Minerals

Three soils, an Alfisol (depth 0–20 cm), Oxisol (depth 17–35 cm), and a Ultisol (depth 11–40 cm) were collected, respectively, from Xiaogan, Hubei Province,

Dongshansi, Hainan Province, and Xianning, Hubei Province in China. Soils were rinsed in distilled water and dispersed by adding 0.01 M NaOH solution. Clay fractions of the soils < 2 μm were obtained by the sedimentation method after ultrasonic dispersion (Jackson, 1969; He et al., 1998).

Kaolinite was purchased from Wako Chemical Industries, Ltd., Japan and its < 2 μm fraction was separated by the sedimentation method after sonication and pH adjustment with 0.01 M NaOH (Jackson, 1969).

After saturation with Na^+ by 1 M NaCl solution, the clays were washed with deionized water and, when necessary (e.g. for goethite), dialyzed until Cl^- free. The pH of the suspensions was adjusted to pH 5.6 with 0.1 M HCl or NaOH and equilibrated for several days until the pH remained unchanged. A few drops of toluene were added to the suspensions to inhibit microbial growth. The final suspensions contained 10 g of clay per liter.

Other aliquots of the < 2 μm clay fractions of the soils and kaolinite separated by the sedimentation method were flocculated by the addition of 0.5 M $CaCl_2$ solution. The Ca-saturated samples were washed with deionized water until Cl^- free, air dried, and then ground to pass a 100-μm sieve before further study.

The clays were analyzed by x-ray diffraction analysis (XRD) using a Rigaku Geigerflex D/Max B diffractometer (Rigaku Company, Tokyo, Japan). The XRD pattern of the K-saturated soil clays at 20°C, or heated at 300°C and 550°C for 2 h, and Mg-saturated and glycerol-solvated, were obtained using CuK_α radiation generated at 35 kV and 20 mA. The clay minerals in the soils were mainly kaolinite, illite, and 1.4 nm intergrade mineral (chloritized vermiculite or smectite). Synthetic goethite showed characteristic peaks at 0.482, 0.436, 0.244, 0.228, 0.204, and 0.104 nm.

Duplicates of the soil samples were extracted separately with ammonium oxalate (0.2 mM, pH 3.0, in darkness, 4 h) for determination of short-range ordered Fe precipitates and sodium dithionite-citrate-bicarbonate (DCB) (pH 7.0, 80°C for 15 min, 2 times) (Jackson et al., 1986). The amounts of Fe in the extracts were determined by inductively coupled plasma spectroscopy. Organic matter (OM) was determined by dichromate oxidation (Jackson, 1969; Jackson et al., 1986; He et al., 1998).

Point of zero charge (PZC) was measured using the modified salt potentiometric titration method (Sakurai et al., 1988). Surface area was evaluated by the retention of ethylene glycol monoethyl ether (Carter et al., 1965). Selected chemical and physicochemical properties of the minerals and soil clays are shown in Table 21.1.

Goethite was synthesized in the laboratory following the method described by Atkinson et al. (1967). This procedure consisted of slow addition under vigorous stirring of 200 mL of 2.5 M NaOH to a solution containing 50 g of $Fe(NO_3)_3 \cdot 9H_2O$ solution in 825 mL of deionized water. The final suspension was aged six days at 60°C and then washed free of salts by dialysis, air dried, and lightly ground to pass a 100-μm sieve. An aliquot was left in suspension

Table 21.1: Selected properties of soil clays and minerals

Samples	OM ($g\ kg^{-1}$)	Fe_o ($g\ kg^{-1}$)	Fe_d ($g\ kg^{-1}$)	PZC	Surface area $m^2\ g^{-1}$	Clay mineral composition
Alfisol	20.9	4.8	36.2	2.96	205	Kaolinite, illite, 1.4 nm mineral
Oxisol	14.7	2.2	46.8	3.85	84	Kaolinite, oxides
Ultisol	18.7	5.6	33.8	3.35	116	Kaolinite, illite, 1.4 nm mineral, oxides
Kaolinite				3.61	54	
Goethite				8.27	121	

OM: Organic matter
Fe_o: iron solubilized by ammonium oxalate at pH 3.0
Fe_d: iron solubilized by dithionite-citrate-bicarbonate
PZC: point of zero charge

after dialysis. The XRD patterns of a randomly oriented sample showed XRD peaks at 0.418, 0.269, and 0.245 nm, characteristic of goethite.

2.2 Chemicals

Acid phosphatase (E.C. 3.1.3.2, Type I, from wheat germ, 0.4 units mg^{-1}) was purchased from Sigma Co., St. Louis, MO (USA). The other chemicals were reagent grade from Analar, BDH, Poole, UK.

2.3 Sorption Isotherms of Acid Phosphatase

Twenty mg of air-dried samples or two mL of the suspensions containing 10 g soil clays, kaolinite, or goethite per liter were mixed with solutions containing suitable amounts of acid phosphatase in the presence of 0.01 M KCl, 0.01–0.1 M acetate, citrate, or phosphate buffer whose pH was previously adjusted to pH 5.6. The final suspensions (3 mL) were then equilibrated for 1 h at 20°C and agitated every 5 min. with a vortex mixer. After this reaction period the suspensions were centrifuged at 10,000 *g* for 30 min.

Selected experiments were carried out in the presence of 0.01–0.1 M KCl, or 0.01–0.1 M acetate by using greater amounts of sorbents (up to 100 mg). The pH of the suspensions (20 mL) was kept constant by adding 0.02 M HCl through an automatic titrator VIT 90 in conjunction with a 5 mL ABU93 syringe burette (Radiometer, Copenhagen).

The enzyme concentration in the supernatants was determined directly at 280 nm by a Jasco Ubest-50 Uv/Vis spectrophotometer using acid phosphatase as the standard (Simpson and Hughes, 1977; Huang et al., 1995). Acid phosphatase was also used for preparation of the standard calibration curve.

The amount of enzyme sorbed was calculated as the difference between the milligrams of enzyme initially added and the amounts recovered in the supernatants.

2.4 Sequence of Addition of Components on Enzyme Sorption

Some experiments were carried out by adding high concentrations of buffer solutions (0.1 M acetate, phosphate, or citrate) at pH 5.6 and phosphatase to the sorbents as a mixture (*ligand + enzyme* systems) or by introducing ligands before enzyme (*ligand before enzyme* systems) as described below:

Ligands and enzyme added as a mixture: 3 mL of 0.1 M acetate, phosphate, or citrate buffer (pH 5.6) containing 0.6–3.6 mg of enzyme were added to 20 mg air-dried minerals or soil clays in centrifuge tubes. The final suspensions were shaken at 20°C for 1 h as described above.

Ligands added before enzyme: 20 mg of each air-dried sample were mixed with 1.5 ml 0.1 M acetate, phosphate, or citrate buffer (pH 5.6) and agitated at 20°C for 1 h. Acetate, phosphate, or citrate buffer (1.5 ml 0.1 M) containing 0.6–3.6 mg enzyme was added and the solution shaken for an additional hour. Enzyme amounts in the supernatants were determined as described above.

Sorption experiments of acid phosphatase on kaolinite, goethite, and the three soil clays were also carried out at pH 5.6 in the presence of increasing concentrations of phosphate or citrate (ranging from 0.001–0.05 M). The ligands the enzyme were added to the sorbents as a mixture.

3 RESULTS AND DISCUSSION

3.1 Sorption Isotherms of Acid Phosphatase

Figure 21.1 shows the sorption isotherms of phosphatase at pH 5.6 on selected samples air dried (Fig. 21.1A) or in suspension (Fig. 21.1B) in 0.01 M acetate. Similar experiments carried out in the presence of 0.01 M KCl by keeping pH suspensions constant at pH 5.6 showed no significant differences in sorption of phosphatase in the presence of KCl or acetate (data not shown), indicating that acetate has a little ability to prevent phosphatase sorption (as discussed below).

The sorption isotherms can be described by the Langmuir equation:

$$S = S_m K c/(1 + Kc)$$

where S is the amount of enzyme taken up per unit mass of soil colloids or minerals, S_m the maximum amount of enzyme that may be bound, c the equilibrium concentration of enzyme, and K a constant related to sorption energy. Most of the sorption isotherms appeared to be L-class (Giles et al., 1960).

Table 21.2 reports the values of S_m and K constants for the Ca-saturated and air-dried samples, as defined in the Langmuir equation. S_m values ranged from 175 g kg^{-1} for the Alfisol to 115–118 g kg^{-1} for kaolinite and goethite. Ultisol

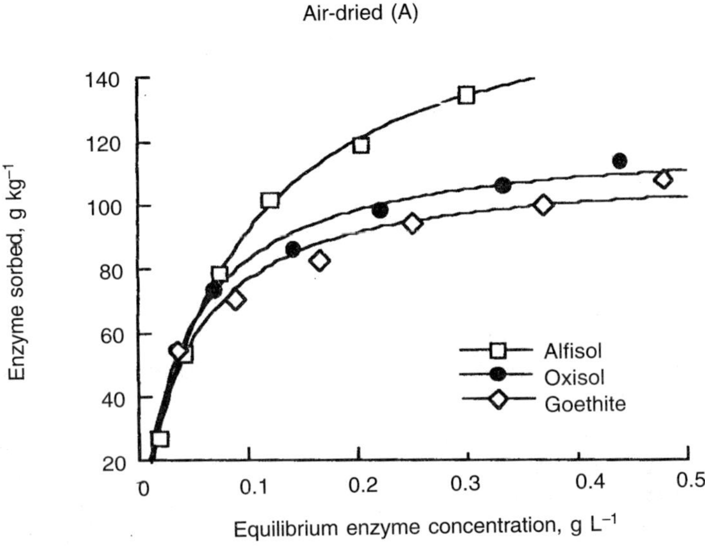

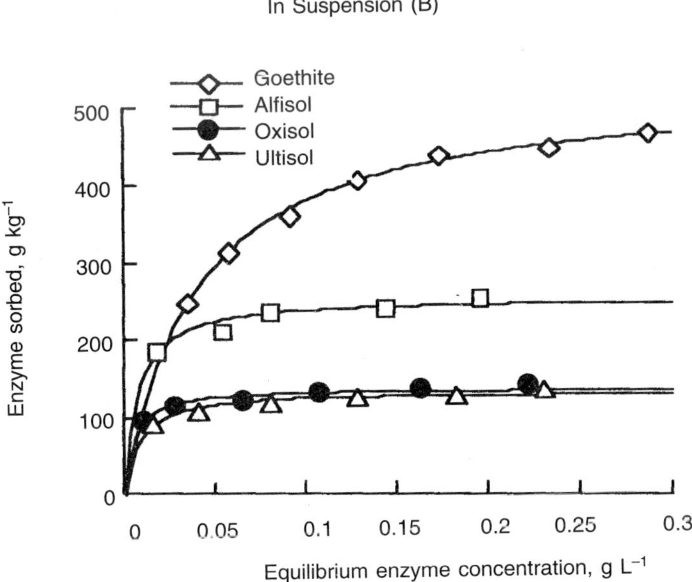

Fig. 21.1: Sorption isotherms of acid phosphatase at 20°C and pH 5.6 on Alfisol, Oxisol, Ultisol or goethite. The soil clays and goethite were air dried (A) or kept in suspensions (B).

Table 21.2: Langmuir parameters for sorption of acid phosphatase on selected air-dried soil clays and minerals in 0.01 M acetate buffer at pH 5.6

Samples	S_m, g kg^{-1}	K	r^2
Alfisol	175	11.4	0.998***
Oxisol	125	20.0	0.996***
Kaolinite	115	21.8	0.992***
Goethite	118	17.0	0.996***

*** Significant at 0.001 probability level.

showed a sorption capacity similar to that of Oxisol, both when air dried and when kept in suspension (Fig. 21.1B).

At a given enzyme equilibrium concentration (e.g. 0.2–0.5 mg mL^{-1}), the amounts of enzyme sorbed on the minerals and soil clays as g per kg (Fig. 21.1A and Table 21.2) were usually in the order Alfisol > Oxisol ≈ Ultisol > kaolinite ≈ goethite. The higher capacity of Alfisol clay in enzyme sorption could be mainly due to the larger surface area after drying as well as the high content of organic matter (Table 21.1). Many studies have shown that humic substances adsorb great amounts of enzyme molecules on their surfaces (Burns, 1986; Boyd and Mortland, 1990; Nannipieri et al., 1996; Rao et al., 1996, 1998).

However, if we consider the amount of enzyme sorbed on the air-dried samples per m^2 (Tables 21.1 and 21.2), the sequence is completely different: kaolinite > Oxisol > Ultisol > goethite > Alfisol. This behavior indicates that physicochemical properties, including surface area, PZC, mineralogy of the sorbents (Table 21.1), and nature of cations saturating the clays play a very important role in enzyme sorption. The PZC of the clay fraction of Alfisol, Ultisol, Oxisol, and kaolinite was 2.96, 3.35, 3.85, and 3.61 respectively, while that of goethite was 8.27 (Table 21.1) (Theng, 1979; Staunton and Quiquampoix, 1994; De Cristofaro et al., 1999; Violante and Gianfreda, 2000 and references therein). Leprince and Quiquampoix (1996) found that the isoelectric point of two acid phosphatases from an ectomycorrhizal fungus was 6.6 and 7.1 respectively. Then, electrostatic force, more than some other interaction was probably involved at pH 5.6 in the sorption of enzyme molecules on the soil clays and kaolinite (Theng, 1979; Violante et al., 1995; De Cristofaro et al., 1999; Violante and Gianfreda, 2000 and references therein). Conversely, acid phosphatase on the goethite surface at pH 5.6 was sorbed mainly by ligand exchange (Naidja et al., 1995, 1997; Violante and Gianfreda, 2000).

Air-dried clays showed a much lower sorption capacity to fix enzyme molecules than did clays kept in suspensions (compare the isotherms of Alfisol, Oxisol, and goethite in Fig. 21.1A with those in Fig. 21.1B). This behavior must be attributed to the strong aggregation of the clay particles after air drying. The nature of the sorbents before and after drying changed tremendously. In fact, goethite, kept in suspension, sorbed larger amounts of enzyme than all the other sorbents, clearly because the exposed surface area of this Fe-oxide in particular was much greater before than after drying (Fig. 21.1).

3.2 Sorption of Acid Phosphatase in Presence of Phosphate and Citrate

Sorption isotherms of acid phosphatase on selected air-dried minerals and soil clays in the presence of 0.01 M acetate, phosphate, or citrate buffer are shown in Figures 21.2A–C. All the sorption models of the enzyme in these systems also conformed to the Langmuir pattern. The amounts of enzyme sorbed on minerals and soil clays in phosphate or citrate systems were smaller than those in 0.01 M acetate systems (Figs. 21.1 and 21.2). The decrease was greater in the presence of citrate than in the presence of phosphate. To evaluate the ability of phosphate or citrate to decrease enzyme sorption on the minerals and soil clays, the percent efficiency of each ligand was calculated according to the expression of Deb and Datta (1967):

Efficiency of phosphate [or citrate] (%) = 1 – [enzyme sorbed in the presence of phosphate (or citrate)/enzyme sorbed in the absence of phosphate (or citrate)] × 100

At an enzyme concentration of 0.3 mg mL^{-1}, the efficiency of phosphate in preventing phosphatase sorption on the Alfisol (Fig. 21.2A), kaolinite, Oxisol (Fig. 21.2B), and goethite (Fig. 21.2C), with reference to the amount of enzyme sorbed in acetate systems, was respectively of 7, 20, 30, and 64%, whereas the efficiency of citrate in inhibiting enzyme sorption was 41, 42, 70, and 90% respectively (Fig. 21.2). These results indicate that inhibition of phosphate and citrate was particularly high on goethite and soil clays containing greater amounts of oxides (e.g. Oxisol; Table 21.1). We also found that phosphate and citrate strongly precluded phosphatase sorption (more than 60%) on a synthetic ferrihydrite, a short-range Fe-oxyhydroxide (data not shown).

Naidja et al. (1995), studying the sorption of tyrosinase onto montmorillonite and hydroxy-aluminum-montmorillonite complexes containing different levels of OH-Al coatings, likewise noted that the presence of phosphate did not significantly affect the amount of tyrosinase sorbed onto montmorillonite, but substantially reduced the sorption of this enzyme onto $Al(OH)_x$-montmorillonite complexes. These authors found that the higher the content of $Al(OH)_x$ species in the system, the greater the inhibition of phosphate on tyrosinase sorption. Gianfreda et al. (1991) also demonstrated that the pH-sorption profiles of invertase on a montmorillonite, an $Al(OH)_x$-montmorillonite complex, and a noncrystalline Al-oxide differed when citrate-phosphate-borate, acetate (pH 4.0–5.5), or phosphate (pH 4.0–8.0) solutions were employed as buffer. Much more enzyme was sorbed in the acetate than in the citrate-phosphate-borate systems on the minerals used.

A possible explanation of these findings is that citrate and phosphate strongly compete with phosphatase for sorption sites of soil clays and minerals.

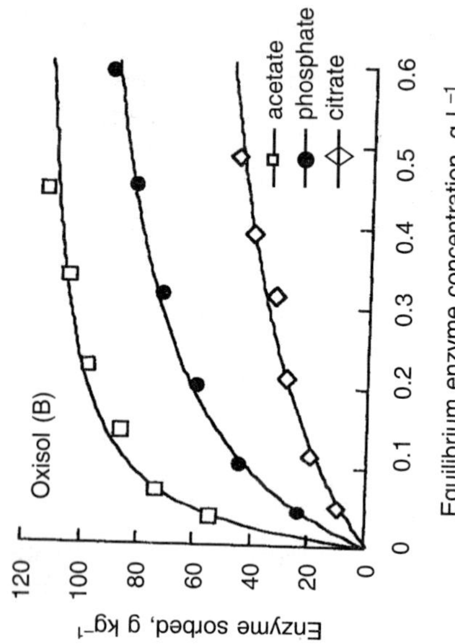

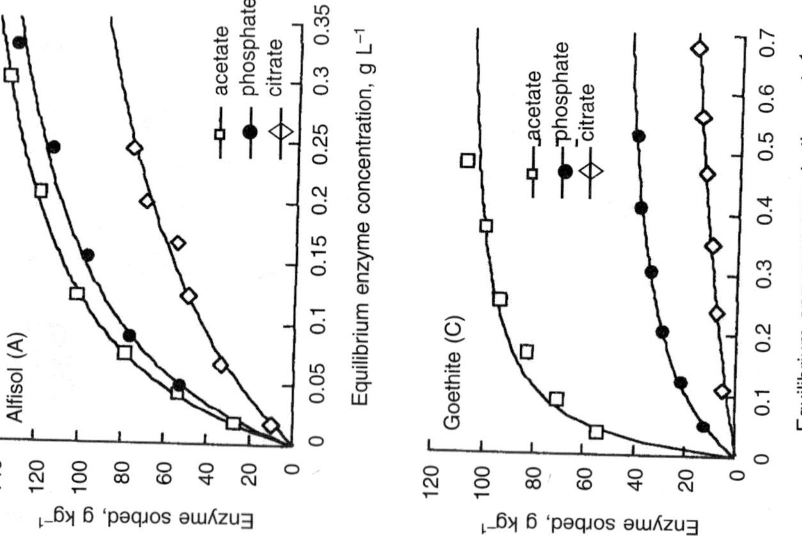

Fig. 21.2: Sorption isotherms of acid phosphatase at 20°C and pH 5.6 on air-dried Alfisol (A), Oxisol (B), and goethite (C) in 0.01 mol L^{-1} acetate, phosphate, or citrate.

In fact, these ligands form inner-sphere complexes on the surfaces of sorbents, especially on variable-charge minerals such as oxides of Fe, Al or Mn, kaolinite, chlorite-like minerals, and short-range aluminosilicates (e.g. allophane, imogolite) (Barrow, 1985; Sparks, 1995; Naidja et al., 1995; Violante et al., 2002b). Conversely, chloride and acetate do not strongly compete with the enzyme molecule because they are weakly sorbed on the surfaces of clay minerals by forming outer-sphere complexes (Sparks, 1995). However, the greater capacity of citrate than phosphate to prevent phosphatase sorption is surprising because phosphate usually shows a sorption capacity similar to or even greater than citrate. On variable-charge minerals, greater quantities of phosphate than citrate (or other LMMOLs, such as oxalate, malate, or succinate) are usually sorbed on soils and clay minerals (Liu et al., 1999; He et al., 1999; Violante and Gianfreda, 2000; Violante et al., 2002a). Clearly, some other processes (e.g. hindrance) may also occur in the reactions among soil sorbents, biopolymers (enzymes), and low molecular mass organic and inorganic ligands that may inhibit or promote removal of biopolymers from the solution (as discussed below).

3.3 Sorption of Enzyme as Influenced by Ligand Concentrations

Sorption of phosphatase in the presence of increasing concentrations of citrate or phosphate was studied on both samples kept in suspension and those that were air dried.

Figure 21.3 shows the sorption of acid phosphatase (in percent with reference to the amounts sorbed in the absence of citrate) on goethite and Oxisol clay

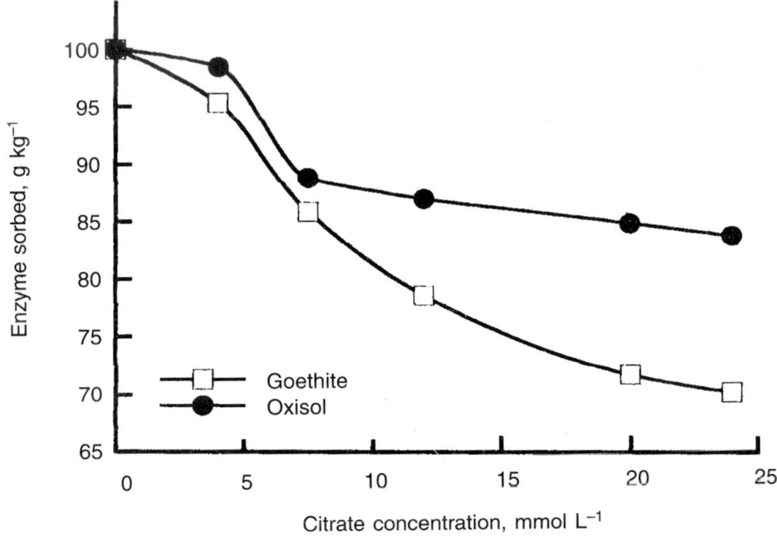

Fig. 21.3: Amounts (%) of acid phosphatase sorbed on Oxisol clay and goethite (kept in suspension) in the presence of increasing concentrations of citrate.

(in suspension) in the presence of increasing concentrations of citrate (up to 0.025 M). It is evident that citrate precluded sorption of phosphatase more on goethite than on the Oxisol.

We also found that at 0.01 M phosphate, the efficiency of phosphate in reducing enzyme sorption was around 6% for air-dried Alfisol, Oxisol, and kaolinite and 17% for goethite. As the concentration of phosphate was increased to 0.05 M, the efficiency of phosphate increased to 36, 41, 38, and 70% for Alfisol, Oxisol, kaolinite, and goethite, respectively (data not shown).

3.4 Sorption of Enzyme as Affected by High Concentrations of Buffer Solutions and Sequence of Components Addition

In the presence of high concentrations of buffer solutions (0.1 M or more) containing phosphate or citrate, we have often found that, surprisingly, the removal of phosphatase from the solutions was particularly high and sometimes greater than that observed in acetate or KCl solutions or in 0.01–0.05 M phosphate or citrate buffer solutions.

Figure 21.4 shows the sorption of phosphatase fixed on the Ultisol clay in 0.01 M acetate and in 0.01 and 0.1 M citrate. Sorption of the enzyme in the presence of 0.1 M citrate was greater than in the presence of 0.01 M acetate or citrate. Similar results were also found using other sorbents (in particular goethite, ferrihydrite, and Oxisol clay) in the presence of both 0.01 and 0.1 M citrate or phosphate (data not shown).

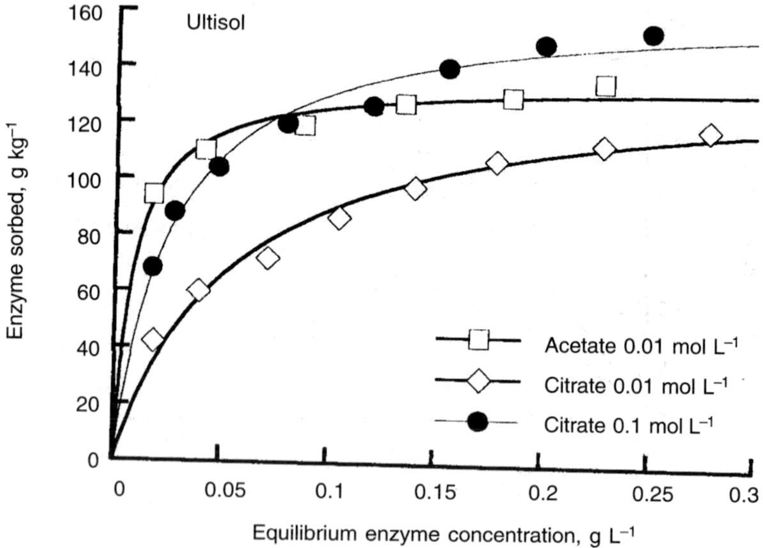

Fig. 21.4: Sorption isotherms of acid phosphatase at 20°C and pH 5.6 on Ultisol clay (kept in suspension) in 0.01 mol L^{-1} acetate or citrate and in 0.1 mol L^{-1} citrate.

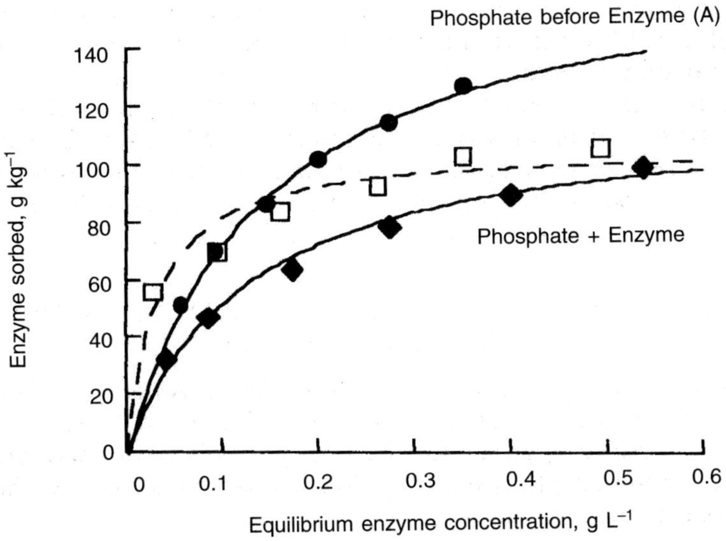

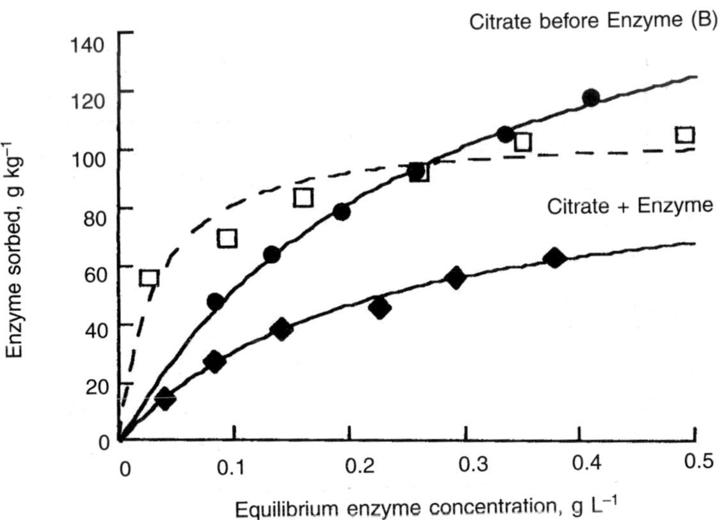

Fig. 21.5: Sorption isotherms of acid phosphatase at 20°C and pH 5.6 on kaolinite (air dried) in 0.1 mol L^{-1} phosphate (A) and in 0.1 mol L^{-1} citrate (B) by adding phosphate and citrate before or with phosphatase. The symbol □ indicates the amounts of acid phosphatase sorbed in the presence of acetate.

Furthermore, we found that the removal of phosphatase from the solutions was usually greater with the addition of 0.1 M citrate or phosphate buffer before the enzyme than when phosphatase was added as a mixture with phosphate or citrate anions to the sorbents (Fig. 21.5). In competitive sorption studies between two different ligands, it has been demonstrated that the order of anion addition affects the sorption of each ligand on the sorbents, with the sorption of the anion added first being greater than when the ligands that compete for sorption sites are added together (Violante and Gianfreda, 1993; Liu et al., 1999; He et al., 1999). In these experiments (Fig. 21.5) the opposite was true, probably because the addition of excessively high concentrations of phosphate or citrate profoundly modified the surface properties of the sorbents (as discussed below).

A complete understanding of the mechanisms of sorption of proteic molecules (as well as other biopolymers and anions or cations) on soil components and soils in the presence of inorganic or organic ligands is especially problematical because different processes may occur simultaneously in these reactions, such as (i) the kind of surface complexes formed by inorganic and organic ligands and biopolymers when added in different amounts and sequences to the sorbents, (ii) the change in surface charge after sorption, (iii) the possible formation of precipitates (e.g. Al- or Fe-phosphates), or (iv) dissolution reactions, which may create new sorption sites when large amounts of organic and inorganic ligands are added to the sorbents. For example, large amounts of phosphate added to the sorbents may promote the formation of precipitates on their surfaces (precipitation versus adsorption) (Sparks, 1995) and, consequently, the nature, surface area, PZC, and surface charge of the sorbent could be completely altered. Furthermore, high concentrations of strongly chelating organic ligands such as citrate not only modify the surface properties of the sorbents, but may promote dissolution, releasing OH-Al or OH-Fe species that may act as new sorbents for phosphatase. These "artifacts" may explain the greater amounts of enzyme molecules removed from solution when concentrations of the buffer solutions containing ligands with high affinity for sorbent surfaces were too high and, mainly, when the ligands were added before the enzyme.

Much evidence seems to demonstrate that the influence of foreign ligands on the sorption/desorption of biopolymers onto soil components must be related to the nature and concentration of the ligand(s) present in the systems, physicochemical properties and mineralogy of sorbent, the sorbent/solution ratio, the initial pH, and so forth. These studies merit close attention.

4 CONCLUSIONS

Organic and inorganic ligands such as citrate and phosphate, which are strongly sorbed on the surfaces of clay minerals and soil clays by forming strong inner-sphere complexes, drastically inhibited the sorption of phosphatase on goethite, kaolinite and soil clays, clearly because competition for sorption sites occurred between the enzymatic molecules and the organic and inorganic ligands. With an increase in citrate or phosphate concentrations, sorption of enzyme decreased steadily. Inhibition of phosphate and citrate was particularly strong on metal oxides and soil clays containing greater amounts of oxides (e.g. Oxisol). The inhibitory effect of citrate on enzyme sorption was greater than that of phosphate. Conversely, organic and inorganic ligands, such as acetate and chloride, which are weakly adsorbed on the surfaces of minerals and oxides, exerted no influence on enzyme sorption.

However, in the presence of high concentrations of phosphate or citrate (especially when ligands were added before the enzyme), removal of phosphatase from the solution was surprisingly high, probably due to alteration of the clay surfaces after ligand addition as well as formation of precipitates (Al- or Fe-phosphates) or solubilization in the presence of citrate of OH-Al or OH-Fe species, which promoted enzyme precipitation (Rao et al., 1996).

Certainly, the competitive effects of various organic and inorganic ligands on the sorption/desorption of enzymes on/from various soil components and soils are of paramount importance in environmental sciences and warrant special attention. In fact, studies on the sorption of biopolymers (e.g. proteins, DNA, RNA) on clay minerals and soils, on determination of activity of enzymes immobilized on clay minerals, or of soil enzymes are carried out in buffer solutions (Alef and Nannipieri, 1995), but the nature and concentration of natural and synthetic ligands present in these solutions are not usually taken into consideration. That the nature and concentration of foreign ligands added to systems in the study of sorption or activity of enzymes must be critically considered is clearly demonstration.

Our study also has implications for rhizosphere chemistry. Rhizosphere carbon flow has been estimated to account for a major fraction, up to 40%, of plant primary production. The rhizosphere is a favorable habitat for acid-producing bacteria. Microorganisms also release many biomolecules, so that the amount of biomolecules in the rhizosphere is much higher than in the bulk soil. Oxalic, acetic, formic, malic, succinic, citric, tartaric, and malonic acids are the most abundant aliphatic acids present in the rhizosphere.

Interactions among plants, soil colloids, organics (including proteins) and microorganisms in the rhizosphere are still very poorly known. Our study demonstrates that the presence of root exudates at the soil-plant interface may strongly influence not only the sorption of enzymes on soil colloids, but may probably also affect their residual activity. The different activity of soil enzymes present in the rhizosphere and bulk soil might also be due to the different concentration and nature of organic ligands.

Aknowledgments

The research was supported partly by the National Natural Science Foundation of China (40271064) and the International Foundation for Science (IFS, C12527-2). This paper represents Journal Series no. 0063 from DiSSPA.

References

Alef K. and Nannipieri P. 1995. *Methods in Applied Soil Microbiology and Biochemistry*. Acad. Press, London, UK.

Atkinson R.T., Posner A.M., and Quirk J.P. 1967. Adsorption of potential-determining ions at the ferric oxide aqueous electrolyte interface. *J. Phys. Chem.* 71: 550–558.

Barrow N.J. 1985. Reactions of anions and cations. *Adv. Agro.* 38: 183–230.

Bowden J.W., Nagarajah S., Barrow N.J., Posner A.M., and Quirk J.P. 1980. Describing the adsorption of phosphate, citrate and selenite on a variable-charge mineral surface. *Austr. J. Soil Res.* 18: 49–60.

Boyd S.A. and Mortland M.M. 1985. Urease activity on a clay-organic complex. *Soil Sci. Soc. Amer. J.* 49: 619–622.

Boyd S.A. and Mortland M.M. 1990. Enzyme Interactions with clays and clay-organic matter complexes. In: *Soil Biochemistry.* J.M. Bollag and G. Stotzky (eds.). Marcel Dekker, New York, NY, vol. 6, pp. 1–28.

Burns R.G. 1986. Interactions of enzyme with soil mineral and organic colloids. In: *Interactions of Soil Minerals with Natural Organics and Microbes.* P.M. Huang and M. Schnitzer (eds.). Soil Sci. Soc. Amer., Madison, WI (USA), pp. 429–451.

Carter D.L., Heilman M.D., and Gonzalez C.L. 1965. Ethylene glycol monoethyl ether for determining surface area of silicate minerals. *Soil Sci.* 100: 356–360.

Curl E.A. and Truelove B. 1986. *The Rhizosphere.* Adv. Series Agric. Sci. 15. Springer-Verlag, New York, NY.

De Cristofaro A., Colombo C., Gianfreda L., and Violante A. 1999. Effect of pH, exchange cations and hydrolytic Species of Al and Fe on formation and properties of montmorillonite-Protein Complexes. In: *Effect of Mineral-Organic-Microorganism Interactions on Soil and Freshwater Environments.* J. Berthelin et al. (eds.). Plenum Publ. Co., London, pp. 278–286.

Deb D.L. and Datta N.P. 1967. Effect of associating anions on phosphorus retention in soil: I. Under variable phosphorus concentration. *Plant Soil* 26: 303–316.

Dinkelaker B. and Marschner H. 1992. In vivo demonstration of acid phosphatase activity in the rhizosphere of soil-grown plants. *Plant Soil* 144: 199–205.

Fusi P., Ristori G.G., Calamai L., and Stotzky G. 1989. Adsorption and binding of protein on "clean" (homoionic) and "dirty" (coated with Fe oxyhydroxides) montmorillonite, illite and kaolinite. *Soil Biol. Biochem.* 21: 911–920.

Gianfreda L., Rao M.A., and Violante A. 1991. Invertase (β-fructosidase): Effects of montmorillonite, Al-hydroxide and AL(OH)x-montmorillonite complex on activity and kinetic properties. *Soil Biol. Biochem.* 23: 581–587.

Gianfreda L., Rao M.A., and Violante A. 1992. Adsorption, activity and kinetic properties of urease on montmorillonite, aluminum hydroxide and Al(OH)x-montmorillonite complexes. *Soil Biol. Biochem.* 24: 51–58.

Giles C.H., McEwan T.H., Nakhwa S.N., and Smith D. 1960. Studies in adsorption. XI. A system with classification of solution adsorption isotherms, and its use in diagnosis of adsorption mechanisms and in measurement of specific surface area of solids. *J. Chem. Soc.* 786: 3973–3993.

Harter R.D. and Stotzky G. 1973. X-ray diffraction, electron microscopy, electrophoretic mobility, and pH of some stable smectite-protein complexes. *Soil Sci. Soc. Amer. Proc.* 37: 116–123.

Haubling M. and Marschner H. 1989. Organic and inorganic soil phosphates and acid phosphatases activity in the rhizosphere of 80-years old Norway spruce (*Picea abies* L. Karst.) trees. *Biol. Fertil. Soils.* 8: 128–133.

He J.Z., Li X.Y., and Violante A. 1998. Sequential extraction of aluminum and iron from acidic soils by chemical selective dissolution methods. *Pedosphere* 8: 37–44.

He J.Z., De Cristofaro A., and Violante A. 1999. Comparison of adsorption of phosphate, tartrate, and oxalate on hydroxy aluminum montmorillonite complexes. *Clays Clay Miner.* 47: 226–233.

Huang P.M. and Germida J.J. 2002. Chemical and biological processes in the rhizosphere: Metal pollutants. In: *Interactions between Soil Particles and Microorganisms: Impact on the Terrestrial Ecosystem.* P.M. Huang, J.-M. Bollag, and N. Senesi (eds.). John Wiley & Sons Ltd., Chichester, England, pp. 381–438.

Huang Q., Shindo H., and Goh T.B. 1995. Adsorption, activity and kinetics of acid phosphatase as influenced by montmorillonite with different interlayer material. *Soil Sci.* 159: 271–278.

Jackson M.L. 1969. *Soil Chemical Analysis—Advanced Course.* Author published. Univ. Wisconsin, Madison, WI (USA) (2nd ed.).

Jackson M.L., Lim C.H., and Zelazny L.W. 1986. Oxides, hydroxides and aluminosilicates. In: *Methods of Soil Analysis*, Part 1. A. Klute (ed.). Soil Sci. Soc. Amer., Madison, WI, pp. 159–221 (2nd ed.).

Kobayashi Y. and Aomine S. 1967. Mechanism of inhibitory effect of allophane and montmorillonite on some enzymes. *Soil Sci. Plant Nutr.* 13: 189–194.

Leprince F. and Quiquampoix H. 1996. Extracellular enzyme activity in soil: Effect of pH and ionic strength on the interaction with montmorillonite of two acid phosphatases secreted by the ectomycorrhizal fungus *Hebeloma cylindrosporum. Eur. J. Soil Sci.* 47: 511–522.

Liu F., He. J.Z., Colombo C., and Violante A. 1999. Competitive adsorption of sulfate and oxalate on goethite in the absence or presence of phosphate. *Soil Sci.* 164: 180–189.

Marschner H. 1995. *Mineral Nutrition of Higher Plants*. Acad. Press Limited, London, UK (2nd ed.).

McLaren A.D., Peterson G.H., and Barshad I.. 1958. The adsorption and reactions of enzymes and proteins on clay minerals. IV. Kaolinite and montmorillonite. *Soil Sci. Soc. Amer. Proc.* 22: 239–244.

Naidja A., Violante A., and Huang P.M. 1995. Adsorption of tyrosinase onto montmorillonite as influenced by hydroxyaluminum coatings. *Clays Clay Miner.* 43: 647–655.

Naidja A., Huang P.M., and Bollag J.-M. 1997. Activity of tyrosinase immobilized on hydroxyaluminum-montmorillonite complexes. *J. Molec. Catal. A: Chemical* 115: 305–316.

Nannipieri P., Sequi P., and Fusi P. 1996. Humus and enzyme activity. In: *Humic Substances in Terrestrial Ecosystems.* A. Piccolo (ed.). Elsevier, Amsterdam, Netherlands, pp. 293–328.

Quiquampoix H. 1987. A stepwise approach to the understanding of extracellular enzyme activity in soil, I. Effects of electrostatic interactions on the conformation of a β-D-glucosidase adsorbed on different mineral surfaces. *Biochimie* 69: 753–763.

Quiquampoix H. 2000. Mechanisms of protein adsorption on surfaces and consequences for extracellular enzyme activity in Soil. In: *Soil Biochemistry*, vol. 10. J.-M. Bollag and G. Stotzky (eds.). Marcel Dekker, New York, NY, pp. 171–206.

Rao M.A., Violante A., and Gianfreda L. 1998. Interactions between tannic acid and acid phosphatase. *Soil Biol. Biochem.* 30: 111–112.

Rao M.A., Gianfreda L., Palmiero F., and Violante A. 1996. Interactions of acid phosphatase with clays, organic molecules and organo-mineral complexes. *Soil Sci.* 161: 751–760.

Sakurai K., Ohdate Y., and Kyuma K. 1988. Comparison of salt titration and potentiometric titration methods for the determination of zero point of charge. *Soil Sci. Plant Nutr.* 34: 171–182.

Sepelyak R.J., Feldkamp J.R., Moody T.E., White J.L., and Hem S.L. 1984. Adsorption of pepsin by aluminum hydroxide, I: sorption mechanism. *J. Pharm. Sci.* 73: 1514–1517.

Simpson G.H. and Hughes J.D. 1977. Arylsulphatase-clay interactions, I. Absorption of arylsulphatase by kaolinite and montmorillonite. *Austt. J. Soil Res.* 16: 27–33.

Sparks D.L. 1995. *Environmental Soil Chemistry*. Acad. Press, San Diego, 267 pp.

Staunton S. and Quiquampoix H. 1994. Adsorption and conformation of bovine serum albumin on montmorillonite: Modification of the balance between hydrophobic and electrostatic interactions by protein methylation and pH variation. *J. Colloid Interf. Sci.* 166: 89–94.

Theng B.K.G. 1979. *Formation and Properties of Clay-Polymer Complexes.* Elsevier, New York, NY, pp. 362.

Violante A. and Gianfreda L. 1993. Competition in adsorption between phosphate and oxalate on an aluminum hydroxide montmorillonite complex. *Soil Sci. Soc. Amer. J.* 57: 1235–1241.

Violante A. and Gianfreda L. 2000. Role of biomolecules in the formation of variable-charge minerals and organo-mineral complexes and their reactivity with plant nutrients and organics in soil. In: *Soil Biochemistry,* Vol. 10. J.-M. Bollag and G. Stotzky (eds.). Marcel Dekker, New York, NY, pp. 207–270.

Violante A., De Cristofaro A., Rao M.A., and Gianfreda L. 1995. Physicochemical properties of protein-smectite and Protein-AL$(OH)_x$-smectite complexes. *Clay Miner.* 30: 325–336.

Violante A., Huang P.M., Bollag J.-M., and Gianfreda L. (eds.). 2002a. *Soil Minerals–Organic Matter-Microorganism Interactions and Ecosystem Health. Ecological Significance of the Interactions among Clay Minerals, Organic Matter and Soil Biota.* vol. II. Developments in Soil Science 28B, Elsevier, Amsterdam, Netherlands, 432 pp.

Violante A., Krishnamurti G.S.R., and Huang P.M. 2002b. Impact of organic substances on the formation of metal oxides in soil environments. In: *Interactions between Soil Particles and Microorganisms and Their Impact on the Terrestrial Environment.* P.M. Huang et al. (eds.). John Wiley & Sons, Inc., New York, NY, pp. 133–188.

Index

L

M

N

T

U

V

W

X

Z

About the Editors

P.M. Huang received his Ph.D. degree in Soil Science at the University of Wisconsin, Madison, in 1966. He is Professor Emeritus of Soil Science at the University of Saskatchewan, Saskatoon, Canada. His research work has significantly advanced the frontiers of knowledge on the nature and surface reactivity of mineral colloids and organomineral complexes of soils and sediments and their role in the dynamics, transformations, and fate of nutrients, toxic metals, and xenobiotics in terrestrial and aquatic environments. His research findings, embodied in over 300 refereed scientific publications, including research papers, book chapters, and 16 books, are fundamental to the development of sound strategies for managing land and water resources.

Dr Huang developed and taught courses in soil physical chemistry and mineralogy, soil analytical chemistry, and ecological toxicology. He has successfully trained and inspired M.Sc. and Ph.D. students and postdoctoral fellows, and received visiting scientists from around the globe. He has served on numerous national and international scientific and academic committees. He has also served as a member of many editorial boards such as the *Soil Science Society of America Journal, Geoderma, Chemosphere, Water, Air and Soil Pollution,* and *Soil Science and Plant Nutrition.* He has served as a titular member of the Commission of Fundamental Environmental Chemistry of the International Union of Pure and Applied Chemistry and is the founding Chairman of the Working Group MO "Interactions of Soil Minerals with Organic Components and Microorganisms" of the International Union of Soil Sciences. He received the Distinguished Researcher Award from the University of Saskatchewan and the Soil Science Research Award from the Soil Science Society of America. He is a Fellow of the Canadian Society of Soil Science, the Soil Science Society of America, the American Society of Agronomy, the American Association for the Advancement of Science, and the World Innovation Foundation.

Antonio Violante is Professor of Agricultural Chemistry at the University of Naples (Italy). He received his Ph.D. in Chemistry at the University of Naples in 1969. He was awarded postdoctoral fellowships from the University of Wisconsin, USA (1976–1977) and the University of Saskatchewan, Canada (1981–1982) and was invited Visiting Professor in the Department of Soil Science, University of Saskatchewan, Canada in 1985, 1992, and 2003.

Dr Violante was Head of the Dipartimento di Scienze Chimico-Agrarie and is Coordinator of the *Doctoral School in Agrobiology and Agrochemistry* of the University of Naples Federico II. He has served on many committees of the Italian Society of Soil Science (President of the Session Soil Chemistry), and Italian Society of Agricultural Chemistry. He is vice-president and liaisons officer of Gruppo Italiano AIPEA. He was the scientific chairman and chief organizer of International and National Congresses.

Dr Violante has contributed to promote research on the interface between soil chemistry and mineralogy and soil biology. The areas of research include the formation mechanisms of Al-hydroxides and oxyhydroxides, the surface chemistry and reactivities of short-range ordered precipitation products of Al and Fe, the influence of biomolecules on the sorption/desorption of nutrients and xenobiotics on/from variable charge minerals and soils and on the factors which influence the sorption and residual activity of enzymes on phyllosilicates, variable charge minerals, organo-mineral complexes, and soils. Dr Violante is the author or co-author of 141 research articles and book chapters. He presented papers at many scientific Congresses and Symposia and gave invited lectures at universities and research institutes worldwide. Dr Violante has international research/teaching experience in Canada, the US, Europe, China and Chile. He has trained students for Masters and Ph.D. Degrees and postdoctoral fellows and received visiting scientists from worldwide. He serves on the editorial board of three international journals. He is a Fellow of the Soil Science Society of America and the American Society of Agronomy.

Jean-Marc Bollag is Professor Emeritus of Soil Biochemistry and Past Director of the Center for Bioremediation and Detoxification, the Environmental Resources Research Institute at the Pennsylvania State University. Dr Bollag has advised more than 100 co-workers-graduate students, post-doctoral scholars and visiting scientists. He has published over 250 research articles and book chapters, and presently he is editor of the book series *Soil Biochemistry.* He served on the editorial board of five international journals. He is a frequent lecturer at conferences and seminars throughout the world.

Dr Bollag is a recipient of the Julius Baer Fellowship, the Gamma Sigma Delta Research Award, and the Badge of Merit from the Polish Ministry of Agriculture. He is a Fallow of the American Academy of Microbiology, the Soil Science Society of America and the American Society of Agronomy. He is also recipient of the Environmental Quality Research Award from the American Society of Agronomy.

Dr Bollag received his Ph.D. degree from the University of Basel (Switzerland, and conducted postdoctoral work at the Weizmann Institute of Science, Rehovoth, Israel, and at Cornell University, Ithaca, New York. He was also a Visiting Scientist in the Biochemistry Section of Agrochemicals at Ciba-Geigy, Basel, Switzerland. Most of his research is related to the fate of pollutants in the environment and to bioremediation problems (incorporation of pollutants into soil organic matter as a detoxification method and application of enzymes for pollution control).

Pama Vityakon is an Associate Professor of Soil Science in the Department of Land Resoruces and Environment, Faculty of Agriculture, Khon Kaen University in northeast Thailand. She received her B.Sc. and M.Sc. in Soil Science from Lincoln College, New Zealand and her doctorate from the University of Hawaii, USA. She has been on the staff of Khon Kaen University since 1979. Her research interests include soil organic matter (SOM) changes under different types of organic inputs, land use and management, the effects of trees on soil fertility and SOM, agroforestry systems, land degradation as indicated by soil erosion and SOM, and farming systems research as a means to identify farmers's problems and to develop technology appropriate for farmers and nutritional quality of food crops as affected by soil nutrients. Her major teaching responsibility has been soil fertility at both undergraduate and graduate levels. She has developed teaching materials designed to incorporate concern with trees into agricultural education in Thailand. She is an author of approximately 50 publications in the forms of books, book chapters, journal articles, proceedings papers and research reports. She belongs to the World Association for Soil and Water Conservation (representative) Thailand Network for Agroforestry Education (committee member), and the Soil Science Society of America. She was the East-West Center distinguished alumni fellow in 1991, a participant in the JSPS Thailand-Japan scientific exchange program (1998–2000), and has been a visiting researcher at the Faculty of Agriculture of Kyoto University, Japan, on several occasions.